ETHYLENE
Agricultural Sources and Applications

ETHYLENE
Agricultural Sources and Applications

Edited by

Muhammad Arshad
University of Agriculture
Faisalabad, Pakistan

and

William T. Frakenberger, Jr.
University of California
Riverside, California

Kluwer Academic / Plenum Publishers
New York, Boston, Dordrecht, London, Moscow

Library of Congress Cataloging-in-Publication Data

Arshad, Muhammad, 1956–
    Ethylene: agricultural resources and applications/by Muhammad Arshad and William
T. Frakenberger, Jr.
     p.   cm.
    Includes bibliographical references (p.).
    ISBN 0-306-46666-X
     1. Ethylene. I. Frakenberger, W. T. (William T.), 1952– II. Title.

QK898.E8 A78 2002
631.8′9—dc21

2001041356

ISBN 0-306-46666-X

©2002 Kluwer Academic / Plenum Publishers, New York
233 Spring Street, New York, New York 10013

http://www.wkap.nl/

10 9 8 7 6 5 4 3 2 1

A C.I.P. record for this book is available from the Library of Congress

Printed in the United States of America

PREFACE

With an ever-increasing demand for more food supply, agricultural scientists will have to search for new ways and technologies to promote food production. In recent decades, plant growth regulators (PGRs) have made great strides in promoting plant growth and development. PGRs are organic compounds which have the ability to dramatically affect physiological plant processes when present in extremely low concentrations (in the range of micro- to picograms). Although all higher plants have the ability to synthesize PGRs endogenously, they do respond to the exogenous sources most likely due to not having the capacity to synthesize sufficient endogenous phytohormones for optimal growth and development under given climatic and environmental conditions. In recent years, PGRs have established their position as a new generation of agrochemicals after pesticides, insecticides and herbicides. Interest in the commercial use of PGRs for improving plant growth and crop yields is also increasing because of their non-polluting nature. The use of PGRs in the post-harvest technology is well established and many new breakthroughs have recently been revealed.

Ethylene (C_2H_4) is one of the five major classes of PGRs. Although C_2H_4 is a simple gas, it has a very strong multifaceted role in plant physiology. Initially, known as the "ripening hormone", C_2H_4 is now considered to be involved in numerous processes from seed ontogeny to senescence and even in post-harvest technology. Results of several studies warrant that C_2H_4 of exogenous sources could be used for the betterment of agricultural production. The exogenous sources include natural production by soil microbiota and synthetic C_2H_4 releasing compounds. A commercial product "ethephon" has already provided a breakthrough discovery against lodging in cereals. The importance of C_2H_4 as a potent plant growth regulator has led us to compile the most recent scientific literature into this book.

The biochemistry of endogenous C_2H_4 biosynthesis and its role in physiology of higher plants have been compiled in the form of several books and reviews. However, very little attention has been given to the soil microbiota and their potential exogenous source of C_2H_4. The ecological importance of microbially produced C_2H_4 was evident after the dramatic effects on plant growth were published in the 1990s. We have pooled this information into a comprehensive book. In this book, we provide eight chapters including an introduction of C_2H_4 as a hormone; its role in plant physiology; sources and factors affecting microbial production of C_2H_4; biochemistry of C_2H_4 production by microorganisms; production in soil; the role of C_2H_4 in symbiosis and pathogenesis; and its commercial application in agriculture. Overall, this book is extremely valuable in that it covers all the aspects of C_2H_4 sources, in-depth biochemistry and applications.

Our comprehensive effort in providing this resource will benefit agricultural scientists including physiologists, microbiologists, soil scientists, botanists, ecologists, and horticulturalists to develop new technologies based on the release of C_2H_4 for the betterment of agricultural production.

Special acknowledgement is given to Louise DeHayes for typing this entire book and to Linda Bobbitt for her assistance in the illustrations of the figures. We also wish to thank Ben Johanson and Tariq Siddique for their assistance in proofreading the text.

CONTENTS

To my wife Talat, son Haaris
and daughters Habiba and Fatima
Muhammad Arshad

To my wife, Margaret E. Beebe-Frankenberger
William T. Frankenberger

1

THE PLANT HORMONE, ETHYLENE

1.1. INTRODUCTION

It is now well accepted that normal plant growth and development are controlled by compounds produced by the plant itself (referred to as endogenous plant hormones). These compounds are organic in nature, other than nutrients, that affect the physiological processes of growth and development in plants when present in extremely low concentrations. The well defined groups of these naturally occurring plant growth regulatory substances (PGRs), also termed as plant hormones or phytohormones, fall into five classes: auxins, gibberellins, cytokinins, ethylene (C_2H_4) and abscisic acid. Almost all plants can synthesize these PGRs, however, they also respond when they are exposed to exogenous sources of these substances most likely due to not having the capacity to synthesize sufficient endogenous phytohormones for optimal growth and development under sub-optimal climatic and environmental conditions. The general thinking is that exogenously supplied PGRs produce their effects by changing the endogenous level(s)/balance of the naturally occurring hormones, allowing a modification of growth and development depending on age and the physiological state of plant development, endogenous level of hormones, state of nutrition, and environmental conditions. Moreover, in plant tissues, excessive amounts of these exogenously supplied PGRs may be stored as conjugates and released when and where the plant needs.

It is well established that these PGRs are also produced by soil microbiota which can act as a potential exogenous source in the rhizosphere for plant uptake and create a physiological response after being taken up by plant roots (Frankenberger and Arshad, 1995; Arshad and Frankenberger, 1992, 1993, 1998). Additionally, compounds with plant growth regulating activity are produced synthetically by man and many of these are used commercially to improve the quality and quantity of agricultural commodities (Nickell, 1982; Abeles et al., 1992; Lürssen, 1991).

The role of these five established classes of plant hromones in plant physiology has been documented by several authors (Davies, 1995; Mattoo and Suttle, 1991; Abeles et al., 1992; Moore, 1989; Takahashi, 1986). Ethylene, a simple gas with profound growth regulating activity, is the major focus of this book with our main emphasis on microbial production of C_2H_4 in soil and its subsequent effects on plant growth, and its role in symbiosis and pathogenesis.

1.2. ETHYLENE AS A GAS

Ethylene is a two-carbon symmetrical compound with one double bond. Its molecular weight is 28.05, with freezing, melting, and boiling temperatures of $-181°C$, $-169.5°C$, and $-103.7°C$, respectively, having a relative density to air of 0.978. It is flammable and colorless, and has a sweet, ether-like odor. Its diffusion coefficient in air is approximately 10,000 times of that in water (Jackson, 1985). Its solubility in water is approximately 140 ppm at $25°C$ and 760 mm Hg pressure. It is ~14 times more soluble in lipids than in water (Eger and Larson, 1964) and the water/gas partition coefficient for C_2H_4 is 0.081 (Spencer, 1969).

1.3. BIOLOGICAL ACTIVITY OF ETHYLENE

Ethylene has nearly full biological activity at 1 μl liter^{-1} (Abeles, 1973) which corresponds to 6.5 x 10^{-9} M at $25°C$ (Abeles et al., 1992). In coordinating with other plant hormones, it plays a crucial role in controlling the way in which a plant grows and develops. In the beginning, C_2H_4 was called the "ripening" hormone, but later investigations revealed that C_2H_4 merits equal status with other classes of plant hormones because of its diverse and effective role in plant growth and development. The major areas of plant physiology in which C_2H_4 has been demonstrated to influence plant growth and development include:
- Release of dormancy
- Shoot and root growth and differentiation
- Adventitious root formation
- Leaf and fruit abscission
- Flower induction in some plants
- Induction of femaleness in dioecious flowers
- Flower ripening
- Flower and leaf senescence
- Fruit ripening

Details on the role of C_2H_4 in the above-listed aspects of plant physiology can be found in many reviews (Pratt and Goeschl, 1969; Abeles et al., 1992; Osborne, 1978; Lieberman, 1979; Sembdner et al., 1980; Yang and Hoffman, 1984; Beyer et al., 1984; Imaseki, 1986; McKeon and Yang, 1987; Reid, 1987; Moore, 1989; Mattoo and Suttle, 1991).

The physiological significance, and evolutionary advantage of C_2H_4 as a gaseous hormone is difficult to assess. One important aspect of C_2H_4 is that degradation is not essential to remove excessive amounts of this hormone, offering an advantage in the regulation of C_2H_4 levels. According to Burg (1962), "C_2H_4 is unique in that it is a vapor at physiological temperature, so that its production, accumulation, transport and functions involve special problems not encountered in other areas of hormone physiology." The diffusion of C_2H_4 through intracellular spaces may act as a signal of damage, stress, or physical contact and is important in coordinating ripening of tissues in fruit (Reid, 1987). Unlike other plant hormones, metabolic processes are not required to detoxify higher C_2H_4 concentrations, since C_2H_4 diffuses passively into the ambient atmosphere, depending on the concentration gradient between the inside and outside of the tissue. Ethylene apparently does not undergo direct transport, but serves a hormonal integrative function by diffusing rapidly through the tissues. It rapidly reaches equilibrium between the rate of production and rate of diffusion. In soil, it is consumed by microorganisms and in air, C_2H_4 is photooxidized by UV light or ozone. It is difficult to envision the evolutionary sequences of events, or selection pressures that made an unsaturated two-carbon gas a hormone. Reconstruction of the forces and events which led to the acquisition of biological activity by C_2H_4, or other hormones, has not been actively pursued by plant taxonomists or physiologists (Abeles et al., 1992).

Other gaseous analogues of C_2H_4 also exhibit a weak, but definite C_2H_4-like activity in bioassays. Crocker et al. (1935) examined the relative effectiveness of various unsaturated hydrocarbons in the stimulation of epinasty in tomato plants. Later, Burg and Burg (1967) compared various hydrocarbons for their activity in suppression of pea curvature, whereas Abeles and Gahagan (1968) compared their activity in abscission. Each of these studies confirms that C_2H_4 is the most active among all the related compounds tested. Second to C_2H_4 in activity, propylene (nearest analogue) was 60- to 100-fold less active. A comparison of biological activity among C_2H_4 and some other hydrocarbon gases in evoking a physiological response is given in Table 1.1. From a detailed analysis of the structural requirement for C_2H_4-like action in the pea-curvature bioassay, Burg and Burg (1965, 1967) postulated that (i) only unsaturated aliphatic compounds are active; (ii) double bonds confer much greater activity than triple bonds; (iii) biological activity is inversely related to molecular size; (iv) activity requires a terminal carbon atom adjacent to an unsaturated carbon; (v) substitutions that result in lower electron density in the double bond and cause delocalization, reduce the biological activity; (vi) the terminal carbon atom must be electrophilic and not positively charged; and (vii) the biological activity of a substituted C_2H_4 analogue is highly dependent on the size of the substituted group and the ability of the substituted group to reduce electron density in the double bond.

1.4. HISTORICAL PERSPECTIVE

The recognition of C_2H_4 as a plant hormone originated from observations of premature shedding of street trees, geotropism of etiolated pea seedlings when exposed to an illuminating gas, early flowering of pineapples treated with smoke, and ripening of oranges exposed to gas from kerosene combustion. Neljubow (1901) is credited to have been the first to demonstrate the involvement of C_2H_4 in the tropistic responses of plants. Subsequent studies involving the use of synthetic gas in evoking a physiological response confirmed the growth regulatory properties of C_2H_4. The introduction of gas chromatography in the late 1950s for the analysis of trace amounts of biological substances has established irrefutably that C_2H_4 is a naturally occurring endogenous plant hormone. It has numerous but complex interactions with other plant hormones, particularly with auxins.

Table 1.1. Compounds giving an ethylene response.

Structure	Compound	Conc. for a 1/2 maximum response	
		$\mu l/l$ gas phase	M (in water)
$H_2C=CH_2$	Ethylene	0.1	4.8×10^{-10}
$CH_3\text{-}CH=CH_2$	Propylene	10	5.2×10^{-8}
$CH_3\text{-}CH_2\text{-}CH=CH_2$	1-Butene	27,000	1.3×10^{-4}
$C\equiv O$	Carbon monoxide	270	5.4×10^{-8}
$HC\equiv CH$	Acetylene	280	1×10^{-6}
$CH_3\text{-}N\equiv C$	Methyl isocyanide	2	--
$H_2C=C=CH_2$	Allene	2,900	5.6×10^{-5}
C_4H_4O	Furan	9,000	1.7×10^{-3}

After Burg and Burg (1967); Sisler and Yang (1984); Sisler and Wood (1988).
Source: Sisler (1991).

Girardin (1864) reported premature shedding of leaves and defoliation of trees near faulty gas mains (leaking gas). Molisch (1884) speculated that abnormal and geotropic responses and horizontal growth of seedlings of etiolated pea seedlings were related to smoke and traces of illuminating gas present in air. Ethylene action on plants was noted for the first time by Neljubow (1901). He identified C_2H_4 as a component of illuminating gas and concentrations of C_2H_4 were sufficient to cause a curvature response similar to that observed by illuminating gas alone. Later, C_2H_4 was identified by Crocker and Knight (1908) and Knight and Crocker (1913) as an active constituent of both illuminating gas and tobacco smoke, causing early flowering of carnations. Cousins (1910) speculated that the premature ripening of bananas stored near oranges was related to the production and release of some unidentified volatile agent (gas) from oranges. Sievers and True (1912) reported that fumes of incompletely combusted kerosene promoted ripening of citrus fruits. Biological responses produced by paper burning were attributed to C_2H_4 (Harvey, 1913). Similarly, early fading and closure of blossoms near traces of a gas escaping from a glasshouse were observed by flower growers (Knight and Crocker, 1913).

Ethylene was found to be the active substance in kerosene fumes causing ripening of lemons. Smoke initiated and synchronized the flowering of pineapple and mango (Denny, 1924a,b; Rodriguez, 1932). Many workers demonstrated that exogenously applied C_2H_4 was effective in inducing fruit ripening, flowering, and leaf abscission, and in breaking seed dormancy (Denny, 1924a,b; Rosa, 1925; Zimmerman et al., 1930; Rodriguez, 1932). Elmer (1932) reported that potatoes stored in containers with ripe apples or pears inhibited shoot growth. The effect was attributed to a volatile product from the fruits. A volatile substance produced by ripe apples stimulated the so-called "climacteric" of unripe fruit (Huelin, 1933; Kidd and West, 1933). Botjes (1933) reported that a volatile substance from apples causing epinasty in tomato seedlings was removed from air by an C_2H_4 absorbent. Gane (1934, 1935) provided the indisputable chemical evidence for C_2H_4 production by plants. It was suggested that C_2H_4 is a natural product of fruit ripening and its release stimulates ripening of fruit on exposure.

Later on, C_2H_4 production was demonstrated not only by ripening fruit, but also by other plant organs such as shoots, flowers, seeds, leaves, and roots (Denny and Miller, 1935; Nelson and Harvey, 1935). Crocker et al. (1935) proposed that C_2H_4 should be considered as a plant hormone. Zimmerman and Wilcoxon (1935) found that auxin treatment of plant tissues sometimes stimulated C_2H_4 production. It was suggested that the physiological effects of auxin and C_2H_4 may overlap and some scientists were critical in accepting C_2H_4 as a plant hormone because of the strong interaction between auxin and C_2H_4. It was suggested that the physiological effects of C_2H_4 were, in fact, due to alterations in auxin content (Went and Thimann, 1937). After development of various methods for qualitative analysis of C_2H_4 in plants, C_2H_4 was detected in numerous plants (Pratt et al., 1948; Niederl and Brenner, 1938; Niederl et al., 1938; Gibson and Crane, 1963; Burg, 1962; Ward et al., 1978). In the middle of this century, C_2H_4 was accepted by fruit physiologists as an important natural self-regulator of fruit ripening (Biale, 1950). Extremely sensitive analytical techniques involving gas chromatography were introduced in the field of C_2H_4 research, which promoted rapid progress in this field and finally led to the acceptance of C_2H_4 as an endogenous plant hormone (Burg and Stolwijk, 1959; Huelin and Kennett, 1959; Meigh, 1959).

Two additional discoveries have brought a surge in C_2H_4 studies. The first was the discovery of compounds which blocked C_2H_4 biosynthesis and the second was of compounds which blocked C_2H_4 action. Owens et al. (1971) reported the inhibition of C_2H_4 production by rhizobitoxin, a phytotoxin produced by the root nodule bacterium, *Rhizobium japonicum*. Rhizobitoxin and its analog, aminoethoxyvinylglucine (AVG) were shown to be inhibitors of pyridoxal phosphate-based enzymes such as ACC synthase. AVG and other compounds, such as α-aminooxyacetic acid (AOA) and L-canaline, were used to block C_2H_4 biosynthesis in hundreds of studies (Amrhein and Wenker, 1979). In 1976, Beyer reported that silver, Ag(I), was a potent anti-ethylene agent (Beyer, 1976a,b). Beyer's discovery was startling because the ability of a substance to inhibit the action of a plant hormone was unparalleled in plant biology. Three years later, Sisler (1979) introduced the use of various volatile unsaturated ring compounds as inhibitors of C_2H_4 action, which had a profound effect on C_2H_4 research. The modern search for the metabolic pathway of C_2H_4 biosynthesis started with the identification of methionine as an C_2H_4 precursor (Lieberman and Mapson, 1964). Adams and Yang (1977) showed that C_2H_4 was derived from S-adenosylmethionine (SAM). The search for the C_2H_4 pathway culminated in 1979 with the discovery that 1-aminocyclopropane-1-carboxylic acid (ACC) was the immediate precursor of C_2H_4 (Adams and Yang, 1979). Lürssen et al. (1979) identified ACC as the precursor of C_2H_4 by screening thousands of chemicals. The last two decades have seen a virtual explosion in research on the biology and biochemistry of C_2H_4 as a plant hormone.

1.5. SOURCES OF ETHYLENE AND APPLICATIONS

The effect of non-plant produced C_2H_4 on the biochemistry of plants appears to be the same as that which they synthesize (Bunch et al., 1991). Most of the functions attributed to C_2H_4 have been studied by exogenous applications of this hormone, implying that C_2H_4 sources could be of geat importance with respect to altering the growth and development of plants and subsequently agricultural production. These sources can be divided into two main categories:

(i) Agricultural Sources: These refer to sources which may have a direct or indirect affect on plants due to the hormonal activity of C_2H_4 . These include C_2H_4 production by higher and lower plants, soil microbiota, biological and nonbiological accumulation of C_2H_4 in the soil atmosphere and commercially available C_2H_4 gas or C_2H_4-releasing compounds such as ethephon, Ethrel and Retprol. The agricultural sources include both synthetic and natural sources of C_2H_4 , and have great applications in agriculture.

(ii) Non-Agricultural Sources: These refer to the sources which contribute to C_2H_4 as a pollutant to the atmosphere and include industry emission, fires and automobiles.

Among the natural sources, soil microbiota are considered the most potent C_2H_4 producers in addition to C_2H_4 biosynthesis by higher plants. In contrast to plants, however, there is no conceivable physiological role attributed to C_2H_4 in microorganisms, even though the amount they produce can be substantial. Biochemical and physiological studies of microbial C_2H_4 have led to its classification as being a secondary metabolite (Bunch et al., 1991). The biochemistry and enzymology of C_2H_4 biosynthesis in higher plants is reviewed in Chapter 2, while the biochemistry and factors affecting microbial biosynthesis of C_2H_4 are revealed in Chapters 3 and 4, respectively. The accumulation of physiologically active concentrations of C_2H_4 in the soil atmosphere is discussed in Chapter 5 with special reference to biological and nonbiological generation of C_2H_4 in soil with response to various organic and inorganic amendments. The role of C_2H_4 in extremely important plant-microbe interactions such as symbiosis (nodulation, mycorrhizae and lichens) and pathogenesis, is critically reviewed in Chapters 6 and 7, respectively. Among the synthetic sources, the commercially available C_2H_4 gas or C_2H_4-releasing compounds such as ethephon/Ethrel and Retprol are the most important and are used in agriculture in various parts of the world. The agricultural uses of these commercial sources of C_2H_4 are highlighted in Chapter 8.

1.6. FUTURE PROSPECTS

There is no doubt that C_2H_4 along with other plant hormones regulate the growth and development of plants. The agricultural industry can benefit from the use of agricultural sources of C_2H_4. Recent work has demonstrated that the potential of soil microflora as a source of exogenous C_2H_4 in the rhizosphere, as well as various commercial C_2H_4-releasing products can be successfully used to improve crop yields. Moreover, recent developments in the molecular and genetic aspects of C_2H_4 have opened up opportunities to regulate C_2H_4-mediated plant responses in favor of desired directions. According to Nickell (1982), "although use of PGRs is presently small in volume and value compared to their more important pesticide relatives – herbicides, insecticides, and fungicides – predictions are that the rate of dollar volume will increase much faster for PGRs than for pesticides during the next several years, thus being the most rapidly expanding segment of the agricultural chemical business."

1.7. REFERENCES

Abeles, F. B., 1973, *Ethylene in Plant Biology*, Academic Press, New York and London.

Abeles, F. B. and Gahagan, H. E., 1968, Abscission: The role of ethylene, ethylene analogues, carbon dioxide and oxygen, *Plant Physiol.* **43**:1255-1258.

Abeles, F. B., Morgan, P. W., Saltveit, M. E., Jr., 1992, *Ethylene in Plant Biology*, 2nd ed., Academic Press, San Diego, CA.

Adams, D. O., and Yang, S. F., 1977, Methionine metabolism in apple tissues. Implication of S-adenosylmethionine as an intermediate in the conversion of methionine to ethylene, *Plant Physiol.* **60**:892-896.

Adams, D. O., and Yang, S. F., 1979, Ethylene biosynthesis: Identification of 1-aminocyclopropane-1-carboxylic acid as an intermediate in the conversion of methionine to ethylene, *Proc. Natl. Acad. Sci. USA* **76**:170-174.

Amrhein, N., and Wenker, D., 1979, Novel inhibitors of ethylene production in higher plants, *Plant Cell Physiol.* **20**:1635-1642.

Arshad, M., and Frankenberger, W. T., Jr., 1992, Microbial biosynthesis of ethylene and its influence on plant growth, *Adv. Microbial. Ecol.* **12**:69-111.

Arshad, M. and Frankenberger, W. T., Jr., 1993, Microbial Production of Plant Growth Regulators, in: *Soil Microbial Ecology*, B. F. Metting, ed., Marcel Dekker, New York, NY, pp. 307-347.

Arshad, M., and Frankenberger, W. T., Jr., 1998, Plant growth-regulating substances in the rhizosphere: Microbial production and functions, *Adv. Agron.* **62**:45-151.

Beyer, E. M., Jr., 1976a, Silver ion: A potent antiethylene agent in cucumber and tomato, *HortSci.* **11**:195-196.

Beyer, E. M., Jr., 1976b, A potent inhibitor of ethylene action in plants, *Plant Physiol.* **58**:268-271.

Beyer, E. M., Jr., Morgan, P. W., and Yang, S. F., 1984, Ethylene, in: *Advanced Plant Physiology*, M. B. Wilkins, ed., Pitman Publishing, London, pp. 111-126.

Biale, J. B., 1950, Postharvest physiology and biochemistry of fruits, *Annu. Rev. Plant. Physiol.* **1**:183-206.

Botjes, J. O., 1933, Aethyleen aus vermoedelijk oorzaak von de groeivemmende werking van rijpe appels, *Plantonzick* **39**:207-211.

Bunch, A. W., McSwiggan, S., Lloyd, J. B., and Shipston, N. F., 1991, Ethylene synthesis by soil microorganisms: Its agricultural and biotechnological importance, *Agrofoodindustry Hi-Tech.* **2**:21-24.

Burg, S. P., 1962, The physiology of ethylene formation, *Annu. Rev.* Plant *Physiol.* **13**:265-302.

Burg, S. P., and Burg, E. A., 1965, Ethylene action and the ripening of fruits, *Science* **148**:1190-1196.

Burg, S. P., and Burg, E. A., 1967, Molecular requirements for the biological activity of ethylene, *Plant Physiol.* **42**:144-152.

Burg, S. P., and Stolwijk, J. A. A., 1959, A highly sensitive katharometer and its application to the measurement of ethylene and other gases of biological importance, *J. Biochem. Microbiol. Technol. Eng.* **1**:245-259.

Cousins, H. H., 1910, *Annual Report of the Jamaican Department of Agriculture* **7**:15.

Crocker, W., and Knight, L. I., 1908, Effect of illuminating gas and ethylene upon flowering carnations, *Bot. Gaz.* **46**:259-276.

Crocker, W., Hitchcock, A. E., and Zimmerman, P. W., 1935, Similarities in the effects of ethylene and the plant auxins, *Contrib. Boyce Thompson Inst.* **7**:231-248.

Davies, P. J., 1995, The plant hormones: Their nature, occurrence, and functions, in: Plant Hormones and Their Role in Plant Growth and Development, P. J. Davies, ed., Martinus Nijhoff, Dordrecht, pp. 1-11.

Denny, F. E., 1924a, Hastening the coloration of lemons, *J. Agric. Res.* **27**:757-768.

Denny, F. E., 1924b, Effect of ethylene upon respiration of lemons, *Bot. Gaz.* **77**:322-329.

Denny, F. E., and Miller, L. P., 1935, Production of ethylene by plant tissue as indicated by the epinastic response of leaves, *Contrib. Boyce Thompson Inst.* **7**:97-102.

Eger, E. I., and Larson, C. P., 1964, Anesthetic solubility in blood tissues. Values and significance, *Brit. J. Anaesthesia* **36**:140-149.

Elmer, O. H., 1932, Growth inhibition of potato sprouts by the volatile products of apples, *Science* **75**:193.

Frankenberger, W. T., Jr., and Arshad, M., 1995, *Phytohormones in Soils: Microbial Production and Functions*, Marcel Dekker, New York, NY.

Gane, R., 1934, Production of ethylene by some ripening fruit, *Nature* **134**:1008.

Gane, R., 1935, The formation of ethylene by plant tissues and its significance in the ripening of fruits, *J. Pomol. Hort. Sci.* **13**:351-358.

Gibson, M. S., and Crane, F. L., 1963, Paper chromatography method for identification of ethylene, *Plant Physiol.* **38**:729-730.

Girardin, J. P., 1864, Einflub des Leuchtgases auf die Promnaden und Strabenblaume, *Jahresber. Agric. Chem.* **7**:199-200.

Harvey, E. M., 1913, The castor bean plant and laboratory air, *Bot. Gaz.* **56**:439-442.

Huelin, F. E., 1933, Effects of ethylene and of apple vapours on the sprouting of potatoes, *Rep. Food Invest. Board Gr. Br.*, pp. 51-53.

Huelin, F. E., and Kennett, B. H., 1959, Nature of the olefines produced by apples, *Nature* **184**:996.

Imaseki, H., 1986, Ethylene, in: Chemistry of Plant Hormones, N. Takahashi, ed., CRC Press, Boca Raton, FL, pp. 249-264.

Jackson, M. B., 1985, Ethylene and responses of plants to soil waterlogging and submergence, *Annu. Rev. Plant Physiol.* **36**:145-174.

Kidd, F., and West, C., 1933, The influence of the composition of the atmosphere upon the incidence of the climacteric in apples, *Gt. Brit. Dept. Sci. Ind. Res. Food Invest. Bd. Rept.*, pp. 51-57.

Knight, L. I., and Crocker, W., 1913, Toxicity of smoke, *Bot. Gaz.* **55**:337-371.

Lieberman, M., 1979, Biosynthesis and action of ethylene. *Annu. Rev. Plant Physiol.* **30**:533-591.

Lieberman, M., and Mapson, L. W., 1964, Genesis and biogenesis of ethylene, *Nature* **204**:343-345.

Lürssen, K., 1991, Ethylene and agriculture, in: *The Plant Hormone Ethylene*, A. K. Mattoo and J. C. Suttle, eds., CRC Press, Boca Raton, FL, pp. 315-326.

Lürssen, K., Nauman, K., and Schroder, R., 1979, 1-Aminocyclopropane-1-carboxylic acid--an intermediate of the ethylene biosynthesis in higher plants, *Z. Pflanzenphysiol. Bd.* **92**:285-294.

Mattoo, A. K., and Suttle, J. C., eds., 1991, *The Plant Hormone Ethylene*. CRC Press, Boca Raton, FL.

McKeon, T. A., and Yang, S.-F., 1987, Biosynthesis and metabolism of ethylene, in: *Plant Hormones and Their Role in Plant Growth and Development*, P. J. Davies, ed., Martinus Nijhoff, Dordrecht, pp. 94-112.

Meigh, D. F., 1959, Nature of the olefines produced by apples, *Nature* **184**:1072-1073.

Molisch, H., 1884, Setzungsber, *Kais. Acad. Wiss. (Wein)* **90**:111-196.

Moore, T. C., 1989, Ethylene: *Biochemistry and Physiology of Plant Hormones*, 2nd ed., Springer-Verlag, New York, pp. 228-259.

Neljubow, D. N., 1901, Uber die horizontale Nutation der Stengel von *Pisum sativum* und einiger anderen Pflanzen, *Beih. Bot. Zentralbl.* **10**:128-138.

Nelson, R. C., and Harvey, R. B., 1935, The presence in self-blanching celery of unsaturated compounds with physiological action similar to ethylene, *Science* **82**:133-134.

Nickell, L. G., 1982, *Plant Growth Regulators. Agricultural Uses*, Springer-Verlag, Berlin, Heidelberg, Germany.

Niederl, J. B., and Brenner, M. W., 1938, Micro-method for the identification and estimation of ethylene in ripening fruits, *Mikrochemie* **24**:134-145.

Niederl, J. B., Brenner, M. W., and Kelly, J. N., 1938, The identification and estimation of ethylene in the volatile products of ripening bananas, *Am. J. Bot.* **25**:357-360.

Osborne, D. J., 1978, Ethylene, in: *Phytohormones and Related Compounds -- A Comprehensive Treatise*, Vol. 1, D. S. Letham, P. B. Goodwin, and T. J. V. Higgins, eds., Elsevier/North Holland Biomedical Press, Amsterdam, pp. 265-294.

Owens, L. D., Lieberman, M., and Kunishi, A., 1971, Inhibition of ethylene production by rhizobitoxin, *Plant Physiol.* **48**:1-4.

Pratt, H. K., and Goeschl, J. D., 1969, Physiological roles of ethylene in plants, *Annu. Rev. Plant Physiol.* **20**:541-584.

Pratt, H. K., Young, R.E., and Biale, J. B., 1948, The identification of ethylene as a volatile product of ripening avocados, *Plant Physiol.* **23**:526-531.

Reid, M. S., 1987, The functioning of hormones in plant growth and development. E1, Ethylene in plant growth, development and senescence, in: *Plant Hormones and Their Role in Plant Growth and Development*, P. J. Davies, ed., Martinus Nijhoff, Dordrecht, pp. 257-279.

Rodriguez, A. G., 1932, Influence of smoke on ethylene on the fruiting of pineapple (*Ananas satius* Shult), *J. Dept. Agric. Puerto Rico* **16**:5-18.

Rosa, J. T., 1925, Shortening the rest period of potatoes by ethylene gas, *Potato News Bull.* **2**:363-365.

Sembdner, G., Gross, D., Liebisch, H. W., and Schneider, G., 1980, Biosynthesis and metabolism of plant hormones, in: *Hormonal Regulation of Development. I. Molecular Aspects of Plant Hormones, Encyclopedia of Plant Physiology*, New Series, Vol. 9. J. MacMilan, ed., Springer-Verlag, Berlin, p. 281-444.

Sievers, A. F., and True, R. H., 1912, A preliminary study of the forced curing of lemons as practiced in California, *U.S. Dept. Agric. Bur. Plant Ind. Bull.* **232**:1-38.

Sisler, E. C., 1979, Measurement of ethylene binding in plant tissues, *Plant Physiol.* **64**:538-542.

Sisler, E. C., 1991, Ethylene-binding components in plants, in: *The Plant Hormone Ethylene*, A. K. Mattoo and J. C. Suttle, eds., CRC Press, Boca Raton, FL, pp. 81-99.

Sisler, E. C., and Yang, S. F., 1984, Ethylene, the gaseous plant hormone, *Bioscience* **34**:234-238.

Sisler, E. C., and Wood, C., 1988, Computation of ethylene for unsaturated compounds for binding and action in plants, *Plant Growth Regul.* 7:181-191.

Spencer, M. S., 1969, Ethylene in nature, in: *Progress in the Chemistry of Organic Natural Products*, L. Zechmeister, ed., Springer-Verlag, Wien and New York, pp. 32-80.

Takahashi, N., 1986, *Chemistry of Plant Hormones*, CRC Press, Boca Raton, FL.

Ward, T. M., Wright, M., Roberts, J. A., Self, R., and Osborne, D. J., 1978, Analytical procedures for the assay and identification of ethylene, in: *Isolation of Plant Growth Substances*, J. R. Hillman, ed., Cambridge University Press, Cambridge, pp. 135-151.

Went, F. W., and Thimann, K. V., 1937, *Phytohormones*, Macmillan, New York, p. 294.

Yang, S. F., and Hoffman, N. E., 1984, Ethylene biosynthesis and its regulation in higher plants, Annu. Rev. Plant Physiol. **35**:155-189.

Zimmerman, P. W., and Wilcoxon, F., 1935, Several chemical growth substances which cause initiation of roots and other responses in plants, *Contrib. Boyce Thompson Inst.* 7:209-229.

Zimmerman, P. W., Crocker, W., and Hitchcock, A. E., 1930, The response of plants to illuminating gas, *Proc. Am. Soc. Hort. Sci.* **27**:53-56.

2

ETHYLENE IN PLANT PHYSIOLOGY

2.1. INTRODUCTION

Ethylene (C_2H_4) is a plant hormone that is involved in the regulation of many physiological responses (Abeles et al., 1992; Mattoo and Suttle, 1991; Reid, 1995). Initially designated as a "ripening hormone", C_2H_4 is involved in almost all growth and developmental processes ranging from germination of seeds to senescence of various organs and in many responses to environmental stress. Ethylene production occurs in all plant organs, including roots, stems, leaves, buds, tubers, bulbs, flowers, and seeds, but the magnitude of C_2H_4 production varies from organ to organ and is dependent on growth and developmental processes. Recent scientific progress has increased the understanding of biosynthetic pathways and enzymes involved in C_2H_4 production including genetic control, leading to the development of several ways to manipulate C_2H_4 production by genetic alteration of plants (Kende, 1993; Fluhr and Mattoo, 1996; Zarembinski and Theologis, 1994; Stella et al., 1996; Woltering and de Vrije, 1995).

2.2. ENDOGENOUS PRODUCTION OF ETHYLENE

The endogenous level of C_2H_4 in plants is controlled primarily by its rate of production. Measurements of the rate of C_2H_4 released per unit amount of tissue provides information on the relative changes of C_2H_4 in cellular concentrations. Ethylene is produced by various parts/organs of plants growing under normal conditions; however, any kind of biological, chemical or physical stress strongly promotes endogenous C_2H_4 synthesis by plants (Abeles et al., 1992; Hyodo, 1991). For instance, in germinating seeds, high rates of C_2H_4 production usually occur when the radical starts to penetrate the seed coat and during periods when the seedling forces its way through soil (Beyer et al., 1984).

The plant hormone, auxin is known to have a strong interaction with C_2H_4 and a higher rate of C_2H_4 production in young tissues often correlates with high auxin concentrations (Kende, 1993; Peck and Kende, 1995).

2.3. BIOSYNTHETIC PATHWAY

   Ethylene was recognized as a plant hormone over 50 years ago, yet the biosynthetic pathway of C_2H_4 in plants remained elusive until the key intermediate, 1-aminocyclopropane-1-carboxylic acid (ACC) was identified as the immediate precursor of C_2H_4 in methionine (MET)-dependent C_2H_4 production in plant tissues. The formation of C_2H_4 from MET proceeds through two major intermediates, S-adenosylmethionine (SAM) and ACC (Fig. 2.1). The credit of discovering the C_2H_4 biosynthetic pathway in plants primarily goes to Shang Fa Yang and his co-workers. Their work provides the basis for all subsequent biochemical and molecular genetic analysis of the pathway (Yang and Hoffmann, 1984; Beyer et al., 1984). Several discoveries in the late 1960s and 1970s resulted in understanding C_2H_4 physiology and unraveling the biosynthetic pathway of C_2H_4. These include the discovery of MET as the precursor of C_2H_4 (Lieberman et al., 1966), the discovery of rhizobitoxine and its analog aminoethoxyvinylglycine (AVG) as specific inhibitors of C_2H_4 biosynthesis (Owens et al., 1971), the use of silver salts (Beyer, 1976) and cyclic olefins (Sisler, 1977) as inhibitors of C_2H_4 action, and finally, the breakthrough discovery of ACC as the intermediate between MET and C_2H_4 (Adams and Yang, 1979; Lurssen et al., 1979). The existence of this MET-ACC pathway of C_2H_4 biosynthesis in plants has been confirmed by the use of inhibitors, precursors, and labeled compounds (Yang and Hoffman, 1984).

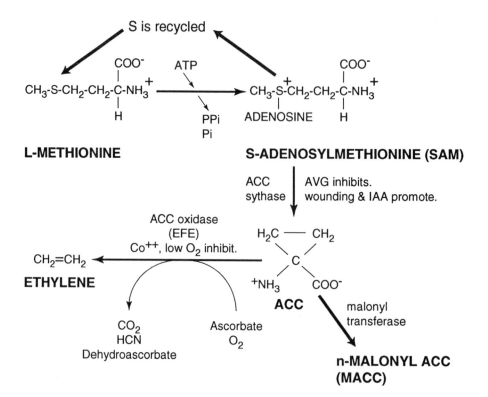

Fig. 2-1.   Biosynthetic pathway of C_2H_4 . Source: Gaspar et al. (1996).

2.3.1. Establishment of Methionine as a Physiological Precursor of Ethylene

Chemical model systems were initially used to study C_2H_4 synthesis because cell-free homogenates from plants did not evolve C_2H_4 (Lieberman and Mapson, 1964). A number of precursors were identified that produced C_2H_4 through various chemical reactions (Mapson, 1969; Yang, 1974). Methionine, however, was the only candidate that produced C_2H_4 when fed to apples (Lieberman et al., 1966) and other plant tissues (Burg and Clagget, 1967; Lieberman, 1979). Although MET carbons 3 and 4 form C_2H_4 and carbon 1 yields CO_2 in both chemical and plant tissue, the mechanisms of C_2H_4 formation are different. Experiments with labeled MET unequivocally established that MET is the physiological precursor of C_2H_4 in higher plants. The steps involved in conversion of MET into C_2H_4 are discussed in the following sections.

2.3.2. Conversion of Methionine into S-Adenosylmethionine (Step I)

The pathway starts with the conversion of MET into SAM and other intermediates (Burg, 1973; Murr and Yang, 1975). The pathway of MET to C_2H_4 requires O_2 and is inhibited by 2,4-dinitrophenol, a decoupler of oxidative phosphorylation, implying the involvement of an energy-dependent step in the biosynthesis of C_2H_4 from MET which led to recognition of SAM in the pathway (Burg and Clagget, 1967; Burg, 1973; Murr and Yang, 1975). Adams and Yang (1977) confirmed this proposal by demonstrating that labeled $[^{35}S']$-MET and $[^3H]$-methylmethionine released labeled 5'-methythioadenosine (MTA) and its hydrolysis product, 5-methylthioribose (MTR) upon its conversion to C_2H_4 in apple tissues. They concluded that MET must be converted into SAM before C_2H_4 is released. The activation of MET into SAM is demonstrated in Fig. 2.2 (Stella et al., 1996).

The conversion of SAM into ACC occurs at such a low rate that it may not result in depletion of steady state SAM levels, so the enzyme responsible for SAM formation (methionine adenosyltransferase) is unlikely to become a rate-limiting enzyme in C_2H_4 biosynthesis (Yu and Yang, 1979). However, the continuous flux of the substrates of intermediary metabolism does not allow unambiguous experimentation to prove that SAM levels do not significantly change during plant metabolism (Fluhr and Mattoo, 1996). In this regard, the operation of a MET cycle (see Section 2.2.5) results in recycling of MTA to regenerate MET and thereby SAM, providing a salvage pathway to maximize the availability of SAM. On the other hand, studies with selenomethionine-stimulated C_2H_4 synthesis (Konze and Kende, 1979b) and those that used inhibitors of either C_2H_4 or polyamine biosynthesis (Roberts et al., 1984) indicate that SAM levels may indeed regulate C_2H_4 production (Fluhr and Mattoo, 1996). Moreover, the sensitivity of methionine adenosyltransferase to SAM implies that this enzyme may play a regulatory role in C_2H_4 biosynthesis (Giovanelli et al., 1980). SAM can also be converted to S-adenosylmethylthiopropylamine by the enzyme, S-adenosylmethionine decarboxylase (SAM-decarboxylase) (Icekson et al., 1985). Treatment of pea seedlings with 0.3 μl L^{-1} C_2H_4 resulted in a reduction of SAM-decarboxylase activity.

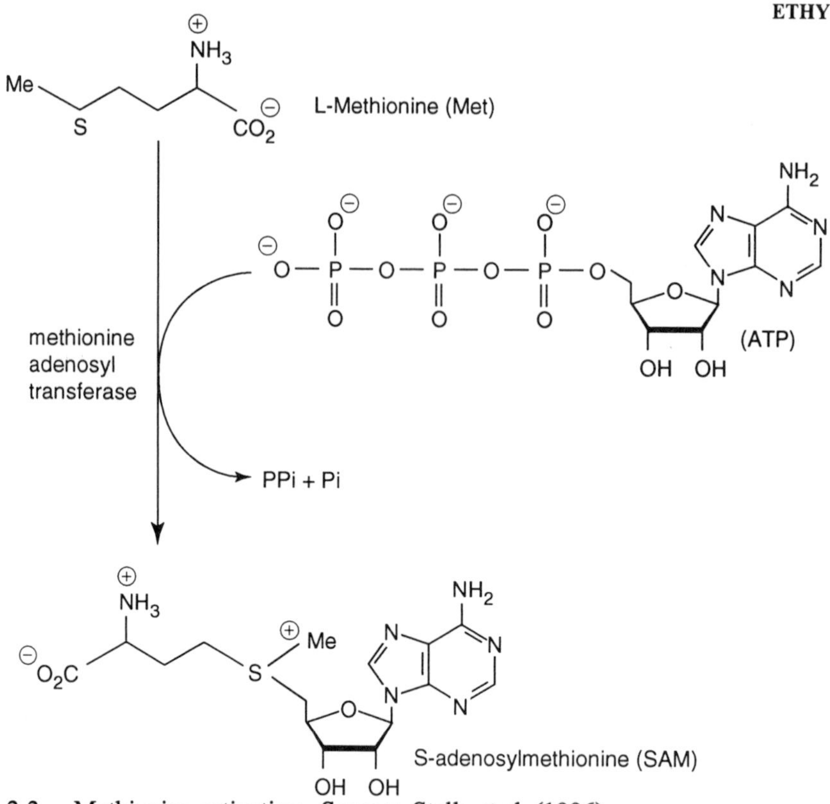

Fig. 2-2.   Methionine activation. Source: Stella et al. (1996).

2.3.2.1. Biochemical and Molecular Aspects of Methionine Adenosyltransferase

The enzyme, methionine adenosyltransferase (ATP-methionine S-adenosyl-transferase, EC 2.5.1.6) catalyzes the conversion of MET and ATP into SAM. This enzyme has been partially purified with a reported K_m for MET of 0.4 mM and for ATP of 0.3 mM (Aarnes, 1977). This enzyme has been classified as an important "house-keeping" enzyme and is ubiquitous in plant tissues (Thomas and Surdin-Kerjan, 1991). It is present in multiple molecular forms and encoded by a multi-gene family (Peleman et al., 1989; Larsen and Woodson, 1991).

2.3.3. Conversion of S-Adenosylmethionine into 1-Aminocycloprpane-1-carboxylic Acid (Step II)

The field of C_2H_4 physiology and biochemistry progressed rapidly ever since Adams and Yang (1979) identified ACC as the immediate precursor of C_2H_4 in higher plants. The formation of ACC from SAM is mediated by a pyridoxal enzyme, ACC synthase (EC 4.4.1.14) (Adams and Yang, 1979). The low concentration and instability of ACC synthase has made its isolation and characterization difficult.   Moreover, the immunological properties of this enzyme isolated from different plant tissues varies. For example, ACC synthase isolated from ripening apples (Yang et al., 1990a) did not cross-react with a tomato monoclonal antibody (Bleecker et al., 1986) or with winter squash polyclonal antibodies (Nakajima et al., 1988). Both internal cues and external inducers cause *de novo* synthesis of ACC synthase (Kende, 1989; Yang and Hoffman, 1984). Exposure of a climateric fruit to C_2H_4 induced ACC synthase (Mita et al., 1999), but not in nonclimateric fruits (Shiomi et al., 1999). During ripening of mume fruit, Mita et al. (1999) noted marked increased expression of ACC synthase prior to that of ACC oxidase and an increase in C_2H_4 biosynthesis. The levels of the mRNAs for the genes corresponded closely to the levels of activity of the C_2H_4 –biosynthetic enzymes (Mita et al., 1999). Taken together, ACC synthase represents the key regulator enzyme in the pathway. It is strongly inhibited by AVG (a competitive inhibitor) and aminoisobutyric acid (an inhibitor of pyridoxal phosphate-mediated enzyme reactions) (Rando, 1974). The overall biochemistry of SAM conversion into ACC is demonstrated in Fig. 2.3 (Stella et al., 1996). By using SAM with stereospecific deuterium labeled on carbon atoms 3 and 4, Wiesendanger et al. (1986) have proved that the mechanism of formation of ACC is via direct nucleophillic substitution of the sulfonium group accompanied by a complete inversion of the configuration of carbon atom 4.

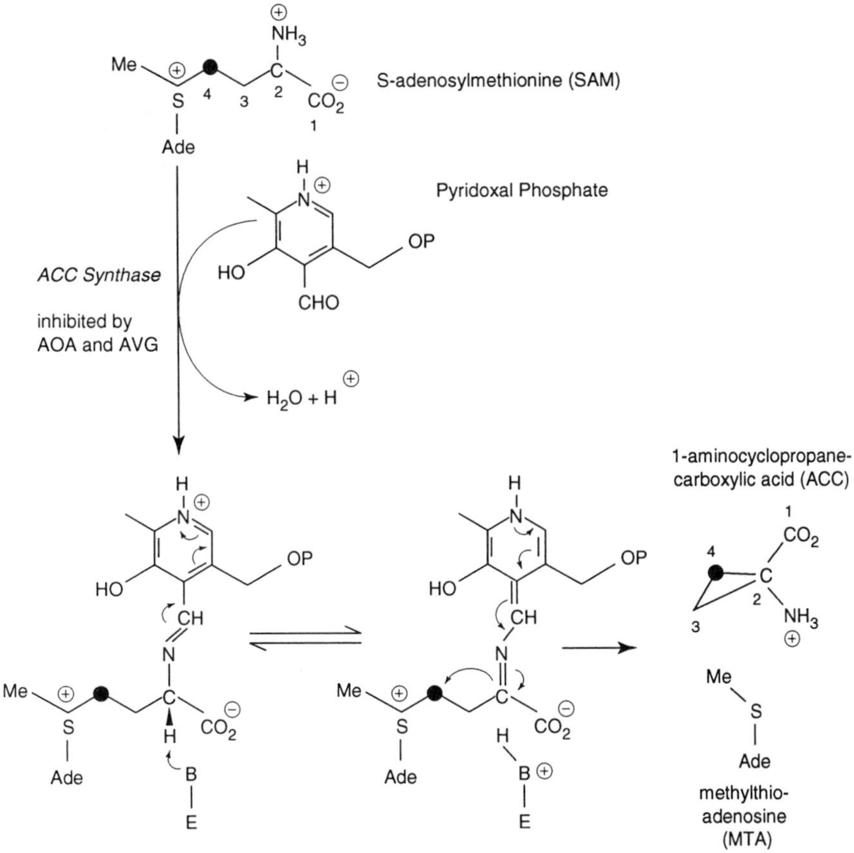

Fig. 2-3.   The formation of 1-aminocyclopropane-1-carboxylic acid (ACC) from *S*-adenosylmethionine (SAM). Source: Stella et al. (1996).

2.3.3.1. Biochemical and Molecular Aspects of ACC Synthase

ACC synthase is a rate-limiting, cytosolic enzyme which catalyzes the first committed step in the C_2H_4 biosynthetic pathway. This enzyme catalyzes the elimination of 5'-methylthioadenosine (MTA) from SAM to yield ACC. The K_m for SAM is between 13 and 60 µM (Boller et al., 1979; Burns and Evensen, 1986; Privalle and Graham, 1987; Yu et al., 1979; Yoshii and Imaseki, 1981). ACC synthase has a molecular weight of 45 to 58 kDa (Yang, 1980; Acaster and Kende, 1983; Bleecker et al., 1986) and a pH optimum of 8.5 (Boller et al., 1979; Yu et al., 1979). The K_i for AVG is around 10 µM (Privalle and Graham, 1987). Pure ACC synthase has an activity of 2-4 x 10^5 units per mg protein (1 unit = 1 nmole ACC produced per hour at 30°C) (Kende, 1989). The enzyme functions as a homodimer (Capitani et al., 1999; Li et al., 1997). ACC synthase is labile *in vitro* and *in vivo*. Its half-life is short; the $t_{1/2}$ of tomato ACC synthase is 58 minutes (Kim and Yang, 1992). The reaction of the isolated enzyme with SAM was linear for only 30 minutes (Yoshii and Imaseki, 1981). At the normal endogenous level of 40 µM SAM, the catalytic activity of ACC synthase lasted 24 minutes (Satoh and Esashi, 1986). The rate of the reaction with (*S,S*)-SAM is diffusion-controlled and exhibits a limiting k_{cat}/K_m value of 1.2 x 10^6 M^{-1} s^{-1} at pH 8.3 (Li et al., 1997).

ACC synthase catalyzes the formation of ACC and MTA from SAM seemingly by an α,γ-elimination reaction mechanism (Adam and Yang, 1979; Ramalingham et al., 1985). It has suicidal inhibition during catalysis when the substrate SAM undergoes a β,γ-elimination reaction to form vinylglycine and the resulting alkylation irreversibly inactivates the enzyme (Satoh and Yang, 1988, 1989; Yip et al., 1990). The active site of apple ACC synthase has been shown to contain an amino acid sequence that participates in binding pyridoxal phosphate. This sequence was covalently linked to the 2-aminobutyrate portion of SAM during inactivation (Yip et al., 1990). Yang (1996) has postulated a scheme as shown in Fig. 2.4 to describe the mechanism based on inactivation of ACC synthase by SAM.

Amino acid profiling (Mehta and Christen, 1994) and X-ray analysis (Capitani et al., 1999) showed that ACC synthase is a member of the α family of pyridoxal 5'-phosphate-dependent enzymes. The roles of several active-site residues surrounding the pyridoxal 5'-phosphate binding pocket have been clarified by site-directed mutagenesis investigations (White et al. 1994; Li et al., 1997). Random mutagenesis *in vivo* was also used in an attempt to identify the functional residues (Tarun et al., 1995). It is noteworthy that the first four amino acids of this sequence are characteristic of all enzymes of the known sequence functioning with pyridoxal phosphate as a coenzyme (Stella et al., 1996).

L-Vinylglycine (L-VG) has been shown to be a mechanism-based inhibitor of ACC synthase (Satoh and Yang, 1989) as well as other pyridoxal phosphate-dependent enzymes. Recently, Feng and Kirsch (2000) demonstrated that L-VG is primarily an alternative substrate for ACC synthase. The L-VG deaminase activity of ACC synthase yields the products, α-ketobutyrate and ammonia with a k_{cat} value of 1.8 s^{-1} and a K_m of 1.4 mM.

cDNA sequences encoding ACC synthase have been isolated from various tissues including fruit of zucchini (Sato and Theologis, 1989), winter squash (Satoh and Yang, 1989; Sato et al., 1991; Nakajima et al., 1990), tomato (van der Straeten et al., 1990) and

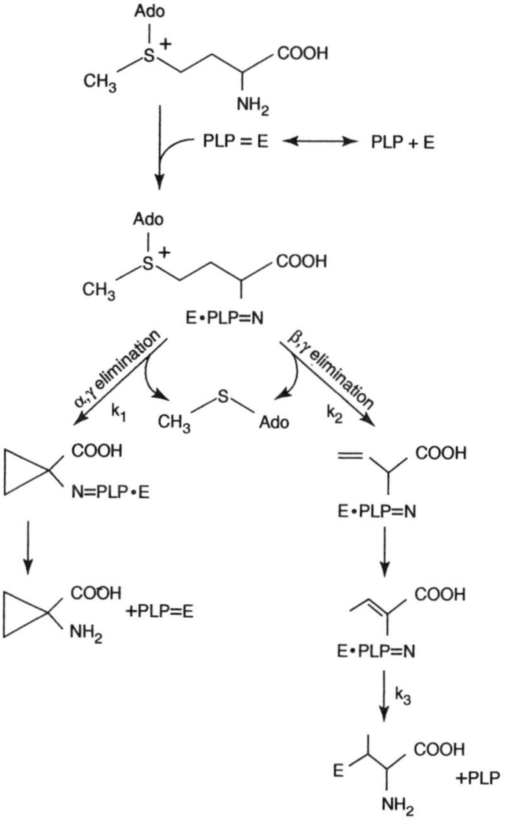

Fig. 2-4. A postulated scheme based on the inactivation of ACC synthase by SAM. Ado, E, PLP = E and PLP = N-R stand for adenosine moiety, apo-ACC synthase, internal aldimine and external aldimine, respectively. The values for k_1, k_2 and k_3 were estimated to be 300, 0.01 and 0.1 min^{-1}, respectively. Source: Yang (1996).

apple (Dong et al., 1991). The identity of the clone was confirmed by the ability of the cloned sequence to direct the synthesis of ACC synthase by *Escherichia coli* and yeast (Satoh and Theologis, 1989; Nakajima et al., 1990). The ACC synthases from apple and tomato have been expressed in *E. coli* and purified to homogeneity with yields ranging from 1 to 8 mg L^{-1} (White et al., 1994; Zhou et al., 1998). *In vivo* studies using the ACC cDNA as a probe showed that the ACC synthase gene was induced by a diverse group of inducers, including wound (Satoh and Yang, 1989; Nakajima et al., 1990; van der Straeten et al., 1990; Mita et al., 1999), ripening (Dong et al., 1991), lithium ions and IAA (Satoh and Yang, 1989). Arteca and Arteca (1999) reported that the ACC synthase gene (ACS6) in *Arabidopsis* is also induced by wounding, and by treatment with LiCl, NaCl, CuCl$_2$, auxin, cycloheximide, aminooxyacetic acid and C$_2$H$_4$ in addition to touch treatment. ACC levels were increased in response to each of these treatments with the exception of cycloheximide and aminooxyacetic acid which resulted in a decrease and no effect, respectively. In tomato, ACC synthase is encoded by six genes (Olson et al., 1991;

Rottmann et al., 1991; Yip et al., 1992; Lincoln et al., 1993); one of these, LE-ACS2, is expressed during fruit ripening and is induced by treatment with exogenous C_2H_4. LE-ACS4 appears to be expressed at a lower level during these stages (Rottmann et al., 1991). It was suggested that endogenous, auxin-induced and wounding-induced ACC synthase were coded by different genes (Nakajima et al., 1990; van der Straeten et al., 1990).

Cloning and characterization of genes encoding the two key enzymes, ACC synthase and ACC oxidase, has been carried out in many species (Sato and Theologis, 1989; Van der Straeten et al., 1990; Nakajima et al., 1990; Lay-Yee and Knighton, 1995). ACC synthase and ACC oxidase are encoded by a multigene family whose various members are differentially expressed in response to many factors (Olson et al., 1991; Rottmann et al., 1991; Lincoln et al., 1993; Lasserre et al., 1996). Expression of ACC synthase and ACC oxidase may be developmentally regulated, show tissue specificity, and be controlled by signal transduction pathways that respond to C_2H_4 (Kim et al., 1997). Itai et al. (1999) have used two ACC synthase genes (pPPACS1 and pPPACS2) and an ACC oxidase gene (pPPAOX1) as probes on 33 Japanese pear cultivars expressing different levels of C_2H_4 in ripening fruit. They reported that the cultivars that produce high levels of C_2H_4 possess at least one additional copy of pPPACS1 and those producing moderate levels of C_2H_4 have at least one additional copy of pPPACS2.

2.3.4. Conversion of 1-Aminocyclopropane-1-carboxylic Acid into Ethylene (Step III)

The last step in the MET-dependent C_2H_4 production pathway is the conversion of ACC to C_2H_4. Peiser et al. (1984) have demonstrated with labeled ACC that the carboxyl carbon of ACC is liberated as CO_2, whereas carbon 1 of ACC yields hydrogen cyanide (HCN) which is rapidly metabolized to yield β-cyanoalanine and aspargine. Dong et al. (1992) have shown that for each mole of ACC consumed, one mole of O_2 is utilized and equimolar quantities of dehydroascorbate, C_2H_4, HCN, and CO_2 are formed. The overall reaction stoichiometry of ACC oxidase activity is shown in Fig. 2.5 (Brunhuber et al., 2000). The oxidation of ACC to C_2H_4 is catalyzed by ACC oxidase, previously named as C_2H_4 –forming enzyme (EFE) (Dong et al., 1992), which requires Fe^{2+} for its activity (Ververidis and John, 1991) and is CO_2 activated (Dong et al., 1992). This enzyme has been studied in melon (Smith et al., 1992), avocado (McGarvey and Christoffersen, 1992), apple (Kuai and Dilley, 1992; Fernandez-Maculet and Yang, 1992; Dong et al., 1992; Dupille et al., 1993), winter squash (Hyodo et al., 1993), pear (Vioque and Castellano, 1994) and banana (Moya-Leon and John, 1995). The enzyme was puified to homogeneity from apple (Dong et al., 1992; Dupille et al., 1993) and banana fruits (Moya-Leon and John, 1995). The expression of the ACC oxidase gene was studied in broccoli (Pogson et al., 1995), melon (Yamamoto et al., 1995), tomato (Nakatsuka et al., 1997), pear (Lelievre et al., 1997), and kiwifruit (Whittaker et al., 1997). *In vivo* analysis of ACC oxidase activity showed that CO_2 promoted activity in leaves of sunflower and corn, while it had little or no effect on ACC oxidase activity in roots of these plants (Finlayson et al., 1997). *In vitro* analysis revealed that substrate K_m, Kco_2 and response to pH were different for the same enzyme extracted from the two organs (leaves and roots), implying that ACC oxidase may exist as organ specific isoenzymes which are tailored to the environmental and physiological status of each organ (Finlayson et al.,

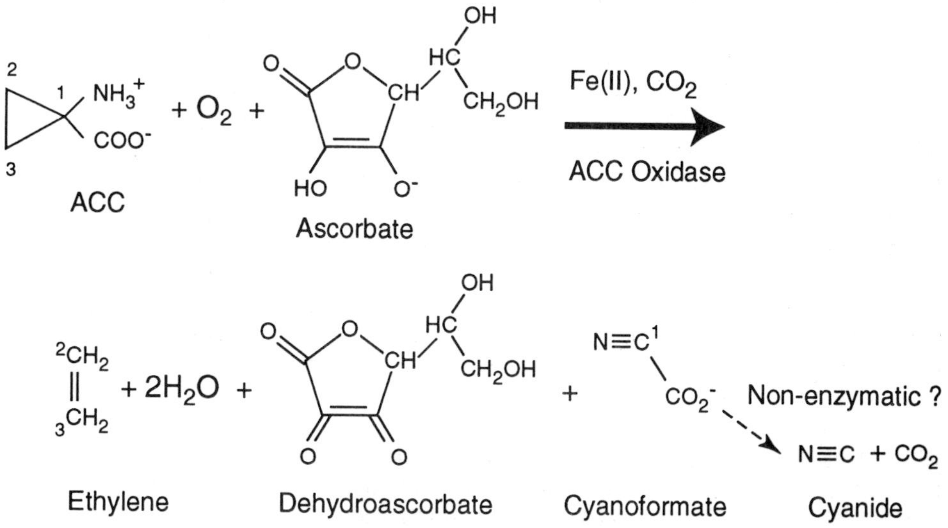

Fig. 2-5. Reaction stoichiometry of ACC oxidase. Source: Brunhuber et al. (2000).

1997). Several types of chemicals, including analogs of ACC, uncouplers of oxidative phosphorylation, free-radical scavengers, metal chelators, sulfhydryl reagents and heavy metal ions inhibit C_2H_4 production *in vivo* by their action on the conversion of ACC to C_2H_4 (Yang and Hoffman, 1984). It is well documented that cobalt ions inhibit the oxidative deamination of ACC to C_2H_4 , in both *in vivo* and *in vitro* systems. It was reported that Co^{2+} at 10 μM inhibits *in vitro* pear ACC oxidase activity by about 70% (Vioque and Castellano, 1994). α-Aminoisobutyric acid, a structural analog of ACC, acts as a competitive inhibitor of ACC oxidase *in vivo* (Satoh and Esashi, 1983) and *in vitro* (Vioque and Castellano, 1994; Fernandez-Maculet and Yang, 1992; Nijenhuis-DeVries et al., 1994).

ACC oxidase activity is not as highly regulated as that of ACC synthase. It is constitutive in most vegetative tissues (Yang and Hoffman, 1984) and it is induced during fruit ripening (McGarvey et al., 1990), wounding (Callahan et al., 1992), senescence

(Wang et al., 1982) and by fungal elicitors (Spanu et al., 1991). Exposure to exogenous C_2H_4 induces ACC oxidase activity in ripening mume (climacteric) and cucumber (nonclimacteric) fruits (Mita et al., 1999; Shiomi et al., 1999). It was suggested that the induction of ACC oxidase activity was regulated by endogenous C_2H_4 production during senescence (Kasai et al., 1996).

2.3.4.1. Biochemical and Molecular Aspects of ACC Oxidase

The enzymology of ACC oxidase has been studied extensively (Ververidis and John, 1991; Dong et al., 1992; McKeon et al., 1995; Prescott and John, 1996; John, 1997). ACC oxidase is a bisubstrate enzyme since it requires both O_2 and ACC. The K_m for ACC was only slightly dependent on O_2 concentration, but varied greatly among different plant tissues. It ranged from 8 μM in apple to 120 μM in etiolated wheat leaves. ACC oxidase belongs to the 2-oxoacid-dependent dioxygenase (2-ODD) family of enzymes (Prescott, 1993), and like other members of this group it requires Fe^{2+} and ascorbate for its *in vitro* activity. However, most other 2-ODDs use 2-oxoglutarate as a cosubstrate, whereas ACC oxidase uses ascorbate, oxidizing it stoichiometrically to dehydroascorbate (Dong et al., 1992). Also unique among 2-ODDs, ACC oxidase activity has an absolute requirement for CO_2 (Fernandez-Maculet et al., 1993; Smith and John, 1993). Molecular and biochemical studies have confirmed that this enzyme requires cofactors including iron and ascorbate and CO_2 for its activity (Hamilton et al., 1990; Ververidis and John, 1991; Dong et al., 1992; Fernandez-Maculet et al., 1993; McGarvey et al., 1992; Pirrung et al., 1993). It has been hypothesied that ACC oxidase is a non-heme Fe(II) enzyme (Prescott, 1993; Feig and Lippard, 1994). While O_2 is necessary for ACC oxidase activity in tissues, its elimination during extraction is critical for *in vitro* activity. The presence of 30 mM ascorbate and 0.1 mM $FeSO_4$ was required for stability.

The subcellular location of ACC oxidase is still a point of controversy. The primary sequence suggests ACC oxidase to be a cytosolic enzyme since it does not contain putative membrane-spanning domains or a signal peptide (Hamilton et al., 1991). However, there is a large body of data indicating that the enzyme is either associated with the plasma membrane (Kende, 1993) or that it is apoplastic or tonoplastic (Rombaldi et al., 1992; Ayub et al., 1992; Latche et al., 1992; Bouzayen et al., 1990; Konze and Kende, 1979a). The plasmalemma ACC oxidase converted apoplastic ACC to C_2H_4 more readily and was more sensitive to high osmotica than the tonoplast ACC oxidase.

Like ACC synthase, ACC oxidase is also an unstable enzyme; the $t_{1/2}$ of apple ACC oxidase is 2 h (Pirrung et al., 1993). The enzyme is particularly labile, undergoes inactivation during catalysis and has low catalytic power (Smith et al., 1994; Barlow et al., 1997; Vioque and Castellano, 1998). Castellano and Vioque (2000) reported that ACC oxidase from pear (*Pyrus communis* L. cv. *Blanquilla*) fruit tissues had a maximum activity at pH 6.5. The enzyme eluted as a single hydrophobic protein with an estimated molecular weight of 37.9 kDa. Kato and Hyodo (1999) reported that ACC oxidase, isolated from pear fruit at the climacteric stage, had a molecular mass of 40 kDa. The apparent K_m values for their substrates, ACC and O_2 were 42.2 μM and 0.53%, respectively, while the apparent K_m for ascorbate (co-substrate) was estimated to be 1.9 mM. The concentration of Fe^{2+} as a cofactor for maximum activity was 25 μM. These enzymatic properties were found to be similar to ACC oxidase derived from other plants (Kato and Hyodo, 1999). Kasai et al. (1998) isolated and purified ACC oxidase from

Escherichia coli cells transformed with cDNA of broccoli ACC oxidase. The recombinant bacterial enzyme was very similar to that of a native broccoli enzyme in its enzymatic properties. The molecular mass of the purified enzyme from the transformed cells was estimted to be 37 kDa (Kasai et al., 1998).

The *in vivo* ACC binding site has proved to be chiral, with an absolute configuration of the *1R, 2S* substrate being generally accepted (Yang and Hoffman, 1984). Based on the strong analogy in amino acid sequence between ACC oxidase and isopenicillin *N* synthase (another Fe^{2+}-dependent oxidase) (Christoffersen et al., 1992; Ming et al., 1990), it has been proposed that the active site of ACC oxidase is very confined (Lay et al., 1996), being buried within a "jelly-roll" motif and lined by hydrophobic residues, where Fe^{2+} does not belong to a heme group (Roach et al., 1995). Stella et al. (1996) postulated a hexacoordinated metal center for the working hypothesis of ACC oxidase, with three histidine residue ligands, one site for cysteine thiol, one site for O_2 (apparently occupied by a carboxylate group of an aspartate residue), and the presence near the active site of a tryptophan residue (Fig. 2.6). In addition, a hypothetical hexacoordinated active

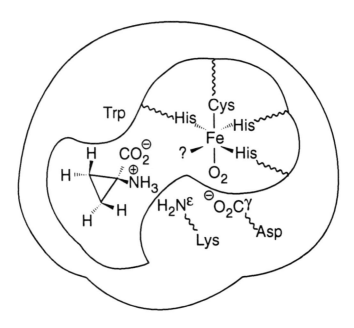

Fig. 2-6.   ACC oxidase active site: the working hypothesis.
Source: Stella et al. (1996).

site Fe^{2+} center, where two histidines (H177 and H234) and an aspartate (D179) residue are the likely Fe^{2+} binding sites, has been suggested (Shaw et al., 1996). In this context, Castellano and Vioque (2000) have observed that those metal ions (Co^{2+}, Cu^{2+}, and Zn^{2+}) whose cationic radius are similar to that of Fe^{2+} (0.74 Å) were more effective in inhibiting ACC oxidase activity than others (Mn^{2+}, Ni^{2+}, and Mg^{2+}). These inhibitory metals might act by replacing Fe^{2+} and forming inactive enzyme-metal complexes (Mathooko et al., 1993). Pirrung and Mcgeehan (1986) postulated that ACC could interact in the catalytic complex at three points: the amino and carboxylic groups and one of the hydrogen atoms on C2. Recently, Zhang et al. (1997) indicated that the ACC carboxylate binds to the enzyme through the side chain of the R244 and S246 residues, while the amino group is orientated towards the hexacoordinated Fe^{2+} complex by the CO_2 activator, which is iron ligated as bicarbonate.

The proposed mechanism for the reaction catalyzed by ACC oxidase involves free-radical intermediates (Vioque, 1986; Stella et al., 1996). However, in a generic sense, free-radical scavengers did not inhibit the *in vitro* pear ACC oxidase activity (Table 2.1). Free-radical intermediates generated during the catalysis would thus remain in the active site and not pass to the solution. In this context, a confined active site within a hydrophobic cavity contributes to their isolation from the external environment. Sulfhydryl reagents such as *p*-chloromercuriphenylsulfonate, *N*-ethylmaleimide and 2,4,6-trinitrobenzenesulfonate strongly reduced the *in vitro* ACC oxidase activity, indicating that sulfhydryl groups are involved in the reaction mechanism. Similar results have also been reported for the enzymes extracted from melon (Smith et al., 1992) and citrus (Dupille and Zacarias, 1996). The fact that iodoacetic acid is more effective than

Table 2.1. Action of inhibitors of *in vivo* ethylene production on *in vitro* pear ACC oxidase activity.

Inhibitor	Concen-tration	ACC oxidase activity nmol C_2H_4/g/h	%
None		12.94 ± 1.04	100
Sodium benzoate	1.0 mM	13.04 ± 1.69	101
n-Propyl gallate	0.5 mM	0.00 ± 0.00	0
CCCP	1.0 mM	4.14 ± 0.45	32
DNP	1.0 mM	12.37 ± 0.23	96
p-CMPS	1.0 mM	2.07 ± 0.05	16
N-EM	1.0 mM	4.66 ± 0.45	36
TNBS	0.5 mM	3.75 ± 0.10	29
Salicylic acid	1.0 mM	2.98 ± 0.11	23
Cycloheximide	0.1 mM	10.74 ± 0.24	83

CCCP, *m*-chlorophenylhydrazone; DNP, dinitrophenol; *p*-CMPS, *p*-chloromercuriphenylsulfonate; N-EM, *N*-ethylmaleimide; TNBS, 2,4,6-trinitrobenzinesulfonate.
Source: Castellano and Vioque (2000).

iodoacetamide in inhibiting tomato ACC oxidase activity was interpreted as a sign that the thiol group must not be near a negatively charged residue, e.g., aspartate or glutamate (Mathooko et al., 1993).

A cDNA clone (pTOM13), isolated from ripening and wounded tomato tissue (Smith et al., 1986), corresponded to an mRNA encoding a protein involved in the conversion of ACC to C_2H_4 (Hamilton et al., 1990). The predicted amino acid sequence of the pTOM13 polypeptide was similar to that of flavanone-3-hydroxylase (EC 1.14.11.9). Garcia-Pineda and Lozoya-Gloria (1999) also isolated cDNA of ACC oxidase gene from pepper and found that 3-4 members of the ACC oxidase gene family were similar to genes reported in other plants. The reported amino acid sequences for several ACC oxidases indicate that Cys 165 is a good candidate for being part of the active site, perhaps as a ligand for Fe^{2+} (Lay et al., 1996).

2.3.5. Conjugation of 1-Aminocyclopropane-1-carboxylic Acid

Conjugation of ACC with malonate to form 1-(malonylamino)cyclopropane-1-carboxylic acid (MACC) was discovered by Amrhein et al. (1981). The natural presence of MACC has since been well documented (Hoffman et al., 1982; Kende, 1993; Tan and Bangerth, 2000). A rapid decline in the rate of C_2H_4 production can result from decreased ACC synthesis or from the conjugation of ACC into MACC. The later reaction is catalyzed by the enzyme, ACC malonyltransferase (Yang et al., 1990b). The presence of malonyltransferase may serve as an alternative means of converting high levels of ACC into an inactive product (Hoffman et al., 1983). Once formed, MACC can be transported and stored into a vacuole (Bouzayen et al., 1988, 1989). Because of this, MACC is a sink for ACC only when the levels of ACC are high or ACC oxidase activity is saturated. The increase in malonyltransferase may account for autoinhibition of C_2H_4 production. Ethylene can increase the levels of malonyltransferase in various tissues (Gupta and Anderson, 1989; Liu et al., 1985a,b; Philosoph-Hadas et al., 1985). Treatment of green tomato fruits with C_2H_4 caused a rise in malonyltransferase within an hour (Liu et al., 1985b). Phytochrome-mediated enhancement of ACC to MACC has been observed in etiolated wheat (Jiao et al., 1987). Abscisic acid (ABA) increased the level of MACC nine-fold in spinach leaf discs (Philosoph-Hadas et al., 1989). IAA caused a transient rise in ACC and C_2H_4 production, with a steady increase in MACC content in mung bean (Riov and Yang, 1989).

A new conjugate, 1-(γ-L-glutamylamino)cyclopropane-1-carboxylic acid (GACC) of ACC has been identified in tomato (Martin et al., 1995). The enzyme activity catalyzing the formation of GACC from ACC and glutathione has been detected in cell extracts. Thus, more than one acid-hydrolyzable, amide-linked conjugate of ACC is present in tomato. It remains to be determined what regulates the accumulation of these conjugates and to what level they influence the ability of a tissue to make available ACC for C_2H_4 production.

Conjugation of ACC may be one way of preserving the precursor to prevent its accumulation and conversion to C_2H_4 . In this regard, it is interesting to note that, under certain stress conditions, MACC serves as a source of ACC (Jiao et al., 1986; Hanley et al., 1989). The overall processes involved in biosynthesis of MET-dependent C_2H_4 , and its regulation along with enzymes involved is summarized in Fig. 2.7 (Fluhr and Mattoo, 1996).

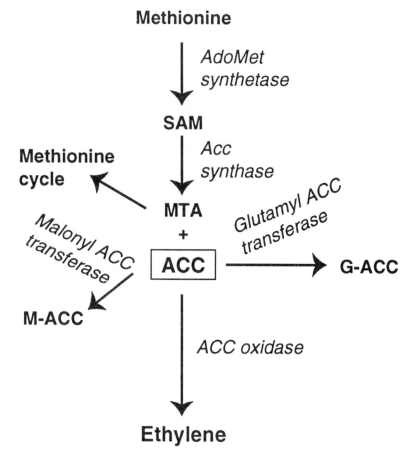

Fig. 2-7.   The major pathway of C_2H_4 biosynthesis in higher plants
and the enzymes involved. Also shown are routes to the MET cycle
and conjugation reactions that conserve ACC being directly available
for C_2H_4 production. Source: Fluhr and Mattoo (1996).

2.3.5.1. Biochemical Aspects of ACC N-Malonyltransferase

The enzyme ACC N-malonyltransferase, which catalyzes malonylation of ACC with
malonyl CoA has been purified from mung bean hypocotyls (Guo et al., 1992; Benichou
et al., 1995) and tomato fruit pericarp (Martin and Saftner, 1995; Amrhein etal., 1987; Su
et al., 1985). ACC N-malonyltransferase may exist as isoforms. One form of the enzyme
from mung bean is a monomer of 50 to 55 kDa with K_m values of 0.5 mM for ACC and
0.2 mM for malonyl CoA with a pH optimum of 8.0 (Guo et al., 1992), while the second
form is a monomer of 36 kDa with much lower K_m values for ACC (0.068 mM) and
malonyl CoA (0.074 mM) being inhibited by D-phenylalanine (Benichou et al., 1995).
The K_m for malonyltransferase was 170 mM compared to 66 µM for ACC-oxidase
(Mansour et al., 1986). The enzyme from tomato fruit is a monomer of 38 kDa and has

similar kinetic parameters as that of the larger form of mung bean ACC *N*-malonyltransferase (Martin and Saftner, 1995). The tomato fruit enzyme is developmentally regulated, stimulated by C_2H_4 and not subject to inhibition by millimolar concentrations of L- or D-amino acids as is the mung bean enzyme.

2.3.6. Methionine Cycle

Methionine pools are too low in plant tissue to sustain normal rates of C_2H_4 production, implying that there must be regeneration of MET to continue its supply. This recycling of MET has been confirmed as illustrated in Fig. 2.8. As discussed in Section 2.3.2, during the formation of ACC from SAM, MTA is released as a by-product which is recycled back to MET via 5'-methylthioribose (MTR), MTR-1-phosphate (MTR-1-P), and 2-keto-4-methylthiobutyrate (KMBA).

In plants and in many microorganisms, MTA is rapidly hydrolyzed to MTR and adenosine by MTA nucleosidase (Adams and Yang, 1977, 1979; Baxter and Coscia, 1973; Ferro et al., 1976; Guranowski et al., 1981; Kushad, 1990; Yu et al., 1979). MTR is converted into MTR-1-P by MTR kinase and subsequently into KMBA by uncharacterized enzymes and finally into MET by KMBA-transaminase (Kushad et al., 1983; Kushad, 1990; Guranowski, 1983; Guranowski et al., 1981). MTR kinase which catalyzes the ATP-dependent phosphorylation of MTR to MTR-1-P has been identified in plants (Guranowski, 1983; Kushad et al., 1983). This enzyme requires the presence of a divalent metal ion (Mg^+, Mn^{2+}) and has K_m values of 4.3 and 8.3 µM for MTR and ATP, respectively. The conversion of MTR-1-P into KMBA is believed to occur in three successive steps, catalyzed by yet three uncharacterized enzymes. First, MTR-1-P is isomerized to methylthioribulose-1-P in the presence of a protein fraction, followed by two unidentified intermediates formed in the course of a reaction catalyzed by a second protein and lastly, conversion of two intermediates into KMBA with the consumption of oxygen by the third protein fraction (Stella et al., 1996). This conversion involves the release of formic acid (Miyazaki and Yang, 1987). Finally, KMBA is transaminated to MET (Backlund et al., 1982). Radiotracer studies (Wang et al., 1982; Yung et al., 1982; Giovanelli et al., 1983; Yang and Hoffman, 1984) demonstrated that MTA furnishes both the methylthio group and the ribose moeity for incorporation into MET. Once MET is present, the system runs as long as there is a supply of ATP. This MTR salvage pathway occurs widely in plant tissues. The overall result of the Yang cycle (Fig. 2.8) is that the 4-carbon portion of MET, from which ACC is derived, is ultimately furnished from the ribose portion of ATP via SAM, while the CH_3S- group of MET is conserved for continued regeneration of MET (Yang et al., 1990b). Thus, given a constant pool of CH_3-S groups and available ATP, a high rate of C_2H_4 production can be achieved without high intracellular concentrations of MET, a less abundant amino acid (Zarembinski and Theologis, 1994).

2.4. REGULATION OF ETHYLENE PRODUCTION

The rate of C_2H_4 synthesis is controlled by endogenous as well as exogenous biotic and abiotic factors (Abeles et al., 1992; Mattoo and Suttle, 1991). Plants have evolved a highly regulated system for production of C_2H_4. It is supported by the fact that

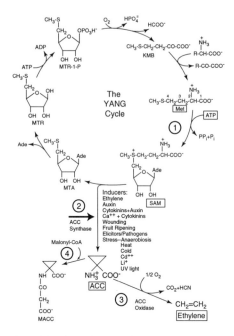

Fig. 2-8.   The ethylene biosynthetic pathway of higher plants. SAM, S-adenosyl-L-methionine; ACC, 1-aminocyclopropane-1-carboxylic acid; KMB, 2-keto-4-methylthiobutyrate; MACC, malonyl-ACC; MTA, 5'-methylthioadenosine; MTR, 5'-methylthioribose; MTR-1-P, MTR-1-phosphate. Source: Yang and Hoffman (1984).

multiple genes encode key enzymes involved in C_2H_4 biosynthesis (e.g., ACC synthase and ACC oxidase) and whose transcripts are differentially regulated (Fluhr and Mattoo, 1996). The formation of ACC by ACC synthase and its conversion to C_2H_4 by ACC oxidase regulates C_2H_4 synthesis; however, in many cases, the presence of sufficient ACC oxidase implies that the conversion of SAM to ACC is the rate-limiting reaction. The fact that the amount of ACC synthase usually parallels the endogenous levels of ACC and the rate of C_2H_4 production confirms this premise.

Ethylene itself stimulates a tissue's ability to convert ACC to C_2H_4 as well as to sequester ACC as MACC (Mattoo and White, 1991), a phenomenon of autoregulation. Positive feedback regulation of C_2H_4 biosynthesis is a characteristic feature of ripening fruits and senescing flowers because a massive increase in C_2H_4 production is triggered by exposure to exogenous C_2H_4 with activation of ACC synthase and/or ACC oxidase (Liu et al., 1985c; Inaba and Nakamura, 1986; Wang and Woodson, 1992; Suttle and Kende, 1980). Ethylene also induces ACC oxidase activity in a number of vegetative (Chalutz et al., 1984; Kim and Yang, 1994) and reproductive tissues (Liu et al., 1985c; Bufler, 1986; Dominguez and Vendrell, 1993; McKeon et al., 1995). This stimulation is correlated with the accumulation of transcripts of ACC oxidase (Nadeau et al., 1993; Drory et al., 1993; Kim and Yang, 1994; Tang et al., 1994; Peck and Kende, 1995; Avni et al., 1994; Barry et al., 1996), suggesting transcriptional regulation. Interestingly, ACC synthase transcripts also accumulate in tissues treated with C_2H_4 or those tissues that copiously produce C_2H_4 (Rottmann et al., 1991; Li et al., 1992; Lincoln et al., 1993; Park et al., 1992; Henskens et al., 1994). The accumulation of ACC oxidase transcripts in several cases occurs earlier than that of ACC synthase, implying that these transcripts are coordinately induced by C_2H_4, causing a very intense autocatalytic C_2H_4 production (Fluhr and Mattoo, 1996). This is best exemplified by results obtained with floral organs during pollination-induced senescence (Nadeau et al., 1993; Drory et al., 1993; Porat et al., 1994; Tang et al., 1994). Using 1-methylcyclopropene (MCP), a new inhibitor of C_2H_4 action (Serek et al., 1994), Nakatsuka et al. (1997) demonstrated that the regulation and expression in tomato fruit of the ACC synthase and oxidase genes (at the transcriptional level) is under a positive feedback control mechanism even at the stage of massive C_2H_4 production.

In some instances, C_2H_4 mediates inhibition of its own synthesis. Negative feedback has been recognized in a number of fruit and vegetative tissues (Riov and Yang, 1982; Vendrell and McGlasson, 1971; Zeroni et al., 1976; Hyodo et al., 1985; Aharoni et al., 1979; Yoshii and Imaseki, 1982). In such cases, exogenous C_2H_4 significantly inhibits endogenous C_2H_4 production induced by ripening, wounding, and/or treatment with auxins. This autoinhibitory effect (Vendrell and McGlasson, 1971; Zauberman and Fuchs, 1973; Zeroni et al., 1976; Saltveit and Dilley, 1978) seems more directed toward limiting the availability of ACC (Riov and Yang, 1982; Yoshii and Imaseki, 1982). In the presence of AVG (Owens et al., 1971), an inhibitor of ACC synthase, IAA-induced extractable ACC synthase activity in mung bean hypocotyls increased several-fold (Yoshii and Imaseki, 1982). In winter squash, autoinhibition of C_2H_4 production was found correlated with the suppression of an ACC synthase transcript accumulation (Nakajima et al., 1990), suggesting that some of the ACC synthase genes may be negatively regulated by C_2H_4 . Another facet to negative regulation of C_2H_4 biosynthesis is the finding of a gene, E8, whose expression leads to inhibition of C_2H_4 production in tomato (Lincoln and Fischer, 1988; Penarruba et al., 1992).

Induction of C_2H_4 by IAA in etiolated pea seedlings also occurs via a rapid increase in the accumulation of ACC oxidase transcript and the induction of ACC oxidase transcripts is preceded by that of ACC synthase (Peck and Kende, 1995). However, in such a case there is a reasonable constitutive level of ACC oxidase that may be sufficient to start the autocatalytic stimulation of C_2H_4 production.

2.5. METABOLISM

The gaseous nature of C_2H_4 has led some scientists to believe for many years that C_2H_4 was not metabolized in plant tissues because of its diffusion out of the tissue. Others believed that C_2H_4 metabolism within the plant tissues had a direct role in the mechanisms of action of the gas. Although C_2H_4 metabolism by plants was considered for some time to be an artifact (Abeles, 1972), it has now been unambiguously established that higher plants can metabolize C_2H_4 . However, whether C_2H_4 metabolism carries any physiological role is not yet clear. In some plant tissues C_2H_4 is oxidized to CO_2, in others it is incorporated into the tissue by conversion to ethylene oxide (C_2H_4O) and ethylene glycol ($C_6H_6O_2$) and, in certain plants, both processes occur simultaneously (Sanders et al., 1989). While working with developed cotyledons of *Vicia faba*, Jerie and Hall (1978) found that C_2H_4 metabolism occurred at high rates at physiological concentrations of C_2H_4 with a major product namely C_2H_4O and minor amounts of CO_2 and other metabolites.

It has been suggested that C_2H_4 metabolism and C_2H_4 action may be integrally related (Beyer et al., 1984; Smith and Hall, 1984). The mode of action of C_2H_4 may be through a metabolite of C_2H_4 such as C_2H_4O or $C_6H_6O_2$ (Beyer, 1975; Smith and Hall, 1984). The similarity between affinities of C_2H_4 for binding sites on *Phaseolus* spp. and oxidized products (e.g., C_2H_4O) in *Vicia* spp. have led Bengochea et al. (1980) to propose that the two activities of a common system are involved in the regulation of the plant's response to C_2H_4.

In most instances, the rate of metabolism of C_2H_4 is nearly first order with respect to C_2H_4, even at fairly high levels of C_2H_4 (>40 ml L^{-1}). In the pea, the concentration of C_2H_4 giving a half-maximal rate for C_2H_4 metabolism is approximately 1000 times the concentration necessary for the half-maximal response in the pea growth test (Beyer et al., 1984). However, the K_m for C_2H_4 in tissue incorporation by *Vicia faba* corresponds closely to the levels evoking a physiological response (Sanders et al., 1989). By using the inhibitors of C_2H_4 action, Ag(I) and CO_2, it was demonstrated that they also inhibit C_2H_4 metabolism in different plant tissues (Beyer, 1977; Beyer and Sudin, 1978; Beyer, 1979b), and the extent of inhibition of metabolism closely paralleled the effect of the inhibitor in antagonizing C_2H_4 effects upon development (Beyer, 1979a). These correlations taken together seemed to indicate an association between C_2H_4 metabolism and sensitivity to C_2H_4. Further work with pea by studying the effects of inhibitors of C_2H_4 action upon C_2H_4 metabolism, the effects of elicitors of C_2H_4 upon C_2H_4 metabolism, and the effects of inhibitors of C_2H_4 metabolism upon C_2H_4 action (Sanders et al., 1989; Sisler and Yang, 1984) revealed that there is no correlation between the characteristics of C_2H_4 metabolism and C_2H_4 action in peas. Hall (1991) concluded that in most plants, C_2H_4 metabolism does not have a significant effect on endogenous C_2H_4 concentration; it is related to the mode of action of C_2H_4. Although it was first demonstrated that oxygen deprivation and Ag^+ treatment similarly inhibited C_2H_4 metabolism and action (Beyer et al., 1984), later work demonstrated that the physiological effects of C_2H_4 analogues and antagonists do not correlate with their metabolism or their effect on C_2H_4 metabolism (Sanders et al., 1989; Raskin and Beyer, 1989). Ethylene metabolism thus has no role in affecting C_2H_4 action, as discussed in detail in Section 2.6.

It is likely that C_2H_4 metabolism is a nonessential consequence of C_2H_4 action, resulting from high levels of C_2H_4 production and an endogenous hydrocarbon oxidation system.

2.6. MECHANISM OF ACTION AND ETHYLENE RECEPTORS

A number of possibilities have been hypothesized to explain the mechanism(s) of C_2H_4 action. These include that C_2H_4 might act either by serving as a cofactor in some reaction, or by being oxidized to some essential component and being incorporated into tissue, or binding to a receptor, providing some essential function, and then either diffusing away or being destroyed (as is the case with other hormones). By using deuterium-labeled C_2H_4, the first "cofactor" hypothesis was ruled out (Beyer, 1972; Abeles et al., 1972). As opposed to C_2H_4 metabolism (the second hypothesis), most of the experimental evidence points to C_2H_4 binding with receptors (the third hypothesis) as the general mediator of C_2H_4 action. Firstly, the products of C_2H_4 metabolism do not evoke C_2H_4-like responses; i.e., C_2H_4O has no C_2H_4-like effects (Sanders et al., 1989).

Secondly, C_2H_4 effects can be mimicked by some hydrocarbons (Burg and Burg, 1967) and counteracted by certain olefin antagonists (Sisler and Blankenship, 1993) at levels which correlate to their ability to compete with C_2H_4 binding to the silver ion. Thirdly, there are several inorganic compounds, such as phosphorus trifluoride, which cannot be metabolized to products related to any of the metabolic products of C_2H_4, yet they evoke C_2H_4 effects (Sisler, 1977; Sisler and Blankenship, 1993). Finally, carbon disulfide (CS_2), a potent inhibitor of C_2H_4 oxidation, has no effect on C_2H_4 action (Abeles, 1984). Thus, evidence at the present time favors the theory that C_2H_4 most probably binds to a receptor and brings about some essential change and then diffuses away unchanged, but the function of this binding is not fully understood. According to Woltering and deVrije (1995), C_2H_4 presumably acts by binding, most probably in a reversible way, to a receptor protein (Fig. 2.9). The C_2H_4 receptor complex alters the activity of signal transduction reactions, leading to the transcription of specific genes and the synthesis and/or activation of enzymes responsible for the physiological effects (Sisler and Goren, 1981; Lawton et al., 1990). The changes caused by C_2H_4 in enzymes of treated plant tissues (Table 2.2) support the premise that C_2H_4 receptors are involved in its action.

The C_2H_4 evoked responses can be classified into two categories, namely concentration response and sensitivity response. The former involves the changes in concentration of cellular C_2H_4, while the latter involves the increase in tissue sensitivity to C_2H_4. Studies on both of these C_2H_4 evoked resonses led to the premise that C_2H_4 must bind to some components of the cell to mediate its physiological effects (Abeles et

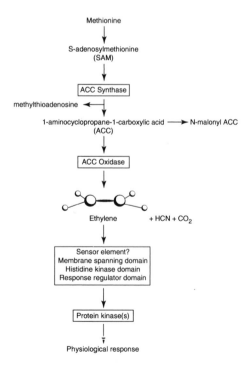

Fig. 2-9. Summary of ethylene biosynthesis and action. Source: Woltering and deVrije (1995).

Table 2.2.   Enzyme activity affected by C_2H_4 treatment.

Enzyme	Plant	Reference
Increased levels of enzyme activity		
4-Coumarate:CoA ligase	Carrot root	Ecker and Davis (1987)
α-Amylase	Rice	Smith et al. (1987)
Arginine decarboxylase	Rice	Cohen and Kende (1986)
ATPase	Rubber	Gidrol et al. (1988)
β-1,3-Glucanase	Tobacco	Felix and Meins (1987)
β-Cyanoalanine synthase	Pea seedling	Goudey et al. (1989)
β-Galactosidase	Pepper fruit	Gross et al. (1986)
Chalcone synthase	Carrot root	Ecker and Davis (1987)
Chitinase	Bean leaf	Broglie et al. (1989)
Chlorophyllase	Citrus fruit	Purvis (1980)
Cinnamic acid 4-hydroxylase	Pea seedling	Hyodo and Yang (1971)
Cellulase	Sunflower	Kawase (1979)
DNA polymerase	Potato tuber	Apelbaum et al. (1984)
Ethylene-forming enzyme	Avocado fruit	Sitrit et al. (1986)
Hevein	Rubber tree	Broekaert et al. (1990)
Invertase	Tomato fruit	Iki et al. (1978)
Lysine decarboxylase	Tomato fruit	Bakanashvili et al. (1987)
Malonyltransferase	Tomato fruit	Liu et al. (1985b)
Nitrate reductase	*Agrostemma*	Schmerder and Borriss (1986)
Peroxidase	Cucumber	Abeles et al. (1988)
Phenylalanine ammonia-lyase	Carrot root	Ecker and Davis (1987)
Polygalacturonase	Tomato	Jeffrey et al. (1984)
Polyphenol oxidase	Sweet potato	Buescher et al. (1975)
Proteinase	Tomato plants	Vera and Conejero (1989)
RNA polymerase	Potato tuber	Apelbaum et al. 1984)
RNase	Potato tuber	Isola and Franzoni (1989)
S-Adenosylmethionine decarboxylase	Rice	Cohen and Kende (1986)
Sucrose phosphate synthase	Muskmelon	Hubbard et al. (1989)
Superoxide dismutase	Bean leaf	Henry and Ananich (1985)
Uronic acid oxidase	Citrus leaf	Huberman and Goren (1982)
Decreasd levels of enzyme activity		
Acid invertase	Melon fruit	Hubbard et al. (1989)
ACC synthase	Squash fruit	Hyodo et al. (1985)
Phenylalanine ammonia-lyase	Mustard	Buhler et al. (1978)
β-Amylase	Sweet potato	Buescher et al. (1975)
Arginine decarboxylase	Pea seedling	Palavan et al. (1984)
S-Adenosylmethionine decarboxylase	Pea seedling	Icekson et al. (1985)
Citrate synthase	Tomato fruit	Jeffrey et al. (1984)
Nitrate reductase	Purslane	Stacewicz-Sapuncakis et al. (1973)

Source: Abeles et al. (1992).

al., 1992). Most dose-response studies have shown that C_2H_4 saturates near 10 to 100 μl L^{-1} and higher levels have no additional effects. These observations suggest that all receptors (binding sites) are saturated under these conditions and the additional C_2H_4 has no other toxic effects on the cell, supporting the premise that C_2H_4 action is mediated by a receptor (Bleeker et al., 1988; Sanders et al., 1989; Sisler and Blankenship, 1993).

Payton et al. (1996) suggested that changes in C_2H_4 sensitivity are mediated by modulation of receptor levels during plant development. However, some workers are of the view that the oxidation of C_2H_4 likely does involve in the interaction of C_2H_4 with a receptor or enzyme (Evans et al., 1984; Smith and Hall, 1984; Smith et al., 1985).

Attempts to locate the C_2H_4 binding sites in various plant tissues have been made. The C_2H_4 -binding protein has been found associated with cellular membranes (Hall, 1986; Sisler, 1980; Thomas et al., 1984). The location of C_2H_4 binding sites in developing bean cotyledons was shown to be the endoplasmic reticulum and protein body membranes. Endomembrane systems and protein bodies had a high affinity for C_2H_4 , confirming earlier work with cell fractionation techniques (Evans et al., 1982). *In vitro* studies have shown that both membrane and cytosolic fractions contain measurable amounts of C_2H_4 binding sites (Harpham et al., 1996).

Sisler (1991) summarized the early studies demonstrating the presence of C_2H_4 binding in various plant tissues (Table 2.3). Ethylene binding has also been estimated by following the amount of $[^{14}C]$- C_2H_4 released from tobacco leaves. Leaves were exposed to labeled C_2H_4 in the presence of varying amounts of unlabeled C_2H_4 and C_2H_4 analogues. After a short air flush, the amount of labeled C_2H_4 released was measured (Sisler, 1979). Using this technique, values close to that predicted by dose-response curves, were obtained. Similar values were obtained in later studies with other plants (Goren and Sisler, 1986). A summary of the properties of C_2H_4 -binding sites from *Phaseolus vulgaris* is given in Table 2.4 (Hall et al., 1987). Harpham et al. (1996) reported that virtually all higher plants studied contain at least two classes of C_2H_4

Table 2.3.   Ethylene-binding sites in plant vegetative tissue.

Tissue	K_d	Apparent concentration of receptor (mol/kg fresh weight)	Reference
Tobacco leaves	0.30	3.5×10^{-9}	Sisler (1979)
Bean leaves	0.14	2.0×10^{-9}	Goren and Sisler (1986)
Citrus leaves	0.15	5.7×10^{-9}	Goren and Sisler (1986)
Ligustrum leaves	0.31	6.8×10^{-9}	Goren and Sisler (1986)
Mung bean sprouts	--	Identified	--
Tomato leaves	0.30	1.9×10^{-9}	Sisler (1982a)
Tomato fruit	0.30	7.0×10^{-11}	Sisler (1982a)
Carnation petals	0.10	6.0×10^{-9}	Sisler et al. (1986)
Carnation leaves	0.09	2.0×10^{-9}	Sisler et al. (1986)
Bean cotyledons	0.30	4.8×10^{-9}	Goren and Sisler (1986)
Bean roots	--	Identified[a]	--
Pea epicotyl	0.12 0.63	Identified	Smith et al. (1987)
Wheat germ	--	Identified	--
Apple pulp	0.097	3.2×10^{-11}	Blankenship and Sisler (1989)
Cucumber leaves	--	Identified	--
Mushroom sporophore	--	Not detected	--
Morning glory petals	0.10	3.8×10^{-9}	Blankenship and Sisler (1989)

[a]Not corrected for the amount occupied by endogenous C_2H_4.
Source: Sisler (1991).

Table 2.4.   Properties of the C_2H_4-binding site of *Phaseolus vulgaris*.

	Membrane bound	Triton solubilized and partially purified protein
K_d Ethylene (M)	0.88×10^{-10}	5.5×10^{-10}
K_i Propylene (M)	5.6×10^{-7}	0.9×10^{-7}
K_i Acetylene (M)	1.03×10^{-5}	2.0×10^{-5}
K_i CO_2 (M)	1.15×10^{-3}	
K_1 $(M^{-1} s^{-1})$	2.97×10^3	0.042×10^3
K_{-1} (s^{-1})	1.4×10^{-6}	2.3×10^{-8} (calc)
pH optimum	7.5 - 9.5	4 - 7
Sedimentation coefficient		2.2s
Stoke's radius (nm)		6.1
Frictional ratio		2.4
Molecular weight		52 - 60,000

Source: Hall et al. (1987).

binding sites, one of which fully associates and dissociates in about 2 h and a class of sites that takes up to 20 h to become fully saturated. Although the types of sites differ in their rate constants of association, they have similar and high affinities for C_2H_4 . Most of the C_2H_4 responses are reversible and occur within a matter of minutes to hours (Table 2.5). However, the association and dissociation kinetics of C_2H_4 were on the order of 15 minutes in rice seedlings (Sanders et al., 1990). This suggests that potential C_2H_4 binding proteins should have similar characteristics (Abeles et al., 1992). The binding site from bean cotyledons has been characterized as a 50 kD, hydrophobic, acidic protein (Thomas et al., 1985).

Possible C_2H_4 receptors have been isolated and partially purified from several plant sources including tobacco, beans and tomatoes (Sisler, 1982a,b; Sisler and Blankenship, 1993). The binding sites appear to be a protein, based on heat-sensitivity, protease sensitivity, solubility and chromatographic behavior and sensitivity to sulfhydryl agents (Sisler, 1982a,b; Sisler and Blankenship, 1993). The isolated receptor displays properties similar to those expected from studies of C_2H_4 action *in vivo*. Ethylene binding to the isolated receptor is competitively inhibited by propylene at 128 times and by acetylene at 1013 times the concentration of C_2H_4 (gas phase), corresponding to the *in vivo* activity of these analogues. Moreover, the C_2H_4 antagonist 2,5-norbornadiene is similarly effective in blocking binding of C_2H_4 to the receptor (Sisler and Blankenship, 1993).

The predicted receptor protein contains three domains, a putative membrane spanning domain, a transmitter (histidine protein kinase) domain and a receiver domain (Chang et al., 1993). A copper ion associated with the C_2H_4 –binding domain is required for C_2H_4 –binding activity. It is proposed that binding of C_2H_4 to the transition metal induces a conformational change in the sensor domain that is propagated to the cytoplasmatic transmitter domain of the protein (Rodriguez et al., 1999; Bleecker et al., 1998). However, recently reported findings indicate that histidine kinase activity may not be necessary for an C_2H_4 response, because some receptors lack histidine within the

Table 2.5. Response times for ethylene action.

Time (hours)	Process	Reference
Abscission		
24	100% loss of *Zygocactus* flowers and buds	Cameron and Reid (1981)
2	Acceleration of bean petiole explants	Dela Fuente and Leopold (1969)
2	Acceleration of *Digitalis* flowers	Stead nd Moore (1983)
Enzyme synthesis		
24	Phenylalanine ammonia-lyase from carrot	Chalutz (1973)
12	ATPase from rubber trees	Girdol et al. (1988)
6	Chitinase from bean leaves	Boller et al. (1983)
6	Phenylalanine ammonia-lyase from citrus	Riov et al. (1969)
3	β-Cyanoalanine synthase from peas	Goudey et al. (1989)
1	ACC malonyltransferase from tomato	Liu et al. (1985b)
Epinastry		
1	Intact tomato leaves	Leather et al. (1972)
Inhibition of cell division		
2	Apical hook of pea seedlings	Apelbaum and Burg (1972)
Inhibition of root elongation		
1	Intact pea root	Chadwick and Burg (1967)
0.5	Barley root	Hall et al. (1977)
0.3	Primary radish roots	Jackson (1983)
0.3	Intact pea roots	Rauser and Horton (1975)
Inhibition of shoot elongation		
3	Excised pea epicotyl segments	Nee et al. (1978)
1.5	Intact *Pharbitis* shoots	Abdel-Rahman and Cline (1989)
0.5	Excised etiolated pea stem section	Eisinger et al. (1983)
0.2	Intact pea epicotyls	Goeschl and Kays (1975)
Messenger RNA synthesis		
2	Glucanase from bean	Vogeli et al. (1988)
0.5	Tomato mRNA	Lincoln et al. (1987)
Respiration		
24	Avocado fruit	Eaks (1966)
10	Banana fruit	McMurchie et al. (1972)
5	Carrot roots	Christoffersen and Laties (1982)
2	Parsnip roots	Theologis and Laties (1982)
Rooting		
240	Tobacco	Zimmerman & Hitchcock (1933)
Swelling, cell expansion		
1	pea stem sections	Eisinger et al. (1983)
0.1	Intact pea stems	Eisinger (1983)

Source: Abeles et al. (1992).

kinase domain that is predicted to be phosphorylated (Tiemann and Klee, 1999; Hua et al., 1998).

The question of how the C_2H_4 –receptor complex induces the known physiological changes, is being addressed by two complementary approaches, including biochemical and genetic analyses. Unlike the well-characterized C_2H_4 biosynthesis pathway (Abeles et al., 1992; Kende, 1993; McKeon et al., 1995), less is known about C_2H_4 perception and the subsequent signal transduction pathway (Chang, 1996; Fluhr, 1998; Chang and Stewart, 1998; Theologis, 1998; Bleecker et al., 1998). Leyser (1998) postulated that plant hormones coordinate information from many sources through many points of regulation from biosynthesis to signal transduction, as shown in Fig. 2.10. Ethylene perception requires specific receptors and a signal transduction pathway to coordinate downstream responses (Müller et al., 2000). Molecular studies with mutants varying in C_2H_4 sensitivity have provided information regarding C_2H_4 perception and transduction pathways. Components involved in C_2H_4 perception and signal transduction have first been cloned from *Arabidopsis thaliana* by screening mutants for alteration in sensitivity or response to C_2H_4 . By screening *Arabidopsis* seedlings for elongation in the presence of C_2H_4 , Bleecker et al. (1988) first obtained a class of C_2H_4 insensitive mutants, *etr*, which were lacking in a number of C_2H_4 responses and had reduced C_2H_4 binding. The gene *ETR1* encoded a protein (ETR1) which acts early in the response pathway as an C_2H_4 receptor (Bleecker et al., 1988; Chang et al., 1993; Ecker, 1995). Other C_2H_4 insensitive mutants *ein1* and *ein2* have been developed (Guzman and Ecker, 1990; van der Straeten et al., 1993; Kieber et al., 1993) which have a similar phenotype and may

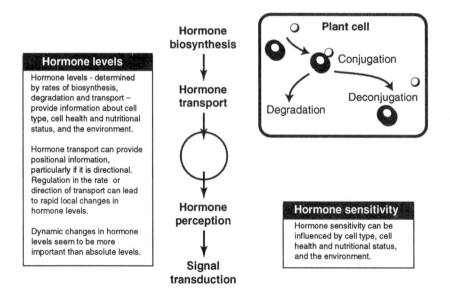

Fig. 2-10. A hypothetical scheme explaining mechanisms of action of hormones. Plant hormones coordinate information from many sources through many points of regulation from biosynthesis to signal transduction. Source: Leyser (1998).

therefore be identical.   The homology studies suggest that the C_2H_4 response is
transmitted by a multicomponent protein phosphorylation pathway (Kieber et al., 1993),
which could explain the involvement of calcium ion in the C_2H_4 transduction pathway
(Raz and Fluhr, 1992).   Double-mutant analysis (epinasis) for an C_2H_4 signal transduction
pathway led Zarembinski and Theologis (1994) to suggest a model for a putative C_2H_4 -
sensing pathway (Fig. 2.11).   According to the proposed genetic model, the ETR1 protein
is a kinase (which is active in the absence of C_2H_4 and inactive in its presence) that
targets CTR1 which is a Raf protein kinase capable of repressing the C_2H_4 signal
transduction pathway.   Activated CTR1 may eventually phosphorylate EIN3, which may
be a transcription factor that is inactivated by phosphorylation and only active when the
ETR1 and CTR1 kinases are switched off in the presence of C_2H_4 (see Fig. 2.11).
Similarly, McGrath and Ecker (1998) have proposed models of the C_2H_4 signal
transduction pathway (Fig. 2.12) which also involves ETR1 and CTR1 protein receptors.

Recently, three more genes, *ETR2*, *ERS2*, and *EIN4*, involved in C_2H_4 signaling in
Arabidopsis were identified (Hua et al., 1998; Sakai et al., 1998).   The receptors
consitutively suppress a downstream signalling pathway, and are inactivated by C_2H_4 ,
leading to the activation of genes necessary for C_2H_4 regulated responses (Hua and
Meyerowitz, 1998; Chang and Stewart, 1998; Theologis, 1998).   The five members of the
putative C_2H_4 receptor gene family appear to belong to two subfamilies.   One subfamily
consists of *ETR1* and *ERS1*, whereas the other contains *ETR2*, *EIN4* and *ERS2* (Hua et
al., 1998).   Likewise, a similar family of two-component C_2H_4 receptors has also been
characterized in tomato, in which five putative receptor genes have now been cloned

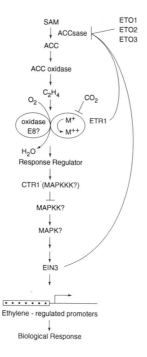

Fig. 2-11.   A putative ethylene-sensing pathway.   Source: Zarembinski
and Theologis (1994).

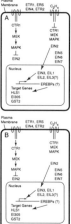

Fig. 2-12. Proposed models of the ethylene signal transduction pathway: A. binding of ethylene by the receptor complex inhibits activity of the MAP kinase cascade, allowing transmission of the ethylene signal; B, the receptor complex stimulates the activity of the MAP kinase cascade to inhibit ethylene responses in the absence of the hormone. ETR1, ERS, EIN4 and ETR2 have several key differences, and their placement together only indicates that they may perform similar functions, either separately or in concert. With the exception of EIN3, the subcellular locations of the components of the signaling cascade are currently unknown. In the cases of the receptor complex and CTR1, subcellular locations are inferred from biochemical data or the locations of similar gene prodcuts in other orgnaisms. Source: McGrath and Ecker (1998).

(Zhou et al., 1996a,b; Payton et al., 1996; Lanahan et al., 1994; Wilkinson et al., 1995; Lashbrook et al., 1998; Tiemann and Klee, 1999). It is conceivable that most plants have several receptors of the *ERS* and/or *ETR1* type. In mung bean seedlings, the level of an *ERS1* homologue was highly inducible by C_2H_4 in all tissues examined, indicating that C_2H_4 positively modulates the expression of its receptor genes (Kim et al., 1999). Müller et al. (2000) studied expression of a gene for the putative C_2H_4 receptor, *RhETR*, in miniature roses (*Rosa hybrida*) during flower senescence. The expression of *RhETR* was distinctly higher in the cultivar with short flower life ("Bronze") than in the long-lasting cultivar ("Vanilla") and modulation of receptor levels was also observed during flower development. Exposure to low C_2H_4 concentrations results in an up-regulation of *RhETR* in flowers of both cultivars. Differences in the expression of the putative C_2H_4 receptor of cultivars with short or long flower life suggest that variation in flower longevity may be due to differences in receptor levels during flower development. These molecular studies convincingly support the view that C_2H_4 action is mediated through specific C_2H_4 binding proteins, the so-called receptors.

2.7. ETHYLENE PRODUCTION BY LOWER PLANTS

Like higher plants, all the lower vascular (water ferns) and non-vascular (liverworts, and horsetail, *Selaginella, Lycopodium, Psilotum*, mosses) plants produce C_2H_4 (Osborne, 1989; Thomas et al., 1983). However, there is compelling evidence that in most cases they do not follow the same biosynthetic pathway operative in higher plants. Added ACC did not promote C_2H_4 production by liverworts (*Marchantia polymorpha, Riella helicophylla*), mosses (*Funaria polymorpha, Polytrichum commune, Sphagnum cuspidatum*), *Psilotum nudum,* lycopods (*Selaginella wildenovii, Lycopodium phleg*maria), and horsetails (*Equisetum hyemale, Equisetum telmateia*). Similarly, ACC did not promote C_2H_4 production in the ferns *Trichomanes speciosum, Ophioglossum reticulatum, Salvinia natans, Azolla caroliniana* (Osborne, 1989). However, ACC and IAA promoted and cobalt ions inhibited C_2H_4 production in the ferns *Pteridium aquilinum, Matteuccia struthioptheris,* and *Polystichum minitum,* suggesting that these plants do have the same biosynthetic pathway as higher plants (Title, 1987). Cookson and Osborne (1978) described the lack of label in C_2H_4 produced by the semi-aquatic fern *Regnellidium diphyllum* when supplied with U-[^{14}C]-1-MET although the [^{14}C] label was readily incorporated into protein and CO_2. Moreover, a blocker of SAM conversion into ACC also did not alter the rates of C_2H_4 production by leaflets, and instead actually increased the production by the petioles. Similarly, application of IAA, a stimulator of C_2H_4 in higher plants or tissue wounding did not result in any change in the rate of C_2H_4 formation (Walters and Osborne, 1979), providing further evidence for a different biosynthetic route from that of the ACC-synthase-regulated pathway in higher plants. Furthermore, ACC was not effective in substituting for C_2H_4 in the promotion of cell elongation in either the petioles of *R. diphyllum* (Osborne, 1982) or in the thallus of the semi-aquatic liverwort, *Riella helicophylla* (Stange and Osborne, 1989). Exogenous addition of ACC had no effect on C_2H_4 production rates by *R. helicophylla.* In a broad survey which included many genera of lower plants, all the species tested in the evolutionary orders up to Gnetales produced C_2H_4 , but none of the lower plants increased C_2H_4 production when supplied with ACC, whether or not they were land, semi-aquatic, or aquatic species (Osborne, 1989, 1991), implying that ACC oxidase activity was apparently absent. In a more recent study, Osborne et al. (1996) provided convincing evidence that lower plants operate via a non-MET-ACC pathway of C_2H_4 formation with very limited conversion of ACC into MACC or other C_2H_4 –release conjugates. Feeding labeled ACC to lower plants (*R. diphyllum* and *R. helicophylla*) demonstrated the inability of these semi-aquatic ferns to derive C_2H_4 from ACC, although a small amount of ACC was conjugated to MACC. Osborne et al. (1996) suggested that although C_2H_4 functions as a regulator of higher plants, the formation of ACC as a precursor of C_2H_4 biosynthesis has arisen relatively late in plant evolution.

Chernys and Kende (1996) confirmed the findings of Osborne et al. (1996) as they also reported the inability of *R. diphyllum* and *Marsilea quadrifolia* to convert labeled ACC into labeled C_2H_4 . Moreover, AVG and AOA did not inhibit C_2H_4 production by these ferns. Although they detected ACC and ACC synthase in both the ferns, they concluded that C_2H_4 in the ferns occurs mainly via an ACC-independent route, even though the capacity to synthesize ACC is present in these lower plants (Chernys and Kende, 1996). No comprehensive study has been undertaken to explore the C_2H_4 biosynthetic pathway in lower plants and ferns, so it is not known whether or not the biosynthetic pathways characterized in microorganisms namely MET $\rightarrow$ KMBA $\rightarrow$ C_2H_4 or glutamate/KGA $\rightarrow$ C_2H_4 are operative in these lower plants. Future research should focus on characterizing the pathway of C_2H_4 biosynthesis in these lower plants.

2.8. CONCLUDING REMARKS

The biochemistry of C_2H_4 biosynthesis in higher plants has been well characterized and the mechanism(s) of action of C_2H_4 to evoke physiological responses is being thoroughly investigated. The understanding of molecular aspects of C_2H_4 biosynthesis has opened up new opportunities for scientists to regulate C_2H_4 production in favor of plant growth and development. Some examples of such applications are discussed in Chapter 8 (Sections 8.3). A better understanding of the mechanism(s) of C_2H_4 action will further facilitate the process of controlling plant growth and development coordinated by C_2H_4.

2.9. REFERENCES

Aarnes, H., 1977, Partial purification and characterization of methionine adenosyltransferase from pea seedlings, *Plant Sci. Lett.* **10**:381-390.

Abdel-Rahman, A. M., and Cline, M. G., 1989, Timing of growth inhibition following shoot inversion in *Pharbitis nil*, *Plant Physiol.* **91**:464-465.

Abeles, F. B., 1972, Biosynthesis and mechanism of action of ethylene, *Annu. Rev. Plant Physiol.* **23**:259-292.

Abeles, F. B., 1984, A comparative study of ethylene oxidation in *Vicia faba* and *Mycobacterium paraffinicum*, *J. Plant Growth Regul.* **3**:85-95.

Abeles, F. B., Morgan, P. W., and Saltveit, M. E., Jr., 1992, *Ethylene in Plant Biology*, 2nd Ed., Academic Press, Inc., San Diego, CA.

Abeles, F. B., Dunn, L. J., Morgens, P., Callahan, A., Dinterman, R. E., and Schmidt, J., 1988, Induction of 33-kD and 60-kD peroxidases during ethylene-induced senescence of cucumber cotyledons, *Plant Physiol.* **87**:609-615.

Abeles, F. B., Ruth, J. M., Forrence, L. E., and Leather, G. R., 1972, Mechanism of hormone action. Use of deuterated ethylene to measure isotopic exchange with plant material and the biological effects of deuterated ethylene, *Plant Physiol.* **49**:669-671.

Acaster, M. A., and Kende, H., 1983, Properties and partial purification of 1-aminocyclopropane-1-carboxylate synthase, *Plant Physiol.* **72**:139-145.

Adams, D. O., and Yang, S. F., 1977, Methionine metabolism in apple tissue: Implication of S-adenosylmethionine as an intermediate in the conversion of methionine to ethylene, *Plant Physiol.* **60**:892-896.

Adams, D. O., and Yang, S. F., 1979, Ethylene biosynthesis: Identification of 1-aminocyclopropane-1-carboxylic acid as an intermediate in the conversion of methionine to ethylene, *Proc. Natl. Acad. Sci. USA* **76**:170-174.

Aharoni, N., Anderson, J. D., and Lieberman, M., 1979, Production and action of ethylene in senescing leaf discs. Effect of indoleacetic acid, kinetin, silver ion, and carbon dioxide, *Plant Physiol.* **64**:805-809.

Amrhein, N., Forreiter, C., Klonka, C., Skorupka, H., and Tophof, S., 1987, Metabolism and its compartmentation of 1-aminocyclopropane-1-carboxylic acid in plant cells, in: *Conjugated Plant Hormones*. K. Schreiber, Schutte, H. R., and Sembdner, G., eds., Institute of Plant Biochemistry, Halle, East Germany, pp. 102.

Amrhein, N., Schneebeck, D., Skorupka, H., Tophof, S., and Stockigt, J., 1981, Identification of a major metabolite of the ethylene precursor 1-aminocyclopropane-1-carboxylic acid in higher plants, *Naturwissen.* **68**:619-620.

Apelbaum, A., and Burg, S. P., 1972, Effect of ethylene on cell division and deoxyribonucleic acid synthesis in *Pisum sativum*, *Plant Physiol.* **50**:117-124.

Apelbaum, A., Vinkler, C., Sfakiotakis, E., and Dilley, D. R., 1984, Increased mitochondrial DNA and RNA polymerase activity in ethylene-treated potato tubers, *Plant Physiol.* **76**:461-464.

Arteca, J. M., and Arteca, R. N., 1999, A multi-responsive gene encoding 1-aminocyclopropane-1-carboxylate synthase (ACS6) in mature *Arabidopsis* leaves, *Plant Molec. Biol.* **39**:209-219.

Avni, A., Bailey, B. A., Mattoo, A. K., and Anderson, J. D., 1994, Induction of ethylene biosynthesis in *Nicotiana tabacum* by a *Trichoderma viride* xylanase is correlated to the accumulation of 1-aminocyclopropane-1-carboxylic acid (ACC) synthase and ACC oxidase transcripts, *Plant Physiol.* **106**:1049-1055.

Ayub, R. A., Rombaldi, C., Petitprez, M., Latche, A., Pech, J. C., and Lelievre, J. M., 1992, Biochemical and immunocytological characterization of ACC oxidase in transgenic grape cells, in: *Cellular and Molecular Aspects of the Plant Hormone Ethylene*, J. C. Pech, Latche, A., and Balague, C., eds., Kluwer Academic Publ., Dordrecht, Netherlands, pp. 98-99.

Backlund, P. S., Jr., Chang, C. P., and Smith, R. A., 1982, Identification of 2-keto-4-methylthiobutyrate as an intermediate compound in methionine synthesis from 5'-methylthioadenosine, *J. Biol. Chem.* **257**:4196-4202.

Bakanashvili, M., Barkai-Golan, R., Kopeliovitch, E., and Apelbaum, A., 1987, Polyamine biosynthesis in *Rhizopus*-infected tomato fruits: Possible interaction with ethylene, *Physiol. Molec. Plant Path.* **31**:41-50.

Barlow, J. N., Zhang, Z., John, P., Baldwin, J. E., and Schofield, C. J., 1997, Inactivation of 1-aminocyclopropane-1-carboxylate oxidase involves oxidative modifications, *Biochem.* **36**:3563-3569.

Barry, C. S., Blume, B., Boyzayen, M., Cooper, W., Hamilton, A. J., and Grierson, D., 1996, Differential expression of the 1-aminocyclopropane-1-carboxylate oxidase gene family of tomato, Plant J. **9**:525-535.

Baxter, C., and Coscia, C. J., 1973, *In vitro* synthesis of spermidine in the higher plant *Vinca rosea, Biochem. Biophys. Res. Commun.* **54**:147-154.

Bengochea, T., Acaster, M. A., Dodds, J. H., Evans, D. E., Jerie, P. H., and Hall, M. A., 1980, Studies on ethylene binding by cell free preparations from cotyledons of *Phaseolus vulgaris* L. II. Effects of structural analogues of ethylene and of inhibitors, *Planta* **148**:407-409.

Benichou, M., Martinez-Reina, G., Romojaro, F., Pech, J. C., and Latche, A., 1995, Partial purification and properties of a 36-kDa 1-aminocyclopropane-1-carboxylate N-malonyltransferase from mung bean, *Physiol. Plant.* **94**:629-634.

Beyer, E., Jr., 1972, Mechanism of ethylene action. Biological activity of deuterated ethylene and evidence against exchange and *cis trans* isomerization, *Plant Physiol.* **49**:672-675 .

Beyer, E. M., Jr., 1975, $^{14}C_2H_4$: Its incorporation and metabolism by pea seedings under aseptic conditions, *Plant Physiol.* **56**:273-278.

Beyer, E. M., Jr., 1976, Silver ion: A potent antiethylene agent in cucumber and tomato, *HortSci.* **11**:195-196.

Beyer, E. M., Jr., 1977, $^{14}C_2H_4$: Its incorporation and oxidation to $^{14}CO_2$ by cut carnations, *Plant Physiol.* **60**:203-206.

Beyer, E. M., Jr., 1979a, [^{14}C]-Ethylene metabolism during leaf abscission in cotton, *Plant Physiol.* **64**:971-974.

Beyer, E. M., Jr., 1979b, Effect of silver ion, carbon dioxide, and oxygen on ethylene action and metabolism, *Plant Physiol.* **63**:169-173.

Beyer, E. M., Jr., and Sundin, O., 1978, $^{14}C_2H_4$ metabolism in morning glory flowers, *Plant Physiol.* **61**:896-899.

Beyer, E. M., Jr., Morgan, P. W., and Yang, S. F., 1984, Ethylene, in: *Advanced Plant Physiology*, M. B. Wilkins, ed., Pitman Publ., London, pp. 111-126.

Blankenship, S. M., and Sisler, E. C., 1989, Ethylene binding changes in apple and morning glory during ripening and senescence, *J. Plant Growth Regul.* **8**:37-44.

Bleecker, E. M., Estelle, M. A., Somerville, C., and Kende, H., 1988, Insensitivity to ethylene conferred by a dominant mutation in *Arabidopsis thalliana*, *Science* **241**:1086-1089.

Bleecker, A. B., Henyon, W. H., Somerville, S. C., and Kende, H., 1986, Use of monoclonal antibodies in the purification and characterization of 1-aminocyclopropane-1-carboxylate synthase, an enzyme in ethylene biosynthesis, *Proc. Natl. Acad. Sci.* **83**:7755-7759.

Bleecker, A. B., Esch, J. J., Hall, A. E., Rodriguez, F. I., and Binder, B. M., 1998, The ethylene-receptor family from *Arabidopsis*: Structure and function, *Philos. Trans. Royal Soc. London* **B353**:1405-1412.

Boller, T., Herner, R. C., and Kende, H., 1979, Assay for and enzymatic formation of an ethylene precursor, 1-aminocyclopropane-1-carboxylic acid, *Planta* **145**:293-303.

Boller, T., Gehri, A., Mauch, F., and Voegeli, U., 1983, Chitinase in bean leaves: Induction by ethylene, purification, properties, and possible function, Planta **157**:22-31.

Bouzayen, M., Latche, A., Allibert, G., and Pech, J.-C., 1988, Intracellular sites of synthesis and storage of 1-(malonylamino)cyclopropane-1-carboxylic acid in *Acer pseudoplatanus* cells, *Plant Physiol.* **88**:613-617.

Bouzayen, M., Latche, A., Pech, J.-C., and Marigo, G., 1989, Carrier-mediated uptake of I-(Malonylamino)cyclopropane-1-carboxylic acid in vacuoles isolated from *Catharanthus roseus* cells, *Plant Physiol.* **91**:1317-1322.

Bouzayen, M., Latche, A., and Pech, J.-C., 1990, Subcellular localization of the sites of conversion of 1-aminocyclopropane-1-carboxylic acid into ethylene in plant cells, *Planta* **180**:175-180.

Broekaert, W., Lee, H. I., Kush, A., Chua, N. H., and Raikhel, N., 1990, Wound-induced accumulation of mRNA containing a hevein sequence in laticifers of rubber tree (*Hevea brassilensis*), *Proc. Natl. Acad. Sci. USA* **87**:7633-7637.

Broglie, K. E., Biddle, P., Cressman, R., and Broglie, R., 1989, Functional analysis of DNA sequences responsible for ethylene regulation of a bean chitinase gene in transgenic tobacco, *The Plant Cell* **1**:599-607.

Brunhuber, N. M. W., Mort, J. L., Christoffersen, R. E., and Reich, N. O., 2000, Steady-state mechanism of recombinant avocado ACC oxidase: Initial velocity and inhibitor studies, *Biochem.* **39**:10730-10738.

Buescher, R. W., Sistrunk, W. A., and Brady, P. L., 1975, Effects of ethylene on metabolic and quality attributes in sweet potato roots, *J. Food Sci.* **40**:1018-1020.

Bufler, G., 1986, Ethylene-promoted conversion of 1-aminocyclopropane-1-carboxylic acid to ethylene in peel of apple at various stages of fruit development, *Plant Physiol.* **80**:539-543.

Buhler, B., Drumm, H., and Mohr, H., 1978, Investigations on the role of ethylene in phytochrome-mediated photomorphogenesis. II. Enzyme levels and chlorophyll synthesis, *Planta* **142**:119-122.

Burg, S. P., 1973, Ethylene in plant growth, *Proc. Natl. Acad. Sci. USA* **70**:591-597.

Burg, S. P., and Burg, E. A., 1967, Molecular requirements for the biological activity of ethylene, *Plant Physiol.* **42**:144-152.

Burg, S. P., and Clagett, C. O., 1967, Conversion of methionine to ethylene in vegetative tissue and fruits, *Biochem. Biophys. Res. Comm.* **27**:125-130.

Burns, J. K., and Evensen, K. B., 1986, Ca^{2+} effects on ethylene, carbon dioxide and 1-aminocyclopropane-1-carboxylic acid synthase activity, *Physiol. Plant.* **66**:609-615.

Callahan, A. M., Morgens, P. H., Wright, P., and Nichols, K. E., Jr., 1992, Comparison of Pch313 (pTOM13 homolog) RNA accumulation during fruit softening and wounding of two phenotypically different peach cultivars, *Plant Physiol.* **100**:482-488.

Cameron, A. C., and Reid, M. S., 1981, The use of silver thiosulfate anionic complex as a foliar spray to prevent flower abscission of *Zygocactus, HortSci.* **16**:761-762.

Capitani, G.,Hohenster, E., Feng, L., Storici, P., Kirsch, J. F., and Jansonius, J. N., 1999, Structure of 1-aminocyclopropane-1-carboxylate synthase, a key enzyme in the biosynthesis of the plant hormone ethylene, *J. Molec. Biol.* **294**:745-756.

Castellano, J. M., and Vioque, B., 2000, Biochemical features and inhibitors of the 1-aminocyclopropane-1-carboxylic acid oxidase isolated from pear fruit, *Eur. Food Res. Technol.* **210**:397-401.

Chadwick, A. V., and Burg, S. P., 1967, An explanation of the inhibition of root growth caused by indole-3-acetic acid, *Plant Physiol.* **42**:415-420.

Chalutz, E., 1973, Ethylene-induced phenylalanine ammonia-lyase activity in carrot roots, *Plant Physiol.* **51**:1033-1036.

Chalutz, E., Mattoo, A. K., Solomos, T., and Anderson, J. D., 1984, Enhancement by ethylene of cellulysin-induced ethylene production by tobacco leaf discs, *Plant Physiol.* **74**:99-103.

Chang, C. 1996. The ethylene signal transduction pathway in *Arabidopsis*: An emerging paradigm? *Trends Biochem. Sci.* **21**:129-133.

Chang, C., and Stewart, R C., 1998, The two-component system, *Plant Physiol.* **117**:723-731.

Chang, C., Kwok, S. F., Bleecker, A. B., and Meyerowitz, E. M., 1993, *Arabidopsis* ethylene-response gene ETR1: Similarity of product to two-component regulators, *Science* **262**:539-544.

Chernys, J., and Kende, H., 1996, Ethylene biosynthesis in *Regnellidium diphyllum* and *Marsilea quadrifolia, Planta* **200**:113-118.

Christoffersen, R.E., and Laties, G. G., 1982, Ethylene regulation of gene expression in carrots. *Proc. Natl. Acad. Sci. USA* **79**:4060-4063.

Christoffersen, R. E., McGarvey, D. J., and Savarese, P., 1992, Biochemical and molecular characterization of ethylene forming enzyme from avocado, in: *Cellular and Molecular Aspects of the Plant Hormone Ethyl*ene. J. C. Pech, A. Latche, and C. Balague, eds., Kluwer Academic Publ., Dordrecht, Netherlands, pp. 65-70.

Cohen, E., and Kende, H., 1986, The effect of submergence, ethylene and gibberellin on polyamines and their biosynthetic enzymes in deepwater-rice internodes, *Planta* **169**:498-504.

Cookson, C., and Osborne, D. J., 1978, The stimulation of cell extension by ethylene and auxin in aquatic plants, *Planta* **144**:39-47.

Dela Fuente, R. K., and Leopold, A. C., 1969, Kinetics of abscission in the bean petiole explant, *Plant Physiol.* **44**:251-254.

Dominguez, M., and Vendrell, M., 1993, Wound ethylene biosynthesis in preclimacteric banana slices, *Acta Hort.* **343**:270-274.

Dong, J. G., Fernandez-Maculet, J. C., and Yang, S. F., 1992, Purification and characterization of 1-aminocyclopropane-1-carboxylate oxidase from apple fruit, *Proc. Natl. Acad. Sci. USA* **89**:9789-9793.

Dong, J. G., Yip, W. K., and Yang, S. F., 1991, Monoclonal antibodies against apple 1-aminocyclopropane-1-carboxylate synthase, *Plant Cell Physiol.* **32**:25-31.

Drory, A., Mayak, S., and Woodson, W. R., 1993, Expression of ethylene biosynthetic pathway mRNAs is spatially regulated within carnation flower petals, *J. Plant Physiol.* **141**:663-667.

Dupille, E., and Zacarias, L., 1996, Extraction and biochemical characterization of wound-induced ACC oxidase from citrus peel, *Plant Sci.* **114**:53-60.

Dupille, E., Rombaldi, C., Lelievre, J.-M., Cleyet-Marel, J.-C., Pech, J.-C., and Latche, A., 1993, Purification, properties and partial amino-acid sequence of 1-aminocyclopropane-1-carboxylic acid oxidase from apple fruits, *Planta* **190**:65-70.

Eaks, I. L., 1966, The effect of ethylene upon ripening and respiratory rate of avocado fruit, *Calif. Avocado Soc. Yearbook* **50**:128-133.

Ecker, J. R., 1995, The ethylene signal transduction pathway in plants, *Science* **268**:667-675.

Ecker, J. R., and Davis, R. W., 1987, Plant defense genes are regulated by ethylene, *Proc. Natl. Acad. Sci. USA* **84**:5202-5206.

Eisinger, W., 1983, Regulation of pea internode expansion by ethylene, *Annu. Rev. Plant Physiol.* **34**:225-240.

Eisinger, W., Croner, L. J., and Taiz, L., 1983, Ethylene-induced lateral expansion in etiolated pea stems, *Plant Physiol.* **73**:407-412.

Evans, D. E., Dodds, J. H., Lloyd, P. C., apGwynn, I., and Hall, M. A., 1982, A study of the subcellular localisation of an ethylene binding site in developing cotyledons *of Phaseolus vulgaris* L.by high resolution autoradiography, *Planta* **154**:48-52.

Evans, D. E., Smith, A. R., Taylor, J. E., and Hall, M. A., 1984, Ethylene metabolism *in Pisum sativum* L. Kinetic parameters, the effects of propylene, silver and carbon dioxide and comparison with other systems, *Plant Growth Regul.* **2**:187-195.

Feig, A. L., and Lippard, S. J., 1994, Reactiions of non-heme iron(II) centers with dioxygen in biology and chemistry, *Chem. Rev.* **94**:759-805.

Felix, G., and Meins, F., Jr., 1987, Ethylene regulation of β-1,3-glucanase in tobacco, *Planta* **172**-386-392.

Feng, L., and Kirsch, J. F., 2000, L-Vinylglycine is an alternative substrate as well as mechanism-based inhibitor of 1-aminocyclopropane-1-carboxylate synthase, *Biochem.* **39**:2436-2444.

Fernandez-Maculet, J. C., and Yang, S. F., 1992, Extraction and partial characterization of the ethylene-forming enzyme from apple fruit, *Plant Physiol.* **99**:751-754.

Fernandez-Maculet, J. C., Dong, J. G., and Yang, S. F., 1993, Activation of 1-aminocyclopropane-1-carboxylate oxidase by carbon dioxide, *Biochem. Biophys. Res. Comm.* **193**:1168-1173.

Ferro, A. J., Barrett, A., and Shapiro, S. K., 1976, Kinetic properties and the effect of substrate analogues on 5'-methylthioadenosine nucleosidase from *Escherichia coli*, *Biochim.* Biophys. Acta **438**:487-494.

Finlayson, S. A., Reid, D. M., and Morgan, P. W., 1997, Root and leaf specific ACC oxidase activity in corn and sunflower seedlings, *Phytochem.* **45**:869-877.

Fluhr, R., 1998, Ethylene perception: From two-component signal transducers to gene induction, *Trends Plant Sci.* **3**:14-146.

Fluhr, R. and Mattoo, A. K., 1996, Ethylene-biosynthesis and perception, *Crit. Rev. Plant Sci.* **15**:479-523.

Garcia-Pineda, E., and Lozoya-Gloria, E., 1999, Induced gene expression of 1-aminocyclopropane-1-carboxylic acid (ACC oxidase) in pepper (*Capsicum annuum* L.) by arachidonic acid, *Plant Sci.* **145**:11-21.

Gaspar, T., Kevers, C., Penel, C., Greppin, H., Reid, D. M., and Thorpe, T. A., 1996, Plant hormones and plant growth regulators in plant tissue culture, *In Vitro Cell. Dev. Biol. Plant* **32**:272-289.

Gidrol, X., Chrestin, H., Mounoury, G., and D'Auzac, J., 1988, Early activation by ethylene of the tonoplast H^+-pumping ATPase in the latex from *Hevea brasiliensis*, *Plant Physiol.* **86**:899-903.

Giovanelli, J., Mudd, S. H., and Datko, A. H., 1980, Sulfur amino acids in plants, in: *Amino Acids and Derivatives. The Biochemistry of Plants: A Comprehensive Treatise*, Vol. 5, B. J. Miflin (ed.). Academic Press, New York, pp. 453-505.

Giovanelli, J., Datko, A. H., Mudd, S. H., and Thompson, G. A., 1983, *In vivo* metabolism of 5'-methylthioadenosine in *Lemna*, *Plant Physiol.* **71**:319-326.

Goeschl, J. D., and Kays, S. J., 1975, Concentration dependencies of some effects of ethylene on etiolated pea, peanut, bean and cotton seedlings, *Plant Physiol.* **55**:670-677.

Goren, R. and Sisler, E. C., 1986, Ethylene-binding characteristics in *Phaseolus*, citrus, and *Ligustrum* plants, *Plant Growth Regul.* **4**:43-54.

Goudey, J. S., F. L. Tittle, and M. S. Spencer. 1989. A role for ethylene in the metabolism of cyanide by higher plants. *Plant Physiol.* **89**:1306-1310.

Gross, K. C., Watada, A. E., Kang, M. S., Kim, S. D., Kim, K. S., and Lee, S. W., 1986, Biochemical changes associated with the ripening of hot pepper fruit, *Physiol. Plant* **66**:31-36.

Guo, L., Arteca, R. N., Phillips, A. T., and Liu, Y., 1992, Purification and characterization of 1-aminocyclopropane carboxylate N-malonyltransferase from etiolated mung bean hypocotyls, *Plant Physiol.* **100**:2041-2045.

Gupta, K., and Anderson, J. D., 1989, Influence of temperature on potentiation of cellulysin-induced ethylene biosynthesis by ethylene, *Plant Cell Physiol.* **30**:345-349.

Guranowski, A., 1983, Plant 5-methylthioribose kinase: Properties of the partially purified enzyme from yellow lupin (*Lupinus luteus* L.) seeds, *Plant Physiol.* **71**:932-935.

Guranowski, A. B., Chiang, P. K., and Cantoni, G. L., 1981, 5'-Methylthioadenosine nucleosidase: Purification and characterization of the enzyme from *Lupinus luteus* seeds, *Eur. J. Biochem.* **114**:293-299.

Guzman, P., and Ecker, J. R., 1990, Exploiting the triple response of *Arabidopsis* to identify ethylene-related mutants, *Plant Cell* **2**:513-523.

Hall, M. A. , 1991, Ethylene metabolism, in: *The Plant Hormone Ethylene.* A. H. Mattoo and Suttle, J. C., eds., CRC Press, Inc., Boca Raton, FL, pp. 65-80.

Hall, M. A., 1986, Ethylene receptors, in: *Hormones, Receptors and Cellular Interactions in Plants,* C. M. Chadwick and J. R. Garrod, eds., Cambridge Univ. Press, Cambridge, pp. 69-89.

Hall, M. A., Kapuya, J. A., Sivakumaran, S., and John, A., 1977, The role of ethylene in the response of plants to stress, *Pestic. Sci.* **8**:217-223.

Hall, M. A., Howarth, C. J., Robertson, D., Sanders, I. O., Smith, A. R., Smith, P. G., Starling, J. J., Tang, Z.-D., Thomas, C. J. R., and Williams, R. A. N., 1987, Ethylene binding proteins, in: *Molecular Biology of Plant Growth Control,* J. E. Fox and M. Jacobs, eds., Alan R. Liss, New York, pp. 335-344.

Hamilton, A. J., Bouzayen, M., and Grierson, D., 1991, Identification of a tomato gene for the ethylene-forming enzyme by expression in yeast, *Proc. Natl. Acad. Sci. USA* **88**:7434-7437.

Hamilton, A. J., Lycett, G. W., and Grierson, D., 1990, Antisense gene that inhibits synthesis of the hormone ethylene in transgenic plants, *Nature* **346**:284-287.

Hanley, K. M., Meir, S., and Bramlage, W. J., 1989, Activity of ageing carnation flower parts and the effects of 1-(malonylamino)cyclopropane-1-carboxylic acid-induced ethylene, *Plant Physiol.* **91**:1126-1130.

Harpham, N. V. J., Berry, A. W., Holland, M. G., Moshkov, I. E., Smith, A. R., and Hall, M. A., 1996, Ethylene binding sites in higher plants, *Plant Growth Regul.* **18**:71-77.

Henry, E. W., and Ananich, M. E., 1985, Ethylene effects on superoxide dismutase in bean abscission zone tissue, *Proc.12th Ann. Meeting Plant Growth Regul. Soc.* Am. Univ. Colorado, Boulder, July 28-Aug. 1, pp. 60-77.

Henskens, J. A. M., Rouwendal, G. J. A., Ten Have, A., and Woltering, E. J., 1994, Molecular cloning of two different ACC synthase PCR fragments in carnation flowers and organ-specific expression of the corresponding genes, *Plant Mol. Biol.* **26**:453-458.

Hoffman, N. E., Fu, J.-R., and Yang, S. F., 1983, Identification and metabolism of 1-(malonylamino)cyclopropane-1-carboxylic acid in germinating peanut seeds, *Plant Physiol.* **71**:197-199.

Hoffman, N. E., Yang, S. F., and McKeon, T., 1982, Identification of 1-(malonylamino)cyclopropane-1-carboxylic acid as a major conjugate of 1-aminocyclopropane-1-carboxylic acid, and ethylene precursor in higher plants, *Biochem. Biophys. Res. Commun.* **104**:765-770.

Hua, J., and Meyerowitz, E. M., 1998, Ethylene responses are negatively regulated by a receptor gene family in *Arabidopsis thaliana, Cell* **94**:261-271.

Hua, J., Sakai, H., Nourizadeh, S., Chen, Q. G., Bleecker, A. B., Ecker, J. R., and Meyerowitz, E. M., 1998, EIN4 and ERS2 are members of the putative ethylene receptor gene family in *Arabidopsis, Plant Cell.* **10**:1321-1332.

Hubbard, N. L., Huber, S. C., and Pharr, D. M., 1989, Sucrose phosphate synthase and acid invertase as determinants of sucrose concentration in developing muskmelon (*Cucumis melo* L.) fruits, *Plant Physiol.* **91**:1527-1534.

Huberman, M., and Goren, R., 1982, Is uronic acid oxidase involved in the hormonal regulation of abscission in explants of citrus leaves and fruits? *Physiol. Plant* **56**:168-176.

Hyodo, H., 1991, Stress/wound ethylene, in: *The Plant Hormone Ethylene*, A. K. Mattoo and J. C. Suttle, eds., CRC Press, Inc., Boca Raton, FL, pp. 43-63.

Hyodo, H., and Yang, S. F., 1971, Ethylene-enhanced formation of cinnamic acid-4-hydroxylase in excised pea epicotyl tissue, *Arch Biochem. Biophys.* **143**:338-339.

Hyodo, H., Tanaka, K., and Yoshisaka, J., 1985, Induction of 1-aminocyclopropane-1-carboxylic acid synthase in wounded mesocarp tissue of winter squash fruit and the effects of ethylene, *Plant Cell Physiol.* **26**:161-167.

Hyodo, H., Hashimoto, C., Morozumi, S., Hu, W., and Tanaka, K., 1993, Characterization and induction of the activity of 1-aminocyclopropane-1-carboxylate oxidase in the wounded mesocarp tissue of *Cucurbita maxima, Plant Cell Physiol.* **34**:667-671.

Icekson, I., Goldlust, A., and Apelbaum, A., 1985, Influence of ethylene on S-adenosylmethionine decarboxylase activity in etiolated pea seedlings, *J. Plant Physiol.* **119**:335-345.

Iki, K., Sekiguchi, K., Kurata, K., Tada, T., Nakagawa, H., Ogura, N., and Takehana, H., 1978, Immunological properties of β-fructofuranosidase from ripening tomato fruit, *Phytochemistry* **17**:311-312.

Inaba, A., and Nakamura, R., 1986, Effect of exogenous ethylene concentration and fruit temperature on the minimum treatment time necessary to induce ripening in banana fruit, *J. Japan Soc. Hort. Sci.* **55**:348-354.

Isola, M. C. and Franzoni, L., 1989, Effect of ethylene on the increase in ribonuclease activity in potato tuber tissue, *Plant Physiol. Biochem.* **27**:245-250.

Itai, A., Kawata, T., Tanabe, K., Tamura, F., Uchiyama, M., Tomomitsu, M., and Shiraiwa, N., 1999, Identification of 1-aminocyclopropane-1-carboxylic acid synthase genes controlling the ethylene level of ripening fruit in Japanese pear (*Pyrus pyrifolia* Nakai), *Mol. Gen. Genet.* **261**:42-49.

Jackson, M. B., 1983, Regulation of root growth and morphology by ethylene and other externally applied growth substances, in: *Growth Regulators in Root Development*, M. B. Jackson and A. D. Stead, eds., Proc. Brit. Plant Growth Regulator Group, Monograph **10**:103-116.

Jeffrey, D., Smith, C., Goodenough, P., Prosser, I., and Grierson, D., 1984, Ethylene-independent and ethylene-dependent biochemical changes in ripening tomatoes, *Plant* **Physiol.** **74**:32-38.

Jerie, P. H., and Hall, M. A., 1978, The identification of ethylene oxide as a major metabolite of ethylene in *Vicia faba* L., *Proc. R. Soc. London Ser. B* **200**:87-94.

Jiao, X.-Z., Yip, W. K., and Yang, S. F., 1987. The effect of light and phytochrome on 1-aminocyclopropane-1-carboxylic acid metabolism in etiolated wheat seedling leaves. *Plant Physiol.* **85**:643-647.

Jiao, X.-Z., Philosoph-Hadas, S., Su, L.-Y., and Yang, S. F., 1986, The conversion of 1-(malonylamino)cyclopropane-1-carboxylic acid to 1-aminocyclopropane-1-carboxylic acid in plant tissues, *Plant Physiol.* **81**:637-641.

John, P., 1997, Ethylene biosynthesis: The role of 1-aminocyclopropane-1-carboxylate (ACC) oxidase, and its possible evolutionary origin, *Physiol. Plant.* **100**:583-592.

Kasai, Y., Hyodo, H., Ikoma, Y., and Yano, M., 1998, Characterization of 1-aminocyclopropane-1-carboxylate (ACC) oxidase in broccoli florets and from *Escherichia coli* cells transformed with cDNA of broccoli ACC oxidase, *Bot. Bull. Acad. Sin.* **39**:225-230.

Kasai, Y., Kato, M., and Hyodo, H., 1996, Ethylene biosynthesis and its involvement in senescence of broccoli florets, *J. Japan. Soc. Hort. Sci.* **65**:185-191.

Kato, M., and Hyodo, H., 1999, Purification and characterization of ACC oxidase and increase in its activity during ripening of pear fruit, *J. Japan. Soc. Hort. Sci.* **68**:551-557.

Kawase, M., 1979, Role of cellulase in aerenchyma development in sunflower, *Amer. J. Bot.* **66**:183-190.

Kende, H., 1993, Ethylene biosynthesis, *Annu. Rev. Plant Physiol. Plant Mol. Biol.* **44**:283-307.

Kende, H., 1989, Enzymes of ethylene biosynthesis, *Plant Physiol.* **91**:1-4.

Kieber, J. J., Rothenberg, M., Roman, G., Feldman, K. A., and Ecker, J. R., 1993, CTR1, a negative regulator of the ethylene response pathway in *Arabidopsis*, encodes a member of the Raf family of protein kinases, *Cell* **72**:427-441.

Kim, J. H., Lee, J. H., Joo, S., and Kim, W. T., 1999, Ethylene regulation of an ERS1 homolog in mung bean seedlings, *Physiol. Plant.* **106**-90-97.

Kim, J. H., Kim, W. T., Kang, B. G., and Yang, S. F., 1997, Induction of 1-aminocyclopropane-1-carboxylic oxidase mRNA by ethylene in mung bean hypocotyls: Involvement of both protein phosphorylation and dephosphorylation in ethylene signaling, Plant J. 11:399-405.

Kim, W.T. and Yang, S. F., 1992, Turnover of 1-aminocyclopropane-1-carboxylic acid synthase in wounded tomato tissues, *Plant Physiol.* **100**:1126-1130.

Kim, W. T., and Yang, S. F., 1994, Structure and expression of cDNAs encoding 1-aminocyclopropane-1-carboxylate oxidase homologs isolated from excised mung bean hypocotyls., *Planta* **194**:223-229.

Konze, J. R., and Kende, H., 1979a, Ethylene formation from 1-aminocyclopropane-1-carboxylic acid in homogenates of etiolated pea seedlings, *Planta* **146**:293-301.

Konze, J. R., and Kende, H., 1979b, Interactions of methionine and selenomethionine with methionine adenosyltransferase and ethylene-generating systems, *Plant Physiol.* **63**:507-510.

Kuai, J., and Dilley, D. R., 1992, Extraction, partial purification and characterization of 1-aminocyclopropane-1-carboxylic acid oxidase from apple fruit, *Postharvest Biol. Technol.* **1**:203-211.

Kushad, M. M., 1990, Recycling of 5'deoxy-5'-methylthioadenosine in plants, in: *Polyamines and Ethylene: Biochemistry, Physiology, and Interactions*, H. E. Flores, R. N. Arteca, and J. C. Shannon, eds., American Soc. Plant Physiologists, MD, pp. 50-61.

Kushad, M. M., Richardson, D. G., and Ferro, A. J., 1983, Intermediates in the recycling of 5-methylthioribose to methionine in fruits, *Plant Physiol.* **73**:257-261.

Lanahan, M. B., Yen, H.-C., Glovannoni, J. J., and Klee, H. J., 1994, The Never Ripe mutation blocks perception in tomato. *Plant Cell* **6**:521-530.

Larsen, P. B., and Woodson, W. R., 1991, Cloning and nucleotide sequence of a S-adenosylmethionine synthetase cDNA from carnation. *Plant Physiol.* **96**:997-999.

Lashbrook, C. C., Tiemann, D. M., and Klee, H. J., 1998, Differential regulation of the tomato ETR family throughout plant development, *The Plant J.* **15**:243-252.

Lasserre, E., Bouquin, T., Hernandez, J. A., Bull, J., Pech, J. C., and Balague, C., 1996, Structure and expression of three genes encoding ACC oxidase homologs from melon (*Cucumis melo* L.), *Mol. Gen. Genet.* **251**:81-90.

Latche, A., Dupille, E., Rombaldi, C., Cleyet-Marel, J. C., Lelievre, J. M., and Pech, J. C., 1992, Purification, characterization and subcellular localization of ACC oxidase from fruits, in: *Cellular and Molecular Aspects of the Plant Hormone Ethylene*, J. C. Pech, A. Latche, and C. Balague, eds., Kluwer Academic Publ., Dordrecht, Netherlands, pp. 39-45.

Lawton, K. A., Raghothama, K. G., Goldsbrough, P. G., and Woodson, W. R., 1990, Regulation of senescence related gene expression in carnation flower petals by ethylene, *Plant Physiol.* **93**:1370-1375.

Lay, V. J., Prescott, A. G., Thomas, P. G., and John, P., 1996, Heterologous expression and site-directed mutagenesis of the 1-aminocyclopropane-1-carboxylate oxidase from kiwi fruit, *Eur. J. Biochem.* **242**:228-234.

Lay-Yee, M., and Knighton, M. L., 1995, A full-length cDNA encoding 1-aminocyclopropane-1-carboxylate synthase from apple, *Plant Physiol.* **107**:1017-1018.

Leather, G. R., Forrence, L. E., and Abeles, F. B., 1972, Increased ethylene production during clinostat experiments may cause leaf epinasty, *Plant Physiol.* **49**:183-186.

Lelievre, J.-M., Tichit, L., Dao, P., Fillion, L., Nam, Y.-W., Pech, J.-C., and Latche, A., 1997, Effects of chilling on the expression of ethylene biosynthetic genes in Passe-Crassane pear (*Pyrus communis* L.) fruits, *Plant Mol. Biol.* **33**:847-855.

Leyser, H. M. O. , 1998, Plant hormones, *Curr. Biol.* **8**:R5-R7.

Li, N., Wiesman, Z., Liu, D., and Mattoo, A. K., 1992, A functional tomato ACC synthase expressed in *Escherichia coli* demonstrates suicidal inactivation by its substrate S-adenosylmethionine, *FEBS Lett.* **306**:103-107.

Li, Y., Feng, L., and Kirsch, J. F., 1997, Kinetic and spectroscopic investigations of wild-type and mutant forms of apple 1-aminocyclopropane-1-carboxylate synthase, *Biochem.* **36**:15477-15488.

Lieberman, M., 1979, Biosynthesis and action of ethylene, *Annu. Rev. Plant Physiol.* **30**:533-591.

Lieberman, M., and Mapson, L. W., 1964, Genesis and biogenesis of ethylene, *Nature* **204**:343-345.

Lieberman, M., Kunishi, A. T., Mapson, L. W., and Wardale, D. A., 1966, Stimulation of ethylene production in apple tissue slices by methionine, Plant Physiol. **41**:376-382.

Lincoln, J. E., and Fischer, R. L., 1988, Regulation of gene expression by ethylene in wild-type and *rin* tomato (*Lycopersicon esculentum*) fruit, *Plant Physiol.* **88**:370-374.

Lincoln, J. E., Cordes, S., Read, E., and Fischer, R. L., 1987, Regulation of gene expression by ethylene during *Lycopersicon esculentum* (tomato) fruit development, *Proc. Natl. Acad. Sci.* **84**:2793-2797.

Lincoln, J. E., Campbell, A. D., Oetiker, J., Rottmann, W. H., Oeller, P. W., Shen, N. F., and Thologis, A., 1993, LE-ACS4, a fruit ripening and wound-induced 1-aminocyclopropane-1-carboxylate synthase gene of tomato (*Lycopersicon esculentum*), *J. Biol. Chem.* **268**:19422-19430.

Liu, Y., Hoffman, N. E., and Yang, S. F., 1985a, Ethylene-promoted malonylation of 1-aminocyclopropane-1-carboxylic acid particpates in autoinhibition of ethylene synthesis in grapefruit flavedo discs, *Planta* **164**:565-568.

Liu, Y., Su, L.-Y., and Yang, S. F., 1985b, Ethylene promotes the capability to malonylate 1-aminocyclopropane-1-carboxylic acid and D-amino acids in preclimacteric tomato fruits, *Plant Physiol.* **77**:891-895.

Liu, Y., Hoffman, N. E., and Yang, S. F. Yang, 1985c, Promotion by ethylene of the capability to convert 1-aminocycloprpane-1-carboxylic acid to ethylene in preclimacteric tomato and cantaloupe fruits, *Plant Physiol.* **77**:407-411.

Lurssen, K., Naumann, K., and Schroder, R., 1979, 1-Aminocyclopropane-1-carboxylic acid-an intermediate of the ethylene biosynthesis in higher plants, *Z. Pflanzenphysiol.* **92**:285-294.

Mansour, R., Latche, A., Vallant, V., Pech, J. C., and Reid, M. S., 1986, Metabolism of 1-aminocyclopropane-1-carboxylic acid in ripening apple fruits, *Physiol. Plant.* **66**:495-502.

Mapson, L. W., 1969, Biogenesis of ethylene, *Biol. Rev.* **44**:155-187.

Martin, M. N., and Saftner, R. A., 1995, Purification and characterization of 1-aminocyclopropane carboxylic acid N-malonyltransferase from tomato fruit, *Plant Physiol.* **108**:1241-1249.

Martin, M. N., Cohan, J. D., and Saftner, R. A., 1995, A new 1-aminocyclopropane carboxylic acid-conjugating activity in tomato fruit, *Plant Physiol.* **109**:917-926.

Mathooko, F. M., Kubo, Y., Inaba, A., and Nakamura, R., 1993, Partial characterization of 1-aminocyclopropane-1-carboxylate oxidase from excised mesocarp tissue of winter squash fruit, *Sci. Rep. Fac. Agric. Okayama Univ.* **82**:49-59.

Mattoo, A. K., and Suttle, J. C., eds., 1991, *The Plant Hormone Ethylene*, CRC Press, Boca Raton, FL. p. 337.

Mattoo, A. K., and White, W. B., 1991, Regulation of ethylene biosynthesis, in: *The Plant Hormone Ethylene*, A. K. Mattoo and J. C. Suttle, eds., CRC Press, Boca Raton, FL, pp. 21-42.

McGarvey, D. J., and Christoffersen, R. E., 1992, Characterization and kinetic parameters of ethylene-forming enzyme from avocado fruit, *J. Biol. Chem.* 267:5964-5967.

McGarvey, D. J., Yu, H., and Christoffersen, R. E., 1990, Nucleotide sequence of a ripening-related cDNA from avocado fruit, *Plant Mol. Biol.* **15**:165-167.

McGrath, R. B., and Ecker, J. R., 1998, Ethylene signaling in *Arabidopsis*: Events from the membrane to the nucleus, *Plant Physiol. Biochem.* **36**:103-113.

McKeon, T. A., Fernandez-Maculet, J. C., and Yang, S.-F., 1995, Biosynthesis and metabolism of ethylene, in: *Plant Hormones Physiology, Biochemistry and Molecular Biology*, P. J. Davies, ed., Kluwer Academic Publ., Dordrecht, The Netherlands, pp. 118-139.

McMurchie, E. J., McGlasson, W. B., and Eaks, I. L., 1972, Treatment of fruits with propylene gives information about the biogenesis of ethylene, *Nature* **237**:235-236.

Mehta, P. K., and Christen, P., 1994, Homology of 1-aminocyclopropane-1-carboxylate synthase, 8-amino-7-oxononanoate synthase, 2-amino-6-caprolactam racemase, 2,2-dialkylglycine decarboxylase, glutamate-1-semialdehyde 2,1-aminomutase and isopenicillin-N-epimerase with aminotransferases, *Biochem. Biophys. Res. Commun.* **198**:138-143.

Ming, L.-J., Que, L., Jr., Kriauciunas, A., Frolik, C. A., and Chen, V. J., 1990, Coordination chemistry of the metal binding site of isopenicillin N synthase, *Inorg. Chem.* **29**:1111-1112.

Mita, S., Kirita, C., Kato, M., and Hyodo, H., 1999, Expression of ACC synthase is enhanced earlier than that of ACC oxidase during fruit ripening of mume (Prunus mume), *Physiol. Plant.* **107**:319-328.

Miyazaki, J. H., and Yang, S. F., 1987, Metabolism of 5-methylthioribose to methionine, *Plant Physiol.* **84**:277-281.

Moya-Leon, M. A., and John. P., 1995, Purification and biochemical characterization of 1-aminocyclopropane1-carboxylate oxidase from banana fruit, *Phytochem.* **39**:15-20.

Muller, R., Lind-Iversen, S., Stummann, B. M., and Serek, M., 2000, Expression of genes for ethylene biosynthetic enzymes and an ethylene receptor in senescing flowers of miniature potted roses. *J. Hort. Sci. Biotechnol.* **75**:12-18.

Murr, D. P. and Yang, S. F., 1975, Inhibition of *in vivo* conversion of methionine to ethylene by L-canaline and 2,4-dinitrophenol, *Plant Physiol.* **55**:79-82.

Nadeau, J. A., Zhang, X. S., Nair, H., and O'Neill, S. D., 1993, Temporal and spatial regulation of 1-aminocyclopropane1-carboxylate oxidase in the pollination-induced senescence of orchid flowers, *Plant Physiol.* **103**:31-39.

Nakajima, N., Nakagawa, N., and Imaseki, H., 1988, Molecular size of wound-induced 1-aminocyclopropane-1-carboxylate synthase from *Cucurbita maxima* Duch. and change of translatable mRNA of the enzyme after wounding, *Plant Cell Physiol.* **29**:989-998.

Nakajima, N., Mori, H., Yamazaki, K., and Imaseki, H., 1990, Molecular cloning and sequence of a complementary DNA encoding 1-aminocyclopropane-1-carboxylate synthase induced by tissue wounding, *Plant Cell Physiol.* **31**:1021-1029.

Nakatsuka, A., Shiomi, S., Kubo, Y., and Inaba, A., 1997, Expression and internal feedback regulation of ACC synthase and ACC oxidase genes in ripening tomato fruit, *Plant Cell Physiol.* **38**:1103-1110.

Nee, M., Chiu, L., and Eisinger, W., 1978, Induction of swelling in pea internode tissue by ethylene, *Plant Physiol.* **62**:902-906.

Nijenhuis-DeVries, M. A., Woltering, E. J., and DeVrije, T., 1994, Partial characterization of carnation petal 1-aminocyclopropane-1-carboxylate oxidase, *J. Plant Physiol.* **144**:549-554.

Olson, D. C., White, J. A., Edelman, L., Harkins, R. H., and Kende, H., 1991, Differential expression of two genes for 1-aminocyclopropane-1-carboxylate synthase in tomato fruits, *Proc. Natl. Acad. Sci. USA* **88**:5340-5344.

Osborne, D. J., 1982, The ethylene regulation of cell growth in specific target tissues of plants, in: *Plant Growth Substances*, P. F. Waring, ed., Academic Press, London, p. 279.

Osborne, D. J., 1989, The control role of ethylene in plant growth and development, in: *Biochemical and Physiological Aspects of Ethylene Production in Lower and Higher Plants*, H. Clijsters et al., eds., Kluwer Academic Publishers, pp. 1-11.

Osborne, D. J., 1991, Ethylene in leaf ontogeny and abscission, in: *The Plant Hormone Ethyelene*, A. K. Matto and J. C. Suttle, eds., CRC Press, Boca Raton, FL, pp. 193-214..

Osborne, D. J., Walters, J., Milborrow, B. V., Norville, A., and Stange, L. M. C., 1996, Evidence for a non-ACC ethylene biosynthesis pathway in lower plants, *Phytochem.* **42**:51-60.

Owens, L. D., Lieberman, M., and Kunishi, A., 1971, Inhibition of ethylene production by rhizobitoxine, *Plant Physiol.* **48**:1-4.

Palavan, N., Goren, R., and Galston, A. W., 1984, Effects of some growth regulators on polyamine biosynthetic enzymes in etiolated pea seedlings, *Plant Cell Physiol.* **25**:541-546.

Park, K. Y., Drory, A., and Woodson, W. R., 1992, Molecular cloning of an 1-aminocyclopropane-1-carboxylate synthase from senescing carnation flower petals, *Plant Mol. Biol.* **18**:377-386.

Payton, S., Fray, R. G., Brown, S., and Grierson, D., 1996, Ethylene receptor expression is regulated during fruit ripening, flower senescence and abscission, *Plant Molec. Biol.* **31**:1227-1231.

Peck, S. C., and Kende, H., 1995, Sequential induction of the ethylene biosynthetic enzymes by indole-3-acetic acid in etiolated peas, *Plant Mol. Biol.* **28**:293-301.

Peiser, G. D., Wang, T. T., Hoffman, N. E., Yang, S. F., Liu, H. W., and Walsh, C. T., 1984, Formation of cyanide from carbon 1 of 1-aminocyclopropane-1-carboxylic acid during its conversion to ethylene, *Proc. Natl. Acad. Sci. USA* **81**:3059-3063.

Peleman, J., Saito, K., Cottyn, B., Engler, G., Seurinck, J., Van Montagu, M., and Inze, D., 1989, Structure and expression analyses of the S-adenosylmethionine synthetase gene family in *Arabidopsis thaliana*, *Gene* **84**:359-369.

Penarruba, L., Aguilar, M., Margossian, L., and Fischer, R. L., 1992, An antisense gene stimulates ethylene hormone production during tomato fruit ripening, *Plant Cell* **4**:681-687.

Philosoph-Hadas, S., Meir, S., and Aharoni, N., 1985, Autoinhibition of ethylene production in tobacco leaf discs: Enhancement of 1-aminocyclopropane-1-carboxylic acid conjugation, *Physiol. Plant.* **63**:431-437.

Philosoph-Hadas, S., Meir, S., Pesis, E., Reuveni, A., and Aharoni, N., 1989, Hormone-enhanced ethylene production in leaves, in: *Biochemical and Physiological Aspects of Ethylene Production in Lower and Higher Plants*, H. Clijsters et al., eds., Kluwer Academic Publ., Netherlands, pp. 135-142.

Pirrung, M. C., and Mcgeehan, G. M., 1986, Ethylene biosynthesis: 6. Synthesis and evaluation of methylaminocyclopropanecarboxylic acid. *J. Org. Chem.* **51**:2103-2106.

Pirrung, M. C., Kaiser, L. M., and Chen, J., 1993, Purification and properties of the apple fruit ethylene-forming enzyme, Biochem. 32:7445-7450.

Pogson, B. J., Downs, C. G., and Davies, K. M., 1995, Differential expression of two 1-aminocyclopropane1-carboxylic acid oxidase genes in broccoli after harvest, *Plant Physiol.* **108**:651-657.

Porat, R., Borochov, A., Halevy, A. H., and O'Neill, S. D., 1994, Pollination-induced senescence of *Phalanopsis* petals. The wilting process, ethylene production and sensitivity to ethylene, *Plant Growth Regul.* **15**:129-136.

Prescott, A., 1993, A dilemma of dioxygenases (or where biochemistry and molecular biology fail to meet), *J. Exp. Bot.* **44**:849-861.

Prescott, A. G., and John, P., 1996, Dioxygenases: Molecular structure and role in plant metabolism, *Annu. Rev. Plant Physiol. Plant Mol. Biol.* **47**:245-271.

Privalle, L. S., and Graham, J. S., 1987, Radiolabeling of a wound-inducible pyridoxal phosphate-utilizing enzyme: Evidence for its identification as ACC synthase, *Arch. Biochem. Biophys.* **253**:333-340.

Purvis, A. C., 1980, Sequence of chloroplast degreening in calamondin fruit as influenced by ethylene and $AgNO_3$, *Plant Physiol.* **66**:624-627.

Ramalingham, K., Lee, K.-M., Woodward, R. W., Bleecker, A. B., and Kende, H., 1985, Stereochemical course of the reaction catalyzed by the pyridoxal phosphate-dependent enzyme 1-aminocyclopropane-1-carboxylate synthase, *Proc. Natl. Acad. Sci. USA* **82**:7820-7824.

Rando, R. R., 1974, Chemistry and enzymology of Kcat inhibitors, *Science* **185**:320-324.

Raskin, I., and Beyer, E. M., Jr., 1989, Role of ethylene metabolism in *Amaranthus retroflexus*, *Plant Physiol.* **90**:1-5.

Rauser, W. E., and Horton, R. F., 1975, Rapid effects of indoleacetic acid and ethylene on the growth of intact pea roots, *Plant Physiol.* **55**:443-447.

Raz, V., and Fluhr, R., 1992, Calcium requirement for ethylene-dependent responses, *Plant Cell* **4**:1123-1130.

Reid, M. S., 1995, Ethylene in plant growth, development and senescence, in: *Plant Hormones, Physiology, Biochemistry and Molecular Biology*, P. J. Davies, ed., Kluwer Academic Publishers, Dordrecht, The Netherlands, pp. 486-508.

Riov, J. and Yang, S. F., 1989, Ethylene and auxin-ethylene interaction in adventitious root formation in mung bean (*Vigna radiata*) cuttings, *J. Plant Growth Regul.* **8**:131-141.

Riov, J., and Yang, S. F., 1982, Autoinhibition of ethylene production in citrus peel discs. Suppression of 1-aminocyclopropane-1-carboxylic acid synthesis, *Plant Physiol.* **69**:687-690.

Riov, J., Monselise, S. P., and Kahan, R. S., 1969, Ethylene-controlled induction of phenylalanine ammonia-lyase in citrus fruit peel,, *Plant Physiol.* **44**:631-635.

Roach, P. L., Clifton, I. J., Fulop, V., Harlos, K., Barton, G. J., Hajdu, J., Andersson, I., Schofield, C. J., and Baldwin, J. E., 1995, Crystal structure of isopenicillin N synthase is the first from a new structural family of enzymes, *Nature* **375**:700-704.

Roberts, D. R., Walker, M. A., Thompson, J. E., and Dumbroff, E. J., 1984, The effects of inhibitors of polyamine and ethylene biosynthesis on senescence, ethylene production, and polyamine levels in cut carnation flowers, *Plant Cell Physiol.* **25**:315-322.

Rodriguez, F. I., Esch, J. J., Hall, A., Binder, B. M., Schaller, G. E., and Bleecker, A. B., 1999, A copper factor for the ethylene receptor *ETR1* from *Arabidopsis*, *Science* **283**:996-998.

Rombaldi, C., Petitprez, M., Cleyet-Marel, J. C., Rouge, P., Latche, A., Pech, J. C., and Lelievre, J. M., 1992, Immunocyclocalisation of ACC oxidase in tomato fruits, in: *Cellular and Molecular Aspects of the Plant Hormone Ethylene*, J. C. Pech, A. Latche, and C. Balague, eds., Kluwer Academic Publ., Dordrecht, Netherlands, pp. 96-97.

Rottmann, W. E., Peter, G. F., Oeller, P. W., Keller, J. A., Shen, N. F., Nagy, B. P., Taylor, L. P., Campell, A. D., and Theologis, A., 1991, 1-Aminocyclopropane-1-carboxylate synthase in tomato is encoded by a multi-gene family whose transcription is induced during fruit and floral senescence, *J. Mol. Biol.* **222**:937-961.

Sakai, H., Hua, J., Chen, Q., Chang, C., Medrano, L., Bleecker, A. B., and Meyerowitz, E. M., 1998, *ETR2* is an *ETR1*-like gene involved in ethylene signaling in *Arabidopsis*, *Proc. Natl. Acad. Sci. USA* **95**:5912-5913.

Saltveit, M. E., Jr., and Dilley, D. R., 1978, Rapidly induced wound ethylene from excised segments of etiolated *Pisum sativum* L. cv. Alaska. II. Oxygen and temperature dependency. *Plant Physiol.* **61**:675-679.

Sanders, I. O., Smith, A. R., and Hall, M. A., 1989, Ethylene metabolism in *Pisum sativum* L., *Planta* **179**:104-114.

Sanders, I. O., Ishizawa, K., Smith, A. R., and Hall, M. A., 1990, Ethylene binding and action in rice seedlings, *Plant Cell Physiol.* **31**:1091-1099.

Sato, T., and Theologis, A., 1989, Cloning the mRNA encoding 1-aminocyclopropane-1-carboxylate synthase, the key enzyme for ethylene biosynthesis in plants, *Proc. Natl. Acad. Sci.* **86**:6621-6625.

Sato, T., Oeller, P. W., and Theologis, A., 1991, The 1-aminocyclopropane-1-carboxylate synthase of *Cucurbita*: Purification, properties, expression in *Escherichia coli* and primary structure determination by DNA sequence analysis, *J. Biol. Chem.* **266**:3752-3759.

Satoh, S., and Esashi, Y., 1983, Alpha-aminoisobutyric acid, propyl gallate and cobalt ion and the mode of inhibition of ethylene production by cotyledonary segments of cocklebur seeds, *Phys. Plant.* **57**:521-526.

Satoh, S., and Esashi, Y., 1986, Inactivation of 1-aminocyclopropane-1-carboxylic acid synthase of etiolated mung bean hypocotyl segments by its substrate, *S*-adenosyl-L-methionine, *Plant Cell Physiol.* **27**:285-291.

Satoh, S., and Yang, S. F., 1988, S-Adenosylmethionine-dependent inactivation and radiolabeling of 1-aminocyclopropane-1-carboxylate synthase isolated from tomato fruits, *Plant Physiol.* **88**:109-114.

Satoh, S., and Yang, S. F., 1989, Specificity of S-adenosyl-L-methionine in the inactivation and labeling of 1-aminocyclopropane-1-carboxylate synthase isolated from tomato fruits, Arch. *Biochem. Biophys.* **271**:107-112.

Schmerder, B., and Borriss, H., 1986, Induction of nitrate reductase by cytokinin and ethylene in *Agrosemma githago* L. embryos, *Planta* **169**:589-593.

Serek, M., Reid, M. S., and Sisler, E. C., 1994, Novel gaseous ethylene binding inhibitor prevents ethylene effects in potted flowering plants, *J. Amer. Soc. Hort. Sci.* **119**:1230-1233.

Shaw, J.-F., Chou, Y.-S., Chang, R. C., and Yang, S. F., 1996, Caracterization of the ferrous ion binding sites of apple 1-aminocyclopropane-1-carboxylate oxidase by site-directed mutagenesis, *Biochem. Biophys. Res. Commun.* **225**:697-700.

Shiomi, S., Nakamoto, J.-I., Yamamoto, M., Kubo, Y., Nakamura, R., and Inaba, A., 1999, Expression of ACC synthase and ACC oxidase genes in different tissues of immature and mature cucumber fruits, *J. Japan. Soc. Hort. Sci.* **68**:830-832.

Sisler, E. C., 1977, Ethylene activity of some N-acceptor compounds, *Tobacco Sci.* **21**:43-45.

Sisler, E. C., 1979, Measurement of ethylene binding in plant tissue, *Plant Physiol.* **64**:538-542.

Sisler, E. C., 1980, Partial purification of an ethylene-binding component from plant tissue, *Plant Physiol.* **66**:404-406.

Sisler, E. C., 1982a, Ethylene binding in normal, *rin* and *nor* mutant tomatoes, *J. Plant Growth Regul.* **1**:211-219.

Sisler, E. C., 1982b, Ethylene-binding properties of a Triton X-100 extract of mung bean sprouts, *J. Plant Growth Regul.* **1**:211-218.

Sisler, E. C., 1991, Ethylene-binding components in plants, in: *The Plant Hormone Ethylene*, A. H. Mattoo and J. C. Suttle, eds., CRC Press, Inc., Boca Raton, FL, pp. 81-99.

Sisler, E. C., and Goren, R., 1981, Ethylene binding-the basis for hormone action in plants? *What's New in Plant Physiol.* **12**:37-40.

Sisler, E. C., and Yang, S. F., 1984, Anti-ethylene effects of *cis*-2-butene and cyclic olefins, *Phytochem.* **23**:2765-2768.

Sisler, E. C., and Blankenship, S. M., 1993, Diazocyclopentadiene (DACP), a light sensitive reagent for the ethylene receptor in plants, *Plant Growth Regul.* **12**:125-132.

Sisler, E. C., Reid, M. S., and Yang, S. F., 1986, Effect of antagonists of ethylene action on binding of ethylene in cut carnations, *Plant Growth Regulation* **4**:213-218.

Sitrit, Y., Riov, J., and Blumenfeld, A., 1986, Regulation of ethylene biosynthesis in avocado fruit during ripening, *Plant Physiol.* **81**:130-135.

Smith, A. R., and Hall, M. A., 1984, Biosynthesis and metabolism of ethylene, in: *The Biosynthesis and Metabolism of Plant Hormones*, A. Crozier and J. H. Hillman, eds., Society for Experimental Biology Semianr 23, Cambridge University Press, New York, pp. 201-229.

Smith, A. R., Evans, D. E., Smith, P. G., and Hall, M. A., 1985, Ethylene metabolism *in Pisum sativum* L. and *Vicia faba* L., in: *Ethylene and Plant Development*, J. A. Roberts and G. A. Tucker, eds., Butterworths, London, pp. 139-145.

Smith, A. R., Robertson, D., Sanders, I. O., Williams, R. A. N., and Hall, M. A., 1987, Ethylene binding sites, in: *Plant Hormone Receptors*, D. Klambt, ed., Springer-Verlag, Berlin, pp. 229-238.

Smith, C. I. S., Slater, A., and Grierson, D., 1986, Rapid appearance of an mRNA correlated with ethylene synthesis encoding a protein of molecular weight 35000, *Planta* **168**:94-100.

Smith, J. J., and John, P., 1993, Maximizing the activity of the ethylene-forming enzyme, in: *Cellular and Molecular Aspects of the Plant Hormone Ethylene*, J. C. Pech, A. Latche, and C. Balague, eds., Kluwer, Dordrecht, pp. 33-38.

Smith, J. J., Ververidis, P., and John, P., 1992, Characterization of the ethylene-forming enzyme partially purified from melon, *Phytochem.* **31**:1485-1494.

Smith, J. J., Zhang, Z. H., Schofield, C. J., John, P., and Baldwin, J. E., 1994, Inactivation of 1-aminocyclopropane-1-carboxylate (ACC) oxidase, *J. Exp. Bot.* **45**:521-527.

Smith, M. A., Jacobsen, J. V. G., and Kende, H., 1987, Amylase activity and growth in internodes of deepwater rice, *Planta* **172**:114-120.

Spanu, P., Reinhardt, D., and Boller, T., 1991, Analysis and cloning of the ethylene-forming enzyme from tomato by functional expression of its mRNA in *Xenopus laevis* oocytes, *EMBO J.* **10**:2007-2013.

Stacewicz-Sapuncakis, M., Marsh, H. V., Jr., Vengris, J., Jennings, P. H., and Robinson, T., 1973, Participation of ethylene in common purslane repsonse to dicamba, *Plant Physiol.* **52**:466-471.

Stange, L. M. C., and Osborne, D. J., 1989, Contrary effects of ethylene and ACC on cell growth in the liverwort riella helicophylla, in: *Biochemical and Physiological Aspects of Ethylene Production in Lower and Higher Plants*, H. Clijsters, deProft, M., Marcelle, R., and vanPoucke M., eds., Kluwer Academic Publishers, Dordrecht, The Netherlands, pp. 341-348.

Stead, A. D. and Moore, K. G., 1983. Studies on flower longevity in *Digitalis*. The role of ethylene in corolla abscission. *Planta* **157**:15-21.

Stella, L., Wouters, S., and Baldellon, F., 1996, Chemical and biochemical aspects of the biosynthesis of ethylene, a plant hormone, *Bull. Soc. Chim. Fr.* **133**:441-455.

Su, L. Y., Liu, Y., and Yang, S. F., 1985, Relationship between 1-aminocyclopropane-carboxylate malonytransferase and D-amino acid malonyltransferase, *Phytochem.* **24**:1141-1145.

Suttle, J. C., and Kende, H., 1980, Ethylene action and loss of membrane integrity during petal senescence in *Tradescantia*, *Plant Physiol.* **65**:1067-1072.

Tan, T., and Bangerth, F., 2000, Regulation of ethylene, ACC, MACC production and ACC-oxidase activity at various stages of maturity of apple fruit and the effect of exogenous ethylene treatment, *Gartenbauwissenschaft* **65**:121-128.

Tang, X., Gomes, A.M.T.R., Bhatia, A., and Woodson, W. R., 1994, Pistil specific and ethylene-regulated expression of 1-aminocyclopropane-1-carboxylate oxidase genes in petunia flowers, *Plant Cell* **6**:1227-1239.

Tarun, A. S., Lee, J. S., and Theologis, A., 1995, Random mutagenesis of 1-aminocyclopropane-1-carboxylate synthase: A key enzyme in ethylene biosynthesis, *Proc. Natl. Acad. Sci. USA* **95**:9796-9801.

Theologis, A., 1998, Ethylene signalling: Redundant receptors all have their say, *Curr. Biol.* **8R**:875-878.

Theologis, A., and Laties, G. G., 1982, Potentiating effect of pure oxygen on the enhancement of respiration in plant storage organs: A comparative study, *Plant Physiol.* **69**:1031-1035.

Thomas, C. J. R., Smith, A. R., and Hall, M. A., 1984, The effect of solubilization on the character of an ethylene-binding site from *Phaseolus vulgaris* L. cotyledons, *Planta* **160**:474-479.

Thomas, C. J. R., Smith, A. R., and Hall, M. A., 1985, Partial purification of an ethylene-binding site from *Phaseolus vulgaris* L. cotyledon, *Planta* **164**:272-277.

Thomas, D., and Surdin-Kerjan, Y., 1991, The synthesis of the two *S*-adenosylmethionine synthetases is differently regulated in *Saccharomyces cerevisiae, Mol. Gen. Genet.* **226**:224-232.

Thomas, R. J., Harrison, M. A., Taylor, J., and Kaufman, P. B., 1983, Endogenous auxin and ethylene in *Pellia* (*Bryophyta*), *Plant Physiol.* **73**:395-397.

Tieman, D., and Klee, H. J., 1999, Differential expression of two novel members of the tomato ethylene-receptor family, *Plant Physiol.* **120**:165-172.

Tittle, F. L. , 1987, Auxin-stimulated ethylene production in fern gametophytes and sporophytes, *Physiol. Plant.* **70**:499-502.

Van der Straeten, D., Djudzman, A., Van Caeneghem, W., Smalle, J., and Van Montagu, M., 1993, Genetic and physiological analysis of a new locus in *Arabidopsis* that confers resistance to 1-aminocyclopropane-1-carboxylic acid and ethylene and specifically affects the ethylene signal transduction pathway, *Plant Physiol.* **102**:401-408.

Van der Straeten, D., Van Wiemeersch, L., Goodman, H. M., and Van Montagu, M., 1990, Cloning and sequence of two different cDNAs encoding 1-aminocyclopropane-1-carboxylate synthase in tomato, *Proc. Natl. Acad. Sci.* **87**:4859-4863.

Vendrell, M., and McGlasson, W. B., 1971, Inhibition of ethylene production in banana fruit tissue by ethylene treatment, *Aust. J. Biol. Sci.* **24**:885-895.

Vera, P., and Conejero, V., 1989, The induction and accumulation of the pathogenesis-related P69 proteinase in tomato during *Citrus exocortis* viroid infection and in response to chemical treatments, *Physiol. Mol. Plant Path.* **34**:323-334.

Ververidis, P., and John, P., 1991, Complete recovery *in vitro* of ethylene-forming enzyme activity, *Phytochemistry* **30**:725-727.

Vioque, B., 1986, Ethylene biosynthesis in higher plants, *Grasas Aceites* **37**:156-167.

Vioque, B., and Castellano, J. M., 1994, Extraction and biochemical characterization of 1-aminocyclopropane-1-carboxylic acid oxidase from pear, *Physio. Plant.* 90:334-338.

Vioque, B., and Castellano, J. M., 1998, In vivo and in vitro 1-aminocyclopropane-1-carboxylic acid oxidase activity in pear fruit: Role of ascorbate and inactivation during catalysis, *J. Agric. Food Chem.* **46**:1706-1711.

Vogeli, U., Meins, F., Jr., and Boller, T., 1988, Coordinated regulation of chitinase and β-1,3-glucanase in bean leaves, *Planta* 174:364-372.

Walters, J., and Osborne, D. J., 1979, Ethylene and auxin-induced cell growth in relation to auxin transport and metabolism and ethylene production in the semi-aquatic plant, *Regnellidium diphyllum, Planta.* **146**:309-

Wang, H., and Woodson, W. R., 1992, Nucleotide sequence of a cDNA encoding the ethylene-forming enzyme from petunia corollas, *Plant Physiol.* 100:535-536.

Wang, S. Y., Adams, D. O., and Lieberman, M., 1982, Recycling of 5'-methylthioadenosine-ribose carbon atoms into methionine in tomato tissue in relation to ethylene production, Plant Physiol. **70**:117-121.

White, M. F., Vasquez, J., Yang, S. F., and Kirsch, J. F., 1994, Expression of apple 1-aminocyclopropane-1-carboxylate synthase in *Escherichia coli*: Kinetic characterization of wild-type and active-site mutant forms, *Proc. Natl. Acad. Sci. USA* **91**:12428-12432.

Whittaker, D. J., Smith, G. S., and Gardner, R. C., 1997, Expression of ethylene biosynthesis genes in *Actinidia chinensis* fruit. *Plant Mol. Biol.* **34**:45-55.

Wiesendanger, R., Martinoni, B., Boller, T., and Arigoni, D., 1986, Biosynthesis of 1-aminocyclopropane-1-carboxylic acid: Steric course of the reaction at the C-4 position, Experientia 42:207-209.

Wilkinson, J. Q., Lanahan, M. B., Yen, H. C., Giovannoni, J. J., and Klee, H. J., 1995, An ethylene inducible component of signal transduction encoded by Never Ripe, *Science* **270**:1807-1809.

Woltering, E. J., and deVrije, T., 1995, Ethylene: A tiny molecule with great potential, *BioEssays.* **17**:287-290.

Yamamoto, M., Miki, T., Ishii, Y., Fujinami, K., Yanagisawa, Y., Nakagawa, H., Ogura, N., Hirabayashi, T., and Sato, T., 1995, The synthesis of ethylene in melon fruit during the early stage of ripening, *Plant Cell Physiol.* **36**:591-596.

Yang, S. F. , 1974, The biochemistry of ethylene: Biogenesis and metabolism, in: The chemistry and Biochemistry of Plant Hormones, V. C. Sondheimer and E. Walton, eds., *Recent Adv. Phytochem.*, 7:131-178, Academic Press, NY.

Yang, S. F., 1980, Regulation of ethylene biosynthesis, HortSci. **15**:238-243.

Yang, S. F., 1996, ACC synthase, the key enzyme in biosynthesis of the plant hormone ethylene: From enzymology to biotechnology, *Proc. Czech-Taiwan (R.O.C.) Symp. Biotechnology,* Prague, Czech Republic, June 5-8, 1995, pp. 25-33.

Yang, S. F., and Hoffman, N. E., 1984. Ethylene biosynthesis and its regulation in higher plants, *Annu. Rev. Plant Physiol.* **35**:155-189.

Yang, S. F., Yip, W. K., and Dong, J. G., 1990a, Mechanism and regulation of ethylene biosynthesis, in: *Polyamines and Ethylene: Biochemistry, Physiology, and Interaction.* H. E. Flores et al., eds., Amer. Soc. Plant Physiol., pp. 24-35.

Yang, S. F., Yip, W. K., Satoh, S., Miyazaki, J. H., Jiao, X., Liu, Y., Su, L. Y., and Peiser, G. D., 1990b, Metabolic aspects of ethylene biosynthesis, in: *Plant Growth Substances*, R. P. Pharis and S. B. Road, eds., Springer-Verlag, Berlin, pp. 291-299.

Yip, W. K., Dong, J. G., Kenny, J. W., Thompson, G. A., and Yang, S. F., 1990, Characterization and sequencing of the active site of 1-aminocyclopropane-1-carboxylate synthase, *Proc. Natl. Acad. Sci. USA* **87**:7930-7934.

Yip, W. K., Moore, T., and Yang, S. F., 1992, Differential accumulation of transcripts for tomato 1-aminocyclopropane-1-carboxylate synthase homologs under various conditions, *Proc. Natl. Acad. Sci. USA* **89**:2475-2479.

Yoshii, H., and Imaseki, H., 1981, Biosynthesis of auxin-induced ethylene. Effects of indole-3-acetic acid, benzyladenine and abscissic acid on endogenous levels of 1-aminocyclopropane-1-carboxylic acid (ACC) and ACC synthase, *Plant Cell Physiol.* **22**:369-379.

Yoshii, H., and Imaseki, H., 1982, Regulation of auxin-induced ethylene biosynthesis. Repression of inductive formation of 1-aminocyclopropane-1-carboxylate synthase by ethylene, *Plant Cell Physiol.* **23**:639-649.

Yu, Y. B., and Yang, S. F., 1979, Auxin-induced ethylene production and its inhibition by aminoethoxyvinylglycine and cobalt ion, *Plant Physiol.* **64**:1074-1077.

Yu, Y. B., Adams, D. O., and Yang, S. F., 1979, 1-Aminocyclopropane-carboxylate synthase, a key enzyme in ethylene biosynthesis, *Arch. Biochem. Biophys.* **198**:280-286.

Yung, K. H., Yang, S. F., and Schlenk, F., 1982, Methionine synthesis from 5-methylthioribose in apple tissue, Biochem. Biophys. Res. Commun. **104**:771-777.

Zarembinski, T. I., and Theologis, A., 1994, Ethylene biosynthesis and action: A case of conservation, Plant Mol. Biol. **26**:1579-1597.

Zauberman, G. and Fuchs, Y., 1973, Ripening process in avocado stored in ethylene atmosphere in cold storage, *J. Am. Soc. Hort. Sci.* **98**:477-480.

Zeroni, M., Galil, J., and Ben-Yehoshua, S., 1976, Autoinhibition of ethylene formation in nonripening stages of the fruit of sycamore fig (*Ficus sycomorus* L.), *Plant Physiol.* **57**:647-650.

Zhang, Z., Barlow, J. N., Baldwin, J. E., and Schofield, C. J., 1997, Metal catalyzed oxidation and mutagenesis studies on the iron(ii) binding site of 1-aminocyclopropane-1-carboxylic oxidase, *Biochem.* **36**:15999-16007.

Zhou, D., Mattoo, A., and Tucker, M., 1996a, Molecular cloning of a tomato cDNA encoding an ethylene receptor, *Plant Physiol.* **110**:1435-1436.

Zhou, D., Kalaitzis, P., Mattoo, A. K., and Tucker, M. L., 1996b, The mRNA for an ethylene receptor in tomato is constitutively expressed in vegetative and reproductive tissues, *Plant Molecu. Biol.* **30**:131-133.

Zhou, H., Huxtable, S., Xin, H., and Li, N., 1998, Enhanced high-level expression of soluble 1-aminocyclopropane-1-carboxylase synthase and rapid purification by expanded bed adsorption, *Protein Expression Purification* **14**:178-184.

Zimmerman, P. W., and Hitchcock, A. E., 1933, Initiation and stimulation of adventitious roots caused by unsaturated hydrocarbon gases, *Contrib. Boyce Thompson Inst.* **5**:351-369.

3

BIOCHEMISTRY OF MICROBIAL
PRODUCTION OF ETHYLENE

3.1. INTRODUCTION

The ability of microorganisms to derive C_2H_4 from a variety of compounds has made the biochemistry of microbial production of C_2H_4 very complex. The literature indicates that C_2H_4 -producing microorganisms do not follow the same pathway operative in higher plants (MET → SAM → ACC → C_2H_4). This subject has been reviewed elsewhere (Fukuda and Ogawa, 1991; Fukuda et al., 1993; Arshad and Frankenberger, 1993, 1998; Frankenberger and Arshad, 1995). However, many microorganisms do synthesize C_2H_4 from methionine. This chapter is dedicated to the microbial biosynthesis pathways involving substrates, enzymes, inhibitors and intermediary metabolism.

3.2. SUBSTRATES/PRECURSORS/STIMULATORS OF C_2H_4

A wide range of compounds, including alcohols, sugars, organic acids, amino acids, Krebs cycle intermediates, methionine analogues, and humic phenolic compounds have been reported as possible precursors for C_2H_4 generation by different microbial isolates (Wang et al., 1962; Jacobsen and Wang, 1968; Gibson, 1964a,b; Gibson and Young, 1966; Chalutz et al., 1977; Chalutz and Lieberman, 1978; Primrose, 1976b; Primrose and Dilworth, 1976; Lynch, 1974; Chou and Yang, 1973; Owens et al., 1971; Dasilva et al., 1974; Swart and Kamerbeek, 1977; Fergus, 1954; Sprayberry et al., 1965; Thomas and Spencer, 1977; Considine et al., 1977; Fukuda et al., 1986, 1989a,b,c; Ince and Knowles, 1985, 1986; Goto et al., 1985, Arshad and Frankenberger, 1989). However, L-methionine (L-MET) has been the most extensively investigated substrate. Arshad and Frankenberger (1989) and Primrose (1976a) compared the efficiency of L-MET with several other compounds for their effectiveness as substrates of C_2H_4 production by *Acremonium falciforme* and *Escherichia coli*, respectively (Tables 3.1 and 3.2) and found that L-MET or its analogues were the best substrates of C_2H_4 generation by these microbial isolates.

Similarly, Goto et al. (1985) and   Considine et al. (1977) investigated the effect of various amino acids, organic acids, and carbohydrates on C_2H_4 production by *Pseudomonas syringae* pv. *phaseolicola* and strains of *Penicillium cyclopium* and *P. crustosum* (Table 3.3 and 3.4) and found much variability in C_2H_4 production.

Table 3.1. Utilization of various substrates in ethylene formation by *A. falciforme*.

Substrate	Structure	Efficiency relative to L-methionine (%)
L-Methionine	CH$_3$-S-CH$_2$-CH$_2$-CH-COOH $\mid$ NH$_2$	100.0
L-Ethionine	CH$_3$-CH$_2$-S-CH$_2$-CH$_2$-CH-COOH $\mid$ NH$_2$	14.8
L-Cysteine	HS-CH$_2$-CH-COOH $\mid$ NH$_2$	0
L-Homocysteine	HS-CH$_2$-CH$_2$-CH-COOH $\mid$ NH$_2$	4.1
L-Homoserine	CH$_2$OH-CH$_2$-CH-COOH $\mid$ NH$_2$	2.7
L-Glutamic acid	COOH-CH$_2$-CH$_2$-CH-COOH $\mid$ NH$_2$	0
β-Alanine	NH$_2$-CH$_2$-CH$_2$-COOH	0
Fumaric acid	COOH-CH=CH-COOH	0
L-Malic acid	COOH-CH$_2$-CHOH-COOH	0
L-Lactic acid	CH$_3$-CHOH-COOH	0
Pyruvic acid	CH$_3$-CO-COOH	0
Succinic acid	COOH-CH$_2$-CH$_2$-COOH	0
α-Ketoglutaric acid	COOH-CH$_2$-CH$_2$-CO-COOH	0
Acrylic acid	CH$_2$=CH-COOH	0
Acetic acid	CH$_3$COOH	0

Source: Arshad and Frankenberger (1989).

Table 3.2. Ethylene formation by *Escherichia coli* from methionine analogues and precursors.

Substrate	Structure	Efficiency relative to L-methionine (%)
L-Methionine	CH_3-S-CH_2-CH_2-CH-COOH NH_2	100
D-Methionine L-Ethionine	CH_3-CH_2-S-CH_2-CH_2-CH-COOH NH_2	107
L-Cysteine	HS-CH_2-CH-COOH NH_2	85
2-Keto-4-methyl- mercaptobutyric acid	CH_3-S-CH_2-CH_2-CO-COOH	90
L-Homocysteine	HS-CH_2- CH_2-CH-COOH NH_2	92
N-Formyl methionine	CH_3-S-CH_2-CH_2-CH-COOH NH CHO	61
L-Methionine methyl ester	CH_3-S-CH_2- CH_2-CH-CO-O-CH_3 NH_2	123
DL-Methionine hydroxy analogue	CH_3S-CH_2-CH_2-CHOH-COOH	112
L-Norleucine	CH_3-CH_2-CH_2-CH-COOH NH_2	0
α-Ketobutyric acid	CH_3-CH_2-CO-COOH	0
L-Aminobutyric acid	CH_3-CH_2-CH-COOH NH_2	0
α-Ketoglutaric acid	CH_3-CH_2-CH_2-CO-COOH	0

Source: Primrose (1976a, 1979)

Table 3.3. Effect of various substrates on ethylene production by Kudzu strain PK2 of *P. syringae* pv. *phaseolicola*.

Substrate	Ethylene production (-fold over control)	Substrate	Ethylene production (-fold over control)
Control	1.0	Tyrosine	9.1
Glutamate	40.0	Isoleucine	7.5
Glutamine	40.0	Cysteine	7.3
Aspartate	40.0	Valine	6.5
Asparagine	30.9	Ornithine	6.4
Alanine	20.9	Histidine	6.2
Arginine	20.0	Lysine	6.0
Tryptophan	17.3	Glycine	3.6
Serine	16.4	Threonine	3.4
Proline	11.8	Methionine	2.4
Phenylalanine	10.9		
Glucose	14.6	Arabinose	6.6
Fructose	13.6	Xylose	6.6
Sucrose	12.7	Glycerol	3.8
Mannose	8.7		
Succinate	26.4	Pyruvate	13.6
Fumarate	22.7	Oxalate	10.0
Malate	22.7	2-Keto-4-methylthiobutyrate	9.1
α-Ketoglutarate	17.3	Tartarate	4.6
γ-Aminobutyrate	17.3	Acetate	4.0
Citrate	17.3	Glycerate	3.6
Malonate	14.6	Oxalacetate	0.2
Ascorbate	13.6		

[a]Assay mixture: 10 mM phosphate buffer 4.0 ml; substrate 0.5 ml; and bacterial suspension (1 x 10^{10} cells ml^{-1}) 0.5 ml. The rate of ethylene production without a substrate (control) was 1.1 x 10^{-9} nl $cell^{-1}$ h^{-1}. Sugars and organic acids were added to the assay mixture at a concentration of 0.1 M, and amino acids at 0.02 M.

Source: Modified from Goto et al. (1985).

Table 3.4. Effects of various carbon sources on ethylene production ($\mu l^{-1}\ L^{-1}\ 24\ h^{-1}$) by strains of *Penicillium* spp.

Growth substances	*P. cyclopium* strains						*P. crustosum* strains		
	A	B	C	D	E	F	1	2	3
Glucose	27	22	21	30	32	29	287	270	203
Acetate	219	301	249	211	152	218	8	6	8
Citrate		65	70	57	65	60	0	0	0
α-Ketoglutarate	163	115	99	127	108	110	8	11	15
Malic acid	148	129	115	137	126	129	6	10	10
Succinic acid		94	100	52	82	104	0	0	0
Glutamic acid	88	114	75	74	90	87	17	22	27
Vanilic acid	15	11	16	24	15	15	31	21	20
p-Hydroxybenzoic acid	75	88	87	168	74	138	306	125	178
Protocatechuic acid	13	9	17	31	20	10	36	44	54
Sabaroud dextrose	288	158	140	317	182	68	1152	3091	3883

Source: Considine et al. (1977).

Most bacteria and fungi generally produce C_2H_4 only in the presence of MET (Primrose, 1976b; Billington et al., 1979; Lynch, 1972; Ince and Knowles, 1985, 1986; Fukuda et al., 1989a; Arshad and Frankenberger, 1989). However, there are exceptions. Ethylene was not detected when cultures of *Erwinia carotovora* or *Penicillium digitatum* were supplied with MET (Lund and Mapson, 1970; Ketring et al., 1968). Similarly, a partially purified ethylene-forming enzyme isolated from *P. digitatum* utilized α-ketoglutarate (KGA) as an immediate precursor of C_2H_4 , and no C_2H_4 was released when it was substituted with MET (Fukuda et al., 1986; also see Section 3.3.2). The use of MET as substrate by a microbial culture may also be dependent on incubation conditions as Chalutz et al. (1977) found that only shake cultures of *P. digitatum* produced C_2H_4 from MET; whereas, static cultures produced C_2H_4 in the absence of MET, which is contrary to Ketring et al. (1968) findings. Methionine-derived C_2H_4 production by *P. digitatum* was demonstrated by Primrose (1977) and Billington et al. (1979).

Like L-MET, glucose addition to the cultural media has been shown to promote C_2H_4 biosynthesis by some microbial cultures. However, several scientists believe that glucose stimulates microbial production of C_2H_4 as an energy source, and does not directly serve as an C_2H_4 substrate (Lynch and Harper, 1974; Chalutz et al., 1977; Fukuda et al., 1989a; Arshad and Frankenberger, 1989). Similarly, Brown-Skrobot et al.

(1985) reported that of the seven carbohydrates, including glucose tested, none supported production of C_2H_4 by *Septoria musiva*. However, Swart and Kamerbeek (1977) found that a tulip strain of *Fusarium* sp. produced copious amounts of C_2H_4 when grown in a medium in which glucose was the only organic nutrient. Dasilva et al. (1974) reported that an *Aspergillus* sp. and a *Mucor* sp. were capable of deriving C_2H_4 from glucose. The combination of glucose and MET stimulated C_2H_4 evolution even more, but they did not test MET alone as an C_2H_4 precursor. Wang et al. (1962) reported the active conversion of ^{14}C-labeled glucose into C_2H_4 by *P. digitatum*. As shown in Table 3.4, the addition of glucose stimulated C_2H_4 production to a greater extent by strains of *P. crustosum* than by strains of *P. cyclopium*, implying that the suitability of glucose as an C_2H_4 substrate varies with the microorganism under study (Considine et al., 1977). Pazout et al. (1981) found that glucose was a suitable substrate for C_2H_4 formation by *Pseudomonas putida*, especially in cultures with limited aeration. Similarly, Goto et al. (1985) reported that glucose was a better substrate than MET for C_2H_4 generation by *Ps. syringae* pv. *phaseolicola*. Hahm et al. (1992) also reported that C_2H_4 was produced in a *Ps. syringae* pv. *phaseolicola* culture when glucose was used as a carbon source. Experiments with resting cells of *Rhizoctonia solani* showed that C_2H_4 was produced from glucose under conditions of reduced O_2 tension (Brown and Brown, 1981). Fukuda et al. (1988) proposed a biosynthetic pathway for C_2H_4 by *P. digitatum* revealing that glucose is the starting point and is converted into KGA, which serves as the immediate precursor of C_2H_4.

3.3. BIOSYNTHETIC PATHWAYS

Since microbial isolates can generate C_2H_4 from a variety of structurally unrelated compounds, more than one pathway are involved in the microbial biosynthesis of C_2H_4. Although most soil microorganisms can synthesize C_2H_4 from MET, evidence suggests that the pathway followed is different from that of higher plants (Fig. 3.1). A comparison of *in vitro* systems with C_2H_4-forming enzymes (EFE) from microbes and higher plants is given in Table 3.5. The premise that microorganisms do not produce C_2H_4 from MET by a pathway similar to that of higher plants is supported by the following observations:

1. Very few studies have demonstrated intermediates of the plant's pathway of C_2H_4 synthesis. SAM has not yet been isolated in C_2H_4–producing microbial cultures (Coleman and Hodges, 1986; Beard and Harrison, 1992; Jia et al., 1999). Coleman and Hodges (1986) detected ACC in a culture medium but found that *Bipolaris sorokiniana* did not convert ACC in C_2H_4 effectively. A report on preliminary results, published as an abstract only, revealed that a *Rhizobium* spp. was capable of synthesizing C_2H_4 and ACC in a culture medium (Beard and Harrison, 1992). Moreover, exogenous application of ACC resulted in a large burst of C_2H_4 in the rhizobial culture, indicating an active ACC-dependent C_2H_4-forming enzyme system. Since details of this study are not available, the validity of the results cannot be critically reviewed. Very recently, Jia et al. (1999) detected ACC in the culture medium of *Penicillium citrinum* (Fig. 3.2) and purified ACC synthase from the fungal cells; however, they observed that the fungus also

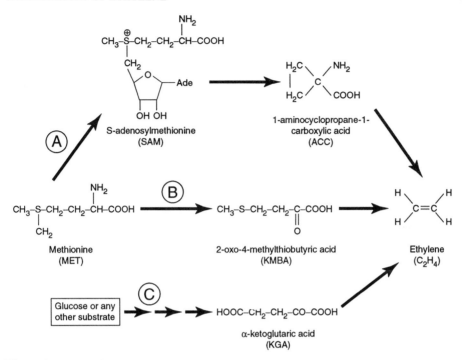

Figure 3-1.   Pathway of C_2H_4 biosynthesis identified in (A) higher plants, (B) *E. coli* and *C. albidus*, and (C) *P. digitatum* and *P. syringae* pv. *phaseolicola*. Source: Arshad and Frankenberger (1998).

Table 3.5. Comparison of *in vitro* systems with ethylene-forming enzymes from microbes and higher plants.

Microbe or plant	Substrate	Co-substrate	Metal ion	Oxygen	Other factors
Penicillium digitatum	2-Oxoglutarate	L-Arginine	Fe(II)	+	Dithiothreitol
Pseudomonas syringae	2-Oxoglutarate	L-Arginine	Fe(II)	+	L-Histidine
Cryptococcus albidus	2-Keto-4-methyl-thiobutyric acid	NADH	Fe(III) + EDTA	+	None
Cucumis melo (melon fruit)	1-Aminocyclopropane -1-carboxylic acid	Ascorbate	Fe(II)	+	None

+ indicates oxygen is essential
Source: Fukuda et al. (1993).

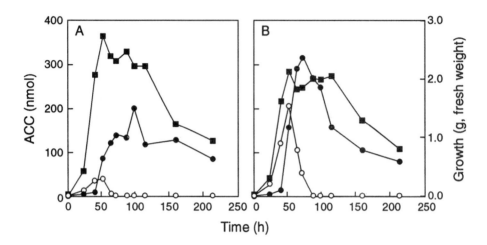

Figure 3-2. Formation of ACC by *P. citrinum*. A, in the absence of L-methionine; B, in the presence of L-methionine. *P. citrinum* was cultured in 100 ml of the medium at 28°C with shaking. The cells were harvested and weighed (■), ACC in cell extract (●) and ACC in medium (○) were measured. Each point shows the mean of four measurements. Source: Jia et al. (1999).

possessed ACC deaminase activity but no ACC-oxidase activity implying that ACC-dependent pathways of C_2H_4 was not functional in this fungus (Jia et al., 1999).

2. Studies have demonstrated the inability of microbial cultures to utilize SAM or ACC as C_2H_4 precursors. An efficient producer of C_2H_4 from MET, *Acremonium falciforme*, grows well on ACC, but does not transform ACC to C_2H_4. Furthermore, it could not utilize SAM, even as a C and N source for growth (Arshad and Frankenberger, 1989). Similarly, Thomas and Spencer (1977) reported that *Saccharomyces cereviseae* was unable to transform SAM into C_2H_4. Both ACC and SAM did not serve as substrates for C_2H_4 formation by cell-free extracts of *E. coli* (Ince and Knowles, 1986) and *Cryptococcus albidus* (Fukuda et al., 1989a). Similarly, C_2H_4 synthesis by the fungi, *Endothia gyrosa* and *Cytospora eucalyticola*, growing on artificial substrates and dead sapwood was MET-dependent, but was unaffected by the addition of ACC (Wilkes et al., 1989). Brown-Skrobot et al. (1985) reported that *Septoria musiva* grew well on SAM, but no C_2H_4 evolution took place. Fukuda et al. (1986) reported that a cell-free C_2H_4 – forming system isolated from *P. digitatum* was unable to derive C_2H_4 from MET, SAM, and ACC. Similarly, the addition of MET or ACC did not have any effect on C_2H_4 formation by living cells of *Ps. syringae* pv. *phaseolicola* (Goto et al., 1985). Although ACC synthase (which derives ACC from SAM) has been purified from cells of the fungus *Penicillium citrinum* (Jia et al., 1999; 2000), with a specific activity of 327 milliunits/mg protein, a molecular mass of 96000 Da, a K_m for SAM of 1.74 mM and K_{cat} 0.56 g^{-1} per monomer, the fungus lacked the ACC oxidase activity. Instead, ACC deaminase activity was observed, indicating the absence of ACC-dependent pathways in this fungus (Jia et al., 1999). These studies clearly provide evidence that the MET-

dependent C_2H_4 biosynthetic pathway of higher plants does not exist in microorganisms. However, the addition of ACC to nonsterile soils results in high emission of C_2H_4 compared with the small amounts of C_2H_4 released after its addition to sterile soils (Frankenberger and Phelan, 1985a,b; Arshad and Frankenberger, 1990; Nazli and Arshad, unpublished data), indicating that soil microbiota can convert ACC into C_2H_4. Despite the fact that numerous soil microbiota can proliferate when ACC is supplied as the sole nitrogen source added to a chemically defined medium, microorganisms have not been isolated that are capable of producing C_2H_4 directly from ACC (Arshad and Frankenberger, unpublished data). Several studies have demonstrated the ability of microorganisms to grow on ACC as a nitrogen source resulting in the induction of the enzyme, ACC deaminase, which catalyzes the conversion of ACC to α-ketobutyrate and NH_3 (Honma, 1993; Shah et al., 1997; Glick et al., 1998; Jacobson et al., 1994; Campbell and Thomson, 1996; Penrose and Glick, 1997). It is very likely that microorganisms may possess ACC deaminase, rather than ACC oxidase; and that is why they can grow on ACC, but do not produce C_2H_4 from ACC (see Section 3.3.1 for details).

3. By the use of inhibitors of C_2H_4 biosynthesis in higher plants, some studies have demonstrated that MET-dependent C_2H_4 production by microbial cultures does not follow the same route. The addition of aminoxyacetic acid (AOA) did not inhibit MET-derived C_2H_4 production by *A. falciforme* (Arshad and Frankenberger, unpublished data). Exogenous application of Co^{2+} as $CoCl_2$, at less than 1.0 mM, suppressed C_2H_4 synthesis by *A. falciforme*, but this inhibition was not a result of inhibiting the activity of the ethylene-forming enzyme (ACC oxidase), since this fungus did not produce C_2H_4 from ACC (Arshad and Frankenberger, 1989). Primrose (1976a, 1977) reported that L-. canaline inhibited MET-induced C_2H_4 synthesis by *E. coli*, which indicates a requirement for pyridoxal phosphate, but Primrose (1977) proposed that an intermediate other than SAM was involved in C_2H_4 generation. In feeding experiments with labeled compounds, Ince and Knowles (1986) found that C_2H_4 produced by *E. coli* was derived from the C-3 and C-4 atoms of MET and 2-oxo-4-methylthiobutyric acid (KMBA) served as an intermediate in this biosynthetic pathway. Fukuda et al. (1989a,c) also confirmed that KMBA serves as an intermediate in MET-derived C_2H_4 production by *Cryptococcus albidus* (see Section 3.3.2). Several studies have demonstrated that *Penicillium digitatum* and *Ps. syringae* pv. *phaseolicola* derive C_2H_4 from KGA, rather than from MET (see Section 5.3.2). It is highly likely that in addition to these two well-characterized pathways, some microorganisms may utilize other pathway(s) to generate C_2H_4 for a given substrate.

4. The isolation, purification and identification of ACC deaminase enzyme from a number of microorganisms further provides evidence of a nonexisting ACC pathway of C_2H_4 biosynthesis in microorganisms. The enzyme metabolizes ACC to NH_3 and α-ketobutyrate.

5. Like bacteria and fungi, the nonexistence of an ACC pathway of C_2H_4 biosynthesis has also been demonstrated in lower plants, the semi-aquatic ferns, *Regnellidium diphyllum* Lindm. and *Mersilea quadrifolia* L. (Chernys and Kende, 1996; Osborne et al., 1996). ACC addition did not stimulate C_2H_4 production from the leaflets of these ferns and AVG and AOA did not inhibit C_2H_4 production. However, ACC and ACC synthase were detected in both the ferns (Cherneys and Kende, 1996).

3.3.1. ACC Deminase

The enzyme, ACC deaminase which has been found only in microorganisms, catalyses ACC to NH_3 and α-ketobutyrate as shown in Fig. 3.3 (Penrose and Glick, 1997). This enzyme was first discovered in1978 and was purified from *Pseudomonas* and a yeast (Honma and Shimomura, 1978) and subsequently has been detected in other bacterial and fungal strains (Klee et al., 1991; Klee and Kishmore, 1992; Honma, 1993; Jacobson et al., 1994; Glick et al., 1995). Klee et al. (1991) screened 600 bacterial strains for their ACC degrading activity and selected two strains with the highest ACC deaminase activity (strains 3F2 aqnd 6G5) for further characterization. Likewise, 103 strains of soil microorganisms were screened by Campbell and Thompson (1996) and they observed that 84 displayed ACC degrading ability with different efficiencies. The seven most efficient were identified as fluorescent pseudomonads. *P. putida* GR12-2 was isolated from the rhizosphere of grass in the Canadian high arctic (Lifshitz et al., 1986). Very recently, Jia et al. (1999) reported the ability of *Penicillium citrinum* to degrade ACC into ammonia and α-ketobutyrate. The purified ACC deaminase had a K_m of 4.8 mM and a specific activity of 4.7 unit/mg protein. Likewise, the presence of ACC deaminase activity in *Enterobacter cloaceae* CAL2 and UW2 was confirmed by Shah et al. (1998). Interestingly, most of the microorganisms known to contain this enzyme have been isolated from soil samples or are microbes typically found in the soil (Table 3.6). Many of these microorganisms were identified by their ability to grow on minimal media containing ACC as the sole nitrogen source.

ACC deaminase was first purified to homogeneity from *Pseudomonas* sp. strain ACP (Honma and Shimomura, 1978) and partially purified from *P. chloroaphis* 6G5 (Klee et al., 1991) and *P. putida* GR12-2 (Jacobson et al., 1994). ACC deaminase purified from all three sources appears similar in molecular mass and form (Penrose and Glick, 1997). However, the enzyme isolated from the yeast, *Hansenula saturnus* had 60-63% amino acid residues identical to those of reported in pseudomonad enzymes (Minami et al., 1998). ACC deaminase activity occurs exclusively in the cytoplasm of *P. putida* GR12-2 (Jacobson et al., 1994). Some properties of this enzyme are summarized in Table 3.7.

ACC deaminase activity was induced in *Ps. putida* GR12-2 as a function of ACC which increased approximately 10-fold over the basal level of activity as the concentration of ACC was increased up to 10,000-fold (Jacobson et al., 1994). The level of ACC deaminase activity observed when *P. putida* GR12-2 was grown on minimal medium plus ammonium sulfate represents a basal level of activity and never exceeded 5% of the total activity measured in extracts grown on minimal medium containing ACC.

The coenzyme, pyridoxal phosphate is a tightly bound cofactor of ACC deaminase in the amount of approximately three mol of enzyme-bound pyridoxal phosphate per mol of enzyme or one mole per trimeric subunit (Honma, 1985). The aldimine-ketimine complex formed between pyridoxal phosphate and substrate has a characteristic absorption spectrum which shows a maximum absorbance at 418 nm (Walsh et al., 1981) and is used to monitor the formation and disappearance of intermediate enzyme-substrate complexes (Lehninger, 1975). However, the ACC deaminase from *Penicillium citrinum* was found dimeric in nature having absorption spectrum of about 420 and 330 nm at pH 7.5, similar to that of the enzyme from *Pseudomonas* sp. ACP (Honma and Shimomura et al., 1978).

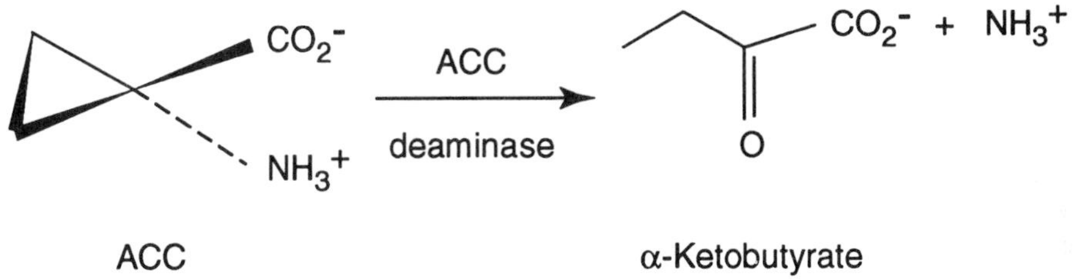

Figure 3-3. The reaction catalyzed by ACC deaminase. Source: Penrose and Glick (1997).

The genes for the enzyme, ACC deaminase have recently been isolated from a few soil bacteria (Klee et al., 1991; Sheehy et al., 1991; Klee and Kishmore, 1992; Campbell and Thompson, 1996; Shah et al., 1998; Grichko et al., 2000; Jia et al., 2000). These studies reveal that there may be similar as well as several different types of deaminase genes in different microbes. The cloned ACC deaminase genes can be expressed in other microorganisms and higher plants (Shah et al., 1998; Campbell and Thompson, 1996; Klee and Kishmore, 1992; Minami et al., 1998; Wang et al., 2000; Grichko et al., 2000). Shah et al. (1998) reported expression of ACC deaminase genes in *Escherichia coli, Ps. putida* and *Ps. fluorescens* enabling these bacteria to hydrolyze ACC and conferred upon these strains the ability to promote root growth of canola seedlings. Similarly, Wang et al. (2000) reported that inoculation with transformed strains of *Ps. fluorescens* with ACC deaminase activity increased root length of canola plants under gnotobiotic conditions. ACC deaminase genes from *Pseudomonas* strains were used to make expression cassettes which allowed expression of ACC deaminase in plant tissue (Klee and Kishmore, 1992). The enzyme, ACC deaminase of *Ps.* sp. strain ACP was detected immunologically in electrophorated tobacco protoplasts and in transformed petunia leaves. Extracts of transformed plant tissue showed ACC deaminase activity and tissue from one of the transformants appeared less able to convert ACC to C_2H_4 (see Section 8.3.3). When plant cells were transformed with a chimeric gene encoding the ACC deaminase gene from *P. chloroaphis* strain 6G5, ACC deaminase was expressed in transformed petunia, tobacco and tomato plants (Klee and Kishmore, 1992). The significance of ACC deaminase in microorganisms or transformed plants in improving plant growth and yield is discussed in detail in Chapter 8.

Table 3.6. Soil microorganisms with ACC deaminase activity.

Genus / Species	Strain	Source	K_m (mmol)	Reference
Enterobacter cloacae	UW2	soil isolate clover Waterloo Canada	nd**	Glick et al. (1995)
Enterobacter cloacae	UW4	soil isolate reeds Waterloo Canada	nd	Glick et al. (1995)
Enterobacter cloacae	CAL2	soil isolate tomato King City Calif USA	nd	Glick et al. (1995)
Enterobacter cloacae	CAL3	soil isolate cotton Fresno Calif USA	nd	Glick et al. (1995)
Hansetzula saturnus		soil isolate	26	Honma and Shimomura (1978); Minami et al. (1998)
Kluyvera ascorbata	SUD165	soil isolate Sudbury, Ontario, Canada.	nd	Burd et al. (1998)
Penicillium citrinum		soil isolate	46	Honma (1993)
Penicillium citrinum			4.8	Jia et al. (1999)
Pseudomonas	ACP	soil isolate	15	Honma and Shimomura (1978)
Pseudomonas chloroaphis	6G5	Drahos collection	90	Klee and Kishmore (1992), Drahos et al. (1988)
Pseudomonas fluorescens	CAL1	soil isolate oats, San Benito Calif USA	nd	Glick et al. (1995)
Pseudomonas nd	3F2	Drahos collection	58	Klee and Kishmore (1992), Drahos et al. (1988)
Pseudomonas nd		Drahos collection		Klee and Kishmore (1992), Drahos et al. (1988)
Pseudomonas putida	GR12-2	soil isolate grass Canadian Arctic	nd	Jacobson et al. (1994), Litshitz et al. (1986)
Pseudomonas putida	UW1	soil isolate bean Waterloo Canada	nd	Glick et al. (1995)
Pseudomonas putida	UW3	soil isolate corn Waterloo Canada	nd	Glick et al. (1995)
Soil isolate*	388	soil isolate St.Charles USA	86	Klee and Kishmore (1992)
Soil isolate	391	soil isolate Malaysia	174	Klee and Kishmore (1992)
Soil isolate	392	soil isolate Peru	71	Klee and Kishmore (1992)
Soil isolate	393	soil isolate St.Charles USA	59	Klee and Kishmore (1992)
Soil isolate	401	soil isolate St.Charles USA	78	Klee and Kishmore (1992)
Soil isolate	T44	soil isolate Tanzania	118	Klee and Kishmore (1992)
Soil isolate	PRI	soil isolate Puerto Rico	41	Klee and Kishmore (1992)

*identification not determined.
**nd, not determined.
Source: updated from Penrose and Glick (1997).

Table 3.7. Properties of microbial ACC deaminase.

Property	Reference
Inducible by ACC	(Honma and Shimomura, 1978; Jacobson et al., 1994; Minami et al., 1998)
Induction rapid (within hours)	(Honma and Shimomura, 1978; Jacobson et al., 1994; Minami et al., 1998)
Induction by aminoisobutyric acid comparable to ACC	(Honma, 1983)
Inducible by other amino acids to lesser extent	(Honma, 1985)
K_m 1.62-174 mM (affinity for ACC is low)	(Klee and Kishmore, 1992; Walsh et al., 1981; Penrose and Glick, 1997; see Table 3-6)
Optimum pH 8.5	(Honma and Shimomura, 1978; Klee and Kishmore, 1992; Honma, 1983; Jacobson et al., 1992; Jia et al., 1999)
pK_a 7.7 and 9.2	(Jacobson et al., 1994)
Optimum temperature 30-35°C	(Jacobson et al., 1994; Jia et al., 1999)
Dimeric, or trimeric	(Honma, 1985; Jia et al., 1999)
Molecular size 68-112 kDa	(Honma and Shimomura, 1978; Honma, 1985; Jacobson et al., 1994)
Subunit mass 35-41 KDa	(Jacobson et al., 1994; Sheehy et al., 1991; Klee et al., 1991; Jia et al., 1999)
Competitive inhibition by L-alanine and L-serine	(Honma, 1985; Minami et al., 1998)
Coenzyme – pyridoxal phosphate	(Honma, 1985)
Absorption spectrum at pH 6 – 416 nm and at pH 4 – 326 nm	(Honma, 1985)

3.4. METHIONINE-DEPENDENT AND INDEPENDENT PATHWAYS

Intensive studies with *E. coli, C. albidus, P. digitatum*, and *Ps. syringae* pv. *phaseolicola* reveal that these microorganisms produce C_2H_4 via two distinct biosynthetic pathways, i.e., MET-dependent and MET-independent pathways. In the MET-dependent pathway, C_2H_4 production proceeds from MET via KMBA and, in the MET-independent pathway, KGA (which may or may not be derived from glucose) serves as an immediate substrate for C_2H_4 formation (Fig. 3.1). However, these two pathways may not be the only biosynthetic route to produce C_2H_4 by pathogenic or nonpathogenic microflora. Nagahama et al. (1992) screened 757 bacterial isolates and found that 225 of these isolates produced C_2H_4 from MET and only one, *Ps. syringae* pv. *phaseolicola*, yielded C_2H_4 from KGA. They suggested that various bacterial strains unable to synthesize C_2H_4 either lacked NADH: Fe(III) EDTA oxidoreductase or lacked MET-uptake activity. They also reported that *Thiobacillus novellus* was unable to produce C_2H_4 in the presence of MET or glutamate, but generated C_2H_4 when incubated with meat extract. They were of the view that meat extract contains growth factors essential for synthesis of C_2H_4 by *T. novellus*. Fukuda and Ogawa (1991) examined 93 fungal strains for KGA-dependent C_2H_4 production. They found that 8 were able to derive C_2H_4 from KGA. Akhtar and Arshad (unpublished data) screened 55 fungal strains isolated from the rhizosphere of wheat, maize, potato and tomato for C_2H_4 production and found that 43 were capable of producing C_2H_4 from L-MET. Out of 55 strains, we tested 16 isolates for their ability to derive C_2H_4 from KMBA and KGA. All produced KMBA-dependent C_2H_4, but none utilized KGA as a physiological precursor of C_2H_4. It is highly likely that in addition to the substrate(s), cofactors and environmental factors (incubation conditions) may play a critical role in C_2H_4 biosynthesis by various microbial cultures.

3.4.1. Methionine-Dependent Pathway

Primrose (1976a,b; 1977) carried out extensive studies with *E. coli* to elucidate the biochemistry of the MET-derived C_2H_4 pathway. Later, the work on C_2H_4 formation by *E. coli* was extended by Billington et al. (1979) and Ince and Knowles (1985, 1986). Fukuda and co-workers (1989a,b,c) conducted extensive work on biosynthetic pathways of C_2H_4 formation by *Cryptococcus albidus,* while Brown-Skrobot et al. (1985) characterized C_2H_4 biosynthesis in *Septoria musiva*. These findings show that *E. coli, C. albidus, and S. musiva* follow a similar MET-dependent pathway of C_2H_4 synthesis. The work in *E. coli, C. albidus*, and *S. musiva* is reviewed here in detail.

3.4.1.1. Escherichia coli

The following observations led Primrose (1976a,b; 1977) to propose a pathway of C_2H_4 biosynthesis by *E. coli*, as illustrated in Figure 3.4.

1. Light stimulated MET-derived C_2H_4 production by *E. coli* and cell-free filtrates (Primrose, 1976a) and the addition of riboflavin to filtrates resulted in a 25-fold increase in the rate of C_2H_4 synthesis (Primrose, 1977) suggesting the involvement of a photosensitive intermediate derived from MET in the MET-dependent C_2H_4 synthesis.

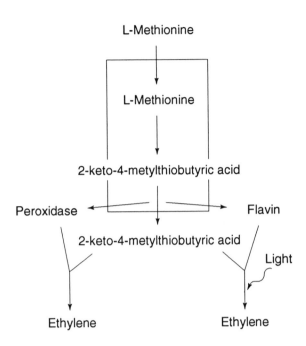

Figure 3-4. Proposed pathway for microbial biosynthesis of
C$_2$H$_4$ Source. Primrose (1977).

2.   Ethylene formation also occurs in the absence of light (Primrose, 1976a),
suggesting another pathway in which intermediates give rise to C$_2$H$_4$. The addition of
catalase, a known inhibitor of peroxidase, suppressed C$_2$H$_4$ formation by *E. coli*
(Primrose, 1977), suggesting that peroxidases are involved in microbial C$_2$H$_4$ synthesis in
the absence of light. The simultaneous loss of peroxidase and the ability to produce C$_2$H$_4$
after washing of *E. coli* cells (Primrose, 1976b) further supports the existence of a second
pathway involving peroxidase in C$_2$H$_4$ synthesis.
      Primrose (1976b) suggested that methional or KMBA were possible intermediates of
the MET-dependent pathway based primarily on their photochemical degradation
properties. However, more evidence supports the involvement of KMBA, rather than
methional, since the pH curves of photochemical degradation for KMBA and the
intermediate of MET-derived C$_2$H$_4$ were superimposed (Primrose, 1977). He also found
that the addition of 2,4-dinitrophenylhydrazine (DNPH) to filtrates of *E. coli* cultures
grown in the presence of MET resulted in the formation of a yellow precipitate
characteristic of DNPH.   No precipitation was formed when MET and DNPH were
added to filtrates of control cultures grown in the absence of MET, revealing that the
intermediate reacts with DNPH to form the precipitate (Primrose, 1977).   Later,
Billington et al. (1979) found that DNPH derivatives of KMBA and the intermediate
have identical infrared (IR), nuclear magnetic resonance (NMR), mass spectra and

identical R_f values in a number of different thin-layer chromatography (TLC) solvent systems. They also detected KMBA as a DNPH derivative in culture fluids of a diverse group of C_2H_4 –producing bacteria as well as some yeast, when supplemented with MET. Billington et al. (1979) described methods for identifying KMBA and showed that this compound is a common metabolic product of microorganisms such as *E. coli, Pseudomonas pisi, Bacillus mycoides, Acinetobacter calcoacteticus, Aeromonas hydrophila, Rhizobium trifolii* and *Corynebacterium* sp., when cells are grown in the presence of MET. Labeled KMBA was detected in bacerial cultures that were fed labeled MET, strongly suggesting a role for KMBA as an intermediate in MET-dependent C_2H_4 formation. Relative to KMBA, the yield of methional from the same cultures was less than 1%, excluding the possibility of methional as a possible intermediate (Billington et al., 1979).

The proposed pathway (Fig. 3.4) reveals that L-MET is deaminated to KMBA, which is then degraded to C_2H_4, either by a peroxidase or photochemically in the presence of a flavin (Primrose, 1977). Primrose (1977, 1979) suggested that MET analogues are converted to MET by *E. coli* before C_2H_4 formation.

Later, Ince and Knowles (1985, 1986) extended the work of Primrose (1977) and extensively investigated the biosynthetic pathway of C_2H_4 in dark-grown *E. coli* by feeding labeled compounds. They confirmed the findings of Primrose (1977) that KMBA serves as an intermediate in MET-dependent C_2H_4 biosynthesis. Ince and Knowles (1986) developed an C_2H_4 -forming system using a cell-free extract of *E. coli* B SPAO. Their system consisted of KMBA, NAD(P)H, Fe(III) chelated to EDTA and O_2, and they discussed the possibility that production of C_2H_4 by many bacteria might follow the same route identified in *E. coli*. They demonstrated that the conversion of MET into KMBA is catalyzed by a soluble transaminase enzyme. 2-Hydroxy-4-methylthiobutyric acid (HMBA) was also a product, but did not serve as an intermediate. Both KMBA and HMBA were formed before optimal production of C_2H_4 occurred. Ethylene formation was also stimulated by increasing the concentrations of Fe(III) when it was chelated to EDTA and by decreasing phosphate concentrations. The KMBA was converted to C_2H_4, methanethiol (methylmercaptan; CH_3SH), and probably CO_2 by a soluble enzyme system requiring the presence of NAD(P)H, Fe(II) chelated to EDTA, and O_2. In the absence of added NAD(P)H, C_2H_4 formation by cell-free extracts from KMBA was stimulated by glucose, which suggests the possibility that more than one enzyme, such as glucose dehydrogenase, might be involved in the conversion of KMBA into C_2H_4. The addition of NAD(P)H did not increase the production of KMBA by cell-free extracts implying that the enzyme involved was not a dehydrogenase or an oxidase but probably a transaminase (Shipston and Bunch, 1989). The enzyme which converts MET to KMBA appeared to be constitutive (Bunch et al., 1991). Ethylene produced was derived from the C-3 and C-4 atoms of MET. The overall scheme for C_2H_4 biosynthesis by *E. coli* is illustrated in Fig. 3.5. Other studies have also demonstrated the conversion of MET into C_2H_4 through KMBA in *E. coli* cultures (Mansouri and Bunch, 1989; Shipston and Bunch, 1989).

3.4.1.2. *Cryptococcus albidus*

In an attempt to elucidate MET-derived C_2H_4 biosynthesis by *C. albidus*, Fukuda et al. (1989a) detected KMBA, a deaminated product of MET, accumulating in the culture filtrate of this fungus. Heat treatment (in boiling water for 5 min) of cells reduced the rate of C_2H_4 formation, indicating inactivation of the enzyme involved in C_2H_4 formation. It seems very likely that *C. albidus* produces C_2H_4 through a pathway similar to that of *E. coli* (MET → KMBA → C_2H_4). They partially purified a cell-free EFE from *C. albidus* IFO0939. All the C_2H_4 –forming activity was found in the supernatant

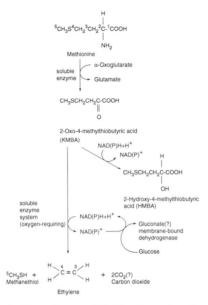

Figure 3-5. Biosynthetic pathway of C_2H_4 formation from methionine by *Escherichia coli* Source. Ince and Knowles (1986).

fraction, which reveals the soluble nature of EFE. This system required KMBA, NADH, EDTA, and Fe(III) for C_2H_4 production, with optimal concentrations of 3 mM, 0.5 mM, 0.4 mM, and 0.4 mM, respectively. Since little formation of C_2H_4 occurred under nitrogen gas, O_2 may be essential for the C_2H_4 –forming reaction. NADH was better than NADPH in stimulation of C_2H_4 formation. In another study, Fukuda et al. (1989c) purified the EFE of *C. albidus* from a cell-free extract. The relative mass of the EFE of *C. albidus* was estimated to be 56,000 by gel filtration and 62,000 by sodium dodecyl sulphate-polyacrylamide gel electrophoresis (SDS-PAGE). These results showed that the enzyme from *C. albidus* IFO0939 is a monomeric protein. It was concluded that this enzyme is an NADH-Fe^{2+}-EDTA oxidoreductase, which reduces 2 mol of Fe(III) EDTA with 1 mol of NADH to give 2 mol of Fe(II) EDTA and 1 mol of NAD+ under anaerobic conditions at pH 6.5 (Fukuda et al., 1989c; Ogawa et al., 1990). From these results, Fukuda et al. (1993) proposed a mechanism for formation of C_2H_4 from MET, as shown in Fig. 3.6. NADH-dependent reduction of Fe(III) to Fe(II), which is catalyzed by an NADH-Fe(III)-EDTA oxidoreductase occurs as a first step. Oxidation of Fe(II)-EDTA by molecular oxygen yields the superoxide radical anion [Fe(II) + O_2 → Fe(III) + O_2^-], which can undergo a dismutation reaction to form hydrogen peroxide ($2O^{2-}$ + $2H^+$ → H_2O_2 + O_2). Hydrogen peroxide, in turn, can react with Fe^{2+} via the Fenton reaction, generating the hydroxyl radical (OH⁻) as shown below:

$$Fe(II) + H_2O_2 \rightarrow Fe(III) + OH^{\cdot} + OH^{-}$$

It had previously been well known that hydroxyl radicals are the reactive species directly responsible for the oxidation of KMBA to C_2H_4 (Diguiseppi and Fridovich, 1980; Tauber and Babior, 1977; Kutsuki and Gold, 1982).   All enzymes that catalyze conversion of Fe(III) to Fe(II) are membrane bound and of the oxidase type.   Since the purified EFE of *C. albidus* IFO0939 is neither membrane bound nor an oxidase, it seems to be a new type of enzyme (Fukuda et al., 1993).   Considering these aspects, Fukuda et al. (1989c) proposed that NADH:Fe(III) oxidoreductase may act as an Fe(III)-Fe(II) recycling enzyme *in vivo*, but the natural electron acceptor is not yet known; however, it is almost certainly not Fe-EDTA. Ogawa et al. (1990) compared the mechanism for the formation of C_2H_4 from KMBA with NADH:Fe(III) EDTA oxidoreductase purified from *C. albidus*.   The characteristics of the reactions catalyzing the NADH:Fe(III) EDTA oxidoreduction and formation of C_2H_4 were almost identical. A chemical C_2H_4 –forming system, composed of Fe(II)EDTA, KMBA, and O_2, was constructed, and the characteristics of the formation of C_2H_4 were compared with those of the enzymatic reaction.   Both the enzymatic and the chemical C_2H_4 –forming reactions were strongly inhibited by scavengers of free radicals, such as benzoic acid, hydroquinone, and catalase, and both were activated by H_2O_2.

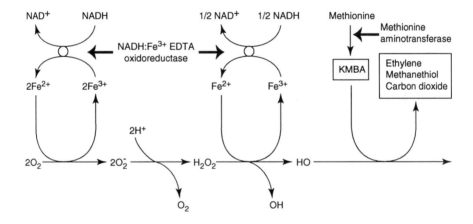

Figure 3-6.   Pathway for formation of C_2H_4 from 2-keto-4-methylthiobutyric acid (KMBA) by the C_2H_4 -forming enzyme from *Cryptococcus albidus* IFO 0939. Source: Fukuda et al. (1993).

Ogawa et al. (1990) concluded that production of C_2H_4 requires the presence of a specific transaminase that catalyzes the formation of KMBA, depending on NADH:Fe(III)EDTA oxidoreductase. To investigate whether or not an NADH-Fe^{3+}-EDTA oxidoreductase is present in microbes that do not produce C_2H_4 even after addition of MET to the medium, Ogawa et al. (1990) measured the activities both of transaminases and of NADH-Fe(III)-EDTA oxidoreductases in bacteria unable to produce C_2H_4 . All strains tested had transaminase activity; while some had NADH:Fe(III)-EDTA oxidoreductase activity others did not. The former type may be negative for MET-uptake activity. It seems that formation of C_2H_4 by microorganisms through a MET-dependent C_2H_4 -forming system, as found in most C_2H_4 producing microorganisms, may occur through the route identified in *C. albidus*, as suggested by Ince and Knowles (1986). It is thus clear why there exist in nature so many C_2H_4 – forming microorganisms that can synthesize C_2H_4 by the MET-dependent pathway. It seems also very likely that *C. albidus* produces C_2H_4 through a pathway similar to that of *E. coli*.

3.4.1.3. *Septoria musiva*

Brown-Skrobot et al. (1985) investigated the possible biosynthetic pathway of C_2H_4 , by *Septoria musiva*, derived from L-MET. To determine whether this fungus follows the same pathway as found in higher plants, SAM was used as a substrate, by substituting it for MET in the medium. The fungus grew well in this medium, but failed to produce C_2H_4 . To determine whether a transaminase reaction may be involved in MET-derived C_2H_4, the MET-glucose medium was amended with isoniazid (isonicotinic acid hydrazide), a known transaminase inhibitor. No C_2H_4 was produced, indicating that transamination of L-MET is involved in C_2H_4 production by the fungus, which results in the formation of an α-keto acid (KMBA) that is converted into C_2H_4 by reactions involving peroxide (Fig. 3.7).

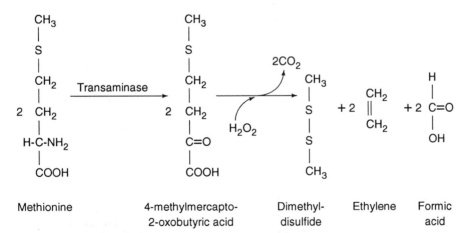

Figure 3-7. Pathway for the conversion of methionine to C_2H_4 by transamination (Source: Brown-Skorobot et al., 1985).

Experiments with methional as a substrate revealed that C_2H_4 was produced from both intact and disrupted cells (25% more with disrupted cells than that of intact cells), implying possible involvement of hydroxyl radicals in the reaction. These findings do not prove that KMBA is converted to C_2H_4 through the action of peroxide, but they do support the premise that MET is converted to C_2H_4 by a transaminase pathway. In this pathway, dimethyldisulfide and formic acid are produced along with C_2H_4. The presence of these products were confirmed by GC-MS and colorimetric tests.

Chandhoke et al. (1992) also reported KMBA-dependent C_2H_4 production by *Gloeophyllum trabeum* and found that the extent of the reaction was influenced by the concentration of chelators (siderophores), iron ($FeCl_3$), manganese ($MnCl_2$) and pH.

3.4.1.4. Summary

There is no doubt that a vast majority of microorganisms generate C_2H_4 from MET via KMBA pathway. Moreover, this transformation has been well-characterized. However, the enzyme involved in the conversion of MET to KMBA has not been well characterized. Wild et al. (1974) suggested that KMBA was produced from D-MET by a D-amino acid dehydrogenase, but Primrose (1979) reported that mutants lacking this enzyme produce C_2H_4 just as readily as the wild-type strains and proposed that an inducible catabolic transaminase, specific for MET, might be involved. Ince and Knowles (1986) found that incubation of cell-free extracts (non-dialyzed) with MET plus NAD^+ or $NADP^+$ did not result in a significant stimulation in the rate of C_2H_4 formation; therefore, it is unlikely that the enzyme involved is an NAD^+ - or $NADP^+$-linked dehydrogenase. However, inclusion of α-keto acids resulted in stimulation of C_2H_4 formation, which suggests that a transaminase enzyme is operative in the transformation of MET into KMBA. However, it is not known whether this transaminase is specific for MET, or if it has a broad specificity for other amino acids. The pathway proposed for C_2H_4 synthesis by *E. coli* may also be applicable to other microorganisms.

3.4.2. Methionine-Independent Pathway

Several studies have unequivocally demonstrated that some microorganisms do not use the MET $\rightarrow$ KMBA pathway for C_2H_4 production and evidence suggests that α-KGA serves as an immediate precursor of C_2H_4 in most of these MET-independent C_2H_4 production systems. Microorganisms which have been reported in utilizing the KGA pathway are *Penicillium digitatum, Pseudomonas cyclopium, Pseudomonas syringae* pv. *phaseolicola, Chaetomium globosum, Phycomyces nitens* and *Fusarium oxysporum* f.sp. *tulipae* (Fukuda et al., 1986, 1988, 1989a,b, 1993; Chou and Yang, 1973; Nagahama et al., 1991a,b; 1992; Goto and Hyodo, 1987; Goto et al., 1985; Hottiger and Boller, 1991; Fukuda and Ogawa, 1991). The biochemistry of KGA-dependent C_2H_4 production by *P. digitatum*, and *Ps. syringae* pv. *phaseolicola* will be reviewed here in detail.

3.4.2.1. *Penicillium digitatum*

The biochemistry of C_2H_4 biosynthesis by *P. digitatum* has been studied more than any other microbial species. Several studies have demonstrated that MET does not serve as an C_2H_4 precursor for this fungus (Ketring et al., 1968; Jacobsen and Wang, 1968;

Chou and Yang, 1973; Fukuda et al., 1986). Wang et al. (1962) reported the active conversion of ^{14}C-labeled glucose, glycine, alanine, glutamic acid, and aspartic acid into C_2H_4 by P. digitatum. They suggested that the transformation of glucose to C_2H_4 occurs via glycolysis and the Krebs cycle. Glutamate is first converted to fumaric acid, which donates the middle carbons to form C_2H_4. Jacobsen and Wang (1965) suggested that acrylic acid may be derived from fumaric acid and gives rise to C_2H_4 by decarboxylation. However, in another study, Jacobsen and Wang (1968) showed that C-2 of acrylic acid was incorporated preferentially into C_2H_4 in comparison with C-3, and that its conversion efficiency into C_2H_4 was lower than that of [2-^{14}C]acetate.

Other workers have demonstrated the involvement of the Krebs cycle in C_2H_4 biosynthesis. Gibson (1964a) and Gibson and Young (1966) found that the C-3 of pyruvate and the C-2 of acetate were incorporated into C_2H_4, but the C-1 of both substrates was not. Ketring et al. (1968) continued this line of work and found that [2-^{14}C] acetate and [2,4-^{14}C] citrate were comparable as precursors in short-time incubation experiments. Their data suggested that C_2H_4 production by this fungus is associated closely with the Krebs cycle. They further showed that monofluoroacetic acid, which effectively blocks the conversion of citrate to isocitrate, markedly inhibited C_2H_4 formation and incorporation of labeled acetate or citrate into C_2H_4. The addition of isocitrate restored C_2H_4 production. Since other members of the Krebs cycle, KGA, succinate, malate, cis-aconitate, or citrate, were ineffective in reversing the inhibition, isocitrate was suggested to be the branching point from which C_2H_4 is derived. Later, Chou and Yang (1973) conducted a comprehensive study by feeding radiolabeled Krebs cycle intermediates and glutamate to the fungus. They concluded that both KGA and glutamic acid are the most efficient precursors of C_2H_4 biosynthesis by P. digitatum, and the carbon atoms in C_2H_4 are derived from C-3 and C-4 of glutamate or the corresponding 2-KGA. However, they could not establish unequivocally whether KGA or glutamate was the direct precursor of C_2H_4. Chou and Yang (1973) concluded that KGA must be the branching point at which the pathway of C_2H_4 biosynthesis breaks off from the Krebs cycle, as illustrated in the proposed biosynthetic pathway (Fig. 3.8).

Fukuda and co-workers (1986; 1988; 1989a,b) further advanced the work of Chou and Yang (1973). They reported that partially purified cell-free extracts of P. digitatum IFO 9372 did not use MET to produce C_2H_4 (Fukuda et al., 1988). They succeeded in constructing an in vitro system using the partially purified enzyme of P. digitatum IFO9372 and found that the immediate precursor of C_2H_4 was KGA (Fukuda et al., 1986). The cell-free system required KGA, L-arginine, Fe(III), and dithiothreitol (DTT), at concentrations of 1 mM, 0.5 mM, 0.75 mM, and 1 mM, respectively, for optimal production of C_2H_4. Labeled experiments showed that L-arginine did not yield C_2H_4 by serving as a precursor. A time-course study on C_2H_4 formation by the cell-free system (in vitro) and that by living cells (in vivo) of the fungus strongly suggested that the cell-free C_2H_4 –forming system operates in living cells of the fungus. Later, they suggested a biosynthetic pathway (Fig. 3.9) of C_2H_4 from glucose that operates by its conversion into KGA and then into C_2H_4 in the presence of L-arginine, Fe(II), DTT (iron-reducing factor), and O_2 by a growth-suppressed mutant of P. digitatum (Fukuda et al., 1988). In another study, Fukuda et al. (1989b), for the first time, purified the EFE from P. digitatum by polyacrylamide disc-gel electrophoresis in an electrophoretically homogenous state, that catalyzed the transformation of KGA into C_2H_4.

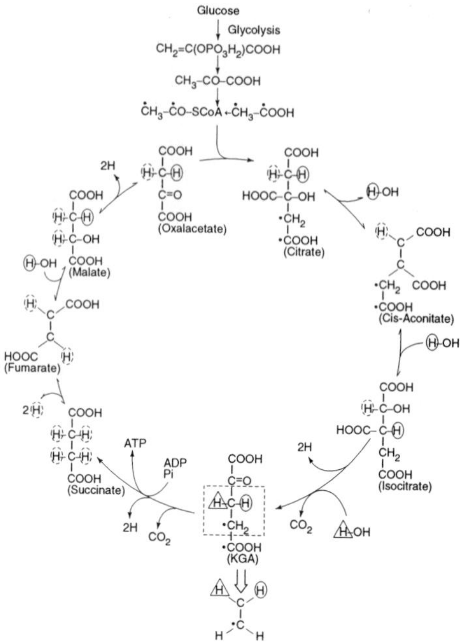

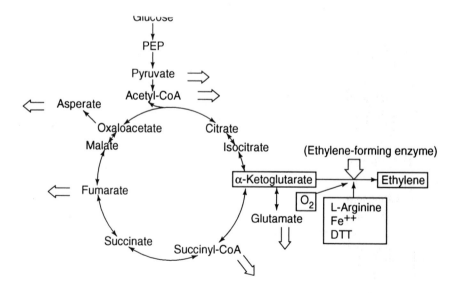

Figure 3-8. Biosynthetic pathway of C_2H_4 production by *Penicillium digitatum* in relation to the Krebs cycle Source. Chou and Yang (1973).

Figure 3-9. Biosynthetic pathway of C_2H_4 from glucose in *Penicillium digitatum* Source. Fukuda et al. (1988).

The enzyme was confirmed to be homogenous, consisting of a single polypeptide and had a relative mass of about 42 kDa, suggesting that the enzyme activity is associated with a single monomeric protein. The presence of KGA, L-arginine, Fe(II), DTT and O_2 were essential for the enzymatic reaction. The EFE has a very high specificity for the substrate, KGA, and cofactor, L-arginine. The effects of various reagents on EFE activity of *P. digitatum* listed in Table 3.8 imply that the activation of O_2 occurs by direct coordination of Fe(II), which may be loosely bound to the enzyme through SH-groups. The enzyme was strongly inhibited by the chelating reagents EDTA and Tiron. This suggests that some kind of Fe(II) complex might be required for catalytic activity. The addition of divalent transition metal ions such as Co(II), Cu(II) and Mn(II) strongly reduced the C_2H_4 -forming activity most likely by competing with Fe(II) for formation of such a complex. The enzyme was inhibited by 5,5'-dithio-bis(2-nitrobenzoate) indicating that SH-groups may have an important role in the catalytic activity. The involvement of superoxide and H_2O_2 and hydroxyl radicals seemed unlikely. Other inhibitors of the activities of free radicals, such as propylgallate, diazibicyclo-octane and hydroquinone, suppressed the reaction considerably probably by acting as chelating agents for Fe(II) and thus suppressing the reaction in a similar manner to EDTA.

Fukuda and Ogawa (1991) proposed a mechanism for C_2H_4 formation by *P. digitatum* (Fig. 3.10). According to their hypothesis, three of the six iron(II) coordination numbers may combine with EFE, two with KGA, and one with oxygen. L-Arginine may act in the reaction between Fe(II) and the enzyme protein. The intermediate (3), shown in Fig. 3.10, is very unstable, especially the –O-O- bonds and, as a result C_2H_4 and CO_2 are released from KGA. The intermediate (4) produces 2 mol of CO_2 and then returns to the original state (1). Consequently, 1 mol of C_2H_4 and 3 mol of CO_2 are produced from 1 mol of KGA, as shown Fig. 3.10. It was observed that the molar ratio between C_2H_4 formation and CO_2 release was 1:3.6 using [U-^{14}C]KGA. Later, Fukuda et al. (1992a) proposed a model for the simultaneous formation of C_2H_4 and succinate from KGA by the enzymes from *P. digitatum* and *Ps. syringae* (Fig: 3.11). Under optimum conditions, the concentrations of L-arginine, Fe(II) and KGA are too low for these chemical species to interact with each other in bulk solution. They must be concentrated within a certain domain of the enzyme in some special way. Thus, the active site of the enzyme can be considered hypothetically to be an Fe(II) complex bound to the enzyme through appropriate ligand atoms L1 and L2 (see also Section 3.4.2). The complex also involves reaction of L-arginine and KGA as a Schiff-base structure to form an intermediate. This intermediate reacts with oxygen to form an unstable Fe(II) complex of the peroxo type (see construct E2 in Fig. 3.11), the KGA moiety of which is decomposed to C_2H_4 and three molecules of CO_2.

The EFE of *P. digitatum* has properties similar to those of KGA-dependent dioxygenase; however, these dioxygenases do not use KGA as a substrate. Similarly, the EFE of *P. digitatum* resembles higher plant EFE in O_2 requirements and reaction to various inhibitors and other reagents except for L-arginine (Fukuda et al., 1989b). Ogawa et al. (1992) reported a nonlinear dependence of the rate of C_2H_4 formation on the amount of EFE present, purified from *P. digitatum*; however, when catalase and bovine serum were added to the reaction mixture, the rate of formation was directly proportional to the amount of enzyme present. They suggested that the nonlinearity of the reaction, in the absence of these additives, is probably due to the hydroxyl radical ions (OH$^+$)

Table 3.8. Effect of various reagents on the ethylene-forming enzyme of *P. digitatum* IFO9372 and *P. syringae* pv. P*haseolicola* PK2.

	Relative ethylene-forming activity (%)	
Reagent	*P. digitatum**	*P. syringae* pv. *phaseolicola***
Control	100	100
Respiratory inhibitors		
KCN*	22	nt
NaN$_3$	62	90
Free radical reagents		
Superoxide dismutase	121	120
Mannitol	74	90
Sodium benzoate	92	100
n-Propylgallate	0	1
1,4-Diazabicyclo[2,2,2]octane (DABCO)	17	nt
Hydroquinone	18	60
Uric acid	nt	90
Transition metals and SH reagents		
CoCl$_2$	3	20
CuSO$_4$·5H$_2$O	15	50
MnCl$_2$·4H$_2$O	11	6
5,5'-Dithio-bis(2-nitrobenzoate) (DTNB)	10	0.7
Chelating reagents		
EDTA (Na)	0	1
Tiron	0	0.8
Others		
Catalase	145	200
H$_2$O$_2$	1	0.7
Ascorbate	128	130

All concentrations were 1 mM, except for superoxide dismutase (150 U ml^{-1} reaction mixture), DABCO (10 mM), hydroquinone (10 mM, catalase (0.05 mg ml^{-1} reaction mixture) and H$_2$O$_2$ (0.03%).
*Source: Fukuda et al. (1989b).   **Source: Nagahama et al. (1991b).

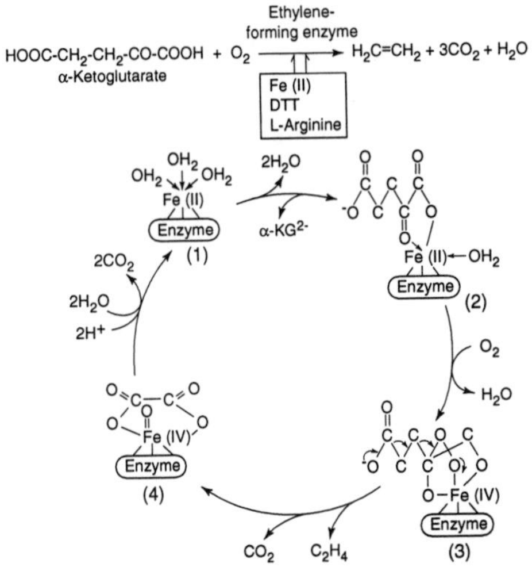

Figure 3-10. A proposed pathway for C_2H_4 formation from α-ketoglutarate by the ethylene-forming enzyme of *Penicillium digitatum* IFO 9372. Source: Fukuda and Ogawa (1991).

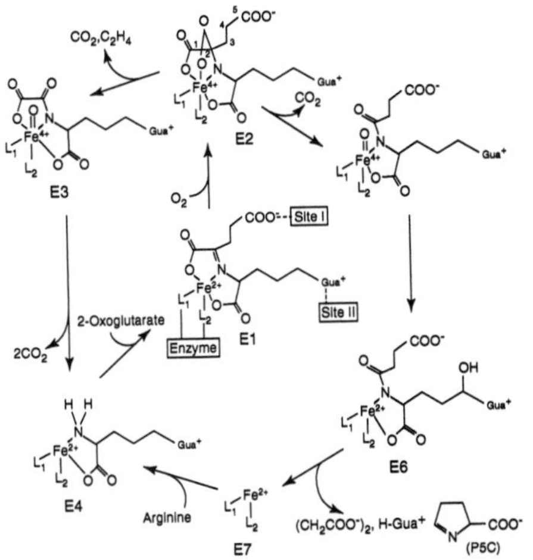

Figure 3-11. A dual-circuit mechanism proposed for the simultaneous formation of C_2H_4 and succinate from 2-KGA by the enzymes from *Penicillium digitatum* and *Pseudomonas syringae* pv. phaseolicola PK2. L1 and L2 are ligands on the enzyme and sites 1 and 11 are binding sites on the enzyme. For simplicity, coordinated water molecules are not shown. All of the iron ions are assumed to be hexa coordinated. Gua^+ indicates a protonated guanidine group; P5C, L-Δ'-pyrroline-5-carboxylate. Source: Fukuda et al. (1992a).

produced by Fenton's reaction that occurs in the reaction mixture when Fe(II) ions and O_2 are present together under reducing conditions (Ogawa et al., 1992).

On the other hand, some studies have demonstrated that *P. digitatum* is also capable of using MET as a C_2H_4 precursor (Primrose, 1977; Chalutz et al., 1977; Chalutz and Lieberman, 1978; Billington et al., 1979). It was shown that *P. digitatum* produces C_2H_4 by two different pathways, depending on whether the fungus is cultured under static or shake conditions (Chalutz et al., 1977; Chalutz and Lieberman, 1978). Glutamate and KGA serve as C_2H_4 precursors in static cultures, whereas MET serves as a precursor and inducer of C_2H_4 production in shake cultures. Chalutz et al. (1977) further indicated that MET is the precursor of C_2H_4 produced by living fungal cells, but not by filtrates. They speculated that the fungus immobilizes MET from solution and, presumably, releases a metabolite into the medium which, in turn, is converted to C_2H_4 . However, the studies conducted by Fukuda et al. (1986, 1989b) did not support the view of Chalutz et al. (1977) and Chalutz and Lieberman (1978), because they found O_2 was essential for KGA or glutamate dependent C_2H_4 production by *P. digitatum*.

3.4.2.2. Pseudomonas syringae pv. phaseolica

Ethylene production by *Ps. syringae* pv. *phaseolicola* has been investigated by many workers (Goto et al., 1985; Goto and Hyodo, 1987; Nagahama et al., 1991a,b; Fukuda et al., 1992a). Goto et al. (1985) reported that the Kudzu strain of this bacterium isolated from *Pueraria lobata* produced C_2H_4 at a rate several times higher than that of *P. digitatum* and *P. solanacearum*. The presence of living cells was essential for C_2H_4 production by the Kudzu strain. The bacterium effectively produced C_2H_4 from amino acids, such as glutamate, aspartate, and their amides. Methionine and ACC did not promote C_2H_4 production by this bacterium. Inhibitors of C_2H_4 biosynthesis in higher plants did not have any effect on C_2H_4 production by *Ps.. syringae* pv. *phaseolicola* (Table 3.9). In another study, Goto and Hyodo (1987) constructed a cell-free C_2H_4 – forming system of *Ps. syringae* pv. *phaseolicola* (Kudzu strain). Ethylene was most effectively produced from KGA at 0.5 mM, followed by glutamate, and then histidine at 5-10 mM. The presence of $FeSO_4$ (0.5 mM) was essential for functioning of this cell-free system. Dithiothreitol and histidine greatly stimulated C_2H_4 production; the latter could be substituted, to some extent, by its analogues. Ethylene formation from KGA was inhibited in the presence of carbonates or organic acids of the Krebs cycle, whereas that from glutamate was inhibited in the presence of ammonium salts. The ethylene-forming reactions were inhibited completely by 11 mM *n*-propylgallate and 1 mM *p*-chloromercuribenzoic acid (PCMB) and partly by coenzymes, such as pyridoxal-1-phosphate, folic acid, and flavin mononucleotide, at 5 mM. The complete cell-free system for the highest C_2H_4 production consisted of 0.5 mM KGA, 50 mM *N*-2-hydroxyethylpiperazine-*N*'-2-ethanesulfonic acid (HEPES; pH 7.0), 5 mM DTT, 0.5 mM $FeSO_4$, and 10 mM histidine. The amount of C_2H_4 produced in this system was equivalent to 40-50% of that produced by living cells. Ethylene production in this reaction mixture appears to arise from enzyme activity, because production was completely terminated when the extract was heated at 100°C, or when PCMB was added to the reaction mixture.

Table 3.9. Inhibitory effects of aminoethoxyvinylglycine (AVG), aminooxyacetic acid (AOA), n-propylgallate (*n*-PG), sodium benzoate and EDTA on C_2H_4 production by *P. syringae* pv. *phaseolicola* PK2.

Inhibitor	Concentration (m*M*)	Rate of C_2H_4 production (%)
None		100
AVG	1.0	117
	0.1	99
AOA	10.0	77
	5.0	72
	2.5	75
	1.0	69
Na-benzoate	1.0	78
	0.1	94
n-PG	2.0	15
	1.0	14
	0.5	16
	0.1	12
	0.05	17
EDTA	2.0	48
	1.0	52
	0.5	87
	0.1	95

Assay mixture: 10 mM phosphate buffer (pH 6.8) 3.5 ml; inhibitor 0.5 ml; substrate (glutamate) 0.5 ml; bacteria (1×10^{10} cells ml^{-1}) 0.5 ml. The rate of C_2H_4 production without an inhibitor (None) was 1.1×10^{-9} nl $cell^{-1} h^{-1}$.
Source: Goto et al. (1985).

The optimal temperature for C_2H_4 production was 25°C. The C_2H_4 –forming activity was completely lost when the cell-free extracts were dialyzed against 10 mM phosphate buffer (pH 7.0) for 24 h at 4°C implying that some reactants are essential for enzymatic activity.

Nagahama et al. (1991a) also studied the formation of C_2H_4 *in vitro* by an extract *of Ps. syringae* pv. *phaseolicola* PK2. The complex system for the formation of C_2H_4 under aerobic conditions *in vitro* required 0.25 mM KGA, 0.2 mM $FeSO_4$, 2 mM DTT, 10 mM L-histidine, and 0.2 mM L-arginine. Dialysis of the cell-free extract against potassium phosphate buffer for 40 h at 4°C resulted in almost complete loss of enzyme activity. The system showed high specificity for the cofactors, L-arginine or L-histidine. The components of this system, with the exception of the stimulatory effects of L-histidine, were similar to those of a system derived from the ethylene-producing plant pathogenic fungus, *P. digitatum*, which also produced C_2H_4 *in vitro* in a reaction dependent on KGA. By assuming that the EFE from *P. syringae* pv. *phaseolicola* PK2 and *P.*

digitatum act in a closely related manner, Nagahama et al. (1991a) proposed a hypothetical reaction mechanism involving the EFE, KGA, L-arginine and an imidazol compound (Fig. 3.12). According to this hypothesis, the presence of a Fe(II) complex bound to the enzyme through an approrpiate ligand provides an active site for the enzymatic reaction. The complex also binds L-arginine to KGA, as a Schiff-base structure, to form an intermediate. The imidazole ring of L-histidine may possibly coordinate with the intermediate Fe(II) complex so that it becomes easier to bind O_2. The subsequent reaction of O_2 leads to the formation of an unstable Fe(IV) complex of the peroxo-type and the KGA moeity is subsequently decomposed to C_2H_4 and three molecules of CO_2. During these oxidation processes, L-arginine, L-histidine, and Fe(II) are regenerated and, thus they can be regarded as cofactors from a mechanistic point of view. The chemical behavior of the enzyme isolated from *P. digitatum* does not exclude this hypothetical mechanism, except that little relevant information is available on the imidazole ring included in the proposed scheme (Nagahama et al., 1991a).

In another study, Nagaham et al. (1991b) purified the EFE from a cell-free extract of *Ps. syringae* pv. *phaseolicola* PK2 that catalyses the formation of C_2H_4 from KGA. The purified enzyme had a specific activity of 660 nmol C_2H_4 per minute per mg of protein . A comparison between the EFE of *Ps. syringae* pv. *phaseolicola* with that of *P. digitatum* showed no homology between the NH_2-terminal amino acid sequence of the two enzymes, but the two enzymes have many properties in common (Table 3.10). The presence of KGA, L-arginine, Fe(II), and O_2 is essential for the enzymatic reaction. The enzymes are highly specific for KGA as a substrate and L-arginine as a cofactor (Table 3.11). The strong inhibition by chelating reagents, such as EDTA and Tiron, suggests that some kind of complex with Fe(II) is involved in the enzymatic reaction (Nagahama et al., 1991a). Some SH-groups in the enzyme may be important for the activity, given the inhibitory effect of DTNB. Superoxide and H_2O_2 are not involved in the reaction, since superoxide dismutase and catalase were not inhibitors. The activation of the reaction by catalase implies the formation of H_2O_2 during the reaction. The involvement of hydroxyl radicals is also unlikely, since the effects of mannitol and sodium benzoate were small. These results imply coordination with Fe(II), which may be loosely bound to the enzyme through its SH-group. Other scavengers of free radicals (e.g., propylgallate and hydroquinone) inhibited the reaction considerably, most likely by forming complexes of Fe(II), similar to EDTA.

Later, Fukuda et al. (1992a) found that purified EFE isolated from *Ps. syringae* pv. *phaseolicola* PK2 simultaneously catalyzed two reactions, i.e., formation of C_2H_4 and formation of succinate from KGA at a molar ratio of 2:1 (Fukuda et al., 1992a). The main reaction is formation of one molecule of C_2H_4 and three molecules of CO_2 from dioxygenation of KGA ($C_5H_5O_5 + O_2 \rightarrow C_2H_4 + 3CO_2 + H_2O$). In the subreaction, succinate is formed from KGA ($C_5H_5O_5 + \frac{1}{2} O_2 \rightarrow C_2H_4O4 + CO_2$) and L-arginine is decomposed into L-hydro-oxyarginine ($C_6H_{14}N_4O_2 + O_2 \rightarrow CH_5N_7O_2 + H_2O$), the latter being further transformed to guanidine and L-Δ-pyrroline-5-carboxylate (P5C). The overall reaction for the EFE from *P. syringae* can be represented by the following:

$$3C_5H_6O_5 + 3O_2 + C_6H_{14}N_4O_2 \rightarrow$$
$$2C_2H_4 + 7CO_2 + C_4H_6O_4 + 3H_2O + CH_5N_3 + C_5H_7NO_2$$

Table 3.10. Comparison of some properties of the ethylene-forming enzymes from *Pseudomonas syringae* and *Penicillium digitatum*.

Property	*Pseudomonas syringae*	*Penicillium digitatum*
Molecular mass (kDa)		
Gel filtration	36	42
SDS-PAGE	42	42
pI	5.9	5.9
pH		
Optimum	7.0-7.5	7.0-7.5
Stability at 5°C	6.0-8.0	6.0-8.0
Temperature (°C)		
Optimum	20-25	25
Stability	<30	<30
K_m value (M)		
Fe(II)	5.9×10^{-5}	4.0×10^{-5}
L-Arginine	1.8×10^{-5}	6.0×10^{-5}
KGA	1.9×10^{-5}	3.8×10^{-5}
N-Terminal sequence	MTNLQTFELP-	LAPPAPSNLG-

Source: Fukuda et al. (1993).

Since formation of P5C and guanidine from L-arginine involves hydroxylation at the Δ-carbon of L-arginine, the latter two reactions appear similar to the KGA-dependent dioxygenase reaction (Hayaishi et al., 1975). Considering the ability of the EFE to mediate the above-cited reactions, Fukuda et al. (1993) proposed a dual-circuit mechanism for the entire reaction, in which binding of L-arginine and KGA in a Schiff-base structure generates a common intermediate for the two reactions (Fig. 3.11). According to this model, the enzyme is assumed to act as a bidentate ligand (L1 and L2) with respect to the active site and it forms a complex with Fe(II) giving rise to E1 (Fig. 3.11). L1 and L2 may be histidine residues (Fukuda et al., 1992a). The Fe(II) ion is further coordinated to a tridentate Schiff's base of KGA and L-arginine, whose terminal

Table 3.11. Cofactor specificity of the ethylene-forming enzyme from *Pseudomonas syringae* pv. *phaseolicola* PK2.

Cofactor (0.2 m*M*)	Relative ethylene-forming activity (%)	Cofactor (10 m*M*)	Relative ethylene-forming activity (%)
L-Arginine	100	Imidazole	62
D-Arginine	3	Imidazole-4-acetic acid	37
L -Canavanine sulfate	7	N-Acetylimidazole	57
L -Citrulline	0	L-Histidine	100
L -Ornithine HCl	0	D-Histidine	89
L -Arginyl- L -arginine	0	1-Methyl-L-histidine	59
L -Arginyl- L -glutamic acid	0	Histamine	45
Poly- L -arginine	0	N-Acetylhistamine	50
N^g-Nitro-L-arginine	0	L-Carnosine	11
N^g-Tosyl- L -arginine	0	L-Histidyl-L-glycine	33
N-carbobenzoxy- L -arginine	0	None	50
None	0		

[a]The specific activity of the ethylene-forming enzyme used was 660 units/mg protein.
Source: Nagahama et al. (1991b).

carboxylate and guanidido groups are trapped, respectively, by binding sites I and II on the enzyme. Then, E1 reacts with dioxygen to form a peroxo complex, E2. Following this event, E2 decomposes irreversibly, with cleavage of the unstable oxygen-oxygen bond and simultaneous cleavage of two bonds in the KGA moiety by separate mechanisms. Thus, one reaction involves simultaneous cleavage of C-2-C-3 and C-4-C-5 bonds to generate C_2H_4, CO_2 and E3. The complex E3 comprises a strongly oxidizing Fe(IV) ion with a strongly reducing oxalyl-like ligand. Thus, an intramolecular redox reaction readily occurs to generate E4. The sequence E1 → E2 → E3 → E4 (→ E1) constitutes a major catalytic cycle for formation of C_2H_4 . The other mode of decomposition of E2 involves cleavage of the C-1-C-2 bond of the KGA moiety, to generate the Fe(IV) complex E5 that contains succinyl arginate as a ligand. The highly reactive Fe(IV) ion of E5 hydroxylates the δ-carbon of the arginine moeity of the ligand. The hydroxylated configuration in E6 is then spontaneously degraded to succinate, guanidine, and P5C. The sequences E1 → E2 → E5 → E6 → E7 → E4 (→ E1) constitutes a minor catalytic cycle for formation of succinate. Complexes E4, E1 and E2 are common to the two catalytic cycles, and it is chemically impossible to determine the specific cycle in which they are involved. The ratio of the activities of the two reactions is determined at the stage of decomposition of E2 since this process is essentially irreversible. Formation of C_2H_4 is favored when all of the bonds in the O-O-C-2-C-3-C-4-C-5 configuration in E2 are stretched into a W-like shape, which maximizes the overlap of orbitals that are produced during the decomposition process. By contrast, when the bonds are not stretched out, formation of C_2H_4 does not occur and succinate is formed from the KGA moiety of E2, in a reaction that is controlled by the spatial alignment of binding site 1, that is to say, by the nature of the enzyme itself (Fukuda et al., 1993).

3.4.2.3. Biosynthesis of C 2 H 4 by other fungi

Two more fungi including *P. cyclopium* and *Fusarium oxysporum* f.sp. *tulipae* have been reported to produce KGA-derived C_2H_4 (Fukuda and Ogawa, 1991; Hottiger and Boller, 1991). *P. cyclopium* KIT0229, a soil fungus, was found to exhibit the highest KGA-dependent C_2H_4 -forming activity. The EFE from *Penicillium cyclopium* KIT0229 was partially purified and its properties were found to be similar to those of the EFE from *P. digitatum* (Fukuda and Ogawa, 1991). Hottiger and Boller (1991) reported that *Fusarium oxysporum* f. sp. *tulipae*, a tulip pathogen, produced high amounts of C_2H_4 during its stationary phase. Feeding experiments with labeled compounds indicated that C_2H_4 is derived from C_3 and C_4 of glutamate or KGA. Fe(II) ions markedly stimulated the rate of C_2H_4 formation *in vivo*, whereas Fe(III), Cu(II) or Zn(II) had little or no effect. Ethylene biosynthesis was strongly inhibited by the heavy metal chelator, α,α'-dipyridine which was fully reversed by Fe(II) ions and partially by Cu(II) and Zn(II) ions, but not by the supply of glutamate or KGA, suggesting that a step in the C_2H_4 biosynthesis pathway downstream of KGA is dependent on Fe(II). When stationary phase cultures were supplied with arginine, ornithine, or proline, C_2H_4 production increased dramatically, while the addition of glutamate or KGA had little effect. Tracer studies with [U-^{14}C] arginine or [U-^{14}C] glutamate revealed that specific radioactivity of C_2H_4 was closely similar to the specific radioactivity of the endogenous glutamate pool, indicating that glutamate was part of the pathway between arginine and C_2H_4. The results suggest that a similar enzyme system of *P. digitatum* catalyzes the final step of C_2H_4 biosynthesis in *F. oxysporum*.

3.4.2.4   Summary

It is now clear that some microorganisms produce C_2H_4 from KGA in the presence of O_2, Fe(II), and L-arginine or L-histidine. The KGA transformation into C_2H_4 is an enzyme-mediated reaction and most likely the same enzyme is active in all those microorganisms which derive C_2H_4 from KGA. Properties of the EFE which generate C_2H_4 from KGA are listed in Table 3.10.

3.5   MOLECULAR ASPECTS OF ETHYLENE-FORMING ENZYME OF *PSEUDOMONAS SYRINGAE*

As discussed in Chapter 2, important advances in the molecular biology of C_2H_4 - forming enzymes of higher plants have been made recently. However, there were no reports of isolating a gene that encodes the EFE in microorganisms until 1992. Fukuda et al. (1992b) described the molecular cloning of the EFE of *Ps. syringae* and demonstrated the presence of cccDNA in this bacterium. They reported that the indigenous plasmid DNA (pPSP1) encodes for the EFE of the bacterium (Fukuda et al., 1992b). *Escherichia coli* JM109 was transformed with the resultant plasmids and the transformants were selected as white colonies on a modified Luria-Bertani (LB) medium. Clones of *E. coli* harboring the gene for the EFE were screened by Southern hybridization of the alkaline-extracted plasmid DNAs using mixtures of oligonucleotide probes. Two positive clones were obtained from the 40 colonies selected. Fukuda et al.

(1992b) selected one clone, *E. coli* JM109 (pEFE01), having the highest C_2H_4 -forming activity which was 230 nl C_2H_4 (ml culture medium)$^{-1}$ h^{-1}, while *E. coli* JM109, the original host, did not produce any C_2H_4. The activity of the enzyme was about one-fifth to one-tenth of that of *P. syringae* at the same density of cells. Antibody raised against the C_2H_4 -forming enzymes reacted with the cell-free extract of *E. coli* JM109 cells that harbored pEFE01, and the immuno-reactive protein showed identical electrophoretic mobility with that of the purified enzyme. They concluded that the gene for the EFE in *Ps. syringae* is encoded by this indigenous plasmid pPSPI. Nucleotide sequence analysis of the sub-clones constructed from pEFF01 revealed an open reading frame that encodes 350 amino acids (mol. wt. 39,444). The calculated relative mass is similar to the values of 36,000 obtained by gel filtration and 42,000 obtained by SDS-PAGE (Nagahama et al., 1991b). The *N*-terminal amino acid-residue sequence from the cloned DNA was identical with that of the purified enzyme, as determined by Edman degradation up to the 29th residue (Nagahama et al., 1991b). The C_2H_4-forming activity of *E. coli* JM109 (pEFE01) was not affected by addition of isopropyl-β-D-thiogalactoside, and both clones that harbored pEFE01 and those that harbored a plasmid with the reverse orientation of the gene had the same C_2H_4 -forming activity. Therefore, EFE activity of *E. coli* JM109 (pEFE01) may be controlled by the promoter from *Ps. syringae*, and not by the *lac* promoter of pUC19. The sequence of pEFE02 contains an inverted repeat in the 3' non-coding region. The role of this inverted repeat is not clear, but it may function as a terminator (Freidman et al., 1987) of the transcription of mRNA from the gene coding for the EFE (Fukuda et al., 1993). Later, Nagahama et al. (1994) further investigated the molecular characteristics of the EFE strains of *Ps. syringae*. The C_2H_4-producing activities of nine strains as measured *in vivo* and *in vitro* were similar, except for that of *Ps. syringae* pv. *mori* M5. A polyclonal antibody and a DNA probe for the EFE from *Ps. syringae* pv. *phaseolicola* PK2 were prepared to investigate homologies among the proteins and genes for the EFE. With the exception of *Ps. syringae* pv. *mori* M5, eight strains tested expressed the same antigen as the EFE from *Ps. syringae* pv. *phaseolicola* PK2 and were homologous to DNA sequences on indigenous plasmids. Molecular masses of antigenic proteins from all C_2H_4-producing strains were 40 kDa. The N-terminal amino acid sequence of the purified EFE from *Ps. syringae* pv. *glycinea* KN130 was identical to that of the enzyme from *Ps. syringae* pv. *phaseolicola* PK2. These results show that the C_2H_4-forming enzymes encoded by the indigenous plasmid(s) in the pathogenic bacteria examined were similar (Nagahama et al., 1994).

Strains of *Ps. syringae* (78 strains and 43 pathovars) and other strains (79) of plant and insect origin were examined for their presence of EFE by a polymerase chain reaction (PCR) assay (Sato et al., 1997). The sequence of the EFE gene of *Ps. syringae* pv. *phaseolicola* PK2 was used to design two primer sets for amplification of the gene. In addition to *Ps. syringae* pv. *phaseolicola* (the "Kudzu strain") and *Ps. syringae* pv. *glycinea* which were efficient C_2H_4 producers, several strains of *Ps. syringae* pv. *sesami* and *cannabina* generated PCR products of the predicted size. A DNA probe of the EFE gene, isolated from strain PK2, hybridized to these PCR products indicating homology to the *Ps. syringae* pv. *phaseolicola* EFE gene. PCR restriction fragment length polymorphism analyses suggested that these four pathovars harbor a similar EFE gene. Furthermore, the probe hybridized to an indigenous plasmid of *Ps. syringae* pv. *cannabina* suggesting that the EFE gene could be located on a plasmid in this pathovar, but did not hybridize to plasmids of *Ps. syringae* pv. *sesami* strains. *Ps. syringae* pvs.

sesami and *cannabina* strains produced C_2H_4 in King's medium B at levels similar to those of *Ps. syringae* pvs. *phaseolicola* and *glycinea* (Sato et al., 1997).

Very recently, Watanabe et al. (1998) provided direct evidence to support previous findings on the EFE encoded plasmid isolated from *Ps. syringae*. They isolated the conjugative plasmid harboring the EFE gene (ethylene plasmid) designated pETH2 from *Ps. syringae* pv. *glycinea* MAFF3010683. pETH2 was detected by Southern blot hybridization using the EFE probe, marked with the transposon mini-Tn5-Km1 and transferred into *Ps. syringae* $Ni27^a$ which does not produce C_2H_4. The transconjugant $Ni27^a$ (pETH2) produced C_2H_4 at a level similar to pv. *glycinea* MAFF301683. In addition, the plasmid designated pCOR2, which encodes coronatine biosynthesis was detected in the same strain. Although the molecular size of the plasmid pCOR2 was not easily distinguishable from pETH2, pCOR2 transferred independently into $Ni27^a$ and the transconjugants produced coronatine. These findings suggested that the EFE gene has been horizontally dispersed among pathovars of *Ps. syringae* by plasmid-mediated conjugation in nature (Watanabe et al., 1998).

3.6. STRUCTURES OF ETHYLENE-FORMING ENZYMES

Partial homology between the EFE of *Ps. syringae* and counterpart enzymes from plants, ripening-related proteins, KGA-dependent dioxygenases, and isopenicillin-N synthases has been reported (Fukuda et al., 1993). The amino acid-residue sequence is highly conserved among the EFE of plants, and middling scores were obtained between KGA-dependent oxygenases. In comparison of the EFE from *Ps. syringae* with EFE from plants, the homology score was low (15% sequence identity), but several clusters of invariant residues were found in the middle part of the sequences. Comparison of these sequence segments reveals that the EFE from *Ps. syringae* has a relationship with the counterpart enzyme from plants. When EFE was compared with KGA-dependent dioxygenases, homology scores were also low. All proteins that have partial homology to the EFE from *Ps. syringae* are members of an Fe(II)/ascorbate oxidase superfamily (McGarvey et al., 1992). Only six invariant amino-acid residues (Gly_{34}, His_{189}, $Asp_{19}1$, Gly_{252}, His_{268}, Arg_{277}) are observed in the optimized alignment of the amino acid-residue sequences in the Fe(II)/ascorbate oxidase superfamily (Fukuda et al., 1993). Matsuda et al. (1991) suggested that there are three regions with a relatively high degree of sequence conservation among KGA-dependent dioxygenases, and that these regions might function as KGA-binding sites and as sites for binding of Fe(II) to two histidine residues. The EFE from *Ps. syringae* also contains the sequences for the iron-binding sites, including two histidine residues (His_{189} and His_{268}), but only Gly_{34} is conserved at another homologous site. The function of the *N*-terminal segment around Gly_{34} in catalysis is not clear because of its low degree of homology. Two amino-acid residues (Asp_{191} and Arg_{277}) were conserved in all sequences examined. These charged residues may perform significant roles in the catalytic reaction (e.g., substrate-binding or intermediate trap). The residue Thr_{239} is replaced by a serine residue in four KGA-dependent enzymes and is conserved in the other enzymes examined. The hydroxyl side-chains seem to be important in catalysis.

The hydropathy plot of the EFE from *Ps. syringae*, calculated by the method of Kyte and Doolittle (1982), is similar to those of corresponding enzymes from plants and ripening-related proteins, with the exception of two extra regions (Fig. 3.13). One extra region is relatively hydrophobic (residues 93-114) and the other (residues 209-226) is extremely hydrophilic. L-Arginine is essential for the formation of C_2H_4 by the enzyme from *Ps. syringae*, as are KGA, Fe(II) and oxygen. However, L-arginine is not a cosubstrate for other KGA-dependent enzymes. Sequence alignment revealed that amino-acid residues from Tyr_{207} and Glu_{232} represent an insertion sequence between two highly conserved domains, which include important histidine residues, and it provides an extra segment within the enzyme. In this region, there are four negatively charged glutamic acid residues that may be related to the binding of L-arginine (Fukuda et al., 1993).

The low degree of homology of the entire sequence compared with those of the family of EFEs of higher plants and the similarities between the hydropathy profiles between the EFE from *P. syringae* and the other proteins cannot be explained at this time.

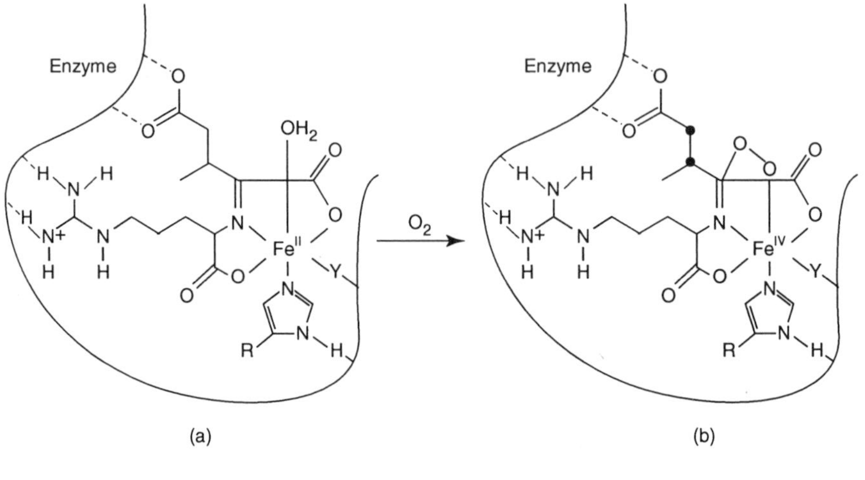

(a)   (b)

Figure 3-12. Model of the proposed intermediate consisting of the EFE, 2-oxoglutarate, L-arginine, and the imidazole compound. (a) The intermediate before reaction with oxygen and (b) after reaction. The carbon atoms from 2-oxoglutarate that are converted to C_2H_4 are indicated (●). Source: Nagahama et al. (1991a).

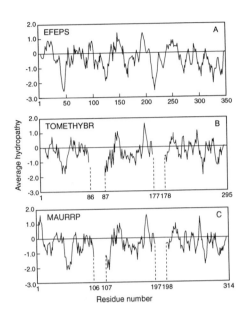

Figure 3-13. Comparison of hydropathy profiles of the EFE from (A)
Pseudomonas syringae pv. *phaseolicola* PK2 (EFEPS) and (B)
TOMETHYBR, the EFE related to the TOM13 gene product of
Lycopersicon esculentum, (C) MAURRP, the ripening-related protein
of *Malus sylvestris*. Hydropathy profiles (n=9) were calculated by the
method of Kyte and Doolittle (1982). Dashed lines indicate boundary
divisions. Source: Fukuda et al. (1993).

3.7. MICROBIAL METABOLISM OF ETHYLENE

In addition to C_2H_4 biosynthesis, a number of bacteria have been found capable of
metabolizing C_2H_4 (Table 3.12). Many of these bacteria are able to utilize C_2H_4 as a
growth substrate for C and energy. Many microorganisms that grow on C_2H_4 (de Bont,
1976; Heyer, 1976) can also grow on other alkenes and alkanes (Hou et al., 1979; Patel et
al., 1983; Higgins et al., 1979; de Bont et al., 1983). The oxidation of C_2H_4 occurs by its
conversion into ethylene oxide via the following reaction:

$$C_2H_4 + 2(H) + O_2 \rightarrow H_2C - CH_2 + H_2O.$$
$$\backslash \; /$$
$$O$$
$$(\text{ethylene oxide})$$

Table 3.12.   Ethylene-utilizing and -oxidizing bacteria.

Arthrobacter sp. B-53	Nikitin and Arakelyan (1979)
Arthrobacter sp. CRL-68	Patel et al. (1983)
Arthrobacter sp. CRL-70	Patel et al. (1983)
Arthrobacter sp. CRLGO	Patel et al. (1983)
Bacillus sp. P-101	Nikitin and Arakelyan (1979)
Brevibacterium fuscum ATCC15993	Patel et al. (1983)
Caulobacter sp. P-104	Nikitin and Arakelyan (1979)
Caulobacter sp. C-6	Nikitin and Arakelyan (1979)
Caulobacter sp. P-103	Nikitin and Arakelyan (1979)
Methylcoccus capsulatus CRLM1 NRRL B11219	Hou et al. (1979)
Methylobacterium organophilum CRL 26NRRL B-11222	Hou et al. (1979)
Methylosinus trichosporium	Higgins et al. (1979)
Methylosinus trichosporium OB 3b	Hou et al., (1979)
Micrococcus sp. M90C	Mahmoudian and Michael (1992)
Microcyclus aquaticus Z2434	Nikitin and Arakelyan (1979)
Microcyclus aquaticus Paq	Nikitin and Arakelyan (1979)
Microcyclus aquaticus sp. P100	Nikitin and Arakelyan (1979)
Mycobacterium sp. CRL-69	Patel et al. (1983)
Mycobacterium E20	de Bont and Harder (1978); de Bont et al. (1979)
Mycobacterium sp. NCIB 11626	Weijers et al. (1988)
Mycobacterium sp. L1	Weijers et al. (1988)
Mycobacterium sp. strain E3	van Ginkel et al. (1987); Habets-Crutzen et al. (1984, 1985); Habets-Crutzen and de Bont (1985); Weijers et al. (1988)
Mycobacterium sp. strain E4	Habets-Crutzen et al. (1984, 1985)
Mycobacterium sp. strain E44	Habets-Crutzen et al. (1984, 1985)
Mycobacterium sp. strain Eu1	Habets-Crutzen et al. (1984, 1985)
Mycobacterium sp. strain Eu3	Habets-Crutzen et al. (1984, 1985)
Mycobacterium sp. strain 12D	Habets-Crutzen et al. (1984, 1985)
Mycobacterium sp. strain 3W	Habets-Crutzen et al. (1984, 1985)
Mycobacterium sp. strain Tu1	Habets-Crutzen et al. (1984, 1985)
Mycobacterium sp. strain 2W	van Ginkel et al. (1987); Habets-Crutzen et al. (1984, 1985); Habets-Crutzen and de Bont (1985); Weijers et al. (1988)
Mycobacterium sp. strain Py1	van Ginkel et al. (1987)
Mycobacterium sp. 11-M	Nikitin and Arakelyan (1979)
Mycobacterium aureum	Hartman et al. (1991)
Mycobacterium rhodochorous ATCC29670	Patel et al. (1983)
Nitrosomonas europaea	Hyman and Wood (1984); Hommes et al. (1998)
Nocardia sp. strain BY1	van Ginkel et al. (1987
Nocardia sp. strain BT1	van Ginkel et al. (1987)
Nocardia sp. strain IP1	van Ginkel et al. (1987)
Nocardia sp. strain H8	van Ginkel et al. (1987)
Pseudomonas sp. CRL-71	Patel et al. (1983)
Pseudomonas fluorescens	Nikitin and Arakelyan (1979)

Table 3.12. (continued)

Renobacter vacuolatum	Nikitin and Arakelyan (1979)
Thermophillic bacterium	Weijers et al. (1988)
Xanthobacter sp. strain Py2	van Ginkel et al. (1986, 1987))
Xanthobacter sp. strain By2	van Ginkel et al. (1986, 1987)
Xanthobacter sp. strain Py3	van Ginkel et al. (1986)
Xanthobacter sp. strain Py7	van Ginkel et al. (1986)
Xanthobacter sp. strain Py10	van Ginkel et al. (1986)
Xanthobacter sp. strain Py11	van Ginkel et al. (1986)
Xanthobacter sp. strain Py17	van Ginkel et al. (1986)

de Bont and Harder (1978) demonstrated that the initial steps in C_2H_4 metabolism in *Mycobacterium* E20 (an isolate from soil) was the oxidation of C_2H_4 to an epoxyalkene by alkene monooxygenase (AMO). Later, by using a labeled substrate, de Bont et al. (1979) confirmed that the initial step in the oxidation of C_2H_4 by *Mycobacterium* E20 was catalyzed by a monooxygenase (Fig. 3.14). The product, ethylene oxide (C_2H_4O) from C_2H_4 is NADPH-dependent in the case of crude cell-free extracts. This enzyme is soluble and has a specificity limited to alkenes (Table 3.13). In another study, they observed that resting cells of *Mycobacterium* Py1 (grown on propene) catalyzed the oxidation of C_2H_4 into epoxyethene (de Bont et al., 1983). Hartmans et al. (1991) detected NADH- or NADPH-dependent AMO activity in cell-free extracts of the C_2H_4 - utilizing *Mycobacterium* E3 and *Mycobacterium aureum* L1. The activity was not linear with protein concentration in the assay suggesting that AMO is a multicomponent enzyme. The inhibition pattern of AMO activity was very similar to the inhibition patterns published for the three-component soluble methane monooxygenases. An C_2H_4-utilizing bacterial soil isolate identified as *Micrococcus* sp. M90C showed a specific activity of 67 nmol/min/mg protein for C_2H_4 and produced epoxyethane with speciific productivity of 785 μmol/h/g dry weight (Mahmoudian and Michael, 1992). The cell-free extract study revealed that epoxidation activity was localized in the soluble fraction as was also observed in the case of whole cells. The formation of epoxyethane by the soluble fraction was dependent on the presence of O_2 and NADH. No activity was observed in the absence of cells or with heat-killed cells indicating the presence of a monooxygenase. The activity was stable in short term (7-10 days) storage at −40°C in the absence of stabilizing agents, but was completely inhibited in the presence of 5 mM β-mercaptoethanol as reported in Table 3.14 (Mahmoudian and Michael, 1992). Higgins et al. (1979) reported that the cell suspension of *Methylosinus trichosporium* OB 3b partially metabolized a wide range of compounds including C_2H_4 which was oxidized to C_2H_4O. Similarly, Hou et al. (1979) found that the resting cell suspensions of methane-growing bacteria, *M. trichosporium* OB 3b, *Methylococcus capsulatus* CRL M, NRRL B11219, *Methylobacterium organophilum* CRL 26 and NRRL B11222 oxidized alkenes to their corresponding 1.2-epoxide which accumulated extracellularly. Ethylene was oxidized to C_2H_4O and *M. capsulatus* was the most efficient in mediating this oxidation reaction (5.5 μmol/h/assay). Eight strains of bacteia isolated from lake water and soil

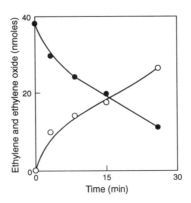

Figure 3-14. The conversion of C_2H_4 to ethylene oxide by cell-free extracts of *Mycobacterium* E20. The reaction was carried out with dialysed cell-free extracts in the presence of 0.2 mM propylene oxide in screw-cap tubes. (●) ethylene; (○) ethylene oxide. Source: deBont et al. (1979).

capable of growing on alkanes catalyzed the epoxidation of alkenes (Patel et al., 1983). The reaction proceeded linearly for up to 60 min at 30°C. The product accumulated extracellularly. The optimum temperature and pH for the reaction were 35°C and 7.0, respectively. Ethylene was oxidized to C_2H_4O at a rate of 0.5 to 2.3 μmol/h/mg protein by the bacteria (Patel et al., 1983). van Ginkel et al. (1987) screened 11 bacteria strains for their ability to grow and oxidize various alkenes. They found that only two strains, *Mycobacterium* strains E3 and 2W, grew on C_2H_4 (Table 3.15), while all except one were capable of oxidizing C_2H_4 at rates ranging from 2 to 50 nmol/min per mg of protein. The alkenes tested were oxidized either to water and CO_2 or to epoxyalkanes. Similarly, the *Xanthobacter* strains isolated on propene or butene oxidized C_2H_4 at a rate of 35-55 nmol per minute per mg protein (van Ginkel et al., 1986). Habets-Crützen et al. (1984) reported that resting cell suspension of all the alkene-utilizing bacteria (14 strains) converted C_2H_4 quantitatively to epoxyethane and C_2H_4 –utilizers had a doubling growth time ranging from 8 to 28 minutes. They also observed that bacteria grown on C_2H_4 predominantly produced the R form of 1,2- epoxypropane from propene and 1,2- epoxybutadien from 1-butene (Habets-Crützen et al., 1985). Similarly, Weijers et al. (1988)found that bacteria grown on 1-alkenes (including C_2H_4) formed epoxyalkanes stereospecifcally, but the enantiomeic composition depended on both the epoxyalkane formed and the organism used.

    A nitrifying bacterium, *Nitrosomonas europaea* has been reported capable of oxidizing C_2H_4 into C_2H_4O (Hyman and Wood, 1984; Hyman et al., 1988; Hommes et al., 1998). This bacterium which oxidizes NH_3 into NO_2^- (nitrification) can also use C_2H_4 as a cosubstrate. Hyman and Wood (1984) reported that incubation of *N. europaea* with C_2H_4 led to the formation of C_2H_4O. Inhibitors of ammonium oxidation

Table 3.13. Oxidation of gaseous hydrocarbons by (a) suspensions of ethylene-grown, ethane-grown, and succinate-grown cells and (b) cell-free extracts of ethylene-grown *Mycobacterium* E20.

(a) Substrate	nmoles substrate utilized min^{-1} mg protein^{-1} [a]		
	ethylene-grown	ethane-grown	succinate-grown
Ethylene	5.2	2.3	1.1
Ethane	1.8	9.0	0.6
Propylene	6.3	--	--
Propane	2.0	--	--
Butylene	4.0	--	--
Butadiene	13.2	--	--
Butane	1.8	--	--
Allene	8.2	--	--

(b) Substrate	Activity (nmoles substrate utilized min^{-1} mg protein^{-1})
Ethylene	0.35
Ethane	0
Propylene	0.37
Propane	0
Butylene	0.14
Butadiene	0.42
Butane	0
Allene	0.73

Source: de Bont et al. (1979).

inhibited the formation of C_2H_4O implying that C_2H_4 is a substrate for ammonium monooxygenase. The presence of C_2H_4 restricts ammonium oxidation by this enzyme ($K_i = 80$ μM). Hyman and Wood (1984) also reported that C_2H_4O is a substrate for this enzyme, but with much lower affinity than C_2H_4 . Later they found that ammonium monooxygenase of *N. europaea* also catalyzes the oxidation of alkanes to alcohols and alkenes including C_2H_4 to epoxides and alcohols in the presence of ammonium revealing a broad substrate specificity (Hyman et al., 1988). Very recently Hommes et al. (1998) confirmed that ammonium monooxygenase catalyzes the transformation of C_2H_4 into

Table 3.14. Production of epoxyethane by whole cells and cell-free extracts of *Micrococcus* sp. M90C.

Fractions	Specific activity[a]	Relative activity (%)
Whole cells	60	100
(S) with NADH without BME	19	32
(S) with NADH and BME	0	0
(S) without NADH and BME	0	0

The soluble fraction (S) was assayed with a total volume of 1 ml containing 5 μmol NADH.
β-mercaptoethanol (BME) was added to a final concentration of 5 mM during the preparation of cell-free extracts.
[a]nmol/min/mg protein.
Source: Mahmoudian and Michael (1992).

C_2H_4O and the presence of C_2H_4 inhibits ammonium oxidation (50% inhibition of NO_2^- production occurred at 0.41 mM C_2H_4). They also reported that oxidation of endogenous compounds most likely provided the reductant for the enzymatic transformation of C_2H_4 into ethylene oxide (Hommes et al., 1998).

3.8. CONCLUDING REMARKS

Recent progress in characterizing biosynthesis of C_2H_4 by microorganisms has been reviewed in this chapter. Currently, two pathways for formation of C_2H_4 in microorganisms have been identified, namely the KMBA and the KGA pathways. By contrast, the ACC pathway is exploited by higher plants. In the KMBA pathway of microbial production of C_2H_4, the final step most likely involves a chemical reaction. The EFE of the KGA pathway has been characterized from an enzymology and molecular-genetic perspective. In comparison with the EFE of higher plants, the homology for the amino acid-residue sequence of the EFE of *Ps. syringae* and the KGA-dependent dioxygenases is relatively low. However, functionally significant regions appear to be conserved.

The presence of ACC deaminase which catalyzes ACC to NH_3 and α-ketobutyrate has been well-characterized in various non-C_2H_4 producing microorganisms. Regulating C_2H_4 synthesis in higher plants with ACC deaminase can have a tremendous impact on agronomic development. Biotechnology can provide a breakthrough in promoting agricultural production. Some applications of this approach are discussed in other chapters of this book.

Table 3.15. Oxidation of ethene, propene, 1-butene, 1,3-butadiene, cis-2-butene, trans-2-butene, 1-pentene, and 1-hexene by washed cell suspensions of alkene-grown bacteria.

Strain	Growth substrate	Substrate oxidation rate (nmol/min per mg of protein) on the following substrate:							
		C_2H_4	C_3H_6	C_4H_8	C_4H_6	cis-C_4H_8	trans-C_4H_8	C_5H_{10}	C_6H_{12}
Mycobacterium sp. strain E3	Ethene	50	17	12	19	20	20	14	13
Mycobacterium sp. strain 2W	Ethene	23	6	6	11	12	13	6	7
Mycobacterium sp. strain Py1	Propene	15	20	17	ND*	ND	ND	8	5
Xanthobacter sp. strain Py2	Propene	50	81	62	17	70	60	20	14
Nocardia sp. strain By1	1-Butene	19	23	26	9	21	17	12	11
Xanthobacter sp. strain By2	1-Butene	45	70	61	24	67	29	24	16
Nocardia sp. strain TB1	2-Butene	2	2	3	3	6	5	1	1
Nocardia sp. strain BT1	Butadiene	18	16	17	57	19	19	ND	ND
Nocardia sp. strain 1P1	Isoprene	11	14	12	13	ND	ND	ND	ND
Pseudomonas sp. strain H1	1-Hexene	0	0	1	1	1	1	3	5
Nocardia sp. strain H8	1-Hexene	16	19	19	ND	ND	ND	16	16
Nocardia sp. strain H8	Propene	17	18	16	ND	ND	ND	17	16

*ND, not determined
Source: van Ginkel et al. (198

3.9 REFERENCES

Arshad, M., and Frankenberger, W. T., Jr., 1989, Biosynthesis of ethylene *by Acremonium falciforme, Soil Biol. Biochem.* **21**:633-638.

Arshad, M., and Frankenberger, W. T., Jr., 1990, Ethylene accumulation in soil in response to organic amendments, *Soil Sci. Soc. Am. J.* **54**:1026-1031.

Arshad, M., and Frankenberger, W. T., Jr., 1993, Microbial production of plant growth regulators, in: *Soil Microbial Ecology*, B. F. Metting, ed., Marcel Dekker, Inc., New York, pp. 307-347.

Arshad, M., and Frankenberger, W. T., Jr., 1998, Plant growth-regulating substances in the rhizosphere: Microbial production and functions, *Adv. Agron.* **62**:45-151.

Beard, R., and Harrison, M. A., 1992, Effect of inoculation on ethylene production in beans, *Plant Physiol.* **99**(Suppl.):66.

Billington, D. C., Golding, B. T., and Primrose, S. B., 1979, Biosynthesis of ethylene from methionine, *Biochem. J.* **182**:827-836.

Brown, S. K., and Brown, L. R., 1981, The production of carbon monoxide and ethylene by *Rhizoctonia solani, Dev. Indust. Microbiol.* **22**:725-731.

Brown-Skrobot, S. K., Brown, L. R., and Filer, T. H., Jr., 1985, Mechanism of ethylene and carbon monoxide production by *Septoria musiva, Dev. Indust. Microbiol.* **26**:567-573.

Bunch, A. W., McSwiggan, S., Lloyd, J. B., and Shipston, N. F., 1991, Ethylene synthesis by soil microorganisms: Its agricultural and biotechnological importance, *Agro-Industry Hi-Tech.* **2**:21-24.

Burd, G. I., Dixon, D. G., and Glick, B. R., 1998, A plant growth-promoting bacterium that decreases nickel toxicity in seedlings, *Appl. Environ. Microbiol.* **64**:3663-3668.

Campbell, B. G. and Thomson, J. A., 1996, 1-Aminocyclopropane-1-carboxylate deaminase genes from *Pseudomonas strains, FEMS Microbiol. Lett.* **138**:207-210.

Chalutz, E. and Lieberman, M., 1978, Inhibition of ethylene production in *Penicillium digitatum, Plant Physiol.* **61**:111-114.

Chalutz, E., Lieberman, M., and Sisler, H. D., 1977, Methionine-induced ethylene production by *Penicillium digitatum, Plant Physiol.* **60**:402-406.

Chandhoke, V., Goodell, B., Jellison, J., and Fekete, F. A., 1992, Oxidation of 2-keto-4-thiomethylbutyric acid (KTBA) by iron-binding compounds produced by the wood-decaying fungus *Gloeophyllum trabeum. FEMS Microbiol. Lett.* **90**:263-266.

Chernys, J., and Kende, H., 1996, Ethylene biosynthesis in *Regnellidium diphyllum* and *Marsilea quadrifolia, Planta* **200**:113-118.

Chou, T. W. and Yang, S. F., 1973, The biogenesis of ethylene in *Penicillium digitatum, Arch. Biochem. Biophys.* **157**:73-82.

Coleman, L. W., and Hodges, C. F., 1986, The effect of methionine on ethylene and 1-aminocyclopropane-1-carboxylic acid production by *Bipolaris sorokiniana, Phytopathol.* **76**:851-856.

Considine, P. J., Flynn, N., and Patching, J. W., 1977, Ethylene production by soil microorganisms, *Appl. Environ. Microbiol.* **33**:977-979.

Dasilva, E. J., Henriksen, E., and Henriksson, L. E., 1974, Ethylene production by fungi, *Plant Sci. Lett.* **2**:63-66.

de Bont, J. A. M., 1976, Oxidation of ethylene by soil bacteria, *Antonie van Leeuwenhoek* **42**:59-71.

de Bont, J. A. M. and Harder, W., 1978, Metabolism of ethylene by *Mycobacterium* E20, *FEMS Microbiol. Lett.* **3**:89-93.

de Bont, J. A. M., Attwood, M. M., Primrose, S. B., and Harder, W., 1979, Epoxidation of short chain alkenes in *Mycobacterium* E20: The involvement of a specific monooxygenase, *FEMS Microbiol. Lett.* **6**:183-188.

de Bont, J. A. M., van Ginkel, C. G., Tramper, J., and Luyben, K. C. A. M., 1983, Ethylene oxide production by immobilized *Mycobacterium* Py1 in a gas/solid bioreactor, *Enzyme Microbiol. Technol.* **5**:55-60.

Diguiseppi, J., and Fridovich, I., 1980, Ethylene from 2-keto-4-thiomethylbutyric acid: The Haber-Weiss reaction, *Arch. Biochem. Biophys.* **205**:323-329.

Drahos, D., Barry, G., Hemming, B., Bradt, F., Skipper, H., Kline, E., Kluepful, D., Hughes, T., and Gooden, D., 1988, Pre-release testing procedures: U.S. field test of a lacZY-engineered soil bacterium, in: *The Release of Genetically-Engineered Microorganisms*, M. Sussman, C. Collins, F. Skinner, and D. Stewait-Tull, eds., Academic Press, New York. pp. 181-191.

Fergus, C. L., 1954, The production of ethylene by *Penicillium digitatum, Mycologia* **46**:543-555.

Frankenberger, W. T., Jr., and Arshad, M., 1995, *Phytohormones in Soils: Microbial Production and Functions*, Marcel Dekker, Inc., New York.

Frankenberger, W. T., Jr., and Phelan, P. J., 1985a, Ethylene biosynthesis in soil. I. Method of assay in conversion of 1-aminocyclopropane-1-carboxylic acid to ethylene, *Soil Sci. Soc. Am. J.* **49**:1416-1422.

Frankenberger, W. T., Jr., and Phelan, P. J., 1985b, Ethylene biosynthesis in soil. II. Kinetics and thermodynamics in the conversion of 1-aminocyclopropane-1-carboxylic acid to ethylene, *Soil Sci. Soc. Am. J.* **49**:1422-1426.

Friedman, D. J., Imperiale, M. J., and Adhya, S. L., 1987, RNA 3' end formation in the control of gene expression, *Annu. Rev. Genetics* **21**:453-488.

Fukuda, H., and Ogawa, T., 1991, Microbial ethylene production, in: *The Plant Hormone Ethylene*, A. K. Mattoo and J. C. Suttle, eds., CRC Press, Boca Raton, FL, pp. 279-292.

Fukuda, H., Fujii, T., and Ogawa, T., 1986, Preparation of a cell-free ethylene-forming system from *Penicillium digitatum*, *Agric. Biol. Chem.* **50**:977-981.

Fukuda, H., Fujii, T., and Ogawa, T., 1988, Production of ethylene by a growth-suppressed mutant of *Penicillium digitatum*, *Biotechnol. Bioengineer.* **31**:620-623.

Fukuda, H., Takahashi, M., Fujii, T., and Ogawa, T., 1989a, Ethylene production from L-methionine by *Cryptococcus albidus*, *J. Ferment. Bioeng.* **67**:173-175.

Fukuda, H., Kitajima, H., Fujii, T., Tazaki, M., and Ogawa, T., 1989b, Purification and some properties of a novel ethylene-forming enzyme produced by *Penicillium digitatum*, *FEMS Microbiol. Lett.* **59**:1-6.

Fukuda, H., Takahashi, M., Fujii, T., Tazaki, M., and Ogawa, T., 1989c, An NADH:Fe(III) EDTA oxidoreductase from *Cryptococcus albidus*: An enzyme involved in ethylene production in vivo? *FEMS Microbiol. Lett.* **60**:107-112.

Fukuda, H., Ogawa, T., Tazaki, M., Nagahama, K., Fujii, T., Tanase, S., and Morino, Y., 1992a, Two reactions are simultaneously catalyzed by a single enzyme: The argenine-dependent simultaneous formation of two products, ethylene and succinate, from 2-oxoglutarate by an enzyme from *Pseudomonas syringae*, *Biochem. Biophysic. Res. Communic.* **188**:483-489.

Fukuda, H., Ogawa, T., Ishihara, K., Fujii, T., Nagahama, K., Omata, T., Inoue, Y., Tanase, S., and Morino, Y., 1992b, Molecular cloning in *Escherichia coli*, expression and nucleotide sequence of the gene for the enthylene-forming enzyme of *Pseudomonas syringae* pv. *phaseolicola* PK2, *Biochem. Biophys. Res. Commun.* **188**:826-832.

Fukuda, H., Ogawa, T., and Tanase, S., 1993, Ethylene production by microorganisms, *Adv. Microbiol. Physiol.* **35**:275-306.

Gibson, M. S., 1964a, Incorporation of pyruvate-C^{14} into ethylene by *Penicillium digitatum*, *Sacc. Arch. Biochem. Biophys.* **106**:312-316.

Gibson, M. S., 1964b, Organic acids as a source of carbon for ethylene production, *Nature* **202**:902-903.

Gibson, M. S., and Young, R. E., 1966, Acetate and other carboxylic acids as precursors of ethylene, *Nature* **210**:529-530.

Glick, B. R., Karaturovic, D. M., and Newell, P. C., 1995, A novel procedure for rapid isolation of plant growth promoting pseudomonads, *Can. J. Microbiol.* **41**:533-536.

Glick, B. R., Primrose, D. M., and Li, J., 1998, A model for lowering of plant ehtylene concentrations by plant growth-promoting bacteria, *J. Theor. Biol.* **190**:63-68.

Goto, M., and Hyodo, H., 1987, Ethylene production by cell-free extract of the Kudzu strain of *Pseudomonas syringae* pv. *phaseolicola*, *Plant Cell Physiol.* **28**:405-414.

Goto, M., Ishida, Y., Takikawa, Y., and Hyodo, H., 1985, Ethylene production by the Kudzu strain of *Pseudomonas syringae* pv. *phaseolicola* causing halo blight in *Pueraria lobata* (Wild) *Ohwi*, *Plant Cell Physiol.* **26**:141-150.

Grichko, V. P., Filby, B., and Glick, B. R., 2000, Increased ability of transgenic plants expressing the bacterial enzyme ACC-deaminase to accumulate Cd, Co, Cu, Ni, Pb and Zn, *J. Biotechnol.* **81**:45-53.

Habets-Crützen, A. Q. H., and de Bont, J. A. M., 1985, Inactivation of alkene oxidation by epoxides in alkene- and alkane-grown bacteria, *Appl. Microbiol. Biotechnol.* **22**:428-433.

Habets-Crützen, A. Q. H., Brink, L. E. S., van Ginkel, C. G., de Bont, J. A. M., and Tramper, J., 1984, Production of epoxides from gaseous alkenes by resting-cell suspension and immobilized cells of alkene-utilizing bacteria, *Appl. Microbiol. Biotechnol.* **20**:245-250.

Habets-Crützen, A. Q. H., Carlier, S. J. N., de Bont, J. A. M., Wistuba, D., Schurig, V., Hartmans, S., and Tramper, J., 1985, Stereospecific formation of 1,2-epoxypropane, 1,2-epoxybutane, and 1-chloro-2,3-epoxypropane by alkene-utilizing bacteria, *Enzyme Microb. Technol.* **7**:17-21.

Hahm, D. H., Kwak, M. Y., Bae, M., and Rhee, J. S., 1992, Effect of dissolved oxygen tension on microbial ethylene production in continuous culture, *Biosci. Biotechnol. Biochem.* **56**:1146-1147.

Hartmans, S., Weber, F. J., Somhorst, D. P. M., and de Bont, J. A. M., 1991, Alkene monooxygenase from *Mycobacterium*: a multicomponent enzyme, *J. Gen. Microbiol.* **137**:2555-2560.

Heyer, J., 1976, Mikrobielle verwetang von Äthylen, *Z. Allg. Microbiol.* **16**:633-637.

Higgins, I. J., Hammond, R. C., Sariaslani, F. S., Best, D., Davies, M. M., Tryhorn, S. E., and Tayol, F., 1979, Biotransformation of hydrocarbons and related compounds by whole organism suspensions of methane-grown *Methylosinus trichosporium* OB 3b, *Biochem. Biophys. Res. Commun.* **89**:671-677.

Hommes, N. G., Russell, S. A., Bottomley, P. J., and Arp, D. J., 1998, Effects of soil on ammonia, ethylene, chloroethane, and 1,1,1-trichloroelkane oxidation by *Nitrosomonas europaea*, *Appl. Environ. Microbiol.* **64**:1372-1378.

Honma, M., 1983, Enzymatic determination of 1-aminocyclopropane-1-carboxylic acid, *Agri. Biol. Biochem.* **47**:617-618.

Honma, M., 1985, Chemically reactive sulfhydryl groups of 1-aminocyclopropane-1-carboxylate deaminase. *Agric. Biol. Chem.* **49**:567-571.

Honma, M. 1993. Stereospecific reaction of 1-aminocyclopropane-1-carboxylate deaminase, in: Cellular and Molecular Aspects of the Plant Hormone Ethylene, J. C. Pech, A. Latche, and C. Balague, eds., Kluwer Academic Publishers, Dordrecht, The Netherlands, pp. 111-116.

Honma, M., and Shimomura, T., 1978, Metabolism of 1-aminocyclopropane-1-carboxylic acid, *Agric. Biol. Chem.* **42**:1825-1831.

Hottiger, T., and Boller, T., 1991, Ethylene biosynthesis in *Fusarium oxysporum* f. sp. *tulipae* proceeds from glutamate/2-oxoglutarate and requires oxygen and ferrous ions *in vivo*, *Arch. Microbiol.* **157**:18-22.

Hou, C. T., Patel, R., Laskin, A. I., Barnabe, N., 1979, Microbial oxidation of gaseous hydrocarbons: Epoxidation of C_2 to C_4 *n*-alkenes by methylotrophic bacteria, *Appl. Environ. Microbiol.* **38**:127-134.

Hyman, M. R., and Wood, P. M., 1984, Ethylene oxidation by *Nitrosomonas europaea*, *Arch. Microbiol.* **137**:155-158.

Hyman, M. R., Murton, I. B., and Arp, D. J., 1988, Interaction of ammonia monooxygenase from *Nitrosomonas europaea* with alkanes, alkenes and alkynes, *Appl. Environ. Microbiol.* **54**:3187-3190.

Ince, J. E., and Knowles, C. J., 1985, Ethylene formation by cultures of *Escherichia coli*, *Arch. Microbiol.* **141**:209-213.

Ince, J. E., and Knowles, C. J., 1986, Ethylene formation by cell-free extracts of *Escherichia coli*, *Arch. Microbiol.* **146**:151-158.

Jacobson, C. B., Pasternak, J. J., and Glick, B. R., 1994, Partial purification and characterization of 1-aminocyclopropane-1-carboxylate deaminase from the plant growth promoting rhizobacterium *Pseudomonas putida* GR12-2, *Can. J. Microbiol.* **40**:1019-1025.

Jacobsen, D. W., and Wang, C. H., 1965, The conversion of acrylic acid to ethylene in *Penicillium digitatum*, *Plant Physiol.* **40**:xix.

Jacobsen, D. W., and Wang, C. H., 1968, The biogenesis of ethylene in *Penicillium digitatum*, *Plant Physiol.* **43**:1959-1966.

Jia, Y.-J., Ito, H., Matsui, H., and Honma, M., 2000, 1-Aminocyclopropane-1-carboxylate (ACC) deaminase induced by ACC synthesized and accumulated in *Penicillium citrinum* intracellular spaces, *Biosci. Biotechnol. Biochem.* **64**:299-305.

Jia, Y.-J., Kakuta, Y., Sugawara, M., Igarashi, T., Oki, N., Kisaki, M., Shoji, T., Kanetuna, Y., Horita, T., Matsui, H., and Honma, M., 1999, Synthesis and degradation of 1-aminocyclopropane-1-carboxylic acid by *Penicillium citrinum*, Biosci. Biotechnol. Biochem. **63**:542-549.

Ketring, D. L., Young, R. E., and Biale, J. B., 1968, Effects of monofluoroacetate on *Penincillium digitatum* metabolism and on ethylene biosynthesis, *Plant Cell Physiol.* **9**:617-631.

Klee, H. J., Hayford, M. B., Kretzmer, K. A., Barry, G. F., and Kishmore, G. M., 1991, Control of ethylene synthesis by expression of a bacterial enzyme in transgenic tomato plants, *Plant Cell* **3**:1187-1193.

Klee, H. J., and Kishmore, G. M., 1992, International Patent Number WO92/12249 (to Monsanto Company), July 23, 1992.

Kutsuki, H., and Gold, M. H., 1982, Generation of hydroxyl radical and its involvement in lignin degradation by *Phanerochaete chrysosporium*, *Biochem. Biophys. Res. Commun.* **109**:320-327.

Kyte, J., and Doolittle, R. F., 1982, A simple method for displaying the hydropathic character of a protein, *J. Mol. Biol.* **157**:105-132.

Lehninger, A. L., 1975, *Biochemistry: The Molecular Basis of Cell Structure and Function*, Worth Publishers, Inc., New York, pp. 1104.

Lifshitz, R., Kloepper, J. W., Scher, F. M., Tipping, E. M., and Laliberte, M., 1986, Nitrogen-fixing pseudomonads isolated from roots of plants grown in the Canadian high arctic, *Appl. Environ. Microbiol.* **51**:251-253.

Lund, B. M., and Mapson, L. W., 1970, Stimulation of *Erwinia carotovora* of the synthesis of ethylene in cauliflower tissue, *Biochem. J.* **119**:251-263.

Lynch, J. M., 1972, Identification of substrates and isolation of microorganisms responsible for ethylene production in soil, *Nature* **240**:45-46.

Lynch, J. M., 1974, Mode of ethylene formation by *Mucor hiemalis*, J. Gen. Microbiol. 83:407-411.

Lynch, J. M., and Harper, S. H. T., 1974, Formation of ethylene by a soil fungus, *J. Gen. Microbiol.* **80**:187-195.

Mahmoudian, M., and Michael, A., 1992. Production of chiral epoxides by an ethene-utilizing *Micrococcus* sp., *J. Indust. Microbiol.* **11**:29-35.

Mansouri, S., and Bunch, A. W., 1989, Bacterial ethylene synthesis from 2-oxo-4-thiobutyric acid and from methionine, *J. Gen. Microbiol.* **135**:2819-2827.

Matsuda, J., Okabe, S., Hashimoto, T., and Yamada, Y., 1991, Molecular cloning of hyoscyamine G beta-hydroxylase, a 2-oxoglutarate-dependent dioxygenase, from cultured roots of *Hyoscyamus niger*, *J. Biol. Chem.* **266**:9460-9464.

McGarvey, D. J., Sirevag, R., and Christoffersen, R. E., 1992, Ripening-related gene from avocado fruit, *Plant Physiol.* **98**:554-559.

Minami, R., Uchiyama, K., Murakami, T., Kawai, J., Mikami, K., Yamada, T., Yokoi, D., Ito, H., Matsui, H., and Honma, M., 1998, Properties, sequence, and synthesis in *Escherichia coli* of 1-aminocyclopropane-1-carboxylate deaminase from *Hansenula saturnus*, *J. Biochem.* **123**:1112-1118.

Nagahama, K., Ogawa, T., Fujii, T., Tazaki, M., Goto, M., and Fukuda, H., 1991a, L-Arginine is essential for the formation *in vitro* of ethylene by an extract of *Pseudomonas syringae*, *J. Gen. Microbiol.* **137**:1641-1646.

Nagahama, K., Ogawa, T., Fujii, T., Tazaki, M., Tanase, S., Morino, Y., and Fukuda, H., 1991b, Purification and properties of an ethylene-forming enzyme from *Pseudomonas syringae* pv. *phaseolicola* PK2, *J. Gen. Microbiol.* **137**:2281-2286.

Nagahama, K., Ogawa, T., Fujii, T., and Fukuda, H., 1992, Classification of ethylene-producing bacteria in terms of biosynthetic pathways to ethylene, *J. Ferment. Bioeng.* **73**:1-5.

Nagahama, K., Yoshino, K., Matsuoka, M., Sato, M., Tanase, S., Ogawa, T., and Fukuda, H., 1994, Ethylene production by strains of the plant pathogenic bacterium *Pseudomonas syringae* depends upon the presence of indigenous plasmids carrying homologous genes for the ethylene-forming enzyme, *Microbiol.* **140**:2309-2313.

Nikitin, D. I., and Arakelyan, R. N., 1979, Utilization of ethylene by soil bacteria, *Biol. Bull. Acad. Sci. USSR* **6**:671-673.

Ogawa, T., Takahashi, M., Fujii, T., Tazaki, M., and Fukuda, H., 1990, The role of NADH:Fe(III) EDTA oxidoreductase in ethylene formation from 2-keto-4-methylthiobutyrate, *J. Ferment. Bioeng.* **69**:287-291.

Ogawa, T., Murakami, H., Yamashita, K., Tazaki, M., Fujii, T., and Fukuda, H., 1992, The stimulatory effect of catalase on the formation *in vitro* of ethylene by an ethylene-forming enzyme purified from *Penicillium digitatum*, *J. Ferment. Bioeng.* **73**:58-60.

Osborne, D. J., Walters, J., Milborrow, B. V., Norville, A., and Stange, L. M., 1996, Evidence for a non-ACC ethylene biosynthesis pathway in lower plants, *Phytochem.* **42**:51-60.

Owens, L. D., Lieberman, M., and Kunishi, A., 1971, Inhibition of ethylene production by rhizobitoxine, *Plant Physiol.* **48**:1-4.

Patel, R. N., Hou, C. T., Laskin, A. I., Felix, A., and Derelanko, P., 1983, Epoxidation of *n*-alkenes by organisms grown on gaseous alkanes, *J. Appl. Biochem.* **5**:121-131.

Pazout, J., Wurst, M., and Vancura, V., 1981, Effect of aeration on ethylene production by soil bacteria and soil samples cultivated in a closed system, *Plant Soil* **62**:431-437.

Penrose, D. M., and Glick, B. R., 1997, Enzymes that regulate ethylene levels -- 1-aminocyclopropane-1-carboxylic acid (ACC) deaminase, ACC synthase and ACC oxidase, *Indian J. Exptl. Biol.* **35**:1-17.

Primrose, S. B., 1976a, Formation of ethylene by *Escherichia coli*, *J. Gen. Microbiol.* **95**:159-165.

Primrose, S. B., 1976b, Ethylene-forming bacteria from soil and water, *J. Gen. Microbiol.* **97**:343-346.

Primrose, S. B., 1977, Evaluation of the role of methional, 2-keto-4-methylthiobutyric acid and peroxidase in ethylene formation by *Escherichia coli*, *J. Gen. Microbiol.* **98**:519-528.

Primrose, S. B., 1979, A review, ethylene and agriculture: the role of microbes, *J. Appl. Bacteriol.* **46**:1-25.

Primrose, S. B., and Dilworth, M. J., 1976, Ethylene production by bacteria, *J. Gen. Microbiol.* **93**:177-181.

Sato, M., Watanabe, K., Yazawa, M., Takikawa, Y., and Nishiyama, K., 1997, Detection of new ethylene-producing bacteria, *Pseudomonas syringae* pvs. *cannabina* and *sesami*, by PCR amplification of genes for the ethylene-forming enzyme, *Phytopathol.* **87**:1192-1196.

Shah, S., Li, J., Moffatt, B. A., and Glick, B. R., 1997, ACC deaminase genes from plant growth-promoting rhizobacteria, in: *Plant Growth-Promoting Rhizobacteria: Present Status and Future Prospects.* A. Ogoshi, K. Kobayashi, Y. Honma, F. Kadama, N. Kondo, and S. Akino, eds., OECD, Paris, France, pp. 320-324.

Shah, S., Li, J., Moffatt, B. A., and Glick, B. R., 1998, Isolation and characterization of ACC deaminase genes from two different plant growth-promoting rhizobacteria, *Can. J. Microbiol.* **44**:833-843.

Sheehy, R. E., Honma, M., Yamada, M., Sasaki, T., Martineau, B., and Hiatt, W. R., 1991, Isolation, sequence and expression in *Escherichia coli* of the *Pseudomonas* sp. strain ACP gene encoding 1-aminocyclopropane-1-carboxylate deaminase, *J. Bacteriol.* **173**:5260-5265.

Shipston, N., and Bunch, A. W., 1989, The physiology of L-methioinine catabolism to the secondary metabolite ethylene by *Escherichia coli*, *J. Gen. Microbiol.* **135**:1489-1497.

Sprayberry, B. A., Hall, W. C., and Miller, C. S., 1965. Biogenesis of ethylene in *Pennicillium digitatum*. *Nature* **208**:1322-1323.

Swart, A. and G. A. Kamerbeek. 1977. Ethylene production and mycelium growth of the tulip strain of *Fusarium oxysporum* as influenced by shaking of and oxygen supply to the culture medium. *Physiol. Plant.* **39**:38-44.

Tauber, A. I., and Babior, B. M., 1977, Evidence for hydroxyl radical production by human neutrophils, *J. Clin. Invest.* **60**:374-379.

Thomas, K. C., and Spencer, M., 1977, L-Methionine as an ethylene precursor in *Saccharomyces cerevisiae*, *Can. J. Microbiol.* **23**:1669-1674.

van Ginkel, C. G.,Welten, H. G. J., and de Bont, J. A. M., 1986, Epoxidation of alkenes by alkene-grown *Xanthobacter* spp., *Appl. Microbiol. Biotech.* **24**:334-337.

van Ginkel, C. G., Welten, H. G. J., and de Bont, J. A. M., 1987, Oxidation of gaseous and volatile hydrocarbons by selected alkene-utilizing bacteria, *Appl. Environ. Microbiol.* **53**:2903-2907.

Walsh, C., Pascal, R. A., Jr., Johnston, M., Raines, R., Dikshit, D., Krantz, A., and Honma, M., 1981, Mechanistic studies on the pyridoxal phosphate enzyme 1-aminocyclopropane-1-carboxylate deaminase from *Pseudomonas* sp., *Biochemistry* **20**:7509-7519.

Wang, C., Knill, E., Czlick, B. R., and Defago, G., 2000, Effect of transferring 1-aminocyclopropane-1-carboxylic acid (ACC) deaminase genes with *Pseudomonas fluorescens* strain CHA0 and its gacA derivative CHA96 on their growth-promoting and disease-suppressive capacities, *Can. J. Microbiol.* **46**:898-907.

Wang, C. H., Persyn, A., and Krackov, J., 1962, Role of the Krebs cycle in ethylene biosynthesis, *Nature* **195**:1306-1308.

Watanabe, K., Nagahama, K., and Sato, M., 1998, A conjugative plasmid carrying the *efe* gene for the ethylene-forming enzyme isolated from *Pseudomonas syringae* pv. *glycinea*, *Phytopathol.* **88**:1205-1209.

Weijers, C. A. G., van Ginkel, C. G., and de Bont, J. A. M., 1988, Enantiomeric composition of lower epoxyalkanes produced by methane-, alkane-, and alkene-utilizing bacteria, *Enzyme Microb. Technol.* **10**:214-218.

Wild, J., Walczak, W., Krajewska-Grynkiewicz, K., and Klopotowski, T., 1974, D-Amino acid dehydrogenase: The enzyme of the first step of D-histidine and D-methionine racemization in *Salmonella tryphimurium*, *Mol. Gen. Genet.* **128**:131-146.

Wilkes, J., Dale, G. T., and Old, K. M., 1989, Production of ethylene by *Endothia gyrosa* and *Cytospora eucalypticola* and its possible relationship to kino vein formation in *Eucalyptus maculata*, *Physiol. Molec. Plant Pathol.* **34**:171-180.

4

FACTORS AFFECTING MICROBIAL
PRODUCTION OF ETHYLENE

4.1. MICROBIAL PRODUCTION

Since the publications of Biale (1940) and Miller et al. (1940) on C_2H_4 production by *Penicillium digitatum*, several workers have screened numerous microorganisms for their ability to produce C_2H_4. These studies revealed that a diverse group of soil microbiota, including pathogens, are very active in producing C_2H_4. According to Primrose (1979), C_2H_4 is synthesized by many species of bacteria and fungi, but is oxidized by only a limited number of microorganisms. Out of 166 strains of fungi, yeast, bacteria, and actinomycetes, 49 produced C_2H_4 (Fukuda et al., 1984). Among them 62% were molds, 20% were yeasts, 21% were bacteria, and 6% were actinomycetes. Out of 228 soil fungi tested by Ilag and Curtis (1968), 58 (25%) produced C_2H_4, whereas 20 unidentified actinomycetes also produced C_2H_4. El-Sharouny (1984) reported that 31% of 80 fungal species isolated from diseased roots were capable of producing C_2H_4. Arshad and Frankenberger (1989) found that corn rhizosphere is quite rich with microbiota capable of producing C_2H_4 derived from L-methionine (MET), and among these, fungi were the most preponderant. Babiker and Pepper (1984) isolated 14 soil fungi that produced C_2H_4 in the presence of MET. Recently, Arshad and Akhter (unpublished data) screened rhizosphere microflora of maize, wheat, tomato and potato and found that fungal isolates predominently produced C_2H_4 from exogenous supplied L-MET or 2-oxo-4-methylthiobutyric acid (KMBA).

After examining 497 bacterial isolates obtained from soil and water, Primrose (1976b) concluded that C_2H_4 –producing bacteria are ubiquitous in nature. He reported that 71 bacterial strains isolated from soil produced C_2H_4 when MET was included in the medium (Primrose, 1976b). Billington et al. (1979) and Swanson et al. (1979) also reported that strains belonging to many bacteria have the ability to produce C_2H_4. Nagahama et al. (1992) screened 757 bacterial isolates and found that 229 were capable of producing C_2H_4. Recently, Weingart and Völksch (1997) screened 20 pathovars *of Pseudomonas syringae* (total 117 strains) and reported that three pathovars including *Ps. syringae* pv. *glycinea*, pv. *phaseolicola* (kudzu) and pv. *pisi* produced C_2H_4 in synthetic media. Similarly, Völksch and Weingart (1997) screened various strains belonging to

P.s. syringae pv. *glycinea* (isolated from soybean) and *Ps. syringae* pv. *phaseolicola* (isolated from kudzu and bean) and found that all 12 strains of pv. *glycinea* and all 10 kudzu strains of pv. *phaseolicola* produced C_2H_4, whereas none of the 8 strains of pv. *phaseolicola* isolated from bean produced C_2H_4. In another study, they observed that out of 75 strains of 21 various *Ps. syringae* pathovars, all five strains belonging to *Ps. syringae* pv. *cannabina* and 10 strains of *Ps. syringae* pv. *glycinea* produced C_2H_4 (Völksch and Weingart, 1998). Fujii et al. (1985) screened 296 isolates belonging to the genus *Mycobacterium* and reported that several isolates were excellent C_2H_4 producers. A comprehensive list of C_2H_4 –producing microorganisms is given in Table 4.1.

It is now well documented that C_2H_4 is produced by many soil-inhabiting microorganisms in the presence of different substrates. It should be emphasized that the majority of these microorganisms produce C_2H_4 from MET or keto-glutaric acid (KGA). Out of 229 C_2H_4 producing bacteria, 225 utilized MET while only one utilized KGA for C_2H_4 production (Nagahama et al., 1992). Three chemolithotroph strains had C_2H_4 forming capacity, while *Thiobacillus novellus* IFO12443 had a novel C_2H_4-forming system that was dependent upon addition of meat extract to the culture medium. Arshad and Akhter (unpublished data) found that all 24 fungal isolates of tomato, potato, wheat and maize rhizosphere produced C_2H_4 when the culture media was supplemented with MET or KMBA, but none produced C_2H_4 from KGA.

Although numerous isolates of fungi, bacteria and actinomycetes have been reported capable of producing C_2H_4, they vary greatly in the quantity of C_2H_4 produced. It is very likely that the ability of individual microbial isolates to produce C_2H_4 is regulated by many physicochemical and environmental factors in addition to its genetic makeup. Weingart et al. (1999) reported that cell-free extracts prepared from strains of *Ps. syringae* pv. *pisi* and *Restonia solanacerum* produced C_2H_4 at rates 20-1000 fold lower than strains of *Ps. syringae* pvs. *cannabina, glycinea, phaseolicola* and *sesami* (Table 4.2) when the medium was supplemented by KGA. However, in absence of KGA, all tested strains produced C_2H_4 at similar rates (Weingart et al., 1999). Weingart and Völksch (1997) reported variation in C_2H_4 production rates ranging from 0-100 (10^{-8} nL h^{-1} $cell^{-1}$) by various *Pseudomonas* pathovars (Table 4.3). Tzeng and DeVay (1984) also observed great differences in C_2H_4 production rates among *Fusarum oxysporum* f. sp. *vasifectum, Colletobichia dematinum* var. *truncatum* and various *Verticillum* species grown in MET amended media. Wilkes et al. (1989) reported that various strains of *Endothia gyrosa* strongly differed in their ability to produce MET-dependent C_2H_4. Likewise, Arshad and Frankenberger (1989) noted much variation in MET-dependent C_2H_4 production potential of various fungi isolated from maize rhizosphere. Babiker and Pepper (1984) reported MET-derived C_2H_4 production by 14 soil fungi in concentrations ranging from 41-833 pmol g^{-1} 48 h^{-1}. Similarly, great differences in C_2H_4 production potential of 166 strains of fungi, yeast, bacteria, and actinomycetes were observed by Fukuda et al. (1984).

4.2. FACTORS AFFECTING MICROBIAL PRODUCTION OF ETHYLENE

Factors such as the presence of substrate(s), carbon source(s), nitrogen and phosphorus amendments, pH of medium, aeration, temperature of incubation, and presence of trace elements affect C_2H_4 formation by various microbial cultures. These factors will be discussed in detail in the following section.

Table 4.1. Ethylene-evolving microorganisms.

Species	Reference
A. BACTERIA	
Acinetobacter calcoaceticus	Billington et al. (1979)
Aeromonas spp.	Billington et al. (1979)
Agrobacterium rhizogenes	Swanson et al. (1979)
A. tumefaciens	Swanson et al. (1979)
Arthrobacter spp.	Primrose (1976b)
Azospirillum spp.	Strzelczyk et al. (1994a)
Bacillus licheniformis	Fukuda et al. (1989a)
B. subtilis	Mansouri and Bunch (1989)
B. mycoides	Billington et al. (1979)
Brevibacterium linens	Mansouri and Bunch (1989)
Chromobacterium violaceum	Mansouri and Bunch (1989)
Citrobacter spp.	Primrose (1976b)
Clostridium butyricum	Pazout et al. (1981)
Corynebacterium spp.	Billington et al. (1979)
C. aquaticum	Fukuda et al., 1984; 1989a)
C. paurometabolum	Fukuda et al. (1989a)
Coryneform bacteria	Billington et al. (1979)
Enterobacter aerogenes	Mansouri and Bunch (1989)
E. cloacae	Primrose (1976b)
Erwinia amylovora	Swanson et al. (1979)
E. carotovora	Swanson et al. (1979)
E. herbicola	Primrose (1976b)
Escherichia coli	Primrose (1976b); Swanson et al. (1979); Shipston and Bunch (1989); Mansouri and Bunch (1989)
Klebsiella ozaenae	Primrose (1976b)
K. pneumoniae oxytoca	Primrose (1976b)
K. pneumoniae	Primrose (1976b)
Micrococcus luteus	Mansouri and Bunch (1989)
Pseudomonas spp.	Primrose (1976b)
P. aeruginosa	Mansouri and Bunch (1989)
P. fluorescens	Pazout et al. (1981); Swanson et al. (1979)
P. fluorescens var. *reptilivoria*	Swanson et al. (1979)
P. syringae	Sato et al. (1997); Swanson et al. (1979)
P. syringae pv. *cannabina*	Sato et al. (1997)
P. syringae pv. *glycinea*	Sato et al. (1987)
P. indigofera	Primrose (1976b)
P. marginata	Swanson et al. (1979)
P. syringae pv. *mori*	Sato et al. (1987)
P. syringae pv. *sesami*	Sato et al. (1997)
P. pisi	Swanson et al. (1979); Billington et al. (1979)
P. polycolor	Swanson et al. (1979)
P. putida	Pazout et al. (1981); Fukuda et al. (1989a); Freebairn and Buddenhagen (1964); Sato et al. (1987)
P. solanacearum	Bonn et al. (1975); Swanson et al. (1979); Freebairn and Buddenhagen (1964); Sato et al. (1987)
P. syringae pv. *phaseolicola*	Goto et al. (1985); Goto and Hyodo (1987); Nagahama et al. (1991a,b)
P. tabaci	Swanson et al. (1979)
Rhizobium trifolii	Billington et al. (1979)

Table 4.1. (continued)

Serratia liquefaciens	Primrose (1976b)
S. marcescens	Mansouri and Bunch (1989)
Staphylococcus aureus	Mansouri and Bunch (1989)
Thiobacillus ferrooxidans	Nagahama et al. (1992)
T. novellus	Nagahama et al. (1992)
Xanthomonas campestris	Mansouri and Bunch (1989)
X. citri	Goto et al. (1980)
X. phaseoli	Swanson et al. (1979)
X. phaseoli var. sojense	Swanson et al. (1979)
X. vesicatoria	Swanson et al. (1979)

B. ACTINOMYCETES

Streptomyces spp.	Dasilva et al. (1974); Ilag and Curtis (1968)

C. FUNGI

Acrasis rosea	Amagi and Maeda (1992)
Acremonium falciforme	Arshad and Frankenberger (1989)
Agaricus bisporus	Lockard and Kneebone (1962); Wood and Hammond (1977); Ward et al. (1978)
Alternaria alternata	El-Sharouny (1984)
A. solani	Ilag and Curtis (1968)
Ascochyta imperfecta	Ilag and Curtis (1968)
Aspergillus spp.	Dasilva et al. (1974); Babiker and Pepper (1984)
A. candidus	Ilag and Curtis (1968)
A. clavatus	Ilag and Curtis (1968); Fukuda et al. (1989a)
A. flavus	Ilag and Curtis (1968)
A. niger	El-Sharouny (1984)
A. ustus	El-Sharouny (1984); Ilag and Curtis (1968)
A. variecolor	Ilag and Curtis (1968)
Blastomyces dermatitidus	Nickerson (1948)
Botrytis spectabilis	Ilag and Curtis (1968)
Cacarisporium parasiticum	Hanke and Dollwet (1976)
Candida vartiovaarai	Arshad and Frankenberger (1989); Babiker andn Pepper (1984)
Cenococcum geophilum	Graham andn Linderman (1980)
Cephalosporium gramineum	Ilag and Curtis (1968)
Ceratobasidium cornigerum	Hanke and Dollwet (1976)
Ceratocystis fimbriata	Chalutz and DeVay (1969)
Chaetomium globosum	Hanke and Dollwet (1976); Fukuda and Ogawa (1991)
C. chlamaloides	Ilag and Curtis (1968)
Colletotrichum musae	Peacock and Muirhead (1974)
Coriolus hirsutus	Tanaka et al. (1986)
Coriolus versicolor	Tanaka et al. (1986)
Cryptococcus albidus	Fukuda et al. (1989a)
C. laurentii	Fukuda et al. (1989a); Fukuda (1984)
Cunninghamellia elegans	Fukuda and Ogawa (1991)
Curvularia spp.	Babiker and Pepper (1984)
C. lunata	El-Sharouny (1984)
C. damatium var. trancatum	Tzeng and DeVay (1984)
Cylindrocarpon spp.	El-sharouny (1984)
Cylindrocladium floridanum	Axelrood-McCarthy and Linderman (1981)
C. scoparium	Axelrood-McCarthy and Linderman (1981)
Cytospora eucalypticola	Wilkes et al. (1989)
Daedalea dickinsii	Tanaka et al. (1986)
Debaryomyces hansenii	Fukuda et al. (1989a)

Table 4.1. (continued)

Dematium pullulans	Ilag and Curtis (1968)
Dictyostelium discoideum	Bonner (1973); Amagai and Maeda (1992)
D. mucoroides	Amagai and Maeda (1992); Amagai (1984)
D. purpureum	Amagai and Maeda (1992)
Endothia gyrosa	Wilkes et al. (1989)
Fistulina hepatica	Tanaka et al. (1986)
Flammulina veltipes	Tanaka et al. (1986)
Fomitopsis pimicola	Tanaka et al. (1986)
Fusarium spp.	Babiker and Pepper (1984)
F. avenaceum	Swart and Kamerbeek (1976)
F. camptoceras	Swart and Kamerbeek (1976)
F. culmorum var. *cereale*	Swart and Kamerbeek (1976)
F. equiseti	El-Sharouny (1984)
F. merismoides var. *crassum*	Swart and Kamerbeek (1976)
F. moniliforme	El-Sharouny (1984); Swart and Kamerbeek (1976)
F. oxysporum	El-Sharouny (1984); Swart and Kamerbeek (1976)
F. oxysporum f. *lycopersici*	Dimond and Waggoner (1953); Swart and Kamerbeek (1976)
F. oxysporum f. *narcissi*	Swart and Kamerbeek (1976)
F. oxysporum f. *tulipae*	Swart and Kamerbeek (1976, 1977); Hottiger and Boller (1991)
F. oxysporum f. sp. *vasinfectum*	Tzeng and DeVay (1984)
F. poae	Swart andn Kamerbeek (1976, 1977)
F. redolens dianthi	Swart and Kamerbeek (1976, 1977)
F. solani	El-Sharouny (1984)
F. solani minus	Swart and Kamerbeek (1976, 1977)
F. sporotrichioides	Swart and Kamerbeek (1976, 1977)
Geotrichum spp.	Babiker and Pepper (1984)
G. candidum	Kapulnik et al. (1983)
Gloeophyllum trabeum	Chandhoke et al. (1992)
Geophillum saepiarum	Tanaka et al. (1986)
Graphium sp.	Hanke and Dollwet (1976)
Hansenula subpelliculosa	Ilag and Curtis (1968)
Hebeloma crustuliniforme	Graham and Linderman (1980)
Hirshiopirus abietinus	Tanaka et al. (1986)
Histoplasma capsulatum	Nickerson (1948)
Humicola grisea	El-Sharouny (1984)
Hymenochaete tabacina	Tanaka et al. (1986)
Laccaria laccata	Graham and Linderman (1980)
Laetiporus sulphureus	Tanaka et al. (1986)
Lenzites betulina	Tanaka et al. (1986)
Macrophomina phaseoli	El-Sharouny (1984)
Monilia geophila	Fukuda et al. (1989a)
Mortierella spp.	Babiker and Pepper (1984)
Mucor spp.	Dasilva et al. (1974)
M. genevensis	Hanke and Dollwet (1976)
M. hiemalis	Lynch (1972); Arshad and Frankenberger (1989); Babiker and Pepper (1984); Hanke and Dollwet (1976)
M. silvaticus	Lindberg et al. (1979)
Mycelium radicis atrovirens	Strzelczyk et al. (1989)
Mycoplana dimorpha	Fukuda (1984)
Mycotypha microspora	Hanke and Dollwet (1976)
Myrothecium roridum	Ilag and Curtis (1968)
Nectria flammea	Fukuda et al. (1989a); Fukuda (1984)

Table 4.1. (continued)

Neurospora crassa	Ilag and Curtis (1968)
Paecilomyces elegans	Fukuda et al. (1989a)
Penicillium spp.	Babiker and Pepper (1984);
	Considine and Patching (1975)
P. corylophilum	Ilag and Curtis (1968)
P. crustosum	Considine et al. (1977); El-Sharouny (1984)
P. cyclopium	Considine et al. (1977); El-Sharouny (1984);
	Fukuda and Ogawa (1991)
P. digitatum	Biale (1940); Fukuda et al. (1984); and others
P. expansum	Kapulnik et al. (1983)
P. italicum	Kapulnik et al. (1983)
P. luteum	Ilag and Curtis (1968)
P. notatum	Ilag and Curtis (1968)
P. patulum	Ilag and Curtis (1968)
P. steckii	El-Sharouny (1984)
P. urticae	Fukuda et al. (1989a); Fukuda (1984)
Pestilotium sp.	Hanke and Dollwet (1976)
Phanerochaete chrysosporium	Tanaka et al. (1986)
Phaeolus schweinitzii	Tanaka et al. (1986)
Phoma herbarum	El-Sharouny (1984)
Photiota adiposa	Tanaka et al. (1986)
Phycomyces blakesleeanus	Russo et al. (1977)
Phycoporus coccineus	Tanaka et al. (1986)
P. nitens	Fukuda and Ogawa (1991)
Phymatotrichum omnivorum	Hill and Lyda (1976)
Picea abies	Blaschke (1977)
Polysphondylium pallidum	Amagi and Maeda (1992)
P. violaceum	Amagi and Maeda (1992)
Pythium butleri	El-Sharouny (1984)
P. ultimum	El-Sharouny (1984)
Rhizoctonia solani	El-Sharouny (1984)
Rhizopus spp.	Babiker and Pepper (1984)
R. japonicus	Fukuda et al. (1989a)
R. nigricans	Babiker and Pepper (1984)
R. stolonifer	El-Sharouny (1984)
Rhodotorula spp.	Babiker and Pepper (1984)
Saccharomyces cerevisieae	Thomas and Spencer (1977)
Schizopyllum commune	Ilag and Curtis (1968)
Schizosaccharomyces octosporus	Fukuda et al. (1989a); Fukuda (1984)
Sclerotium spp.	El-Sharouny (1984)
Sclerotinia laxa	Ilag and Curtis (1968)
Scopulariopsis brevicaulis	Ilag and Curtis (1968);
	Babiker and Pepper (1984)
S. candida	El-Sharouny (1984)
Septoria musiva	Brown-Skrobot et al. (1985)
Sordaria spp.	Hanke and Dollwet (1976)
Syncephalastrum racemosum	Fukuda (1984)
Thamnidium spp.	Hanke and Dollwet (1976)
T. elegans	Ilag and Curtis (1968)
Thielavia alata	Ilag and Curtis (1968)
Trichoderma spp.	Babiker and Pepper (1984)
T. viride	Strzelczyk et al. (1989)
Tulasnella calospora	Hanke and Dollwet (1976)
Tyromyces palustris	Tanaka et al. (1986)
Ulocladium atrum	El-Sharouny (1984)
Verticillium spp.	El-Sharouny (1984)

Table 4.1. (continued)

D. ALGAE

Acetabulans mediterranea	Kevers et al. (1989)
Chorella	Kreslavskii et al. (1988)
Codium latum	Watanabe and Kondo (1976)
Padina arborescens	Watanabe and Kondo (1976)
Porphyra tenera	Watanabe and Kondo (1976)

Table 4.2. Rate of C_2H_4 production by cell-free extracts from strains of *Pseudomonas syringae* pathovars and *Ralstonia solanacearum*.

Strains	Ethylene production (nl/h/mg of protein)	
	With 2-oxoglutarate	Without 2-oxoglutarate
P. syringae pv.		
cannabina GSPB2553	369.2 ± 32.3	0.4 ± 0.1
glycinea 7a/90	472.0 ± 60.1	0.7 ± 0.2
phaseolicola GSPB669	$1{,}165.7 \pm 95.9$	1.3 ± 0.4
sesami 962	497.3 ± 40.7	1.1 ± 0.7
pisi GSPB1206	26.2 ± 3.1	0.2 ± 0.1
pisi GSPB104	1.3 ± 0.4	0.7 ± 0.3
R. solanacearum K60	1.1 ± 0.7	0.4 ± 0.1

Source: Weingart et al. (1999).

Table 4.3. *Pseudomonas syringae* pathovars tested for their ability to produce ethylene.

Pathovar	No. of tested strains	Ethylene production (10^{-8} nl h^{-1}cell^{-1})
aptata	4	0
atrofaciens	3	0
atropurpurea	4	0
cannabina	1	0
coronafaciens	3	0
glycinea	50	5-100
lachrymans	3	0
maculicola	3	0
mori	1	0
morsprunorum	4	0
phaseolicola (from kudzu)	4	30-70
phaseolicola (from bean)	8	0
persicae	1	0
pisi	5	1-3
pisi	1	0
primulae	1	0
savastanoi	3	0
striafaciens	1	0
syringae	5	0
tabaci	4	0
tagetis	1	0
tomato	7	0

Source: Weingart and Volksch (1997).

4.2.1. Substrates/Carbon Sources

Several compounds have been examined by various workers as possible substrate(s) or stimulators of C_2H_4 biosynthesis by microbial cultures. As discussed in Chapter 3, many compounds can act as precursors for microbial biosynthesis of C_2H_4. Among these, MET, with or without glucose, is the most frequently used precursor and is considered as the most favorable substrate for microbial biosynthesis of C_2H_4. Glucose may or may not act as an C_2H_4 precursor. Interestingly, all three strains of a well-known N_2-fixer, *Azospirillum*, isolated from the rhizoplane of different hosts produced C_2H_4 when the media was supplemented with MET or fumarate, but the amount of C_2H_4 produced was less in the latter case (Strzelczyk et al., 1994a). In some cases, the Krebs cycle intermediates, particularly KGA is believed to be the precursor of C_2H_4 formation by some microbial isolates. The spiking of culture media with a variety of compounds including amino acids, organic acids, carbohydrates, proteins and vitamins has been shown to stimulate C_2H_4 synthesis, but it is highly likely that not all of these compounds serve as the true precursors of C_2H_4. These compounds might increase C_2H_4 production by enhancing microbial growth or by serving as co-factors (Tables 4.4 and 4.5).

Kutsuki and Gold (1982) found that the assay of C_2H_4 production is suitable for monitoring the ligninolytic activity of *Phanerochaete chrysosporium* strains because there is a positive correlation between C_2H_4 production from KMBA, and production of $^{14}CO_2$ by these ligninolytic cultures. Tanaka et al. (1986) also found a good correlation between production of C_2H_4 from KMBA and degradation of dimeric model-lignin compounds by wood-dwelling fungi. Fukuda and Ogawa (1991) examined the cellulase activity of *P. digitatum* IFO9372 and investigated production of C_2H_4 from homogenates of recycled paper and Unshiu fruit peel. Both carbon sources were utilized by the fungus for C_2H_4 production. Fukuda et al. (1988) attempted to control the efficiency of *P. digitatum* IFO 9372 in terms of amount of carbon consumed for growth and respiration by developing its mutants that formed smaller colonies than the parent on agar plates. Forty-six growth-suppressed mutants, which formed smaller colonies on agar plates, were isolated from the parent strain. *Penicillium digitatum* IFO 9372-1040-S-6 (Met⁻) was the greatest producer of C_2H_4 among the mutant strains tested. Fukuda et al. (1988) compared the carbon-recovery efficiency for the parent *P. digitatum* IFO 9372 and two mutant strains, 1040 (Met⁻) and S-6 (Met⁻) under different culture conditions. The rate of conversion of carbon in glucose to carbon in C_2H_4 by S-6 (Met⁻) was four-fold and 2.3-fold higher than that by IFO 9372 and 1040 (Met⁻), respectively (Fukuda et al., 1988). This may imply that very efficient C_2H_4 producing strains with respect to carbon recovery can be constructed by using genetic and molecular approaches.

4.2.2. Nitrogen and Phosphorus

There are few reports on the effects of added nitrogen (N) on C_2H_4 biosynthesis by microbial isolates. Ince and Knowles (1985) examined the effect of ammonium chloride (NH_4Cl) on MET-dependent C_2H_4 formation by *Escherichia coli*. They reported that an excess of ammonium (10 μM NH_4Cl) led to early growth termination caused by glucose depletion and residual NH_4Cl accumulation in the stationary phase, repressed C_2H_4 formation. Similarly, Arshad and Frankenberger (1989) found a negative relationship between N added as ammonium nitrate and MET-derived C_2H_4 biosynthesis by *Acremonium falciforme*. It is highly likely that the negative effect of N on C_2H_4 biosynthesis by microbial isolates may be related to the preference of the isolate utilizing N from the added source, rather than N of the MET precursor.

Addition of NH_4Cl and NH_4HCO_3 to a cell-free C_2H_4 –forming extract obtained from *Ps. syringae* pv. *phaseolicola* substantially inhibited C_2H_4 production from glutamate, but not from KGA (Table 4.6). The inhibitory effects of these salts were found in proportion to their concentrations in the range of 0-50 m*M* (Goto and Hyodo, 1987). Mansouri and Bunch (1989) studied the ability of *E. coli* to release C_2H_4 from MET and KMBA as affected by the nature and amount of C and N source used for growth by the bacterium. When glutamate was used as the sole source of C and N for growth, the activity of the C_2H_4 –forming system was reduced to 25% of that observed with cultures grown with glucose and NH_4Cl as C and N sources, respectively. Addition of glutamate (as a C and N source) was not effective in promoting MET-derived C_2H_4 by this bacterium. However, the combined application of glutamate (as a N source) and glucose (as a C source) resulted in enhanced production of C_2H_4 from KMBA as well as from MET.

Table 4.4. Effect of substrates and carbon sources on ethylene production by bacterial cultures.

Microorganism	Substrate or carbon source	Comments	Reference
Azospirillum (3 strains)	Malate, succinate, pyruvate, fumarate (all with or without MET)	All strains produced C_2H_4 in media with MET and malate, succinate, and pyruvate. Only one strain produced C_2H_4 with fumarate and MET. All strains also produced C_2H_4 in media without MET and in the presence of pyruvate, but the amount of C_2H_4 was smaller than the media with MET.	Strzelczyk et al. (1994a)
Eschericia coli	MET, serine, fumarate, pyruvate, acetate, succinate, malate	Only MET added resulted in C_2H_4 production and the final yield of C_2H_4 was proportional to the amount of MET added to the medium.	Primrose and Dilworth (1976)
E. coli	MET, glucose, KMBA, α-oxo acids	Methionine was converted into KMBA, which served as an immediate precursor of C_2H_4. Glucose stimulated KMBA-derived C_2H_4. α-Oxo acids enhanced the conversion of MET into KMBA.	Ince and Knowles (1986)
E. coli	MET, ethionine, cysteine, KMBA, homocysteine, N-formyl MET, MET hydroxy analogue, MET methyl ester, norleucine, KGA, succinate, serine, aminobutyric acid	Cultures in the absence of MET failed to produce detectable amounts of C_2H_4. L-Cysteine, L-homocysteine, MET derivatives, and the sulfur-containing analogues of L-MET acted as precursors of C_2H_4.	Primrose (1976a)
E. coli	L-MET, glucose, KMBA, L-tyrosine, L-alanine, L-aspartate, L-glutamate, L-glutamine, L-isoleucine, L-leucine, L-valine	L-MET was transaminated to KMBA, which was then converted to C_2H_4. Cultures without MET did not produce C_2H_4. Among the amino acids tested, L-tyrosine reduced C_2H_4 yield in batch cultures amended with L-MET. KMBA was more effective than L-MET when the C/N ratio of the media was 20.	Shipston and Bunch (1989)

Table 4.4. (continued)

Organism	Substrates	Comments	Reference
Pseudomonas putida and *Ps. fluorescens*	Glucose, glucose plus MET	C_2H_4 production by *P. putida* in a medium containing glucose alone was several-fold greater than in a medium amended with glucose plus MET under a condition of delayed aeration. *Ps. fluorescens* produced more C_2H_4 in the presence of glucose plus MET than with glucose alone.	Pazout et al. (1981)
Pseudomonas solanacearum	Glucose, peptone, glutamic acid, fumaric acid, gluconic acid	C_2H_4 was not produced from a medium containing glucose alone as a carbon source, but C_2H_4 continued to be released at a slow rate in flasks that contained other carbon sources.	Freebairn and Buddenhagen (1964)
Ps. syringae pv. *phaseolicola*	KGA	The cell-free extract catalyzed the formation of C_2H_4 from KGA. EFE has a high specificity for the substrate (KGA) and cofactor (L-arginine). Presence of Fe(II) and O_2 is also essential for this enzymatic reaction.	Nagaham et al. (1991b)
Ps. syringae pv. *phaseolicola*	KMBA, KGA, glutamate, histidine, histamine, urocanic acid, imidazol, succinate, malate, citrate, fumarate, pyruvate, oxalacetate	A cell-free EFE system of this bacterium produced C_2H_4 most effectively from KGA (0.5 mM) followed by glutamate and then histidine (both 5-10 mM). Presence of organic acids or carbonates inhibited KGA-dependent C_2H_4 formation by the enzyme system. KMBA-derived C_2H_4 was largely nonenzymatic and required DTT and $FeSO_4$.	Goto and Hyodo (1987)
Ps. syringae pv. *phaseolicola*	KGA, glycine, L-alanine, L-valine, L-leucine, L-isoleucine, L-phenyl-alanine, L-proline, L-serine, L-threonine, L-cysteine,	The complete reaction mixture of the cell-free EFE system required for the formation of C_2H_4 under aerobic conditions (*in vitro*) consisted of 0.25 mM KGA, 0.2 mM $FeSO_4$, 2 mM DTT, 10 mM L-histidine and 0.2 mM L-arginine. The system was dependent on co-factors (L-arginine and L-histidine).	Nagahama et al. (1991a)

Table 4.4. (continued)

	L-MET, L-tryptophan, L-tyrosine, L-asparagine, L-glutamine, L-aspartic acid, L-glutamic acid, L-lysine, L-arginine, D-arginine, L-histidine, casamino acids, imidazol, L-canavanine		
Ps. syringae pv. *glycinea* KN38	Glutamate, aspargine, aspartate, arginine, proline, alanine, glucose, sucrose, raffinose, inositol, trigonalline, tryptophan, tyrosine, isoleucine, xylose, fructose, galactose, MET, ACC, homoserine, phenyl-alanine, cysteine, valine, threonine, mannose, glycerol, malonate, citrate, fumarate	The bacterium utilized glutamate, aspargine, aspartate, arginine, proline and alanine efficiently for C_2H_4 production while glucose, sucrose, raffinose, inositol and trigonalline also proved to be good substrates. Tryptophan, tyrosine, isoleucine, xylose, fructose and galactose were less effective for stimulating C_2H_4 production by the bacterium. All other compounds were not utilized by the bacterium to produce C_2H_4.	Sato et al. (1987)
Ps. syringae pv. *phaseolicola* (Kudzu strain)	Glutamate, glutamine, aspartate, asparagine, tryptophan, serine, pyroline, phenylalanine, tyrosine, isoleucine, cysteine, glycine, threonine, MET, valine, ornithine, histidine, lysine, succinate, fumarate, malate, KGA, α-aminobutyric acid, citrate, malonate, ascorbate, pyruvate, oxalate, KMBA, tartarate, acetate, glycerate, oxalacetate, glucose, fructose, sucrose, mannose, arabinose, xylose, glycerol	The presence of living cells was essential to produce C_2H_4 from amino acids, such as glutamate, aspartate, and their amides. Glucose and succinate were also good substrates, but significantly less effective than glutamate. MET had no effect.	Goto et al. (1985)
Xanthomonas citri	MET- or cystein-glucose, peptone-glucose, citrus leaf extract	The bacterium produced C_2H_4 during its early stage of growth, i.e., 3 to 6 hours, and production ceased after 9 hours of inoculation in all the culture media used.	Goto et al. (1980)
11 bacteria	Glucose, MET	All bacterial isolates produced C_2H_4 only when MET was present in the growth medium, suggesting that it is a precursor for C_2H_4.	Primrose (1976b)
20 pathogenic bacteria	MET, glucose	MET (up to 10^{-3} M) added to a basal salt medium enhanced C_2H_4 production in 14 of 20 bacteria tested. However, *Pseudomonas pisi* and *Ps. solanacearum* produced high amounts of C_2H_4 in the absence of MET.	Swanson et al. (1979)

Table 4.5. Effect of substrates and carbon sources on ethylene production by fungal cultures.

Microorganism	Substrate or carbon source	Comments	Reference
Bipolaris sorokiniana	MET^+ leaf blade infusion media of host	C_2H_4 production occurred only in the presence of leaf blade infusion media supplemented with MET. ACC was detected in the culture filtrate. Exogenous addition of ACC was less effective than MET in promoting C_2H_4 production by the fungus.	Coleman and Hodges (1986)
Fusarium oxysporum f. sp. *tulipae*	Glutamate, arginine, proline, ornithine, glucose, KGA, glutamine	Studies with labeled compounds revealed that C-3 and C-4 of glutamate or KGA were incorporated into C_2H_4. Addition of arginine, ornithine, or proline to the stationary phase of the culture increased C_2H_4 production. The enzyme system converting KGA into C_2H_4 in the reaction was dependent on O_2, Fe(II) and arginine.	Hottiger and Boller (1991)
Gloeophyllum trabeum	KMBA	The extent KMBA oxidation into C_2H_4 was influenced by the concentration of chelators and by iron and manganese.	Chandhoke et al. (1992)
Mucor spp.	Glucose, acetate, fructose, glucose plus MET	Glucose plus MET resulted in a 5-fold greater amount of C_2H_4 produced compared with the control.	Dasilva et al. (1974)
M. hiemalis	MET, glucose	MET served as a substrate and glucose as an energy source.	Lynch and Harper (1974)
M. hiemalis, Aspergillus clavatus and *M. racemosus*	Glucose plus MET	Addition of glucose plus MET to the growth medium was required for C_2H_4 biosynthesis. *M. hiemalis* was the most efficient producer.	Lynch (1972)

Table 4.5. (continued)

Penicillium spp.	Vanillic acid, syringic acid, vanillic acid plus glucose or yeast extract, syringic acid plus glucose or yeast extract, yeast extract, glucose, glucose plus MET	In a stationary liquid culture, glucose did not promote C_2H_4 generation, whereas C_2H_4 production in the presence of vanillic or syringic acid was stimulated by the addition of glucose or yeast extract and, usually, the effect was more than additive. In a comparatively more aerated system, glucose became more effective in promoting C_2H_4 synthesis. The combined application of glucose plus MET was no more effective than glucose alone.	Considine and Patching 1975)
P. crustosum (3 strains)	Glucose, glucose plus MET, acetate, citrate, KGA, malic acid, succinate, glutamic acid, vanillic acid, *p*-hydroxybenzoic acid, protocatechuic acid, ferulic acid, vanillic acid plus yeast extract	Little or no C_2H_4 was produced from a mineral salt medium amended with acetate or tricarboxylic cycle intermediates, although these media supported good growth. Phenolic acids were effective in promoting C_2H_4 synthesis. Glucose alone significantly enhanced C_2H_4 production; however, the combined application of glucose plus MET yielded less C_2H_4 than glucose alone, suggesting that unlike glucose, MET did not serve as a substrate for C_2H_4 synthesis.	Considine et al. (1977)
P. cyclopium (6 strains)	Glucose, glucose plus MET, acetate, citrate, KGA, malic acid, succinic acid, glutamic acid, vanillic acid, p-hydroxybenzoic acid, protocatechuic acid, ferulic acid, vanillic acid plus yeast extract	Significantly higher quantities of C_2H_4 were produced when grown on a mineral salt medium amended with acetate or tricarboxylic acid cycle intermediates than on a medium containing glucose as the only carbon source. Phenolic acids had variable effects on C_2H_4 synthesis. Growth was excellent in the presence of glucose. MET plus glucose added to the mineral medium gave the poorest yield of C_2H_4. MET did not seem to be a precursor of C_2H_4.	Considine et al. (1977)
P. digitatum	Glutamate, KGA, MET, MET-sulfoxide, MET-sulfoximine, MET-sulfone	Static cultures of this fungus utilized glutamate and KGA as C_2H_4 precursors. The addition of MET or its analogues inhibited C_2H_4 synthesis. However, shake cultures utilized MET as the C_2H_4 precursor	Chalutz and Lieberman (1978)

Table 4.5. (continued)

Organism	Substrate	Notes	Reference
P. digitatum	Glucose, labeled glutamic acid, acetate, proline, glutamine, ornithine, ethanol, α-amino-*n*-butyric acid, succinyl-semialdehyde, succinic acid and 2-ketoglutaric acid	The origins of C_2H_4 seemed to be associated with the Krebs cycle. KGA and glutamic acid were the most efficient precursors of C_2H_4 that were derived from C-3 and C-4 of these substrates. Succinic acid was an inefficient precursor. 2-Ketoglutaric acid is a branching point at which the pathway of C_2H_4 biosynthesis deviates away from the Krebs cycle. The C-2 unit of acetate was incorporated into C_2H_4. *P. digitatum* produced C_2H_4 when glucose was the only carbon source.	Chou and Yang (1973)
P. digitatum	Labeled succinate, sodium propionate, sodium acetate, MET, malate, fumarate, β-alanine potassium acrylate	The origin of C_2H_4 appears to be associated with the Krebs cycle acids, particularly the middle carbon atoms of dicarboxylic acids. Certain carbon atoms of β-alanine, propionic acid, and MET can be incorporated into the C_2H_4 carbon skeleton, presumably by way of an indirect route by the Krebs cycle. Some evidence also led to the possibility that acrylic acid may be related to the precursor of C_2H_4.	Jacobsen and Wang (1968)
P. digitatum	Acetate, pyruvate, malate	Labeled feeding experiments revealed that C-2 of acetate or C-3 of pyruvate was incorporated into C_2H_4.	Gibson and Young (1966)
P. digitatum	Glucose, MET, glutamate	Glucose did not serve as a precursor for C_2H_4 in shake culture, whereas MET induced concentration-dependent C_2H_4 production. C_2H_4 production was several-fold greater in response to the combined application of glucose plus MET to the growth medium. Under static conditions, glutamate was a more effective substrate for C_2H_4 synthesis than MET.	Chalutz et al. (1977)
P. digitatum (12 cultures)	Glucose, yeast extract, yeast extract components (vitamins, amino acids)	C_2H_4 production was observed in the presence of glucose alone; however, the addition of yeast extract enhanced C_2H_4 production by 15-fold. The stimulatory effect of yeast extract could not be duplicated with amino acids or vitamins, added alone or in combination	Spalding and Lieberman (1965)

Table 4.5. (continued)

P. digitatum (38 cultures)	Sucrose	The fungus grown on a mineral salt medium containing 4 g L^{-1} sucrose and 0.5 g L^{-1} yeast extract yielded detectable amounts of C_2H_4.	Young et al. (1951)
P. digitatum	KGA, L-cysteine, L-threonine, L-serine, L-hydroxyproline, L-isoleucine, L-proline, L-glutamine, L-asparagine, glycine, L-aspartic acid, L-arginine, L-glutamic acid, L-phenylalanine, L-leucine, L-MET, L-alanine, L-tyrosine, casamino acid, ACC, SAM, KMBA, α-aminobutyric acid, succinic semialdehyde, α-ketobutyric acid, DL-α-keto-β-methyl-n-valeric acid, α-ketosocapronic acid, α-ketoadipic acid	A cell-free EFE system utilized KGA as an immediate precursor of C_2H_4. L-Arginine and Fe(II) under reduced conditions promoted this reaction. A time-course study revealed that the same enzyme system also operates in living cells.	Fukuda et al. (1986)
P. digitatum	KGA, L-arginine, dithiothreitol	The presence of KGA as a substrate and L-arginine (cofactor), Fe(II), dithiothreitol, and O_2 were essential for C_2H_4 formation by EFE purified from *P. digitatum*.	Fukuda et al. (1989b)

Table 4.5. (continued)

Organism	Substrate	Notes	Reference
Saccharomyces cerevisiae (wild-type, and mutant, ade⁻, MET⁻)	Glucose, MET, ethionine, S-adenosylmethionine (SAM)	Glucose or MET were required for C_2H_4 synthesis by the yeast when grown on lactate medium and the combined application was several-fold more effective than MET alone. Increasing concentrations of MET up to 5 mM resulted in increased C_2H_4 synthesis. Potassium pyruvate suppressed MET-induced C_2H_4 production. Ethionine also stimulated C_2H_4 synthesis, although yeast growth was inhibited. SAM did not promote C_2H_4 generation. A mutant (ade⁻, MET⁻) produced C_2H_4 only in the presence of MET, whereas adenine plus MET yielded the greatest amount. Labeled feeding experiments revealed that only MET was incorporated into C_2H_4, but not glucose, implying that MET acts as a precursor of C_2H_4 in yeast, whereas glucose appears to be an energy source.	Thomas and Spencer (1977)
S. cerevisiae	Glucose, fructose, mannose	The addition of glucose (1%) to a lactate-grown yeast culture induced C_2H_4 production. However, C_2H_4 production was not proportional to the glucose concentration added. Pyruvate inhibited C_2H_4 generation. Fructose or mannose, when substituted for glucose, had no effect on C_2H_4 production.	Thomas and Spencer (1978)
Septoria musiva	L-MET, L-cysteine, L-aspartic acid, DL-ethionine, L-glutamic acid, DL-homocysteine, L-alanine, L-arginine, L-asparagine, L-histidine, L-isoleucine, L-leucine, L-lysine, DL-norleucine, DL-noraline, L-phenylalanine, L-proline, L-serine, L-threonine, L-tyrosine, and L-valine; glucose, D-arabinose, D-mannose, D-rhamnose, D-ribose, and β-xylene	L-Methionine, L-cysteine, L-aspartic acid, DL-ethionine, L-glutamic acid, and DL-homocysteine stimulated production of C_2H_4. None of the carbohydrates tested stimulated C_2H_4 production.	Brown-Skrobot et al. (1985)

Table 4.5. (continued)

Verticillium dahliae	L-MET, L-glutamate, L-aspartate, DL-ethionine, D-MET, DL-MET, tryptophan, ACC, KMBA, yeast extract	Only those supplements related to MET were stimulatory to C_2H_4 production. ACC did not stimulate C_2H_4 production and AOA had no inhibitory effect on MET-enhanced C_2H_4.	Tzeng and DeVay (1984)
12 ectomycorrhizal fungi	MET	Ectomycorrhizal fungi produced C_2H_4 *in vitro* in a modified Melin-Norkyans liquid medium only if amended with 2.5-10 mM MET. Some of these fungi required renewal of the culture media with fresh MET addition to induce or enhance C_2H_4 production.	Graham and Linderman (1980)
228 fungi	Staley's corn steep liquor, glucose	Out of 228 fungi, 58 were capable of producing C_2H_4.	Ilag and Curtis (1968)
6 mycorrhizal fungi	MET	More C_2H_4 production was reported in media containing MET than in media without the precursor.	Strzelczyk et al. (1994b)

Table 4.6. Inhibitory effect of ammonium salts and carbonates on ethylene production from a cell-free extract of *Pseudomonas syringae* pv. *phaseolicola* amended with KGA or glutamate.

Compound	Ethylene production (%) of control	
	KGA	Glutamate
Control[a]	100.0	100.0
NH₄HCO₃[b]	87.3	17.5
NH₄Cl[b]	121.5	43.7
NaHCO₃[b]	30.1	28.8

[a]Without addition of salts.
[b]Concentration: 50 mM.
Source: Goto and Hyodo (1987).

Concentrations of NH_4Cl up to 5 mM or less were more effective in promoting KMBA-derived C_2H_4 production, whereas higher concentrations ceased C_2H_4 generation after 16 h of growth. Similarly, Shipston and Bunch (1989) reported that cultures of *E. coli* grown with L-MET or KMBA produced C_2H_4 and continuous cultures grown under an ammonium limitation produced both C_2H_4 and KMBA from MET. In contrast, when glucose was limiting, neither of these metabolites was produced. Cells harvested from continuous cultures grown under glucose or ammonium limitations were able to synthesize C_2H_4 from either L-MET or KMBA, although their capacity for C_2H_4 synthesis was optimal under an ammonium limitation (C/N ratio = 20). Ethylene formation in a batch culture was unaffected by the concentration of L-MET in the medium, although increasing concentrations of NH_4Cl resulted in progressively less C_2H_4 formation.

Production of C_2H_4 by shake cultures of *P. digitatum* and other microorganisms is affected by the concentration of inorganic phosphate in the growth medium (Chalutz et al., 1977, 1978, 1980; Chalutz and Lieberman, 1978; Mattoo et al., 1979; Pazout et al., 1982). A reduction in the concentration of phosphate, but not of any other components of the medium, markedly stimulated the production of C_2H_4. This increased production was strongly inhibited by exogenous addition of orthophosphate to a low-phosphate medium. Chalutz et al. (1978) demonstrated that inorganic phosphorus is a potent inhibitor of C_2H_4 production by shake cultures of *P. digitatum*. They also hypothesized that the mechanism of stimulation of MET-induced C_2H_4 production under phosphate-limiting growth conditions differs from that of similar shake cultures grown under conditions of phosphate sufficiency. Mattoo et al. (1979) characterized the phosphate effect on C_2H_4 production *by P. digitatum*.

A low level of phosphate (0.001 mM) was about 200-500 times more effective than a high-phosphate level (100 mM) in stimulating C_2H_4 production, and the stimulation was readily reversed by the addition of phosphate. This phosphate effect did not operate in static cultures. Interestingly, the precursor of C_2H_4 in the low-phosphate promoted system was glutamate, instead of KGA, which is a precursor in static systems. Antibiotics effectively inhibited the low-phosphate:high-C_2H_4-producing system. Alkaline phosphatase and protein kinase activities were higher in the lower-phosphate systems, suggesting that the phosphate level regulates C_2H_4 production by $P.$ $digitatum$ via a phosphorylation or a dephosphorylation reaction of some enzyme system associated with C_2H_4 production. Phosphate-mediated control of C_2H_4 production may also involve the transcriptional and translational machinery of the fungal cell. $P.$ $digitatum$ apparently can produce various levels of C_2H_4 by different pathways, depending on the culture conditions under which it is grown (Mattoo et al., 1979).

An C_2H_4 –producing isolate of a wild-type $P.$ $digitatum$ was compared with a non-C_2H_4 –producing isolate of this fungus for various biochemical parameters (Kapulnik et al., 1983). The intracellular, acid-soluble phosphate content was markedly higher in the non- C_2H_4 – producing isolate than in the C_2H_4 –producing isolate, whether both cultures were grown in the medium containing high (100 mM) or low (0.1 mM) phosphate. Ethylene production by the C_2H_4–producing isolate increased markedly in low-phosphate growth medium and correlated with an increase in acid phosphatase activity, but not with changes observed in the activities of glucose-6-phosphate dehydrogenase, pyruvate kinase, malate dehydrogenase, or succinate dehydrogenase. However, these latter enzymatic activities were significantly different in the two isolates when cultivated in a high-phosphate medium. The data indicate that the ability of fungi to produce C_2H_4 may be related to high-phosphatase activity and to their capacity to maintain low-phosphate intracellularly (Kapulnik et al., 1983).

4.2.3. pH

A wide pH buffer range has often been employed in studies involving C_2H_4 production by microbial isolates. However, the direct effect of pH as a factor in microbial biosynthesis of C_2H_4 has not been investigated thoroughly. Lynch (1974) reported that a change of medium pH from 6 to 12 had a negative effect, whereas a change from 6 to 1 had a positive effect and a change from 6 to 12 to 1 resulted in an outburst of C_2H_4 by an extracellular preparation from $Mucor$ $hiemalis$. It was speculated that this stimulation in C_2H_4 formation with an adjustment of the pH from 6 to 12 to 1 might be due to the release of Fe^{3+} from citrate and hydroxyl precipitates. Primrose (1976a) found that pH 6 was the optimum for C_2H_4 biosynthesis by $E.$ $coli$ in the presence of MET. In another study, Primrose (1977) reported that C_2H_4 production in the dark by $E.$ $coli$ in the presence of MET or KMBA had a pH optimum of 5.5; whereas, photochemical production of C_2H_4 by culture filtrates grown in the presence of MET or KMBA had an optimum of pH 3. Ethylene production by $Gloeophyllum$ $trabeum$ from KMBA incubated at pH 2 was up to ten-fold greater than that produced at pH 4, 6, 8, 10 and 12 (Chandhoke et al., 1992). These authors also found that, in the presence of chelating agents (siderophores), the oxidation rate of KMBA to C_2H_4 at pH 3 was double of that at pH 6.0. A pH range of 6-7 was optimum for MET-dependent

C$_2$H$_4$ biosynthesis by *Acremonium falciforme* (Arshad and Frankenberger, 1989). Goto et al. (1985) noted that glutamate-dependent C$_2$H$_4$ production by *Ps. syringae* pv. *phaseolicola* was remarkably reduced at pH 4.0, but no significant differences were observed between pH 5 and 8. Similarly, in another study, they observed pH 7 as the optimal for glutamate- or KGA-dependent C$_2$H$_4$ production by a cell-free C$_2$H$_4$ –forming system of *Ps. syringae* pv. *phaseolicola* (Goto and Hyodo, 1987). Similarly, Nagahama et al. (1991b) reported an isoelectric point and an optimum pH for the C$_2$H$_4$ -forming enzyme (EFE) purified from *Ps. syringae* pv. *phaseolicola* at 5.9 and about 7.0-7.5, respectively. This EFE system produced C$_2$H$_4$ from KGA. A pH of 7.0 was more favorable for C$_2$H$_4$ production by *Rhizoctonia solani* grown in a glucose-peptone medium under given O$_2$ tension conditions. Lu et al. (1989) investigated C$_2$H$_4$ production by *Pseudomonas solanaceanum* cultured in a synthetic medium maintained at pH 6, 7, and 8. They observed much more C$_2$H$_4$ production at pH 7 than pH 6 and 8 (Fig. 4.1). Bacterial growth was also maximum at pH 7. Strzelczyk et al. (1994b) studied the effect of pH of media on C$_2$H$_4$ production by six mycorrhizal fungi and found pH 6 as the optimum pH for C$_2$H$_4$ production. It is most likely that pH optima for C$_2$H$_4$ production differs with the nature of substrates and microorganisms, and incubation conditions.

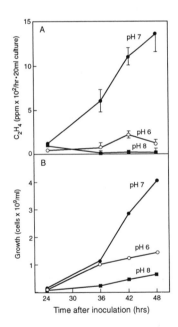

Fig. 4-1. Effect of pH on C$_2$H$_4$ production (A) and growth (B) of a virulent PS61 strain of *Pseudomonas solanacearum* cultured under continuous shaking at 30°C. Data presented are averages of 4 replicates and bars indicate standard deviation. Source: Lu et al. (1989).

4.2.4. Aeration

Aeration is one of the most important critical factors that affects C_2H_4 biosynthesis by microbial cultures. Generally, aerobic conditions are considered more favorable than anaerobic environments for C_2H_4 formation by microbial cultures, as evident from Tables 4.7 and 4.8. This differs from C_2H_4 formation in soil, in which restricted O_2 supply is generally considered more favorable for C_2H_4 release. Interestingly, different compounds (e.g., MET vs. KGA) serve as C_2H_4 precursors for *P. digitatum* under shake versus static incubation (Chalutz et al., 1977; Chalutz and Lieberman, 1978). Microbial EFE systems require aerobic conditions for C_2H_4 biosynthesis as discussed in Section 3.4.2.1. Lu et al. (1989) reported that shaking not only stimulated the growth of *Ps. solanacearum* strain BG1, but also C_2H_4 production by several-fold (Fig. 4.2). The replacement of O_2 with N_2 completely inhibited KGA-derived C_2H_4 production by *Fusarium oxysporum* f. sp. *tulipae* (Hottiger and Boller, 1991). Hahm et al. (1992) studied the effect of dissolved O_2 tensions on C_2H_4 production by *Ps. syringae* pv. *phaseolicola* in continuous culture and reported that a high level of dissolved O_2 (≥ 4.5 mg L^{-1}) was required for efficient production of C_2H_4 , whereas dissolved O_2 tensions below 4 mg L^{-1} restricted C_2H_4 release by this bacterium. Fukuda et al. (1984) tested 80 genera, 150 species, and 178 strains for C_2H_4 production under aerobic conditions. They found that 62% molds, 20% yeasts, 21% bacteria, and 6% actinomycetes were capable of producing C_2H_4 under aerobic conditions. Contrarily, C_2H_4 production in the presence of MET or KMBA by *Verticillium dahliae* was not influenced by shaking, but was stimulated by light (Tzeng and DeVay, 1984).

4.2.5. Temperature

Since temperature affects microbial growth, respiration and enzymatic reactions, it is very likely that C_2H_4 biosynthesis by microbial isolates will be influenced by temperature variations. Primrose (1976a) reported a temperature optimum of 30°C for C_2H_4 formation by *E. coli*, whereas maximum growth was observed at 37°C. A decrease in C_2H_4 production by *P. digitatum* was observed when the temperature was raised beyond 23°C (Mattoo et al., 1977). The Q_{10} and E_a calculated for this system were 5.46 and 113.6 kJ mol^{-1} (between 2° and 23°C), respectively. In static culture, the optimum temperature for C_2H_4 biosynthesis by *Acremonium falciforme* was 30°C (Arshad and Frankenberger, 1989). Strzelczyk et al. (1994b) found that six mycorrhizal fungi produced more C_2H_4 when incubated at a temperature at 26°C than that at 20°C incubation. Swanson et al. (1979) noted a decrease in C_2H_4 production by *Ps. solanacearum* when the temperature was raised above 20°C, although bacterial growth was optimum at 25°C. Lu et al. (1989) found a temperature of 30°C optimum for C_2H_4 production by *Ps. solanacearum* cultures in a synthetic medium while growth was also highest at this temperature (Fig. 4.3). Cultivation at 25°C or 35°C resulted in lowering C_2H_4 synthesis as well as growth. Goto et al. (1985) reported that C_2H_4 production by *Ps. syringae* pv. *phaseolicola* at 30°C was 1.5 times greater than that at 18°C. In another study, they found that increasing the temperature from 1° to 25°C increased C_2H_4 generation by a cell-free extract system of this bacterium (Goto and Hyodo, 1987). Beyond this temperature, the rate of C_2H_4 production decreased. However, with living cells, the optimum temperature was 30°C (Table 4.9). Weingart and Völksch (1997)

Table 4.7. Effect of aeration on ethylene biosynthesis by bacterial cultures.

Microorganism	Incubation conditions	Comments	Reference
Escherichia coli	Aerobic vs. anaerobic	*E. coli* produced C_2H_4 under aerobic conditions only when grown on a MET-containing medium.	Primrose and Dilworth (1976)
E. coli	Aerobic vs. anaerobic	MET-dependent C_2H_4 production occurred only when incubated aerobically.	Primrose (1976a)
E. coli	Aerobic vs. anaerobic	The enzyme system that converted MET-derived KMBA into C_2H_4 required O_2 for this reaction.	Ince and Knowles (1986)
Pseudomonas putida and *Ps. fluorescens*	Immediate aeration (O_2 replenished) or delayed aeration	The highest C_2H_4 accumulation in bacterial cultures was reached under conditions of delayed aeration (i.e., after the O_2 content decreased to 4%). Ethylene production increased immediately after beginning aeration.	Pazout et al. (1981)
Ps. solanacearum	Static vs. shake cultures	Shaking promoted C_2H_4 production as well as cell growth in a synthetic medium.	Lu et al. (1989)
Ps. syringae pv. *phaseolicola*	Flushing the gas atmosphere above the assay medium with fresh air or N_2 gas	The replacement of headspace gas with N_2 decreased the C_2H_4 production rate.	Goto et al. (1985)
Ps. syringae pv. *phaseolicola*	Replacing the gas atmosphere above the assay medium with N_2 gas	Ethylene production was almost completely suppressed under anaerobic conditions (after flushing of air with N_2 gas). When N_2 gas in the flasks was replaced by air, C_2H_4 production started immediately, and the rate of C_2H_4 production in the initial stage was significantly greater.	Goto and Hyodo (1987)

Table 4.7. (continued)

Ps. syringae pv. *phaseolicola*	Shaking	The complete EFE system purified from this bacterium required aerobic conditions to generate C_2H_4 from KGA in the presence of $FeSO_4$, dithiothreitol, L-histidine, and L-arginine.	Nagahama et al. (1991a)
Ps. syringae pv. *phaseolicola* and *Penicillium digitatum*	Shaking	The presence of O_2, Fe(II), and L-arginine was essential for KGA-derived C_2H_4 formation by the EFE system purified from the bacterium and fungus.	Nagahama et al. (1991b)
Ps. syringae pv. *phaseolicola*	Batch and continuous culture; shaking and different dissolved O_2 (DO) tensions	Glucose was the only carbon source. C_2H_4 production was restricted at DO tension below 4 mg L^{-1}, and nearly constant in the broad range of 4.5-8 mg L^{-1}, indicating that a high level of DO was required for efficient production of C_2H_4.	Hahm et al. (1992)

Table 4.8. Effect of aeration on ethylene biosynthesis by fungal cultures.

Microorganism	Incubation conditions	Comments	Reference
Acremonium falciforme	Static vs. shaking	A greater than 7-fold increase in MET-derived C_2H_4 production was observed with shaking over static conditions after 72 h of incubation.	Arshad and Frankenberger (1989)
Cryptococcus albidus	Aerobic vs. anaerobic	The enzyme system that converted MET-derived KMBA into C_2H_4 required O_2 for this reaction.	Fukuda et al. (1989a)
C. albidus	Shaking, or reaction mixture was bubbled with argon gas	The presence of KMBA, NADH, Fe(III), EDTA, and O_2 were essential for the formation of C_2H_4 by the EFE system. EFE is most likely an NADH-Fe(III) EDTA oxidoreductase.	Fukuda et al. (1989c)
Fusarium oxysporum f. sp. *tulipae*	O_2 levels in the gas phase above the assay medium varied between 0 and 20%	The rate of C_2H_4 biosynthesis depended sigmoidally on the O_2 concentration, with a half-maximal value at 2% O_2. Anaerobiosis immediately and completely stopped C_2H_4 biosynthesis.	Hottiger and Boller (1991)
Mucor hiemalis	Chemostat vs. sealed shaking	Low levels of C_2H_4 were observed in sealed shaking flasks compared with a chemostat, which may be partially the result of reduced oxygen levels in the former.	Lynch and Harper (1974)
Penicillium digitatum	Static vs. shaking	Shake cultures produced 20 times as much C_2H_4 as static cultures.	Spalding and Lieberman (1965)
P. digitatum	Headspace gases replaced by air	C_2H_4 production was strongly suppressed in the absence of O_2. The addition of subambient levels of O_2 caused a corresponding increase in the rate of C_2H_4 production.	Considine and Patching (1975)

Table 4.8. (continued)

P. digitatum	Static *vs.* shaking	Static cultures utilized glutamate and KGA as C_2H_4 precursors, whereas shake cultures utilized MET as an C_2H_4 precursor.	Chalutz and Lieberman (1978); Chalutz et al. (1977)
P. digitatum	Shaking	The presence of O_2, Fe(II) and L-arginine (cofactor) was essential for the KGA-dependent C_2H_4 formation by the EFE system purified from this fungus.	Fukuda et al. (1989b)
Saccharomyces cerevisiae	Anaerobic *vs.* aerobic	The conversion of C_2H_4 precursors to C_2H_4 was stimulated by O_2.	Thomas and Spencer (1978)

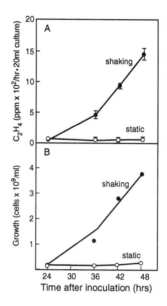

Fig. 4-2. Effect of shaking on C_2H_4 production (A) and growth
(B) of a virulent PS61 strain of *Pseudomonas solanacearum*
growing at 30°C. Data presented are averages of 4 replicates
and bars indicate standard deviation. Source: Lu et al. (1989).

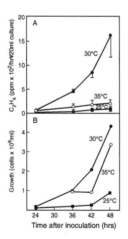

Fig. 4-3. Effect of temperature on C_2H_4 production (A) and
growth (B) of a virulent PS61 strain of *Pseudomonas
solanacearum* cultured under continuous shaking at 30°C.
Data presented are averages of 4 replicates and bars indicate
standard deviation. Source: Lu et al. (1989).

Table 4.9. Effect of temperature on ethylene production by a cell-free extract
and living cells of *Pseudomonas syringae* pv. *phaseolicola*.

Temperature (°C)	Ethylene production			
	Cell-free extract[a]		Living cells[b]	
0	1.1[c]	14.9[d]	1.30[e]	15.8[d]
4	1.7	23.0	2.82	34.3
10	5.2	70.3	4.06	49.3
15	5.6	75.7	4.08	49.6
20	6.8	91.9	4.70	57.1
25	7.4	100.0	7.53	91.5
30	6.4	86.5	8.23	100.
40	1.4	18.9	0.73	8.9
50	0.4	5.4	0.54	6.6

[a]Substrate, α-KG.
[b]Substrate, glutmate.
[c]nl (mg protein)$^{-1}$ h^{-1}.
[d]Percent of the maximal value.
[e] x10^{-8} nl cell^{-1} h^{-1}.
Source: Goto and Hyodo (1987).

reported that C_2H_4 production by various strains of *Ps. syringae* pv. *phaseolicola* or pv. *glycinea* at 28°C was two times higher than that at 18°C.

4.2.6. Trace Elements

Trace elements are known to influence the growth and activity of microorganisms. Their effects on microbial production of C_2H_4 have been examined by many workers. Lynch (1974) found that Cu^{2+} (20 mg L^{-1}) and Mn^{2+} (200 mg L^{-1}) had negative effects on MET-derived C_2H_4 production by an extracellular preparation of *Mucor hiemalis*. Ethylene production was stimulated by the addition of Fe^{3+} (200 mg L^{-1}) and even a greater stimulation (8.6-fold over the Fe^{3+} stimulated effect) was found with Fe^{2+} (200 mg L^{-1}). Ethylene formation by *Cylindrocladium floridantum, C. scoparium* and *P. digitatum* were markedly decreased by the addition of 10 and 20 m*M* $CuSO_4$ (Axelrood-McCarthy and Linderman,1981; Achilea et al., 1985). Arshad and Frankenberger (1989) noted that Co^{2+} (when added at >1 µM) strongly inhibited C_2H_4 production by *A. falciforme*. An EFE system isolated from *P. digitatum* required Fe^{2+} for C_2H_4 formation from KGA (Fukuda et al., 1986, 1988, 1989b; Ogawa et al., 1992). Fukuda et al. (1986) reported that, among the metal ions examined for their effects on KGA-derived C_2H_4 by *P. digitatum*, only Fe(II) had a stimulatory effect, whereas all others had negative effects on C_2H_4 formation (Table 4.10). In another study, they noted that addition of Co(II), Cu(II), and Mn(II) (all applied at 1 mM) had inhibitory effects on KGA-dependent C_2H_4 production by *P. digitatum* (Fukuda et al., 1989b).

Table 4.10. Effects of metals on ethylene formation by a crude enzyme extract of *Penicillium digitatum*.

Metal[a]	Ethylene production rate[b] (nl/mg protein/h)
None	0.26
$FeSO_4 \cdot 7H_2O$	5.21
$Al_2(SO_4)_3$	0.51
$Fe_2(SO_4)_3 \cdot xH_2O$	0.50
K_2SO_4	0.29
$K_2B_4O_7 \cdot xH_2O$	0.26
$(NH_4)_6Mo_7O_{24} \cdot 4H_2O$	0.24
$CaCl_2$	0.23
NaCl	0.23
$MgSO_4 \cdot 7H_2O$	0.17
$CuSO_4 \cdot 5H_2O$	0.15
$MnSO_4 \cdot 4\sim5H_2O$	0.13
$ZnSO_4 \cdot 7H_2O$	0
$CoSO_4 \cdot 7H_2O$	0
$NiSO_4 \cdot 6H_2O$	0

Substrate: KGA 1 mM (final).
[a]Each concentration: 0.1 mM (final).
[b]The protein concentration of the crude enzyme solution was 3.47 mg ml^{-1} reaction mixture.
Source: Fukuda et al. (1986).

Hottiger and Boller (1991) observed that Fe(II) ions markedly stimulated the rate of glutamate- or KGA-derived C_2H_4 production by *Fusarium oxysporum* f. sp. *tulipae*, whereas Fe(III), Cu(II), or Zn(II) (all added at 1.25 mM) had little effect. Since only Fe(II) fully reversed the inhibitory effects of a heavy metal chelator, α,α'-dipyridine, Hottiger and Boller (1991) suggested that a transformation in the C_2H_4 biosynthesis pathway beyond that of KGA is dependent on Fe(II). The addition of Fe(III) (10^{-2} M) stimulated KMBA-derived C_2H_4 by *Gloephyllum trabeum* greater than 13-fold, whereas the same concentration of Mn(II) increased C_2H_4 generation by more than two-fold, while very low levels (10^{-8} and 10^{-5} M) of both trace metals had either little or no effect (Chandhoke et al., 1992). The addition of Fe and Zn metal powder and $CuSO_4$ stimulated C_2H_4 production by *Septoria musiva* grown in a MET-glucose medium, whereas amending the medium with chlorides of Zn, Ni, and Mn resulted in no C_2H_4 production (Brown-Skrobot et al., 1984). Filer et al. (1984) reported that C_2H_4 production by *Fusicladium effusum*, *Pestilotia nucicola*, *Alternaria tenuis* and *Fusarium oxysporum* requires supplemention with iron powder when grown on potato dextrose broth.

Primrose (1977) observed that Fe^{3+}, Mg^{2+} and, to a lesser extent, Ca^{2+} ions are required for MET-dependent C_2H_4 synthesis by *E. coli*. Culture filtrates of *E. coli* did not produce C_2H_4 from KMBA unless the medium was amended with Mn^{2+} and SO_4^{2-} (Primrose, 1977). Ince and Knowles (1985) demonstrated that C_2H_4 formation by *E. coli* was stimulated by increasing the concentration of Fe^{3+} (24-120 μM) chelated with EDTA in the medium. The C_2H_4–forming enzyme system isolated from *E. coli* and

Cryptococcus albidus required Fe^{3+} for synthesizing C_2H_4 from the intermediate, KMBA derived from MET (Ince and Knowles, 1986; Fukuda et al., 1989a,c). Fukuda et al. (1989c) found that Fe(II) chelated to EDTA, added to the reaction mixture containing *C. albidus*, resulted in the nonenzymatic formation of C_2H_4 from KMBA under aerobic conditions. They suggested that EFE is a NADH:Fe(III) EDTA oxidoreductase. In studying C_2H_4 formation by EFE isolated from *Ps. syringae* pv. *phaseolicola*, Nagahama et al. (1991a,b) found that $FeSO_4$ (0.2 mM) was one of the essential cofactors for C_2H_4 synthesis from KGA. They also observed negative effects of Co(II), Cu(II), and Mn(II) (all applied at 1 mM) on C_2H_4 formation from KGA by the EFE system of this bacterium (Nagahama et al., 1991b). Goto and Hyodo (1987) compared the effects of various forms of Fe (0.5 mM) added to the reaction mixture consisting of the C_2H_4 –forming system of *Ps. syringae* pv. *phaseolicola* using KGA as a substrate (Table 4.11); $FeSO_4$ was the most effective stimulator. Ethylene was not produced in the presence of other metals, such as Na(I), Mg(II), Ca(II), Zn(II), Mn(II) and Mo(VI). These studies indicate a critical regulatory role of iron (Fe^{2+} or Fe^{3+}) in C_2H_4 biosynthesis by microbial isolates, as discussed in Section 3.4.2.

4.2.7. Growth Phase

Various studies have demonstrated that production of C_2H_4 by different microbial cultures occurs at their different growth phases. Hahm et al. (1992) indicated that C_2H_4 production by *Ps. syringae* was typically growth associated. Lu et al. (1989) reported that C_2H_4 production by *Ps. solanacearum* increased with an increase in bacterial growth and the highest rate C_2H_4 was detected at the exponential phase (Fig. 4.4). As the bacteria approached their stationary phase, the rate of C_2H_4 production declined rapidly. They also tested various media for growth and C_2H_4 production and found that the media

Table 4.11. Effect of ferrous or ferric compounds on ethylene production from a cell-free extract of Kudzu *Pseudomonas syringae* pv. *phaseolicola* in the presence of KGA.

Ferrous or ferric compound[a]	Ethylene production	
	(nl/mg protein/h)	%
$FeSO_4$	13.0	100
$Fe_2(SO_4)_3$	10.6	81.5
$FeCl_3$	6.4	49.2
Fe-citrate	5.2	40.0
Fe-EDTA	4.0	30.8

Concentration: 0.5 mM.
Source: Goto and Hyodo (1987).

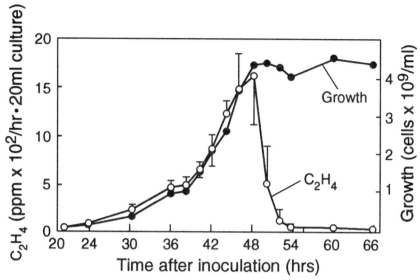

Fig. 4-4. Ethylene production in relation to growth of a
virulent PS61 strain of *Pseudomonas solanacearum* cultured
under continuous shaking at 30°C. Data presented are
averages of 4 replicates and bars indicate standard deviation.
Source: Lu et al. (1989).

which supported maximum growth may not necessarily result in the highest C_2H_4
production (Lu et al., 1989). Bonn et al. (1975) observed that C_2H_4 production by *Ps.*
solanacearum was associated with the growth of the bacterium. Pegg and Cronshaw
(1976) reported that three strains of the pathogen *Ps. solanacearum* had the same growth
curves, but the maximum C_2H_4 released by two strains (K60 and B1) was during the log
phase (21 h) while the third less efficient strain (5228) produced C_2H_4 during the late log
phase (42 h). Goto et al. (1985) reported that C_2H_4 production by *Ps. syringae* pv.
phaseolicola strain PK2 derived from glutamate was proportional to the logarithm of cell
density. The kinetics of C_2H_4 production and bacterial growth of cultures of *Ps. syringae*
pv. *glycinea*, pv. *phaseolicola* and pv. *pisi*, revealed that C_2H_4 production occurred in the
late exponential phase as evident from Fig. 4.5 (Weingart and Völksch, 1997). Weingart
et al. (1999) determined the kinetics of C_2H_4 production and bacterial growth for *Ps.*
syringae pv. *sesami* 962 (KGA-dependent C_2H_4 producer), *Ps. syringae* pv. *pisi*
GSPB1206, and *Restonia solanacearum* K60. They observed that C_2H_4 production by
Ps. syringae pvs. *sesami* and *pisi* was strictly growth associated. The highest rates of
C_2H_4 formation were detected in the late exponential phase. In contrast, C_2H_4 synthesis
by *R. solanacearum* had its maximum production in the early exponential phase.

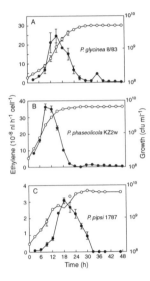

Fig. 4-5. Growth kinetics (o) and C$_2$H$_4$ production (•) by *P. syringae* pv. *glycinea* 8/83 (A), *P. syringae* pv. *phaseolicola* KZ2w (B) and *P. syringae* pv. *pisi* 1787 (C) *in vitro*. The data are the means and standard errors of three independent experiments. Source: Weingart and Volksch (1997).

Hottiger and Boller (1991) reported that *Fusarium oxysporum* f. sp. *tulipae* produced C$_2$H$_4$ from KGA in its stationary phase of growth. Tzeng and DeVay (1984) studied C$_2$H$_4$ production by defoliating and non-defoliating pathotypes of *Verticillium dahliae*. They found that MET-dependent C$_2$H$_4$ production was much greater in 10-d-old cultures of *V. dahliae* than 6-d-old cultures and C$_2$H$_4$ production rates were not apparently related to the fungal growth (cell biomass). The rate of C$_2$H$_4$ production generally followed the growth curve of the fungus, *P. digitatum* and, in general, highest values of production were often associated with the log phase of growth and production decreased as the mycelia matured (Fergus, 1954; Phan, 1962; Spalding and Lieberman, 1965). However, in one study the greatest rate of C$_2$H$_4$ production coincided with sporulation and declined as the mycellium of *P. digitatum* matured (Meheriuk and Spencer, 1964). Other workers observed maximum rates of C$_2$H$_4$ production during the senescence of *P. digitatum* cultures (Spalding and Lieberman, 1965). Time course studies with *P. cyclopium* and *P. velutinum* indicate that the greatest rate of C$_2$H$_4$ production occurs during sporulation, and when the phosphate content of the medium decreases (Pazout and Pazoutova, 1989).

Wilkes et al. (1989) did not find any correlation between growth rate and C$_2$H$_4$ production ability of various strains of *Endothia gyrosa* and *Cyclopora eucalypticola*.

4.3. EFFECT OF ETHYLENE ON MICROBIAL GROWTH

The effect of C$_2$H$_4$ on growth and metabolism of microorganisms has not been extensively studied. However, relatively more studies have been conducted to examine the effect of exogenous C_2H_4 on fungal growth than bacterial growth. The effects of C_2H_4 on microorganisms vary from growth promoting to no effect to growth inhibition, most likely depending upon the type of microorganism, C_2H_4 concentration and environmental factors. It is worth noting that microorganisms strongly differ in their sensitivity to C_2H_4 . Most of the studies have been conducted by exposing the culture to C_2H_4 gas or C_2H_4-releasing compounds.

4.3.1. Bacteria

Ethylene does not appear to play a major role in bacterial growth and development. As discussed comprehensively in Chapter 3 (Section 3.7) a number of bacteria are capable of utilizing C_2H_4 as a carbon source for their growth, particularly the species belonging to *Mycobacterium* which can metabolize C_2H_4 into ethylene oxide. However, the effects of C_2H_4 on bacterial growth and physiology have not been reported, except for a few studies. Recently, Borodina et al. (1997) studied the effect of various levels of C_2H_4 (2×10^{-3} to 2.0%) on the growth of 79 strains of microorganisms belonging to the genera *Micrococcus, Pseudomonas, Bacillus, Flavobacterium, Mycobacterium* and others. The data obtained did not reveal any clear relationship between the response of bacteria and the concentration of C_2H_4. Only 2% C_2H_4 significantly stimulated the growth of microbial cultures and positive responses were found in 18 out of 79 cases. At all other concentrations, there was no effect, indicating that pure C_2H_4 is neutral with respect to bacterial growth. Nikitin and Nikitina (1978) also demonstrated that C_2H_4 at a concentration of 90% in the incubation chamber, had no positive or negative effect on bacterial growth.

4.3.2. Fungi

Ethylene (250 ppm) was reported to promote the germination of *Diplodia natalensis* and *Phomopsis citri* (Brooks, 1944). In the latter case with *Ph. citri*, the effect was not specific and was duplicated by ethyl acetate, ethylene chlorohydrin, and methanol. Brown and Lee (1993) observed that a high level of C_2H_4 (55 μl L^{-1}) significantly increased the growth of *D. natalensis* compared to a control (air) while growth at low levels (2 μl L^{-1}) of C_2H_4 was less than air for the first 2 days of culture, and greater after 3 days. Ethylene applied at concentrations of 10 and 100 μl L^{-1} had no effect on the radial growth of 22 species of fungi cultured on various media (Archer, 1976). Other workers have observed similar results (El-Kazzaz et al., 1983; Elad, 1988; Graham and Linderman, 1981; Lynch, 1975). Spore germination *of Alternaria, Botrytis, Colletotrichum, Monolinia, Penicillium, Rhizopus*, and *Thielaviopsis* was promoted by C_2H_4 (El-Kazzaz et al., 1983). The effects, however, were small and variable. Ethylene was reported to stimulate spore germination of *Botrytis cinerea, Penicillium expansum, Rhizopus nigricans, and Gloeosporium perennans* (Kepczynski and Kepczynska, 1977). The dose-response curve indicated a threshold at 10 μl L^{-1}, saturation at 100 μl L^{-1}, and inhibition at 10,000 μl L^{-1}. Germination and mycelia growth of *Botrytis cinerea* increased in the presence of 3 μl L^{-1} C_2H_4 (Reyes and Smith , 1986). The slime mold, *Physarum polycephalum* showed an increased growth rate under a 75% C_2H_4 atmosphere (Burg, 1973). This enhanced effect was counteracted by CO_2, a well-documented phenomenon in higher plants. In contrast, 18% inhibition of growth of *Gloeosporium album* was observed at 2000 μl L^{-1} C$_2$H$_4$ (Lockhart et al., 1968). Ethylene, however, had no effect on hyphal growth of *Gleosporium album* except at low temperatures (3.3°C), where it was inhibitory (Lockhart et al., 1968). The germination of *Sclerotium rolfsi* sclerotia was inhibited at 1 μl L^{-1} C$_2$H$_4$ (Smith, 1973). Some workers have reported no effect of C$_2$H$_4$ on the germination of spores of *Fusarium*, *Penicillium*, and *Pythium* (Archer, 1973), *Arthobotrys oligospora* (Balis, 1976), *Botrytis cinerea* (Barkai-Golan and Levy-Meir, 1989; Elad, 1988), and *Peronospora trifoliorum* (Johnston et al., 1980).

Few workers have used ethephon in place of C$_2$H$_4$, but they did not take into account the possible effect of the phosphoric acid components of the product. For example, germination of *Glomus mosseae* (Azcon-Aguilar et al., 1981) and *Cylindrocarpon destructans* (Michniewicz et al., 1985) was shown to be inhibited by ethephon. On the other hand, volatiles from ethephon, which are less likely to cause artifactual results, promoted germination of *Urocystis agropyri* (Goel and Jhooty, 1987).

The role of C$_2$H$_4$ as a regulator of fungal growth may be established if the effect is a plant-like response to C$_2$H$_4$ such as: (i) a typical dose response curve; (ii) C$_2$H$_4$ analogue specificity; and (iii) the ability of silver and 2,5-norbornadiene to act as C$_2$H$_4$ action inhibitors. Flaishman and Kolattukudy (1994) reported that the influence of C$_2$H$_4$ on the fungi, *Collectotrichum gloeosporoides* and *C. musae* appears to be plant-like since Ag$^+$ and 2,5-norbonadiene inhibited C$_2$H$_4$ effects and high concentrations of C$_2$H$_4$ reversed the effect of the latter inhibitor. Exposure of these fungi to C$_2$H$_4$ at ≤ 1 μl L^{-1} caused germination , branching of the germ tubes and formation of up to six appressoria from one spore. Ethylene stimulation of germination and appressorium formation appears to be specific for *Collectotrichum* species that normally infects climacteric fruits. Virtually all of the *C. gloeosporiode* and *C. musae* conidia germinated and formed appressoria at 1 μM and 10 μM ethephon, respectively (Table 4.12). Higher concentrations of ethephon did not inhibit appressorium formation. On the other hand, even 100 μM ethephon failed to induce germination and appressorium formation by conidia of *C. trifolii*, *C. pisi*, *C. orbiculare*, and *C. lindemuthianum*, which naturally infect nonclimacteric fruits.

Ishii et al. (1996) found that exposure of vesicular-arbuscular mycorrhizal fungi, *Gigaspora ramisporophora* and *G. mossae* to C$_2$H$_4$ levels ranging between 0.01 and 0.1 ppm stimulated spore germination and hyphal growth of both fungi (Fig. 4.6). At 0.2 ppm and above, the hyphal growth of these fungi was severely inhibited. Michniewicz et al. (1983) reported that application of Ethrel inhibited the growth of fungi, *Fusarium culimorum* grown on Czapek-Dox agar at all concentrations used (1-1000 ppm). The highest concentration (1000 ppm) inhibited the growth of mycelium almost completely. The strongest inhibition of growth caused by Ethrel was observed within the first few days and after 5 days, significant inhibition of growth was found only at the highest concentration of Ethrel, i.e., 100 and 1000 ppm. Similarly, an inhibitory effect of this C$_2$H$_4$ -releasing compound on the growth of *Fusarium* cultivated in the liquid medium (determined after 14 days) was observed at concentrations of 100 and 1000 ppm. Ethrel also inhibited the germination of spores of culture isolates of *Fusarium*; however, they differed in their sensitivity against ethrel (Michniewicz et al., 1983).

The role of C$_2$H$_4$ in the development of *Agricus bisporus* is still not clear since Ward et al. (1978) have shown that exogenous C$_2$H$_4$ had no effect on sporocarp growth.

Table 4.12. Effect of ethephon on germination and appressorium formation by conidia of different *Colletotrichum* species.

| *Colletotrichum* species | Host | Ethephon % of total spores | | | | | | Water | |
| | | 1 μM | | 10 μM | | 100 μM | | | |
		Spores with germ tube only	Spores with appressoria	Spores with germ tube only	Spores with appressoria	Spores with germ tube only	Spores with appressoria	Spores with germ tube only	Spores with appressoria
C. gloeosporioides	Avocado	4 ± 2.1	91±2.0	6 ±2.4	92±2.9	7±2.2	91±3.2	2±1.0	10±2.9
C. musae	Banana	24±7.1	32±4.2	2±1.1	90±2.0	1±1.0	93±4.0	10±5.7	14±10
C. pisi	Peas	1±0.8	4±3.1	2±1.0	3±1.0	2±1.1	6±4.1	1±1.0	5±4.0
C. orbiculare	Cucumber	1±0.4	7±2.6	0	6±1.7	0	3±2.9	0	4±3.1
C. trifolii	Alfalfa	1±1.0	6±4.1	2±1.0	3±3.0	0	4±1.6	0	3±9.1
C. lindemuthianum	Bean	8±4.1	23±7.9	3±2.1	31±4.9	4±2.9	14±7.6	7±4.0	28±4.1

Source: Flaishman and Kolattukudy (1994)

4.3.3. Soil Fungistasis

The role of C_2H_4 as a soil fungistasis is still debatable. Ethylene has been claimed to be the main cause of soil fungistasis (Smith, 1973, 1976; Smith and Cook, 1974), although Archer (1976) could not find any evidence for C_2H_4 –induced inhibition of growth of a wide range of genera and species grown in pure cultures. The evidence for the fungistatic effect was based on air-dried soil studies, and it was claimed that this represented a balance between nutrient-stimulated and C_2H_4 –inhibited germination of the spores (Smith, 1973). Dutta and Deb (1984) found that application of Ethrel caused fungistasis of *Fusarium solani* in sterilized soil, but not in nonsterile soil. Ethrel inhibited spore germination in soil and in aqueous solution and affected soil microbes even at low concentrations (1 μl L^{-1}) (Tables 4.13). It was argued that C_2H_4 –mediated fungistasis is inconsistent with the wide range of ability of microorganisms to synthesize C_2H_4 (Lynch, 1975 and Archer, 1976).

4.4. CONCLUDING REMARKS

The foregoing discussion reveals that although numerous microorganisms have the potential to produce C_2H_4, several factors control this process including incubation conditions as well as availability of substrates, carbon sources, cofactors and trace elements. Diversified cultures (beneficial and pathogenic microorganisms) require different conditions of incubation for optimal C_2H_4 production. Most C_2H_4 producing microflora produce C_2H_4 from MET while some require KGA as a precursor and few produce C_2H_4 independent of these substrates. The ecological significance of microbially produced C_2H_4 in symbiosis, pathogenesis and crop production is discussed in Chapters 6, 7 and 8, respectively.

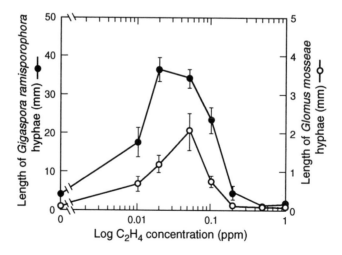

Fig. 4-6. Effect of C_2H_4 on hyphal growth of vesicular-arbuscular mycorrhizal fungi found on agar media. Vertical bars indicate standard error (SE) (n=12). Source: Ishii et al. (1996).

Table 4.13. Effect of Ethrel on spore germination and germ tube length of *Fusarium solani* in sterilized soil and aqueous solution.

Concn. of Ethrel (μl/L)	Spore germination* (%)	Germ tube** length (μm)	Inhibition of spore germination (%)	Inhibition of germ tube length (%)
		Fungistatic effect of Ethrel on:		
AUTOCLAVED SOIL:				
Control	93.92	140.70 ± 10.50	--	--
1	60.84	84.75 ± 9.00	35.22	46.87
10	51.42	79.95 ± 8.10	45.22	50.28
50	42.69	64.95 ± 5.80	54.12	53.12
100	39.37	8.25 ± 1.27	58.08	94.13
AQUEOUS SOLUTION				
Control	83.51	272.24 ± 1.49	--	--
1	72.50	121.18 ± 1.03	13.18	55.5
10	50.00	71.38 ± 0.71	40.12	73.8
50	16.30	66.40 ± 0.89	80.48	75.6
100	4.00	49.80 ± 0.52	95.21	81.7

*Mean of 500 spores.
**Mean of 25 germ tubes.
Source: Dutta and Deb (1984)

4.5. REFERENCES

Achilea, O., Chalutz, E., Fuchs, Y., and Rot, I., 1985, Ethylene biosynthesis and related physiological changes in *Penicillium digitatum*-infected grapefruit (*Citrus paradisi*), *Physiol. Plant Pathol.* **26**:125-134.

Amagai, A., 1984, Induction by ethylene of macrocyst formation in the cellular slime mold. *Dictyostelium mucoroides*, *J. Gen. Microbiol.* **130**:2961-2965.

Amagai, A., and Maeda, Y., 1992, The ethylene action in the development of cellular slime molds: An analogy to higher plants, *Protoplasma* **167**:159-168.

Archer, S. A., 1976, Ethylene and fungal growth, *Trans. Br. Mycol. Soc.* **67**:325-326.

Arshad, M., and Frankenberger, W. T., Jr., 1989, Biosynthesis of ethylene by *Acremonium falciforme*, *Soil Biol. Biochem.* **21**:633-638.

Axelrood-McCarthy, P. E., and Linderman, R. G., 1981, Ethylene production by cultures of *Cylindrocladium floridanum* and *C. scoparium*, *Phytopathol.* **71**:825-830.

Azcon-Aguilar, C., Rodriquez-Navarro, D. N., and Barea, J. M., 1981, Effects of ethrel on the formation and responses to VA mycorrhiza in medicago and triticum, *Plant Soil* **60**:461-468.

Babiker, H. M., and Pepper, I. L., 1984, Microbial production of ethylene in desert soils, *Soil Biol. Biochem.* **16**:559-564.

Balis, C., 1976, Ethylene-induced volatile inhibitors causing soil fungistasis, *Nature* **259**:112-114.

Barkai-Golan, R., and Lavy-Meir, G., 1989, Effects of ethylene on the susceptibility to *Botrytis cinerea* infection of different tomato genotypes, *Ann. Appl. Biol.* **114**:391-396.

Biale, J. B., 1940, Effect of emanations from several species of fungi on respiration and color development of citrus fruits, *Science* **91**:458-459.

Billington, D. C., Golding, B. T., and Primrose, S. B., 1979, Biosynthesis of ethylene from methionine, *Biochem. J.* **182**:827-836.

Blaschke, H., 1977, Der Einfluβ von Temperatur und Feuchtigkeit auf die Bildung von Athylen in Fichtenstreu, *Flora* **166**:203-209.

Bonn, W. G., Sequeira, L., and Upper, C. D., 1975, Technique for the determination of the rate of ethylene production by *Pseudomonas solanacearum*, *Plant Physiol.* **56**:688-691.

Bonner, J. T., 1973, Hormones in social amoebae, in: *Hormonal Control of Growth and Differentiation*, Vol. 2, J. Lobue and A. S. Gordon, eds., Academic Press, New York, NY, pp. 84-98.

Borodina, E. V., Tirranen, L. S., Titova, G. T., and Kalacheva, G. S., 1997, The effects of acetaldehyde and ethylene on growth of reference cultures of microorganisms, *Russian J. Ecol.* **28**:197-199.

Brooks, C., 1944, Stem end rot of oranges and factors affecting its control, *J. Agr. Res.* **68**:363-381.

Brown, G. E., and Lee, H. S., 1993, Interactions of ethylene with citrus stem-end rot caused by *Diplodia natalensis*, *Phytopathol.* **83**:1204-1208.

Brown-Skrobot, S., Brown, L. R., and Filer, T. H., Jr., 1984, Ethylene and carbon monoxide production by *Septoria musiva*, *Dev. Indus. Microbiol.* **25**:749-755.

Brown-Skrobot, S., Brown, L. R., and Filer, T. H., Jr., 1985, Mechanism of ethylene and carbon monoxide production by *Septoria musiva*, *Dev. Indus. Microbiol.* **26**:567-573.

Burg, S. P. , 1973, Ethylene in plant growth, *Proc. Natl. Acad. Sci. USA* **70**:591-597.

Chalutz, E., and DeVay, J. E., 1969, Production of ethylene *in vitro* and *in vivo* by *Ceratocystis fimbriata* in relation to disease development, *Phytopathol.* **59**:750-755.

Chalutz, E., and Lieberman, M., 1978, Inhibition of ethylene production in *Penicillium digitatum*, *Plant Physiol.* **61**:111-114.

Chalutz, E., Lieberman, M., and Sisler, H. D., 1977, Methionine-induced ethylene production by *Penicillium digitatum*, *Plant Physiol.* **60**:402-406.

Chalutz, E., Mattoo, A. K., Anderson, J. D., and Lieberman, M., 1978, Regulation of ethylene production by phosphate in *Penicillium digitatum*, *Plant Cell Physiol.* **19**:189-196.

Chalutz, E., Mattoo, A. K., and Fuchs, Y., 1980, Biosynthesis of ethylene: The effect of phosphate, *Plant Cell Environ.* 3:349-356.

Chandhoke, V., Goodell, B., Jellison, J., and Fekete, F. A., 1992, Oxidation of 2-keto-4-thiomethylbutyric acid (KTBA) by iron-binding compounds produced by the wood-decaying fungus *Gloeophyllum trabeum*, *FEMS Microbiol. Lett.* **90**:263-266.

Chou, T. W., and Yang, S. F., 1973, The biogenesis of ethylene in *Penicillium digitatum*, *Arch. Biochem. Biophys.* **157**:73-82.

Coleman, L. W., and Hodges, C. F., 1986. The effect of methionine on ethylene and 1-aminocyclopropane-1-carboxylic acid production by *Bipolaris sorokiniana*, *Phytopathol.* **76**:851-855.

Considine, P. J., and Patching, J. W., 1975, Ethylene production by microorganisms grown on phenolic acids, *Ann. Appl. Biol.* **81**:115-119.

Considine, P. J., Flynn, N., and Patching, J. W., 1977, Ethylene production by soil microorganisms, *Appl. Environ. Microbiol.* **33**:977-979.

Dasilva, E. J., Henriksson, E., and Henriksson, L. E., 1974, Ethylene production by fungi, *Plant Sci. Lett.* **2**:63-66.

Dimond, A. E., and Waggoner, P. E., 1953, The cause of epinastic symptoms in *Fusarium* wilt of tomatoes, *Phytopathol.* **43**:663-669.

Dutta, B. K., and Deb, P. R., 1984, Sporostatic effect of ethylene and its possible role in soil fungistasis, *Microbios Lett.* **27**:31-35.

Elad, Y., 1988, Involvement of ethylene in the disease caused by *Botrytis cinerea* on rose and carnation flowers and the possibility of control, *Ann. Appl. Biol.* **113**:589-598.

El-Kazzaz, M. K., Sommer, N. F., and Kader, A. A., 1983, Ethylene effects on *in vitro* and *in vivo* growth of certain postharvest fruit-infecting fungi, *Phytopathol.* **73**:998-1001.

El-Sharouny, H. M., 1984, Screening of ethylene producing root infecting fungi in Egyptian soil, *Mycopathologia* **85**:13-15.

Fergus, C. L., 1954, The production of ethylene by *Penicillium digitatum*, *Mycologia* **46**:543-555.

Filer, T. H., Brown, L. R., Brown-Sarobot, S., and Martin, S., 1984, Production of ethylene and carbon monoxide by microorganisms, *J. Mississippi Acad. Sci.* **29**:27-31.

Flaishman, M. A., and Kolattukudy, P. E., 1994, Timing of fungal invasion using host's ripening hormone as a signal, Proc. Natl. Acad. Sci. USA **91**:6579-6583.

Freebairn, H. T., and Buddenhagen, I. W., 1964, Ethylene production by *Pseudomonas solanacearum*, *Nature* **202**:313-314.

Fujii, T., Ogawa, T., and Fukuda, H., 1985, A screening system for microbes which produce olefin hydrocarbons, *Agric. Biol. Chem.* **49**:651-657.

Fukuda, H., 1984, Microbial production of gaseous hydrocarbons and tasks for its practical use (challenge to cope with unstable oil supply forecast in 21st Century), *Chem. Economy Engineer. Review* **16**:18-22.

Fukuda, H., and Ogawa, T., 1991, Microbial ethylene production, in: *The Plant Hormone Ethylene*, A. K. Mattoo and J. C. Suttle, eds., CRC Press, Boca Raton, FL, pp. 279-292.

Fukuda, H., Fujii, T., and Ogawa, T., 1984, Microbial production of C_2-hydrocarbons, ethane, ethylene, and acetylene, *Agric. Biol. Chem.* **48**:1363-1365.

Fukuda, H., Fujii, T., and Ogawa, T., 1986, Preparation of a cell-free ethylene-forming system from *Penicillium digitatum*, *Agric. Biol. Chem.* **50**:977-981.

Fukuda, H., Fujii, T., and Ogawa, T., 1988. Production of ethylene by a growth-suppressed mutant of *Penicillium digitatum*, *Biotechnol. Bioeng.* **31**:620-623.

Fukuda, H., Takahashi, M., Fujii, T., and Ogawa, T., 1989a, Ethylene production from L-methionine by *Cryptococcus albidus*, *J. Ferment. Bioeng.* **67**:173-175.

Fukuda, H., Kitajima, H., Fujii, T., Tazaki, M., and Ogawa, T., 1989b, Purification and some properties of a novel ethylene-forming enzyme produced by *Penicillium digitatum*, *FEMS Microbiol. Lett.* **59**:1-6.

Fukuda, H., Takahashi, M., Fujii, T., Tazaki, M., and Ogawa, T., 1989c, An NADH:Fe(III) EDTA oxidoreductase from *Cryptococcus albidus*: An enzyme involved in ethylene production in vivo? *FEMS Microbiol. Lett.* **60**:107-112.

Gibson, M. S., and Young, R. E., 1966, Acetate and other carboxylic acids as precursors of ethylene, *Nature* **210**:529-530.

Goel, R. K., and Jhooty, J. S., 1987, Stimulation of germination of teliospores of *Urocystis agropyri* by volatiles from plant tissues, *Ann. Appl. Biol.* **111**:295-300.

Goto, M., and Hyodo, H., 1987, Ethylene production by cell-free extract of the Kudzu strain of *Pseudomonas syringae* pv. *phaseolicola*, *Plant Cell Physiol.* **28**:405-414.

Goto, M., Ishida, Y., Takikawa, Y., and Hyodo, H., 1985, Ethylene production by the Kudzu strain of *Pseudomonas syringae* pv. *phaseolicola* causing halo blight in *Pueraria lobata* (Wild) Ohwi, *Plant Cell Physiol.* **26**:14-150.

Goto, M., Yaguchi, Y., and Hyodo, H., 1980, Ethylene production in citrus leaves infected with *Xanthomonas citri* and its relation to defoliation, *Physiol. Plant. Pathol.* **16**:343-350.

Graham, J. H., and Linderman, R. G., 1980, Ethylene production by ectomycorrhizal fungi, *Fusarium oxysporum* f. sp. *pini*, and by aseptically synthesized ectomycorrhizae and *Fusarium*-infected Douglas fir roots, *Can. J. Microbiol.* **26**:1340-1347.

Graham, J. H., and Linderman, R. G., 1981, Effect of ethylene on root growth, ectomycorrhiza formation, and *Fusarium* infection of Douglas-fir, *Can. J. Bot.* **59**:149-155.

Hahm, D. H., Kwak, M. Y., Bae, M., and Rhee, J. S., 1992, Effect of dissolved oxygen tension on microbial ethylene production in continuous culture, *Biosci. Biotechnol. Biochem.* **56**:1146-1147.

Hanke, M., and Dollwet, H. H. A., 1976, The production of ethylene by certain soil fungi, *Sci. Biol. J.* **2**:227-230.

Hill, T. F., Jr. and Lyda, S. D., 1976, Gas exchanges by *Phymatotrichum omnivorum* in a closed, axenic system, *Mycopathologia* **59**:143-147.

Hottiger, T., and Boller, T., 1991, Ethylene biosynthesis in *Fusarium oxysporum* f. sp. *tulipae* proceeds from glutamate/2-oxoglutarate and requires oxygen and ferrous ions, *Arch. Microbiol.* **157**:18-22.

Ilag, L., and Curtis, R. W., 1968, Production of ethylene by fungi, *Science* **159**:1357-1358.

Ince, J. E., and Knowles, C. J., 1985, Ethylene formation by cultures of *Escherichia coli*, *Arch. Microbiol.* **141**:209-213.

Ince, J. E., and Knowles, C. J., 1986, Ethylene formation by cell-free extracts of *Escherichia coli*, *Arch. Microbiol.* **146**:151-158.

Ishii, T., Shrestha, Y. H., Matsumoto, I., and Kadoya, K., 1996, Effect of ethylene on the growth of vesicular-arbuscular mycorrhizal fungi and on the mycorrhizal formation of trifoliate orange roots, J. Japan Soc. Hort. Sci. **65**:525-529.

Jacobsen, D. W., and Wang, C. H., 1968, The biogenesis of ethylene in *Penicillium digitatum*, *Plant Physiol.* **43**:1959-1966.

Johnson, L. B., Davis, L. C., and Stuteville, D. L., 1980, Ethane and ethylene evolution from alfalfa seedlings infected with *Peronospora trifoliorum*, *Physiol. Plant Path.* **16**:155-162.

Kapulnik, E., Mattoo, A. K., Chalutz, E., and Chet, I., 1983, The relationship between the production of ethylene by certain fungi, their enzyme constitution, and their ability to accumulate intracellular phosphate, *Z. Pflazenphysiol.* **109**:347-354.

Kepczynski, J., and Kepczynska, E., 1977, Effect of ethylene on germination of fungal spores causing fruit rot, *Fruit Science Reports (Poland)* **4**:31-35.

Kevers, C., Driessche, T. V., and Gaspar, T., 1989, Relationship between ethylene and ethane production by whole *Acetabularia mediterranea* cells and by microsomal fractions, *Saussurea* **19**:121-123.

Kreslavskii, V. D., Markarov, A. D., Brandt, A. B., Kiseleva, M. I., Mukhin, E. N., Ruzieva, R. Kh., Yakumin, A. F., and Rudendo, T. I., 1988, Effects of red light, phytohormones, and dichlorophenylurea derivatives on ethylene evolution by chlorella, *Sov. Plant Physiol.* **35**:889-896.

Kutsuki, H., and Gold, M. H., 1982, Generation of hydroxyl radical and its involvement in lignin degradation by *Phanerochaete chrysosporium, Biochem. Biophys. Res. Commun.* **109**:320-327.

Lindberg, T., Granhall, U., and Berg, B., 1979, Ethylene formation in some coniferous forest soils, *Soil Biol. Biochem.* **11**:637-643.

Lockard, J. D., and Kneebone, L. R., 1962, Investigation of the metabolic gases produced by *Agaricus bisporus* (Large) Sing, *Mushroom Sci.* **5**:281-299.

Lockhart, C. L., Forsyth, F. R., and Eaves, C. A., 1968, Effect of ethylene on development of *Gloeosporium album* in apple and on growth of the fungus in culture, *Can. J. Plant Sci.* **48**:557-559.

Lu, S.-F., Tzeng, D. D., and Hsu, S.-T., 1989, *In vitro* and *in vivo* ethylene production in relation to pathogenesis of *Pseudomonas solanacearum* (Smith) on *Lycopersicon esculentum* M., *Plant Protection Bull.* **31**:60-76.

Lynch, J. M., 1972, Identification of substrates and isolation of microorganisms responsible for ethylene production in soil, *Nature* **240**:45-46.

Lynch, J. M., 1974, Mode of ethylene formation by *Mucor hiemalis, J. Gen. Microbiol.* **83**:407-411.

Lynch, J. M., 1975, Ethylene in soil, *Nature* **256**:576-577.

Lynch, J. M., and Harper, S. H. T., 1974, Formation of ethylene by a soil fungus, *J. Gen. Microbiol.* **80**:187-195.

Mansouri, S., and Bunch, A. W., 1989, Bacterial ethylene synthesis from 2-oxo-4-thiobutyric acid and from methionine, *J. Gen. Microbiol.* **135**:2819-2827.

Mattoo, A. K., Baker, J. E., Chalutz, E., and Liebermann, M., 1977, Effect of temperature on the ethylene-synthesizing systems in apple, tomato, and in *Penicillium digitatum.* Plant Cell Physiol. 18:715-719.

Mattoo, A. K., Chalutz, E., Anderson, J. D., and Lieberman, M., 1979, Characterization of the phosphate-mediated control of ethylene production by *Penicillium digitatum, Plant Physiol.* **64**:55-60.

Meheriuk, M., and Spencer, M., 1964, Ethylene production during germination of oat seeds and *Penicillium digitatum* spores, *Can. J. Bot.* **42**:337-340.

Michniewicz, M., Czerwinska, E., Rozej, B., and Bobkiewicz, W., 1983, Control of growth and development of isolates of *Fusarium culmorum* (W.G.Sm.) Sacc. of different pathogenicity to wheat seedlings by plant growth regulators. II. Ethylene, *Acta Physiol. Plant.* **5**:189-198.

Michniewicz, M., Czerwinska, E., and Lamparska, K., 1985, Inhibitory effect of ethylene on the growth and development of *Cylindrocarpon destructans* (Zins.) Scholt., *Bull. Pol. Acad. Sci. Biol. Sci.* **33**:125-129.

Miller, E. V., Winston, J. R., and Fisher, D. F., 1940. Production of epinasty of emanations from normal and decaying citrus fruits and from *Penicillium digitatum.* J. Agr. Res. 60:269-278.

Nagahama, K., Ogawa, T., Fujii, T., Tazaki, M., Goto, M., and Fukuda, H., 1991a, L-Arginine is essential for the formation *in vitro* of ethylene by an extract of *Pseudomonas syringae, J. Gen. Microbiol.* **137**:1641-1646.

Nagahama, K., Ogawa, T., Fujii, T., Tazaki, M., Tanase, S., Morino, Y., and Fukuda, H., 1991b, Purification and properties of an ethylene-forming enzyme from *Pseudomonas syringae* pv. *phaseolicola* PK2, *J. Gen. Microbiol.* **137**:2281-2286.

Nagahama, K., Ogawa, T., Fujii, T., and Fukuda, H., 1992, Classification of ethylene-producing bacteria in terms of biosynthetic pathways to ethylene, *J. Ferment. Bioeng.* **73**:1-5.

Nickerson, W. J., 1948, Ethylene as a metabolic product of the pathogenic fungus, *Blastomyces dermatitidis. Arch. Biochem.* **17**:225-233.

Nikitin, D. I., and Nikitina, E. S., 1978, Protsessy samoochishcheniya okruzhayushchei sredy i parazity bakterii (rod *Bdellovibrio*). (Self-purification of the environment and parasites of *Bdellovibrio* bacteria). Nauka, Moscow.

Ogawa, T., Murakami, H., Yamashita, K., Tazaki, M., Fujii, T., and Fukuda, H., 1992, The stimulatory effect of catalase on the formation *in vitro* of ethylene by an ethylene-forming enzyme purified from *Penicillium digitatum, J. Ferment. Bioeng.* **73**:58-60.

Pazout, J., and Pazoutova, S., 1989, Relationship between aeration, carbon source, and respiration yield and the ethylene production and differentiation of static cultures of *Penicillium cyclopium* and *P. velutinum. Can. J. Microbiol.* **35**:619-622.

Pazout, J., Wurst, M., and Vancura, V., 1981, Effect of aeration on ethylene production by soil bacteria and soil samples cultivated in a closed system, *Plant Soil* **62**:431-437.

Pazout, J., Pazoutova, S., and Vancura, V., 1982, Effects of light, phosphate and oxygen on ethylene formation and conidiation in surface cultures of *Penicillium cyclopium, Westling Curr. Microbiol.* **7**:133-136.

Peacock, B. C., and Muirhead, I. F., 1974, Ethylene production by *Colletotrichum musae, Queensl. J. Agric. Anim. Sci.* **31**:249-252.

FACTORS AFFECTING MICROBIAL PRODUCTION

137

Pegg, G. F., and Cronshaw, D. K., 1976, The relationship of *in vitro* to *in vivo* ethylene production in *Pseudomonas solanacearum* infection of tomato, *Physiol. Plant Pathol.* **9**:145-154.

Phan, C. T., 1962, Contribution a L'etude de la production de L'ethylene par le *Penicillium digitatum. Sacc. Rev. Gen. Bot.* **69**:505-543.

Primrose, S. B., 1976a, Formation of ethylene by *Escherichia coli*, *J. Gen. Microbiol.* **95**:159-165.

Primrose, S. B., 1976b, Ethylene-forming bacteria from soil and water, *J. Gen. Microbiol.* **97**:343-346.

Primrose, S. B., 1977, Evaluation of the role of methional, 2-keto-4-methyl-thiobutyric acid and peroxidase in ethylene formation by *Escherichia coli*, *J. Gen. Microbiol.* **98**:519-528.

Primrose, S. B., 1979, A review, ethylene and agriculture: The role of microbes, *J. Appl. Bacteriol.* **46**:1-25.

Primrose, S. B., and Dilworth, M. J., 1976, Ethylene production by bacteria, *J. Gen. Microbiol.* **93**:177-181.

Reyes, A.A., and Smith, R. B., 1986, Controlled atmosphere effects on the pathogenicity of fungi on celery and on the growth of *Botrytis cinerea*, *HortSci.* **21**:1167-1169.

Russo, V. E. A., Halloran, B., and Gallori, E., 1977, Ethylene is involved in the autochemotropism of phycomyces., *Planta* **134**:61-67.

Sato, M., Watanabe, K., Yazawa, M., Takikawa, Y., and Nishiyama, K., 1997, Detection of new ethylene-producing bacteria, *Pseudomonas syringae* pvs. *cannabina* and *sesami*, by PCR amplification of genes for the ethylene-forming enzyme, *Phytopathol.* **87**:1192-1196.

Sato, M., Urushizaki, S., Nishiyama, K., Sakai, F., and Ota, Y., 1987, Efficient production of ethylene by *Pseudomonas syringae* pv. *glycinea* which causes halo blight in soybeans, *Agric. Biol. Chem.* **51**:1177-1178.

Shipston, N., and Bunch, A. W., 1989, The physiology of L-methionine catabolism to the secondary metabolite ethylene by *Escherichia coli*, *J. Gen. Microbiol.* **135**:1489-1497.

Smith, A. M. , 1973, Ethylene as a cause of soil fungistasis, *Nature* **246**:311-313.

Smith, A. M., 1976, Ethylene in soil biology, *Annu. Rev. Phytopathol.* **14**:53-73.

Smith, A. M., and Cook, R. J., 1974, Implications of ethylene production by bacteria for biological balance of soil, *Nature* **252**:703-705.

Spalding, D. H., and Lieberman, M., 1965, Factors affecting the production of ethylene by *Penicillium digitatum*, *Plant Physiol.* **40**:645-648.

Strzelczyk, E., Kampert, M., and Li, C. Y., 1994a, Cytokinin-like substances and ethylene production by *Azospirillum* in media with different carbon sources, *Microbiol. Res.* **149**:55-60.

Strzelczyk, E.,Kampert, M., and Pachlewski, R., 1994b, The influence of pH and temperature on C_2H_4 production by mycorrhizal fungi of pine. *Mycorrhiza* **4**:193-196.

Strzelczyk, E., Pokojska, A., Kampert, M., Michalski, L., and Kowalski, S., 1989, Production of plant growth regulators by non-mycorrhizal fungi associated with the roots of forest trees, in: *Interrelationships between Microorganisms and Plants in Soil*, Proc. Intl. Symp. Liblice. Czech. V. Vancura and F. Kunc (eds.). Acad. Publ., Czech Acad. Sci., Prague, pp. 213-222.

Swanson, B. T., Wilkins, H. F., and Kennedy, B. W., 1979, Factors affecting ethylene production by some plant pathogenic bacteria, *Plant Soil* **51**:19-26.

Swart, A., and Kamerbeek, G. A., 1976, Different ethylene production *in vitro* by several species and formae speciales of *Fusarium*, *Neth. J. Plant Pathol.* **82**:81-84.

Swart, A., and Kamerbeek, G. A., 1977, Ethylene production and mycelium growth of the tulip strain of *Fusarium oxysporum* as influenced by shaking of and oxygen supply to the culture medium, *Physiol. Plant.* **39**:38-44.

Tanaka, H., Enoki, A., and Fuse, G., 1986, Correlation between ethylene production from α- or γ-methylthiobutyric acid and degradation of lignin dimeric model compounds by wood inhibiting fungi, *Mokuzai Gakkaishi* **32**:125.

Thomas, K. C., and Spencer, M., 1977, L-Methionine as an ethylene precursor in *Saccharomyces cerevisiae*, *Can. J. Microbiol.* **23**:1669-1674.

Thomas, K. C., and Spencer, M., 1978, Evolution of ethylene by *Saccharomyces cerevisiae* as influenced by the carbon source for growth and the presence of air, *Can. J. Microbiol.* **24**:637-642.

Tzeng, D. D., and DeVay, J. E., 1984, Ethylene production and toxigenicity of methionine and its derivatives with riboflavin in cultures of *Verticillium*, *Fusarium*, and *Colletotrichum* species exposed to light, *Physiol. Plant.* **62**:545-552.

Völksch, B., and Weingart, H., 1997, Comparison of ethylene-producing *Pseudomonas syringae* strains isolated from kudzu (*Pueraria lobata*) with *Pseudomonas syringae* pv. *phaseolicola* and *Pseudomonas syringae* pv. *glycinea*, *European J. Plant Pathol.* **103**:795-802.

Völksch, B., and Weingart, H., 1998, Toxin production by pathovars of *Pseudomonas syringae* and their antagonistic activities against epiphytic microorganisms, *J. Basic Microbiol.* **38**:135-145.

Ward, T., Turner, E. M., and Osborne, D. J., 1978, Evidence for the production of ethylene by the mycelium of *Agaricus bisporus* and its relationship to sporocarp development, *J. Gen. Microbiol.* **104**:23-30.

Watanabe, T., and Kondo, N., 1976, Ethylene evolution in marine algae and a proteinaceous inhibitor of ethylene biosynthesis from red alga, *Plant, Cell Physiol.* **17**:1159-1166.

Weingart, H., and Völksch, B., 1997, Ethylene production by *Pseudomonas syringae* pathovars *in vitro* and *in planta*, *Appl. Environ. Microbiol.* **63**:156-161.

Weingart, H., Völksch, B., and Ullrich, M. S., 1999, Comparison of ethylene production by *Pseudomonas syringae* and *Ralstonia solanacearum*, *Phytopathol.* **89**:360-365.

Wilkes, J., Dale, G. T., and Old, K. M., 1989, Production of ethylene by *Endothia gyrosa* and *Cytospora eucalypticola* and its possible relationship to kino vein formation in *Eucalyptus maculata*, *Physiol. Mol. Plant Pathol.* 34:171-180.

Wood, D. A. and Hammond, J. B. W., 1977, Ethylene production by axenic fruiting cultures of *Agaricus bisporus*, *Appl. Environ. Microbiol.* **34**:228-229.

Young, R. E., Pratt, H. K., and Biale, J. B., 1951, Identification of ethylene as a volatile product of the fungus *Penicillium digitatum*, *Plant Physiol.* **26**:304-310.

5

ETHYLENE IN SOIL

The presence of ethylene (C_2H_4) in the soil atmosphere may be of great ecological significance because extremely low concentrations (ppb range) in the vicinity of roots could have a profound effect on growth and development of plants (Smith and Russell, 1969; Smith, 1976a; Primrose, 1979; Arshad and Frankenberger, 1988, 1990a,, 1992; Muromtsev et al., 1993, 1995; Bibik et al., 1995; Zahir and Arshad, 1998). Ethylene levels in soil atmosphere can be in the 10 μl L^{-1} and above range (Perret and Koblet, 1984; Meek et al., 1983; Dowdell et al., 1972; Campbell and Moreau, 1979; Smith and Dowdell, 1974) which are physiologically active levels of C_2H_4. The amounts of C_2H_4 in the soil atmosphere can vary greatly from soil to soil depending upon many biotic and abiotic factors (Goodlass and Smith, 1978a; Babiker and Pepper, 1984; Hunt et al., 1982; Arshad and Frankenberger, 1990b,c, 1991; Zechmeister-Boltenstern and Smith, 1998). The amount of C_2H_4 detected in soil represents a net balance of production minus decomposition implying that C_2H_4 production and decomposition take place simultaneously (Arshad and Frankenberger, 1990c; Zechmeister-Boltenstern and Smith, 1998; Nohrstedt, 1983; Hendrickson, 1989). In the following sections, C_2H_4 production, its stability and factors affecting its accumulation in soil will be discussed.

5.1. ETHYLENE PRODUCTION IN SOIL

Several studies have shown that soil is a potential source of C_2H_4 (Smith and Russell, 1969; Campbell and Moreau, 1979; Considine et al., 1977; Dowdell et al., 1972; Goodlass and Smith, 1978a,b; Hunt et al., 1980; Lindberg et al., 1979; Lynch, 1972, 1975; Rovira and Vendrell, 1972; Smith, 1976a; Smith and Cook, 1974; Smith and Dowdell, 1974; Smith and Restall, 1971; Yoshida and Suzuki, 1975; Arshad and Frankenberger, 1988, 1990a,b,c, 1991; Zechmeister-Boltenstern and Smith, 1998). However, the amount of C_2H_4 generated can vary widely in unamended and amended soils (Table 5.1). Both biological and chemical processes seem to contribute to C_2H_4 accumulation in soil, but in most cases, biotic production is the major source of C_2H_4 in soil.

Table 5.1. Ethylene accumulation (nmol kg^{-1} soil 7 days^{-1}) in California soils.

Soil	Unamended (control)	L-Methionine[a]	D-Glucose[b]	L-Methionine[a] + D-glucose[b]	LSD ($p < 0.05$)
				Amendments	
Crownhill	4.6	130.8	83.0	153.7	20.1
Santa Lucia	204.0	126.6	232.5	489.2	86.5
Sheephead	37.0	79.2	193.3	256.5	35.4
Tollhouse	9.3	83.6	115.3	92.7	12.4
Fallbrook	170.0	128.5	344.8	547.2	65.5
Cibo	39.3	40.7	149.0	195.8	36.9
Kitchen Creek	1.1	67.0	9.4	69.3	19.2
Altamont	51.4	52.7	77.6	188.1	33.0
Garey	348.4	211.1	516.9	784.0	115.1
Kimberly	29.0	150.3	207.0	256.6	45.8
Pico	5.0	34.9	154.2	341.1	38.3
Ramona	62.3	72.0	140.2	198.8	43.5
Oildale	31.0	77.4	91.1	302.2	24.7
Hanford	5.2	97.7	119.1	180.6	36.6
Oceano	11.8	78.2	203.3	173.8	47.7
Domino	1.4	34.4	93.0	184.9	31.3
Redding	3.7	16.9	49.8	110.6	16.9
Milham	3.0	50.0	164.3	302.3	22.8
Hesperia	3.0	55.1	99.6	197.3	26.5
LSD ($p < 0.05$)	29.2	11.4	56.1	51.3	

[a]L-Methionine (1 g kg^{-1} soil).
[b]D-Glucose (5 g kg^{-1} soil).
Source: Arshad and Frankenberger (1991).

5.1.1. Abiotic Production

Some studies have demonstrated that a small amount of C_2H_4 found in soil environments might originate from a nonbiological source, because soil sterilization by irradiation or by autoclaving does not completely inhibit C_2H_4 accumulation in soil (Rovira and Vendrell, 1972; Smith and Restall, 1971; Frankenberger and Phelan, 1985a; Harvey and Linscott, 1978; Nakayama and Ota, 1980; Arshad and Frankenberger, 1990b,c). It is highly likely that some compounds present in native organic matter could yield C_2H_4 through nonenzymatic transformations (Smith et al., 1978; Arshad and Frankenberger, 1990b). For example, L-ethionine (an analogue of methionine) can be decomposed nonenzymatically to C_2H_4 in soil since its addition to autoclaved soil results in accumulation of copious amounts of this gas in the soil atmosphere (Arshad and Frankenberger, 1990b). Arshad and Frankenberger (1991) also reported that Fe(II) at ≥ 100 mg g^{-1} soil promotes the abiotic production of C_2H_4 in soil amended with methionine (MET). Nakayama and Ota (1980) investigated the effects of water, compost and MET on C_2H_4 production in sterilized and non-sterilized soil and hypothesized that C_2H_4 production in submerged conditions might be mainly due to the action of microorganisms, while in upland conditions, non-biological evolution of C_2H_4 was the main source. Soil sterilants (autoclaving, treatment with ethylene oxide or selective inhibitors) reduced C_2H_4 production in spruce soil, but C_2H_4 release after the sterilant treatment continued at a low rate (Table 5.2), indicating the potential for low rates of chemical production of C_2H_4 in these spruce soils (Sexstone and Mains, 1990). Many other workers also support the view that some soil C_2H_4 is released by non-microbial or nonenzymatic activity in soil (Nakano and Kuwatsuka, 1979; Ishii and Kadoya, 1984a, 1987). The effect of autoclaving on C_2H_4 evolution from a waterlogged soil containing dead grape leaves is shown in Fig. 5.1 (Ishii and Kadoya, 1984a). It is most likely that the relative contribution of the abiotic sources of the C_2H_4 pool in soil might be dependent on the nature of the soil environment, substrate(s) present in soil (quality and quantity) and the activity of free radicals and peroxides.

5.1.2. Biotic Production

Several studies have reported that the major fractions of soil C_2H_4 comes from biological transformations of either indigenous organic compounds or compounds applied to soil exogenously. The primary evolution of C_2H_4 from soils is caused by microbial activity (Nakano and Kuwatsuka, 1979; Nakayama and Ota, 1980; Pazout et al., 1981; Primrose, 1979; Smith, 1976a,b; Smith and Dowdell, 1974). Non-sterilized soils amended with various organic compounds release C_2H_4 in magnitudes several fold greater than that released when the same soils are sterilized and treated with sterile solutions of these amendments, suggesting a major role of soil microbiota in production of C_2H_4 (Babiker and Pepper, 1984; Frankenberger and Phelan, 1985a; Arshad and Frankenberger, 1990b,c; 1991). Antibiotic treatment of soil reduces C_2H_4 synthesis (Table 5.3) which further confirms that soil living entities are involved in C_2H_4 production in soil (Arshad and Frankenberger, 1990c; Frankenberger and Phelan, 1985b). Ishii and Kadoya (1987) detected copious amounts of C_2H_4 released upon treatment of lipids with a soil enzyme extract compared with traces of C_2H_4 released without treatment, indicating that the major source of C_2H_4 was biotransformation of lipids into C_2H_4 may be an important source. Smith (1976b) proposed that soils should be screened

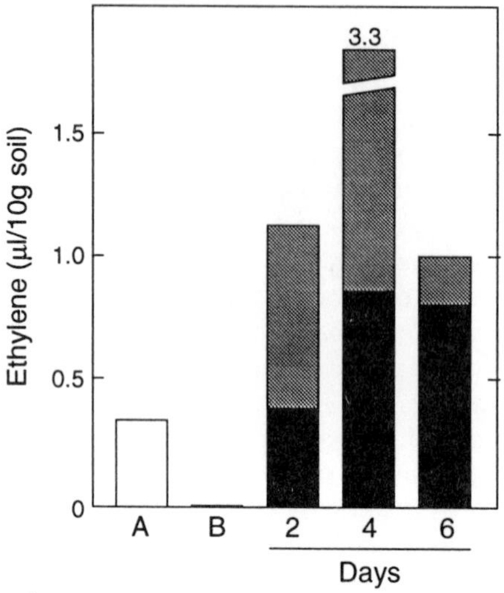

Fig. 5-1. Effect of autoclaving on C_2H_4 evolution from a waterlogged soil containing dead grape leaf (g/10 g soil). (A) immediately after autoclaving, (B) immediately after the air in the vial containing autoclaved soil was replaced with sterilized air (no C_2H_4 was detected), ▩ C_2H_4 evolved from non-autoclaved soil containing dead grape leaf (control), ■ C_2H_4 evolution after the air in the vial containing autoclaved soil was replaced with the air containing no C_2H_4 and then unsterilized soil (20 g) was added to the autoclaved soil. Source: Ishii and Kadoya (1984a).

Table 5.2. Effect of microbial inhibitors on C_2H_4 production in organic spruce forest soils.

	C_2H_4 production rate[a]			
	Spruce Mountain		Gaudineer Knob	
Treatment	L horizon	F horizon	L horizon	F horizon
	(nmol C_2H_4-C kg soil^{-1} h^{-1})			
Control	205.5 (15)	167.2 (29)	166.8[c] (27)	71.8 (3)
Chloramphenicol	191.2 (6)	161.1 (34)	186.6 (9)	78.1 (13)
BES*	200.5 (15)	189.7 (12)	176.6 (7)	78.4 (5)
HgCl	70.1 (47)[b]	23.2 (10)[b]	41.8 (29)[b]	6.0 (46)[b]
Ethylene oxide	7.1 (35)[b]	11.4 (33)[b]	48.5 (17)[b]	21.0 (10)[b]
Autoclaved	32.5 (39)[b]	4.4 (14)[b]	7.8 (26)[b]	0.0[b]

[a]Average of 4 replicate samples except where noted. Values in parentheses are the % coefficient of variation (CV) among replicates. CV = (standard deviation mean) x 100.
[b]Treatment significantly different from control ($p < 0.05$) as determined by Student's t-test.
[c]Average of 3 replicates.
*BES, 2-bromo ethane sulfonic acid.
Source: Sexstone and Mains (1990).

Table 5.3. Effects of various antibiotics on ACC dependent-C_2H_4 formation in soils.

Treatment	Ethylene production after treatment in soil specified[†]		
	Pico	Kitchen Creek	Altamont
Untreated	16.7	5.6	3.1
Amphotericin B	9.0	2.3	1.7
Chloramphenicol	10.5	3.9	2.0
Cycloheximide	12.3	3.8	1.7
Erythromycin	11.9	3.6	1.9
Kanamycin	9.4	3.5	2.3
Nalidixic acid	8.2	3.1	1.3
Nystatin	7.5	2.9	2.1
Penicillin-G	11.9	3.5	2.1
Streptomycin	6.5	1.4	0.6

[†]Expressed in mmol of C_2H_4 released kg^{-1} soil 48 h^{-1}.
Source: Frankenberger and Phelan (1985b).

for their "microbial nutrient status" (measured by O_2 utilization and CO_2 evolution) to which C_2H_4 could be correlated. This assumption was found to hold for six different soils screened for C_2H_4 production by Zechmeister-Boltenstern and Smith (1998). Hodges and Campbell (1998) reported production of C_2H_4 and other gaseous hydrocarbons from calcareous sand inoculated with cyanobacteria, *Nostoc* and *Oscillatoria*. The above discussion supports the view that a major fraction of soil C_2H_4 comes from biological sources.

In addition to soil microbiota, plant roots may also contribute to the biotic fraction of soil C_2H_4, particularly when exposed to stress (Abeles et al., 1992; Hyodo, 1991).

5.1.2.1. Types of Soil Microorganisms Involved

All major groups of soil microbiota, i.e., bacteria, actinomycetes, fungi and algae, are capable of producing C_2H_4, however, their contribution varies greatly with the prevailing soil environments and the nature of substrates present in native soil organic matter or added as soil amendments.

Some workers view soil bacteria as the major contributors to soil C_2H_4 (Smith and Cook, 1974; Cook and Smith, 1977; Smith and Restall, 1971; Sutherland and Cook, 1980; Primrose, 1976; Pazout et al., 1981). Smith and Cook (1974) proposed that the main production of C_2H_4 in soil is carried out by spore-forming bacteria living in anaerobic microsites. Ethylene production in soil decreased as the soil water potential was lowered from saturation to −5 bars or slightly lower (Cook and Smith, 1977). This also supports the premise that soil bacteria are involved mainly in C_2H_4 production because of the inability of fungi to survive under anaerobic conditions. More C_2H_4 was evolved from soil when incubated under anaerobic conditions (Smith and Restall, 1971; Cook and Smith, 1977; Smith and Cook, 1974). This observation suggested that obligate anaerobic or facultative anaerobic bacteria are the main producers of C_2H_4 in soil. Smith

(1976b) also claimed that anaerobic bacteria play a major role in C_2H_4 release in soil. Sutherland and Cook (1980) further supported this theory by reporting that two antibacterial agents, chloramphenical and novobiocin, inhibited C_2H_4 synthesis when they were added to soil; whereas, the antifungal agent cycloheximide had no effect. They suggested that C_2H_4 formation in soil is most likely attributed to facultative and obligate anaerobic bacteria. Numerous studies demonstrated C_2H_4 production in soil incubated under strictly anaerobic conditions, i.e., waterlogged or N_2 environments (see Section 5.2.2) which may exclude the role of aerobic fungi and favor the role of anaerobic bacteria in C_2H_4 accumulation in soil. However, other studies have shown that bacteria isolated from soil are capable of producing C_2H_4 only under aerobic conditions (Primrose, 1976; Pazout et al., 1981).

Many researchers claim that soil fungi are mainly responsible for biotic C_2H_4 production in soil (Lynch and Harper, 1974a,b; 1980; Lynch, 1975; Babiker and Pepper, 1984). Lynch (1983) reported that treatment of an arid soil with an antibacterial agent, novobiocin, had no effect on C_2H_4 production indicating that bacteria are not the sole producers. Similarly, Babiker and Pepper (1984) suggested that *Fusarium* spp. have a greater contribution in C_2H_4 accumulation in a desert soil. Lynch and Harper (1974a), Lynch (1975), and Babiker and Pepper (1984) confirmed that C_2H_4 production can occur under aerobic conditions, but argued that mobilization of the substrate for C_2H_4 production is enhanced under anoxic conditions, whereas, production of C_2H_4 mainly occurs under aerobic conditions. Lynch and Harper (1980) concluded that soil fungi, including yeast, and to a lesser extent, anaerobic bacteria, are responsible for C_2H_4 production in soil. Numerous soil fungi are now known capable of producing C_2H_4 as described in Chapter 4. By using MET as the sole source of N in a basal salt medium, Arshad and Frankenberger (1989) enriched various rhizosphere isolates from corn to investigate their ability to generate C_2H_4. We concluded that both fungi and bacteria are capable of producing C_2H_4 in the presence of MET, with fungi being the predominant microbes. Very recently, Arshad and Akhter (unpublished data) further confirmed this premise as they found that fungal isolates of wheat, maize, tomato and potato rhizosphere dominate among the microorganisms capable of producing C_2H_4 from L-MET or α-keto methylbutyric acid (KMBA).

The use of antibiotics (both antibacterial and antifungal) has demonstrated that both bacteria and fungi contribute actively to soil C_2H_4. Frankenberger and Phelan (1985b) employed nine antibiotics (antibacterial as well as antifungal) to evaluate their effects on 1-aminocyclopropane-1-carboxylic acid (ACC)-derived C_2H_4 synthesis in soil. They observed various degrees of inhibition in C_2H_4 production and concluded that both bacteria and fungi are active in generating C_2H_4 in soil (Table 5.3). Arshad and Frankenberger (1990c) also used antibacterial and antifungal agents (antibiotics) to assess the relative contributions of bacteria and fungi in MET-dependent C_2H_4 production in aerobic (field capacity) and waterlogged soils (Table 5.4) and concluded that the involvement of any specific group of heterotrophs in C_2H_4 generation from soil is highly dependent on the nature of the substrates available and the prevailing environment factors (Arshad and Frankenberger, 1990b,c). Likewise, C_2H_4 production in soils amended with ACC or KMBA was suppressed in response to antibiotics (antifungal and antibacterial) application (Arshad and Nazli, unpublished data) which further confirms the premise that both bacteria and fungi are active in generation of C_2H_4 in soil. Zechmeister-Boltenstern and Smith (1998) hypothesized that C_2H_4 producers belong to physiologically different groups. Cyanobacteria, *Nostoc* and *Oscillatoria* have also been reported capable of producing C_2H_4 when inoculated to a calcareous sand (Hodges and Campbell, 1998). Similarly, lower plants (green algae) including *Regnellidium diphllum, Marsilea quadrifolia* and others have been shown to synthesize C_2H_4 (Cherneys and Kende, 1996; Osborne et al., 1996). Thus, it is very likely that almost all groups of soil microflora contribute to C_2H_4 formation in soil, but their relative efficiency is dependent on the prevailing soil and environmental conditions and the nature of substrates available.

5.2. PHYSICOCHEMICAL PROPERTIES OF SOIL AND ETHYLENE PRODUCTION

Accumulation of C_2H_4 in soil has been found to vary with physico-chemical properties of soils such as organic matter, soil reaction, aeration, temperature, texture, and soil depth. Since these properties are interdependent, it is highly likely that C_2H_4 production in soil is a function of the integrated effects of these physico-chemical and biological parameters. Moreover, as discussed earlier, C_2H_4 catabolism also occurs in soil and the rate of catabolism could differ depending upon soil properties and prevailing environmental factors, thus the amount of C_2H_4 detected is a net balance of both processes. It is difficult to single out the effect of individual soil parameters or environmental conditions on C_2H_4 production.

Table 5.4. Influence of antibiotics on C_2H_4 production (nmol C_2H_4 kg^{-1} soil 7 days^{-1}) in soil.

	Soil conditions	
Treatment	Field capacity	Waterlogged
Unamended control	0	14.8
Streptomycin	5.5	5.5
Amphotericin B	8.1	6.9
LSD (0.05)	2.7	3.0
Relative activity of methionine-dependent C_2H_4 production (%)		
L-methionine	100.0	100.0
L-methionine + streptomycin	102.5	82.7
L-methionine + amphotericin B	86.5	105.4
LSD (0.05)	3.4	14.3
Relative activity of D-glucose-dependent C_2H_4 production (%)		
D-glucose	100.0	100.0
D-glucose + streptomycin	92.1	63.5
D-glucose + amphotericin B	94.6	82.0
LSD (0.05)	8.1	7.3

LSD, least significant difference
Source: Arshad and Frankenberger (1990c).

5.2.1 . Soil Organic Matter

Soil organic matter (SOM) plays a critical role in maintaining the biological activity of soil. SOM can affect the level of C_2H_4 in soil by supporting the activity of C_2H_4–producing soil microbial population and/or by serving as a pool of possible substrate(s) of C_2H_4 . Many scientists believe that high concentrations of C_2H_4 are released from soils rich in organic matter content (van Cleemput et al., 1983; Goodlass and Smith, 1978b; Lynch and Harper, 1980; Babiker and Pepper, 1984). Goodlass and Smith (1978a) reported a significant relationship between organic matter content of fresh and dried soils and their ability to generate C_2H_4 when incubated under anaerobic conditions (Fig. 5.2). Likewise, Smith and Restall (1971) found that increases in C_2H_4 accumulation in six soils were correlated with increases in organic matter content. Among 12 desert soils, two soils with the highest level of organic matter content produced several-fold more C_2H_4 than those with low organic matter (Babiker and Pepper, 1984). Lynch and Harper (1980) indicated a significant correlation ($r = 0.63$, $p < 0.001$) between the amount of uncombined organic matter (nonhumus) and the levels of C_2H_4 accumulated in soil. However, no relation between C_2H_4 production and the amount of humus present in soil was observed (Lynch and Harper, 1980). On the other hand, Zechmeister-Boltenstern and Nikodim (1999) reported that C_2H_4 formation under anaerobic conditions was found to be related to the amount of humus and total nitrogen in soil ($r > 0.90^*$, $p < 0.05$).

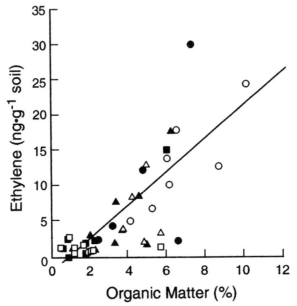

Fig. 5-2. Relation between C_2H_4 evolved from air-dried soils and organic matter content. o, △, □, grassland soils, 0-10, 10-20, 30+ cm depth, respectively. ●, ▲, ■, arable soils, 0-10, 10-20, 30+ cm depth, respectively. Source: Goodlass and Smith (1978a).

They also reported that soils producing high amounts of C_2H_4 were relatively rich in extractable glucose equivalents, implying that availability of low molecular weight carbon sources (reducing sugars) in these soils and C_2H_4 production were positively correlated ($r > 0.70^*$, $p < 0.05$). Three coniferous soils containing 43.6, 35.4, and 14.1% C exhibited C_2H_4 activity of 11.28, 7.99, and 4.09 pmole g^{-1} soil h^{-1} (when incubated under 500 µl acetylene L^{-1} to eliminate C_2H_4 degradation), indicating that the soil with the highest organic C showed highest C_2H_4 production. Rigler and Zechmeister-Boltenstern (1998) reported substantially greater C_2H_4 production rates in an alpine spruce soil (14.2% organic C/humus) compared to a deciduous forest soil (3.7% organic C/humus) and hypothesized that the difference in C_2H_4 production potential of these soils is due to the different organic matter content or Fe content. Ishii and Kadoya (1987) studied the role of lipids present in soil organic matter and C_2H_4 production. They concluded that lipids in soil are important in the release of C_2H_4, and both the quantity and quality of the lipids are related to C_2H_4 evolution in soil. Likewise, Arshad and Nazli (unpublished data) found that C_2H_4 production was much more in organic matter-rich soil than organic matter-poor soils amended with ACC or KMBA and unamended Pakistani soils. This higher accumulation of C_2H_4 in organic matter-rich soils could also be attributed to relatively longer persistence of C_2H_4 due to adsorption of C_2H_4 on organic matter.

Contrary to these findings, some studies have demonstrated no correlation between soil organic matter content and the ability of soil to generate C_2H_4. Arshad and Frankenberger (1991) found no correlation between organic matter content and the C_2H_4-releasing ability of 19 California soils. Similarly, Hunt et al. (1982) tested 21 soils and also found no relation between C_2H_4 accumulation and soil organic matter content alone. However, by using a quadratic surface response equation, Hunt et al. (1982) suggested that maximum C_2H_4 production was correlated with the interaction of soil organic matter, NO_3^- and pH ($R^2 = 0.77$). Out of six soils amended with MET and glucose, the lowest C_2H_4 accumulation was found in soils having the highest organic matter content, i.e., 8.3% (Zechmeister-Boltenstern and Smith, 1998). Tang and Miller (1993) investigated the effects of composted (fermented) and non-composted poultry litter on C_2H_4 production in soil and observed that non-composted poultry litter was more effective in promoting C_2H_4 release than composted litter. The composted litter had 7 times more humic substances than non-composted materials. They argued that higher molecular weight humic substances may not be a good source of C_2H_4 in soil. Smith (1973, 1976a,b) reported that soils high in organic matter content are not always prolific in C_2H_4 production. Smith (1976b) found that soils high in organic matter, but low in available nutrients, produced equal amounts of C_2H_4 when compared to soils with less organic matter and relatively high in available microbial nutrients. It is possible that not only the amount of soil organic matter but its composition may have an important roles in C_2H_4 evolution in soil.

5.2.2. Soil Aeration

Soil aeration is considered one of the most critical factors regulating biological and chemical activity in soil. In many studies, C_2H_4 concentrations detected in the soil atmosphere have been reported without considering its catabolism, thus controversy exists regarding aerobic vs. anaerobic conditions promoting C_2H_4 biosynthesis in soil. Since the first report of C_2H_4 accumulation in an anaerobic soil by Smith and Russell (1969), numerous studies have demonstrated that anaerobiosis is more favorable for C_2H_4 generation in soil. Often more C_2H_4 detection is reported after lowering the O_2 tension, or exposure to waterlogged conditions, or replacing O_2 with other gases, such as N_2, Ar, or He (Lindberg et al., 1979; Smith and Restall, 1971; Smith, 1976b; Smith et al., 1978; Harvey and Linscott, 1978; Smith and Cook, 1974; Cook and Smith, 1977; Dowdell et al., 1972; Smith and Dowdell, 1974; Campbell and Moreau, 1979; Hunt et al., 1980; Pazout et al., 1981). Smith and Restall (1971) reported that appreciable amounts of C_2H_4 did not occur unless the O_2 concentration was below 2% (Fig. 5.3). The analysis of gases collected from soil compacted to various degrees revealed that higher levels of C_2H_4 were detected when CO_2 levels were also high (Zainol et al., 1991). Nakayama and Ota (1980) observed that C_2H_4 production in a submerged soil was 5 to 7 times greater than that of an upland soil. Similarly, the amount of C_2H_4 evolved from a soil containing dead grape leaves under anaerobic (N_2) conditions increased up to 3 times of that detected under aerobic conditions (Ishii and Kadoya, 1984a). They also found that increasing the soil water content resulted in increased C_2H_4 production in soil amended with dead grape leaves (Fig. 5.4). Zechmeister-Boltenstern and Nikodim (1999) examined C_2H_4 production in different soils incubated at various water potentials (-100 to 0 kPa) and detected high levels of C_2H_4 in soils maintained at 0 kPa, i.e., water saturated soils. They speculated that in addition to C_2H_4 production, saturation with water may have caused a release of C_2H_4 or its substrates which had formerly been adsorbed to humus and clay

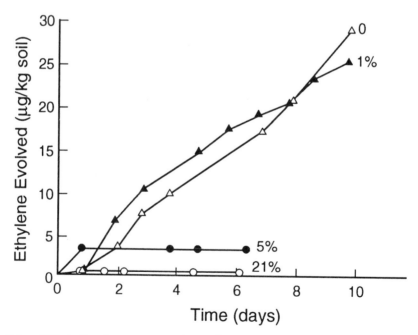

Fig. 5-3.   Effect of various concentrations of oxygen on production of C_2H_4 evolved by suspensions of a clay soil. Source: Smith and Restall (1971).

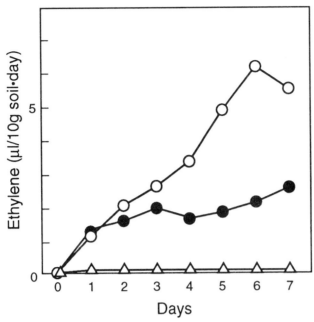

Fig. 5-4. Effect of soil water content on C_2H_4 evolution from the soil containing dead grape leaves (g/10 g soil). △ air-dried, ● 15%, ○ 100% waterlogged. Source: Ishii and Kadoya (1984a).

particles and the death of aerobic microorganisms under anaerobic conditions may have served as a substrate for C_2H_4 production in the water saturated soils. Rigler and Zechmeister-Boltenstern (1998) found that increasing the CO_2 concentrations in the soil atmosphere (0-20%) resulted in more C_2H_4 accumulation in an alpine spruce forest soil while in the presence of acetylene (an inhibitor of C_2H_4 oxidation), CO_2 did not have any affect implying that increases in CO_2 reduced C_2H_4 catabolism and enhanced its accumulation in soil. Conlin and van den Driessche (1996) observed the interactive effects of water treatment and compaction on C_2H_4 accumulation in soil. They reported that C_2H_4 concentrations appeared to substantially rise and then decrease in response to increased compaction at various water table treatments. This may imply that with increased compaction, C_2H_4 production increases until a threshold level of O_2 is available.

Moreover, it has been argued that anaerobic spore-forming bacteria are the primary producers of C_2H_4 in soil (Smith, 1976a,b; Smith and Cook, 1974). Smith (1976a) proposed a self-regulating "O_2-C_2H_4" (Fig. 5.5) cycle which involves anaerobic microsites in an adequately aerated soil where anaerobes proliferate, especially near a rich organic source, and they produce C_2H_4 that diffuses throughout the soil. Diffusion of O_2 into these anaerobic microsites regulates C_2H_4 production, permitting resumption of aerobic growth until the cycle is repeated again. Based upon the observation that Fe(II) stimulates abiotic production of C_2H_4 (see Section 5.5), Smith et al. (1978) modified this

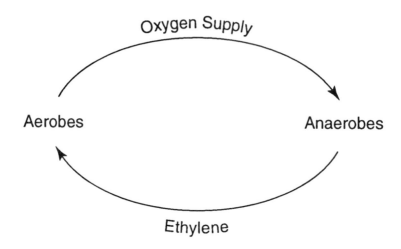

Fig. 5-5. Proposed "oxygen-ethylene" cycle in relation to the
regulation of C_2H_4 formulation in soil. Source: Smith (1976b).

"O_2-C_2H_4" cycle. They proposed that reduced conditions mobilize Fe(II), which triggers
C_2H_4 synthesis chemically which in turn, diffuses out through the soil and inactivates
aerobic microorganisms. The rest of the cycle is the same as previously described.
Primrose (1979) questioned Smith's (1976a,b) proposed "O_2-C_2H_4" hypothesis on the
grounds that (i) C_2H_4 is readily oxidized by aerobic bacteria; (ii) aerobic microorganisms
are the primary producers of C_2H_4; and (iii) C_2H_4 has little effect on the metabolic
activity of aerobic microorganisms. Moreover, Fe(II) added to soil enhances the
nonbiological production of C_2H_4 under fully aerobic conditions (Arshad and
Frankenberger, 1991), which is contrary to the hypothesis of Smith et al. (1978).
However, other workers have suggested that an interactive oxidative-reductive process
may be involved in which anaerobic microsites are conducive for the synthesis or
mobility of C_2H_4 substrates, whereas aerobic conditions favor C_2H_4 production (Lynch,
1975; Lynch and Harper, 1980).

Many studies do not support the modified O_2-C_2H_4 cycle postulated by Smith et al.
(1978). Babiker and Pepper (1984) found significantly greater C_2H_4 levels in autoclaved
soils amended with the aerobic C_2H_4–producing soil isolates, *Mucor hiemalis* and
Fusarium spp., compared with noninoculated, nonautoclaved soils. Moreover, this study
was carried out under conditions in which O_2 was not a limiting factor at field capacity.
Ethylene synthesis in ACC-amended and MET-amended soils revealed that aerobic
conditions are more favorable for these biotransformations (Frankenberger and Phelan,
1985a,b; Arshad and Frankenberger, 1990c). Frankenberger and Phelan (1985a) reported
that anaerobiosis (replacing air with N_2) resulted in 60-85% reduction in ACC-derived
C_2H_4 accumulation in soil. Zechmeister-Boltenstern and Smith (1998) found that C_2H_4

production was highest in sycamore-dominated deciduous woodland and in grassland under aerobic soil conditions as they detected more C_2H_4 in soil maintained at -100 and -5 kPa than at 0 kPa (Fig. 5.6). The soils were amended with glucose and MET. According to Lynch and Harper (1980), soils rich in natural substrates often yield high amounts of C_2H_4 evolution under anaerobic conditions, whereas, substrate-deficient soils are comparable in C_2H_4 production under both aerobic and anaerobic conditions. The addition of MET and glucose resulted in elevated C_2H_4 synthesis under aerobic conditions (Lynch and Harper, 1980).

It is highly likely that reduced conditions might affect chemically generated C_2H_4 while aerobic conditions support the biological release of C_2H_4 in addition to differential degrees of C_2H_4 persistence under these two regimes. Many of the early workers neglected a critical factor of soil acting as a sink for C_2H_4. Soil C_2H_4 is subjected to rapid catabolism under aerobic conditions which could exceed production, whereas, it persists longer under anaerobic conditions with more accumulation of C_2H_4 in anaerobic soils. Hendrickson (1989) found that under aerobic conditions, C_2H_4 production and oxidation occurred simultaneously in organic and inorganic horizons of forest soils. Ethylene oxidation rates in a mineral soil were high relative to other forest soils and exceeded production rates, unless the moisture content was raised to saturation. Jackson (1985) suggested that during the early stages of waterlogging, microorganisms form C_2H_4 when the soil O_2 is still available, but with time the gas becomes entrapped in water into a subsequent anoxia phase and is preserved by slow rates of degradation. Cornforth (1975) claims that C_2H_4 is microbially decomposed in aerobic soils much faster than it is produced in anaerobic soils. He concludes that biologically active concentrations of C_2H_4 will occur in soil for only short periods associated with waterlogging. Similarly, Zechmeister-Boltenstern and Smith (1998) observed that C_2H_4 decomposition rates under

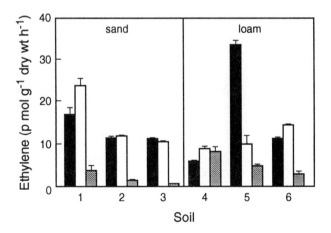

Fig. 5-6. Net C_2H_4 production rates after 24 h under different moisture conditions after the addition of glucose and methionine (ethylene in pmol g^{-1} dry wt h^{-1}). ■ -100 kPa water tension, □ -5 kPa, ▨ 0 kPa. Soils: 1 sycamore, 2 barley, 3 fallow, 4 elm, 5 grassland, 6 spruce. Source: Zechmeister-Boltenstern and Smith (1998).

Fig. 5.6.

aerobic conditions exceeded C_2H_4 production rates by factors of 10 to 100. Rigler and Zechmeister-Boltenstern (1999) found that an increase in CO_2 concentrations (up to 20%) in soil acts negatively upon the function of the soil as a sink for volatile hydrocarbons (C_2H_4 and CH_4), implying that aerobic conditions are more conducive for C_2H_4 catabolism. Arshad and Frankenberger (1990c) investigated C_2H_4 catabolism and stability in soils under field capacity (unsaturated, aerobic) and waterlogged (saturated, anaerobic) conditions. We observed that the amount of C_2H_4 removed in 3 days with a field moist soil was equal to that degraded in 6 days by the same waterlogged soil, indicating greater stability or persistence of C_2H_4 under reduced or waterlogged conditions (Fig. 5.7). We also concluded that C_2H_4 production under these two extremes is dependent on the nature of the organic amendments added to soil. Methionine was a better stimulator of C_2H_4 production in soil maintained at field capacity than under waterlogged conditions, whereas, glucose was equally effective under both moisture regimens (Table 5.5).

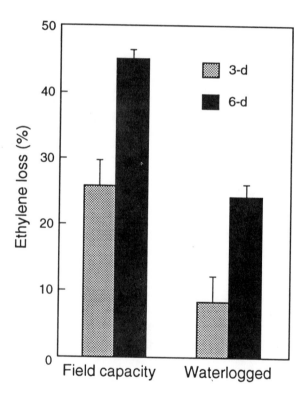

Fig. 5-7. Ethylene stability in soil under field capacity and waterlogged conditions. Source: Arshad and Frankenberger (1990c).

Table 5.5. Ethylene production (nmol kg^{-1} soil) in unamended and amended soil under field capacity and waterlogged conditions.

Soil conditions	Incubation time (h)							
	24	48	72	96	120	144	168	SE[a]
Unamended soil								
Field capacity	0.0	1.2	2.2	3.6	3.8	0.6	0.0	(0.0)
Waterlogged	0.0	0.0	1.9	6.0	12.9	13.3	21.7	(7.5)
L-MET-amended soil								
Field capacity	16.5	26.7	39.0	43.1	61.5	73.8	96.3	(8.2)
Waterlogged	1.6	7.2	10.4	21.0	32.3	44.0	61.8	(15.1)
D-Glucose-amended soil								
Field capacity	0.0	40.1	84.5	95.3	102.6	118.9	129.5	(11.1)
Waterlogged	0.0	14.6	48.1	79.0	88.3	113.9	129.8	(21.9)
L-MET + D-glucose-amended soil								
Field capacity	14.6	53.9	93.8	113.9	114.0	132.0	153.7	(7.1)
Waterlogged	6.0	29.4	122.1	173.6	176.7	199.8	342.5	(22.2)

[a]Standard error in parentheses.
Source: Arshad and Frankenberger (1990c).

It is most likely that greater stability of C_2H_4 under waterlogged conditions contributes to accumulation of high concentrations of C_2H_4 in soils rather than production at high rates. This increased stability under reducing conditions could be due to a less optimal environment for C_2H_4-decomposing microorganisms, because it is known that oxidation is involved in the yield of ethylene oxide as the catabolic product of C_2H_4 (see Chapter 3). Also, under anaerobic conditions, NO_3^- is reduced/denitrified and its suppressive effect on C_2H_4 production is removed (see Section 5.4). Moreover, under anaerobic conditions Fe(III) is reduced to Fe(II) which is a known stimulant of nonenzymatic C_2H_4 production (see Section 5.5). Thus, it is highly likely that these two factors [NO_3^- and Fe(II)] may favor higher C_2H_4 accumulation in soil under anaerobic conditions.

To reach a definitive conclusion regarding C_2H_4 production under aerobic vs. anaerobic conditions, it is necessary to eliminate C_2H_4 degradation by use of inhibitors such as acetylene. Interestingly, Zechmeister-Boltenstern and Smith (1998) found that the addition of 5 mg kg^{-1} glucose and 1 mg kg^{-1} MET not only promoted C_2H_4 production, but also inhibited C_2H_4 decomposition. Since several studies have demonstrated increased C_2H_4 accumulation in soil in response to amendments such as glucose and MET under aerobic conditions (see Section 5.3.1), it is possible that this increase could be partially due to less degradation of produced C_2H_4 in the presence of these substrates.

5.2.3. Soil Reaction

Soil pH directly influences the activity of soil microbiota and enzymatic as well as nonenzymatic reactions in soil; hence, it could have a strong influence on the biological as well as chemical production of C_2H_4 in soil. Several studies have demonstrated a pH-dependent variation in C_2H_4 evolution in soil. In general, acidic pH favors C_2H_4 accumulation. Goodlass and Smith (1978a) observed a significant relationship between enhanced C_2H_4 production and decreasing pH over a pH range of 7.3-5.0 for arid soils (Fig. 5.8). Similarly, Arshad and Frankenberger (1991) found a significant negative linear correlation between soil pH and the amounts of C_2H_4 released in 19 California unamended ($r = 0.54*$) and MET-amended ($r = 0.53*$) soils (Fig. 5.9). While comparing three coarse-textured soils, Babiker and Pepper (1984) reported that a soil having a low pH (6.5) yielded more C_2H_4 than soils having higher pH values (7.6 and 8.0). They speculated that the soil environment of the acidic soil (pH 6.5) was more favorable for the C_2H_4-producing microbiota. Similarly, a negative correlation ($r = -0.926$, $p > 0.008$) was found between C_2H_4 production and the pH of six soils, implying that acidic pH favors C_2H_4 accumulation in soil (Zechmeister-Boltenstern and Smith, 1998). In another study, Zechmeister-Boltenstern and Nikodim (1999) found that C_2H_4 production was negatively correlated with soil pH. However, they speculated that soil pH may have had an indirect affect on C_2H_4 production since decreased pH usually leads to increased availability of Fe(II), which stimulates abiotic C_2H_4 production in soil (see Section 3.5). Ishii and Kadoya (1987) studied lipid-dependent C_2H_4 production by a crude enzyme extract from soil. They found that when pH of the soil extract solution was between 5 to 6, the amount of C_2H_4 evolved was high and decreased with increasing pH.

On the other hand, van Cleemput et al. (1983) argued that the presence of $CaCO_3$ (5.6%) and consequently, the high pH (7.74) made a clayey soil more favorable for C_2H_4 production than two other soils that had a low pH (loamy, pH 6.15; sandy, pH 4.20).

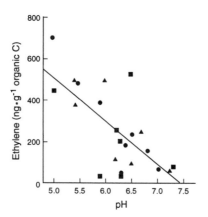

Fig. 5-8.   Relation between C_2H_4 evolved per unit organic C as a function of pH, from air-dried arable soils. ●, 0-10 cm; ▲, 10-20 cm; ■, 30+ cm depth. Source: Goodlass and Smith (1978a).

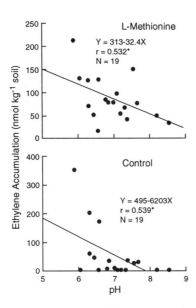

Fig. 5-9. Relation between pH and C_2H_4 accumulation in unamended (control) and L-methionine-amended California soils. Source: Arshad and Frankenberger (1991).

Increased C_2H_4 production in soil in response to the amendment of lime was also observed by Wainwright and Kowalenko (1977). Smith et al. (1978) found that abiotic C_2H_4 production, triggered by Fe(II) applied at 1000 mg kg^{-1} soil was not markedly affected by changing the pH from 4.1 to 5.6. Frankenberger and Phelan (1985a) reported a direct relationship (r = 0.62**) between soil pH of 20, unbuffered surface soils and ACC-enhanced C_2H_4 production. They found that C_2H_4 derived from ACC in three buffered soils was optimum in the alkaline range (Frankenberger and Phelan, 1985a).

From these studies, it is not possible to derive a general conclusion concerning the relationship between soil pH and C_2H_4 production. However, it is likely that a positive or negative effect of soil pH on C_2H_4 production depends primarily on the nature of substrate(s) of C_2H_4 present in soil, the dominating C_2H_4 producing soil microflora and on incubation conditions.

5.2.4. Soil Temperature

Temperature not only affects the biological and enzymatic activity of soil, but also influences chemical transformations. Ethylene accumulation in soil is generally enhanced with increasing temperature in the mesophilic range. Smith and Dowdell (1974) reported that concentrations of C_2H_4 increased logarithmically with soil temperature, nearly 20-fold over the range of 4-11°C. Consistent with enzymatic reactions, an Arrhenius plot of C_2H_4 production was linear over the temperature range of 5 to 45°C for all the soils tested (Sexstone and Mains, 1990). Smith and Restall (1971) indicated that temperatures below 10°C resulted in considerably lower rates of C_2H_4

evolution, and higher temperatures (30-35°C) promoted a 1.6- to 2.0-fold increase. Similarly, Lindberg et al. (1979) reported that no C_2H_4 was produced from root-free humus at <10°C, and incubation above 24°C had a strong stimulation on C_2H_4 formation. At 37°C, C_2H_4 production was about 8-fold greater than that of 24°C. Ethylene released from a waterlogged soil amended with various organic materials was 3 to 4 times greater when incubated at 60°C than that at 30°C, and at 4°C, only a trace amount of C_2H_4 was detected (Ishii and Kadoya, 1984a). In another study, Ishii and Kadoya (1987) detected no C_2H_4 evolution below 4°C from soil lipids upon treatment with crude enzyme extracts of soils. Similarly, Otani and Ae (1993) found a positive correlation between C_2H_4 concentration and soil temperature (Fig. 5.10). They also reported great fluctuations in C_2H_4 concentrations from August to the middle of September, and a sharp decline from October. In ACC-amended soil, C_2H_4 production reached a maximum of 50°C being 2.3- to 3.8-fold greater than under ambient (22°C) conditions (Frankenberger and Phelan, 1985a). Similarly, Sexstone and Mains (1990) reported much more C_2H_4 evolution from L- and F-horizons of Spruce Mountain and Gaudineer Knob soils incubated at 65°C than at 55°C. Heating the soil up to 80°C resulted in increased C_2H_4 evolution in two soils (Smith and Cook, 1974). Lynch (1975) also confirmed that soil heating at 100°C for 10 minutes can stimulate C_2H_4 production. The increase in C_2H_4 content of soil with increasing temperature might be due to enhanced enzymatic activity responsible for biological C_2H_4 production or due to thermal decomposition of organic matter and desorption of adsorbed C_2H_4 (Frankenberger and Phelan, 1985a,b; Sexstone and Mains, 1990). Moreover, temperature variation may have a direct impact on substrate(s) mobilization. This may imply that soils of tropical regions might have higher soil C_2H_4 levels than that of temperate regions.

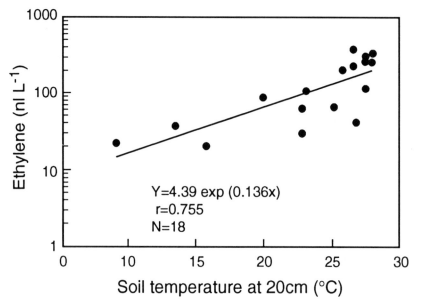

Fig. 5-10. Changes in C_2H_4 concentrations in relation to soil temperature under field conditions. Source: Otani and Ae (1993).

5.2.5. Soil Texture

Soil texture may affect C_2H_4 evolution in soil since it has a strong effect on other soil properties. For instance, soil texture could affect C_2H_4 accumulation in soil by influencing biological activity via controlling the water and gaseous movement as well as substrate mobility. Zechmeister-Boltenstern and Smith (1998) observed that gross C_2H_4 production rates were higher in a loamy soil than in sandy soils. In another study, Zechmeister-Boltenstern and Nikodim (1999) found that fine-textured soils produced 3 to 30 times more C_2H_4 than coarse-textured sandy soils and anaerobic C_2H_4 formation was positively correlated with the clay content (r = 0.952***, p < 0.001). Similarly, C_2H_4 production at a given application rate of either composted or non-composted poultry litter was greater in fine textured Hilleman and Foley soils than in a coarse textured Bosket soil (Tang and Miller, 1993). Many other workers have supported the premise that C_2H_4 production is related to soil clay contents (Smith and Restall, 1971; Lynch and Harper, 1980; Babiker and Pepper, 1984). On the other hand, Arshad and Frankenberger (1991) found no relationship between C_2H_4 production and texture of 19 California soils. Similarly, van Cleemput et al. (1983) did not find any relationship between C_2H_4 production and texture of soil.

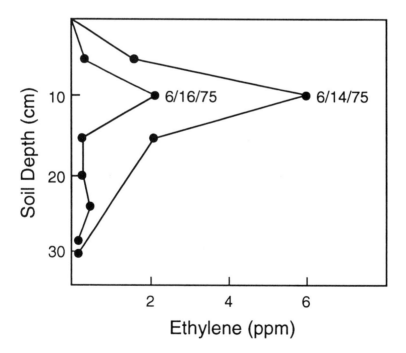

Fig. 5-11. Ethylene concentrations in relation to soil depth on two sampling dates. Source: Campbell and Moreau (1979).

5.2.6. Soil Depth

Available substrates of C_2H_4, microbial populations (density and composition) as well as C_2H_4 and O_2 diffusion may vary with depth. Campbell and Moreau (1979) found the highest level of C_2H_4 at the 10-cm depth (Fig. 5.11). They attributed this observation to the balance between C_2H_4 production and gaseous diffusion, speculating that maximum generation might have occurred at less than 10 cm below the surface, since gaseous diffusion in the surface layer is higher than at lower depths within the profile. Dowdell et al. (1972) determined C_2H_4 levels at 30-, 60-, and 90-cm depths throughout the year, and reported that in general, C_2H_4 concentrations were lower at the 90-cm depth. Similarly, when soil samples were taken at 0- to 10, 10- to 20, and $\geq$30-cm depths and incubated anaerobically, the upper 0- to 10-cm soil samples yielded the most C_2H_4 (Goodlass and Smith, 1978a). Hunt et al. (1980) assessed C_2H_4 evolution from soil samples obtained from different horizons of a sandy loam, incubated at saturation with various O_2 tensions. Higher amounts of C_2H_4 were evolved from the A_p layer than A_2 and B horizons. Syslo et al. (1990) examined the C_2H_4-producing ability of soil samples representing the A_p (0-13 cm), E (13-96 cm) and B_h (96-107 cm) horizons of an Oldsmar soil incubated at field capacity and saturation. They reported that maximum evolution of C_2H_4 occurred from soil samples obtained from the B_h horizon under field-moist conditions.

5.3. ORGANIC AMENDMENTS AND ETHYLENE ACCUMULATION

Exogenous application of organic compounds and materials to soil stimulates C_2H_4 synthesis either directly by serving as possible C_2H_4 precursor(s) and/or indirectly by stimulating the activity of C_2H_4-producing microbial communities and/or by inhibiting C_2H_4 catabolism. Moreover, physical adsorption or desorption of C_2H_4 in soil may also be influenced by organic amendments. Additionally, these amendments may affect the chemical generation of C_2H_4 in soil. Extensive work is needed to understand clearly how these amendments promote C_2H_4 accumulation in soil.

5.3.1. Organic Amendments with Defined Chemical Composition

Many compounds have been tested for their possible effects on C_2H_4 generation in soil including the amino acid, MET which is established as the physiological precursor of C_2H_4 in higher plants (see Chapter 2). Several scientists have reported that the addition of MET promotes C_2H_4 synthesis in soil (Arshad and Frankenberger, 1990a,b,c; 1991; Babiker and Pepper, 1984; Lynch and Harper, 1980), particularly when applied in combination with glucose (Arshad and Frankenberger, 1990c; 1991; Lynch, 1972; Lynch and Harper, 1974a; 1980; El-Sharouny, 1984; Pazout et al., 1981; Hunt et al., 1980; Babiker and Pepper, 1984; Nakayama and Ota, 1980; Ishii and Kadoya, 1987). As discussed in Chapter 3, it is most likely that MET may promote C_2H_4 production and accumulation in soil by serving as a direct precursor, rather than by stimulating the activity of C_2H_4-producing microbiota. Methionine stimulates C_2H_4 synthesis in soil when incubated under aerobic conditions (Arshad and Frankenberger, 1990b,c; Babiker and Pepper, 1984), while Lynch and Harper (1980) reported that MET also enhances C_2H_4 synthesis under anaerobic conditions. Arshad and Frankenberger (1990c) found that MET promoted more C_2H_4 in soil maintained at field capacity than under waterlogged conditions (Table 5.5). Zechmeister-Boltenstern and Smith (1998) reported that the addition of glucose and MET not only promoted C_2H_4 production in soil, but also inhibited C_2H_4 decomposition. They used various inhibitors of C_2H_4 decomposition such as C_2H_2 and CO and compared C_2H_4 generation in soil amended with glucose and MET. They found that C_2H_4 release in the amended soil in the presence or absence of the inhibitors of C_2H_4 was the same, implying that glucose and MET also inhibit the catabolism of C_2H_4.

Many workers that applied MET in combination with glucose reported stimulation of C_2H_4 synthesis more than MET alone (Lynch and Harper, 1980; Pazout et al., 1981; Hunt et al., 1980; El-Sharouny, 1984). Ethylene can be generated from glucose in addition to MET since Arshad and Frankenberger (1990c) found that glucose was equally effective as a soil amendment for C_2H_4 generation at both field capacity as well as waterlogged conditions (Table 5.5). Since α-ketoglutaric acid (α-KGA) has been recognized as one of the precursors of microbial C_2H_4 synthesis (see Chapter 3), which could be derived from glucose, it is most likely that soil microbial communities may contain microbial groups capable of deriving C_2H_4 from MET as well as from glucose under a given set of conditions. Alternatively, glucose may increase the microbial population by serving as a C source which stimulates C_2H_4 production from MET. Ishii and Kadoya (1987) detected C_2H_4 in a waterlogged soil supplied with sugars including glucose, fructose, galactose, D-mannose, D-mannitol, and sucrose. Sugars added to soil, supplied with lipid extracts of various plant materials, promoted C_2H_4 evolution (Ishii et al., 1983). However, further studies are needed with labeled glucose and MET added to both moisture regimes, to resolve the possibility that glucose can serve as a direct precursor of C_2H_4.

ACC, an immediate precursor of MET-derived C_2H_4 in plants, was found very effective in stimulating C_2H_4 generation in soil (Frankenberger and Phelan, 1985a,b; Arshad and Frankenberger, 1990b; Ishii and Kadoya, 1987; Arshad and Nazli, unpublished data) and might also serve as an C_2H_4 precursor. ACC proved to be a much more effective stimulator of C_2H_4 in soil than MET (Arshad and Frankenberger, 1990b; Ishii and Kadoya, 1987; Arshad and Nazli, unpublished data). ACC-derived C_2H_4 is mainly a biotic process since sterilization (autoclaving) reduced the magnitude of C_2H_4 being released (Frankenberger and Phelan, 1985a; Arshad and Frankenberger, 1990b). Efforts to isolate a microorganism from soil capable of producing C_2H_4 from ACC have not yet been successful. The enzyme, ACC oxidase which catalyzes the conversion of ACC into C_2H_4 in plants requires O_2 for activity. Frankenberger and Phelan (1985a,b) and Arshad and Frankenberger (1990b) reported that ACC addition to soil stimulates C_2H_4 production under aerobic conditions. Frankenberger and Phelan (1985a,b) found that anaerobiosis significantly reduced ACC-derived C_2H_4 production in soil. Contrarily, Ishii and Kadoya (1987) demonstrated that ACC can also stimulate C_2H_4 production in soil under waterlogged conditions.

Several other compounds including amino acids, sugars, organic acids, proteins, alcohols and vitamins have been screened by various researchers for their effectiveness to stimulate C_2H_4 production in soil and the majority of them were found effective for this reaction under both aerobic and anaerobic conditions (Arshad and Frankenberger, 1990b; Lynch and Harper, 1980; El-Sharouny, 1980; Goodlass and Smith, 1978a). Arshad and Frankenberger (1990b) screened 63 compounds (19 amino acids, 6 nonprotein amino acids, 10 organic acids, 8 carbohydrates, 6 proteins, 6 alcohols, 5 vitamins, 4 MET analogues) for their efficiency to stimulate C_2H_4 production and found that most of these compounds enhanced C_2H_4 accumulation in soil (Table 5.6). We speculated that the rhizosphere could be the most favorable site for C_2H_4 biosynthesis because most of these compounds are excreted in soil as root exudates. Lynch and Harper (1980) tested MET, glutamic acid, ethanol and glucose applied to soil alone or in combination with glucose and reported that glucose plus MET was the most effective treatment in promoting C_2H_4. Ethanol suppressed glucose-enhanced C_2H_4 production in soil incubated anaerobically. Similarly, El-Sharouny (1984) studied C_2H_4 synthesis in soil amended with amino acids, sugars, organic acids, and vitamins, all applied in combination with glucose. He reported that MET and ethionine induced C_2H_4 synthesis in soil inoculated with four fungal isolates tested.   Ethylene stimulation was promoted by inoculation with *Fusarium oxysporum* and *Phythium ultimum* in soil amended with sucrose or starch and *Penicillium cyclopium* produced high amounts of C_2H_4 with acetate and succinate added. No carbon source enhanced C_2H_4 synthesis in soil inoculated with *Verticillium* (El-Sharouny, 1984). Goodlass and Smith (1978a) screened various organic acids (5), alcohols (4), proteins (2), sugar (1), acetone, and acetaldehyde for their effects on C_2H_4 production in soil. They reported that under anaerobic conditions, C_2H_4 production was enhanced by lower concentrations (2.5 g kg^{-1} soil) of casein, pepsin, glucose, and ethanol and by higher concentrations of lactic and pyruvic acid (25 g kg^{-1} soil), whereas, C_2H_4 production was depressed by n-butyric acid and high concentrations of glucose. Aerobic incubation of the soil resulted in stimulation of C_2H_4 by ethanol, whereas anaerobic conditions suppressed C_2H_4 accumulation in soil (Goodlass and Smith, 1978a).

5.3.2. Organic Amendments Without Defined Chemical Composition

Application of various organic wastes promotes C_2H_4 accumulation in soil either due to a source of C_2H_4 substrates or by stimulating the activity of C_2H_4 producers. Organic amendments can also reduce C_2H_4 catabolism by serving as alternate substrates for C_2H_4 metabolizers, or by adsorption of C_2H_4 on the surfaces of organic particles protecting it from C_2H_4 utilizers. Many studies have been conducted on plant residues with respect to their effects on C_2H_4 production in soil. Harvey and Linscott (1978) tested fresh and dry quack grass rhizomes, oat straw, alfalfa hay, smooth bromegrass leaves and quack grass leaves and reported that of all these amendments increased C_2H_4 production and accumulation in soil with oat straw being the most effective. Quack grass rhizomes were highly effective only under saturated conditions. Hunt et al. (1980) tested bermuda grass and found that C_2H_4 levels in soil containing 28% moisture and incubated under an O_2 atmosphere increased with increasing amounts of rhizomes during the early stages of incubation, and after 24 days, 5 g of bermuda grass kg^{-1} soil yielded the highest level of C_2H_4. Similarly, the addition of hay, wheat straw, and barley straw on C_2H_4 generation were investigated by Goodlass and Smith (1978b). They reported that soil incubated anaerobically yielded 3.5- and 2.6-fold enhancement in response to wheat straw and barley straw additions, respectively, whereas hay had no effect. Anaerobic incubation of soil mixed with crop residues of barley, dwarf bean and cabbage resulted in significantly higher amounts of C_2H_4 accumulation than the control without amendments (Lynch and Harper, 1980). C_2H_4 concentrations in soil maintained at 25% moisture content increased

Table 5.6. Influence of organic amendments on ethylene accumulation in soil.

Amendments	Incubation time, d					C recovered as C_2H_4[a]
	3	5	7	10	14	%
	-------------------nmol kg^{-1} soil-------------------					
Unamended controls						
Sterile (autoclaved)	1.0	2.6	5.0	12.6	13.2 (1.5)b	--
Nonsterile	3.5	1.7	1.6	4.2	1.2 (0.2)	--
Amino acids						
Glycine	0.0	0.0	0.0	15.6	17.4 (2.8)	0.10
L-Aspartic acid	4.2	5.5	10.1	23.0	30.8 (1.2)	0.18
L-Tryptophan	2.4	0.8	4.3	28.8	31.2 (4.4)	0.18
L-Phenylalanine	0.9	2.5	3.0	15.3	39.4 (3.2)	0.24
L-Asparagine	0.8	5.3	16.5	18.9	40.8 (2.5)	0.24
L-Cystine	0.0	1.0	2.0	26.7	62.4 (4.1)	0.36
L-Cysteine	26.2	47.7	52.3	92.3	103.9 (12.4)	0.62
L-Tyrosine	2.1	32.1	53.3	65.0	107.0 (6.6)	0.64
L-Valine	2.5	3.4	10.3	27.6	136.6 (2.1)	0.82
L-Lysine	1.4	0.0	0.0	45.5	138.2 (4.8)	0.83
L-Arginine	9.0	70.6	110.0	139.5	143.7 (7.2)	0.85
L-Glutamic acid	0.0	0.0	3.5	51.7	148.6 (3.0)	0.89
L-Glutamine	0.0	0.0	31.0	95.1	168.7 (11.8)	1.01
L-Alanine	0.0	1.5	43.1	147.2	182.9 (3.8)	1.09
L-Threonine	1.0	2.2	19.9	135.1	183.1 (10.1)	1.09
L-Serine	0.0	23.6	55.9	121.5	189.2 (6.1)	1.13
L-Proline	0.0	12.7	10.6	171.0	202.3 (9.1)	1.21
L-Histidine	9.2	51.6	99.1	140.0	224.1 (4.2)	1.34
L-Methionine (MET)	73.5	125.4	187.5	294.1	358.6 (18.0)	2.14

Table 5.6. (continued).

Non-protein amino acids						
L–Homocysteine	62.1	74.9	85.3	96.3	148.6 (6.3)	0.89
L-Djenkolic acid	10.0	28.3	59.4	114.5	149.1 (7.8)	0.89
L-Ornithine	13.4	82.6	127.7	169.4	177.6 (8.2)	1.06
L-Amino-n-butyric acid	0.0	0.0	19.9	128.9	182.9 (17.2)	1.09
L-Citrulline	0.0	10.9	87.2	161.5	207.4 (5.1)	1.24
1-Aminocyclopropane-1-carboxylic acid	1778.0	2493.0	2582.0	2683.0	2882.0 (82.0)	17.20
Organic acids						
Butyric acid	3.3	6.4	8.2	7.9	9.2 (2.7)	0.05
Acrylic acid	3.0	5.6	10.2	12.9	14.8 (1.7)	0.09
Propionic acid	3.0	5.6	9.1	8.8	19.5 (3.6)	0.12
Maleic acid	1.9	2.4	2.0	48.5	124.5 (8.1)	0.74
L-Malic acid	4.2	18.0	59.1	108.4	129.9 (8.7)	0.77
L-Tartaric acid	3.2	7.6	39.1	113.0	145.9 (6.6)	0.87
Fumaric acid	0.3	21.4	82.3	102.3	147.9 (4.6)	0.88
Succinic acid	3.5	19.6	106.1	134.9	151.2 (9.7)	0.90
Citric acid	0.6	2.4	77.4	121.6	163.4 (11.3)	0.97
Lactic acid	3.6	17.0	124.0	157.0	176.9 (13.4)	1.06
Carbohydrates						
D-Raffinose	1.3	1.8	5.1	14.6	35.0 (5.3)	0.21
Maltose	2.7	4.3	7.4	20.9	62.2 (11.7)	0.37
D-Glucose	1.1	1.2	7.7	22.3	71.9 (12.8)	0.43
D-Xylose	8.8	9.6	17.8	42.1	93.2 (6.5)	0.56
D-Fructose	0.0	0.6	1.8	25.1	113.6 (7.6)	0.68
Sucros	2.6	2.6	8.0	31.3	135.3 (3.5)	0.81
L-Rhamnose	0.0	0.0	1.4	62.9	176.0 (11.1)	1.05
D-Arabinose	0.0	1.3	1.3	53.6	180.5 (8.4)	1.08
Proteins						
Hordenine	1.2	11.1	26.9	35.7	110.4 (7.4)	0.66
Albumin	1.3	7.2	31.1	72.2	143.6 (2.6)	0.86
Zein	5.2	8.4	65.8	78.7	148.9 (17.7)	0.89
Gluten	1.5	36.2	87.9	98.2	149.7 (11.7)	0.89
Casein	6.2	22.3	85.7	100.6	158.8 (4.4)	0.95
Gliadin	3.6	18.2	92.5	101.5	167.7 (0.9)	1.00

Table 5.6. (continued).

Alcohols					
n-Pentanol	0.0	3.3	2.8	33.0 (9.6)	0.20
n-Butanol	0.0	54.6	119.2	155.4 (12.1)	0.93
n-Propanol	0.0	83.9	155.2	166.9 (9.5)	1.00
Ethanol	0.0	97.6	169.3	186.8 (5.3)	1.11
Vitamins					
Pyridoxine	0.0	1.0	2.7	0.0 (0.0)	0.00
Nicotinic acid	0.0	2.1	2.6	3.7 (0.7)	0.02
Thiamine	0.6	2.2	2.5	3.8 (1.1)	0.02
d-Biotin	3.1	4.5	7.7	7.4 (2.2)	0.04
Inositol	0.0	0.0	78.1	168.4 (7.4)	1.00
Methionine analogs					
Lanthionine	2.8	16.9	20.4	55.6 (9.8)	0.33
DL-Norleucine	1.4	8.5	9.2	73.9 (6.2)	0.44
DL-MET-sulfone	6.0	41.1	72.5	74.2 (7.9)	0.44
DL-MET-sulfoxide	26.8	84.4	86.6	107.3 (19.8)	0.64
L-Ethionine	94.3	229.8	302.5	373.1 (18.5)	2.20

[a]Represents the percent fraction of C added as amendments and recovered as C_2H_4 after 14 d of incubation.
[b]Represents the cumulative standard error of C_2H_4 detected after 14 d of incubation.
Source: Arshad and Frankenberger (1990b).

with increasing amounts of chitin, cellulose, and urea; whereas, barley (up to 20 g kg^{-1} soil) promoted C_2H_4 production initially and then a decline in production followed (Wainwright and Kowalenko, 1977). Otani and Ae (1993) detected an increase in C_2H_4 concentration up to 50 μl L^{-1} when a large amount of straw was added to soil under high humidity. Nakayama and Ota (1980) reported that compost application stimulated C_2H_4 release in soil and the amount of C_2H_4 detected was closely related to the amount of compost applied to soil. Application of organic materials including fresh grape leaf, dead citrus leaves, fresh citrus root, dead Japanese pear leaf, dead peach leaf, and persimmon leaf or rice straw also greatly increased C_2H_4 evolution (Table 5.7; Fig. 5.12) in waterlogged soils (Ishii and Kadoya, 1984a). In another study, Ishii and Kayoda (1984b) tested fowl feces, sawdust compost, rice straw, and dead citrus shoot as possible C_2H_4 stimulants in soil and found that the concentrations of C_2H_4 in soil air were sharply increased by the application of all these organic materials except for the redried dead citrus shoots. The role of lipids present in soil organic matter in C_2H_4 accumulation in soil was investigated by Ishii and Kadoya (1987). They reported that enzymatic conversion of lipids into C_2H_4 is dependent on both quality and the quantity of the lipids. In addition to crop materials, the addition of animal waste (animal excreta and poultry litter) and municipal waste (landfill) were also evaluated by scientists for their effects on C_2H_4 accumulation in soil. Burford (1975) found that amendments with both animal dung and urine enhanced C_2H_4 production and accumulation in soil having 20% moisture content. Fresh slurry was more effective than stored slurry. Similarly, soil amendments with both composted and noncomposted poultry litter significantly increased C_2H_4 production under anaerobic flooded conditions (Tang and Miller, 1993). More C_2H_4 was detected per gram of added carbon with noncomposted litter than with the composted litter, most likely due to differences in bioavailability of the carbon (Table 5.8; Fig. 5.13). Similarly, landfill sites were found to contain C_2H_4 in their gas mixture (Ikeguchi and Watanabe, 1991). The C_2H_4 ratio to other gases (hydrogen, acetylene, pentanes, and oxygenated hydrocarbons) decreased as the landfill age increased. The ratio of C_2H_4 to C_2H_6 appeared to have a correlation with the age of the landfill and tended to be a good indicator of aging (Ikeguchi and Watanabe, 1991).

It is obvious from the above discussed studies that organic materials added to soil can stimulate C_2H_4 accumulation in soil and these concentrations can be high enough to be physiologically active on plants. In evaluating the effects of organic amendments on crop growth and yield, enhanced C_2H_4 in soil air should be taken into account to monitor the physiological and developmental changes in crops.

3.4. INORGANIC AMENDMENTS AND ETHYLENE ACCUMULATION IN SOIL

The concentration of C_2H_4 in soil may be regulated by the use of various agricultural practices (Muromtsev et al., 1990). Like organic amendments, addition of inorganic compounds also stimulates C_2H_4 accumulation in soil. Muromtsev et al. (1995) found that the types and doses of fertilizers adminstered to soil affects C_2H_4 accumulation in the soil atmosphere. Addition of dolomitic rock flour containing magnesium along

Table 5.7. Evolution of C_2H_4 from the waterlogged soil containing organic materials.

Organic materials[a] (g/10 g soil)	Ethylene (nl/10 g soil .day)				
	1	2	3	4	5 (days)
Grape 'Muscat of Alexandria'					
d. leaf	1,151.5	1,741.8	2,557.9	3,318.9	4,757.6
f. leaf[b]	2,315.6	2,661.8	2,120.6	2,618.4	1,476.0
d. shoot	3.7	3.7	17.6	9.0	9.0
f. root	0	1.9	3.7	3.7	3.7
Citrus 'Iyokan'					
d. leaf	108.3	264.3	354.3	338.1	275.0
d. shoot	49.4	39.1	18.5	9.0	12.8
f. root	109.6	129.9	132.7	112.4	182.7
Janpanese pear 'Nijisseiki'					
d. leaf	66.1	198.3	359.8	521.6	345.1
f. leaf	3.7	5.5	12.1	23.8	14.7
d. shoot	6.6	12.9	22.0	22.0	11.0
f. shoot	1.9	1.1	3.7	1.9	2.9
f. root	3.7	2.9	1.9	1.9	2.9
Peach 'Ohkubo'					
d. leaf	90.0	186.5	403.8	262.5	161.5
f. leaf	8.5	42.2	31.2	18.3	14.7
d. shoot	5.5	9.2	8.5	9.2	9.2
f. shoot	0	0	3.7	1.9	1.9
f. root	2.9	1.9	1.9	1.9	1.9
Fig 'Masui Dauphine'					
f. leaf	3.7	9.2	16.5	12.9	9.2
d.shoot	9.2	9.2	12.1	16.5	20.2
f. shoot	3.7	7.4	11.1	14.7	18.3
f. root	1.9	5.5	26.8	20.2	22.0
Japanese persimmon 'Fuyu'					
d.leaf	98.7	104.2	162.3	143.4	67.6
f. leaf	10.8	17.6	13.5	10.8	10.8
f. shoot	1.1	1.1	1.9	1.9	1.9
f. root	0	1.1	1.1	1.9	1.1
Rice straw	137.7	275.4	273.5	254.0	224.1
Cattle faeces and rice straw compost (fermented)	3.7	14.4	22.6	24.0	21.4
Rice hull	4.4	12.1	18.3	--	16.5
Fowl feces and sawdust compost	6.4	8.3	7.4	5.9	4.0
Corn straw	3.7	33.1	28.5	19.3	18.3
Digitaria adscendens Henr.	4.9	22.9	26.4	28.3	44.0
Clover	11.0	23.8	19.1	17.6	25.7

[a]f.=fresh, d.=dead.
[b]Delaware leaf
Source: Ishii and Kadoya (1984a).

Table 5.8. Ethylene production per unit of added carbon and recovery of added carbon as C_2H_4.

C source	C added	Ethylene production			Added C recovered as ethylene		
	g C kg^{-1} soil	nmol C_2H_4 g^{-1} C			% × 10^4		
		Hilleman	Foley	Bosket	Hilleman	Foley	Bosket
Non-composted	1.26	528.8	358.2	172.5	13.0	8.6	4.1
Non-composted	2.52	285.0	258.5	126.9	6.8	6.2	3.0
Non-composted	3.77	218.0	179.9	86.7	5.2	4.3	2.1
Composted	1.03	233.8	269.4	159.9	5.6	6.5	3.6
Composted	2.05	165.9	172.7	98.4	4.0	4.1	2.4
Composted	3.08	158.1	143.0	81.6	3.8	3.4	2.0

Source: Tang and Miller (1993).

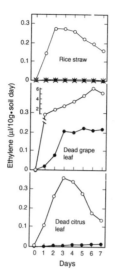

Fig. 5-12. Effect of redrying and fermentation of organic materials (g/10 g soil) on C_2H_4 evolution under waterlogged conditions. o control, ● redried, ✕ fermented. Source: Ishii and Kadoya (1984a).

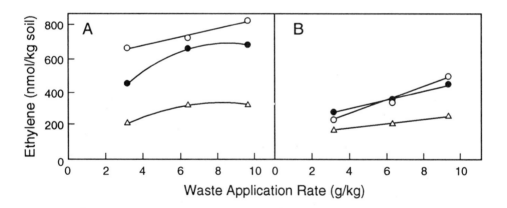

Fig. 5-13. Effect of rate of applied non-composted (A) and composted (B) poultry litter in Hilleman (o), Foley (●), and Bosket (△) soils on C_2H_4 production. The equations describing the response curves in (A) for the Hilleman, Foley and Bosket soils are: $y = 542.3 + 26.0x$, $y = 12.4 + 157.8x - 9.5x^2$, and $y = 8.2 + 75.9x - 4.6x^2$, respectively, where y is C_2H_4 production in nmol kg^{-1} and x is waste application rate in g kg^{-1}. The equations describing the response curves in (B) for the same three soils are $y = 95.3 + 39.3x$, $y = 74.4 + 26.2x$ and $y = 105.2 + 14.4x$. Source: Tang and Miller (1993).

with calcium yielded the best results. Similarly, trace-element fertilizers (molybdenum and cobalt) increased C_2H_4 formation in soil most effectively. Rigler and Zechmeister (1998, 1999) observed more C_2H_4 accumulation in spruce forest soil in response to NH_4^+ [$(NH_4)_2SO_4$] and NO_3^- (KNO_3) addition compared to unamended soils.

After collecting data over a period of 1984-1987, a group of co-workers at the All-Union Scientific Research Institute of Agricultural Microbiology of the Lenin All-Union Academy of Agricultural Science (VASKHNIL), the All-Union Scientific Research Institute of Agricultural Biotechnology of VASKHNIL, and Ukranian Scientific Research Institute of Irrigation Farming suggested that calcium carbide stimulates C_2H_4 accumulation in the soil atmosphere (Muromtsev et al., 1988). According to their hypothesis, calcium carbide (CaC_2) breaks up into acetylene in response to its interaction with water and the acetylene released is readily reduced to C_2H_4 in soil by microorganisms (Muromtsev et al., 1988):

$$\begin{array}{c} \text{soil} \\ CaC_2 \text{ -------------->} Ca(OH)_3 + HC\equiv CH \end{array} \quad \begin{array}{c} \text{soil microorganisms} \\ \text{------------------------->} H_2C=CH_2 \end{array}$$

$$\begin{array}{c} \text{water} \end{array} \qquad\qquad\qquad\qquad \begin{array}{c} \text{enzyme (nitrogenase)} \end{array}$$

Laboratory experiments unequivocally demonstrated the participation of soil microorganisms in the reduction of acetylene to C_2H_4 since no C_2H_4 was detected by gas

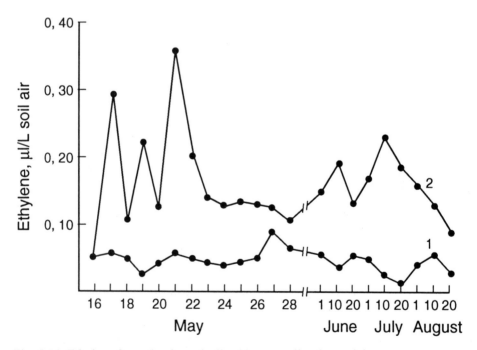

Fig. 5-14. Ethylene formation in podzolized brown soils when calcium carbide was applied: 1) background ($N_{350}P_{250}K_{120}$); 2) background + calcium carbide (150 kg/ha). Source: Muromtsev et al. (1991).

chromatography in sterile soil. The addition of CaC_2 to a non-sterile soil stimulated C_2H_4 formation in soil air by a factor of 45- to 146-fold (1.36 – 4.40 µl L^{-1}). Likewise, greenhouse and field studies over 4 years also supported these findings in the laboratory, with a prolonged and significant increase in the concentration of C_2H_4 in soil of different climatic zones receiving CaC_2 as an amendment (Fig. 5.14) (Muromtsev et al., 1990, 1991). A formulated CaC_2-based product under the trade name of "Retprol" (Muromtsev et al., 1993, 1995; Bibik et al., 1995) was developed as an C_2H_4-releasing compound. Retprol was claimed to provide prolonged decompostion of CaC_2 in soil, being safe in storage and unlike CaC_2 did not absorb moisture from soil air. Long-term comparative studies have demonstrated more release of C_2H_4 from Retprol than CaC_2 (Muromtsev et al., 1993). Retprol is also claimed to be cheaper and ecologically safer than other commercially used C_2H_4–releasing compounds. Muromtsev et al. (1993) have conducted extensive studies over a period of more than a decade to test the effectiveness of Retprol for improving crop production which is discussed in Section 8.3.2.2.

5.5. NITRATE AS A SUPPRESSOR OF ETHYLENE ACCUMULATION IN SOIL

Nitrogenous fertilizers are commonly used all over the world to maximize crop yields. Most of these fertilizers carry N as NO_3^- or NH_4^+ (which is naturally oxidized to NO_3^-). High levels of NO_3^- in soil have been shown to interfere with C_2H_4 accumulation in soil most likely either by suppressing C_2H_4 synthesis or by enhanced C_2H_4 catabolism/degradation (Table 5.9).

Smith and Restall (1971) showed inhibition of C_2H_4 production in anaerobic soils when treated with NO_3^- at 2,000 and 10,000 mg kg^{-1} soil, which was later confirmed by other workers (Smith, 1975, 1976b; Smith and Cook, 1974; Hunt et al., 1980; van Cleemput et al., 1983; Goodlass and Smith, 1978a; Frankenberger and Phelan, 1985b). Goodlass and Smith (1978a) applied NO_3^- at 500, 1000 and 2000 mg kg^{-1} soil and estimated a reduction of 67, 77 and 85%, respectively, in C_2H_4 accumulation. Similarly, addition of NO_3^- at 200 mg kg^{-1} soil caused significant reduction in C_2H_4 accumulation in soil (Adams and Akhter, 1994). However, low levels of NO_3^- (<100 mg kg^{-1} soil) have little to no effect on C_2H_4 accumulation in soil (van Cleemput et al., 1983; Goodlass and Smith,1978a; Kapulnik et al., 1985). Hunt et al. (1980) found that NO_3^- addition (25-100 mg kg^{-1} soil) suppressed C_2H_4 accumulation during the first week, but as denitrification proceeded, the NO_3^- effect diminished. Similarly, van Cleemput et al. (1983) reported that 100 mg NO_3^- kg^{-1} soil had almost no influence on C_2H_4 production, whereas, 300 mg kg^{-1} soil markedly reduced C_2H_4 production. Nitrate up to 10-20 mg kg^{-1} soil enhanced MET-derived C_2H_4 production in soil, while 1000 mg NO_3^- kg^{-1} soil suppressed its synthesis (Kapulnik et al., 1985). Tang and Miller (1993) investigated C_2H_4 accumulation in soils differing in NO_3^- content (6–78 mg kg^{-1} soil) amended with poultry litter and incubated under waterlogged conditions. They reported that differences in NO_3^- content did not affect C_2H_4 evolution in soil appreciably because of the ease with which NO_3^- was reduced upon submergence. It is also likely that the overall NO_3^- levels in these soils were too low to affect C_2H_4 accumulation in soil.

Table 5.9. Effect of nitrate content on C_2H_4 production in soil.

NO_3^--N (mg kg^{-1} soil)	Effect on C_2H_4 generation	Reference
25-100	Suppressed C_2H_4 accumulation during the first week, but as denitrification proceeded, the NO_3^- effect diminished.	Hunt et al. (1980)
20-200	C_2H_4 production was observed only after a period of denitrification or not at all.	Smith and Cook (1974), Smith (1976b)
50 and 100	A transient effect on the evolution of C_2H_4 was noted, with maximum production occurring after a 1- to 2-day delay.	Goodlass and Smith (1978a)
100 and 300	100 mg NO_3^- kg^{-1} soil had almost no influence on C_2H_4 production, whereas 300 mg kg^{-1} markedly reduced C_2H_4 production.	van Cleemput et al. (1983)
500, 1000, and 2000	These NO_3^--N concentrations reduced the total amount of C_2H_4 evolved by 69, 77, and 85%, respectively.	Goodlass and Smith (1978a)
10-20 and 10,000	MET amendment enhanced C_2H_4 production in soils low in NO_3^- (10-20 mg kg^{-1} soil) when compared with a high NO_3^- content (10,000 mg kg^{-1} soil).	Kapulnik et al. (1985)
2,000 and 10,000	These concentrations of NO_3^--N greatly reduced the rate of C_2H_4 production, but never suppressed it completely.	Smith and Restall (1971)

Source: Frankenberger and Arshad (1995).

Since NO_3^- serves as an alternate electron acceptor for facultative anaerobic bacteria in the absence of O_2, many workers believe that NO_3^- inhibits C_2H_4 synthesis in soil by increasing the redox potential of the soil (Smith, 1976a,b; Hunt et al., 1980; Smith and Cook, 1974; Smith et al., 1978). Smith et al. (1978) demonstrated that the addition of Fe(II) at ≥ 100 mg kg^{-1} soil enhanced abiotic production of C_2H_4 in soil under reduced conditions. They postulated that in a biologically active soil with an adequate supply of NO_3^-, a hydrogen donor, and suitable species of bacteria, the redox potential will be poised sufficiently high to stop electrochemical reactions, including Fe(III) reduction that occurs at a low redox potential, leading to inhibition of C_2H_4 synthesis in soil. They concluded that the requirement of Fe(II) for C_2H_4 generation in soil also explains the potency of NO_3^- as an inhibitor of C_2H_4 production. Frankenberger and Phelan (1985a) confirmed that, in the presence of KNO_2 and KNO_3 (0.5 mM kg^{-1} soil), ACC-dependent C_2H_4 production was inhibited by 30-65% and 12-22%, respectively, but they could not confirm that this effect was due to an increased redox potential, since ACC conversion to C_2H_4 was more efficient under high O_2 tensions. Arshad and Frankenberger (1991) reported that Fe(II) at concentrations ≥ 100 mg kg^{-1} soil promoted abiotic C_2H_4 synthesis in soil under highly aerobic conditions, ruling out the belief that Fe(II) effects on C_2H_4

synthesis are related to the redox potential, as suggested by Smith et al. (1978). It is possible that NO_3^- might interfere with the enzymes involved in the biotic production of C_2H_4 in soil. The presence of NO_3^- may result in shifting of the C_2H_4 producing microflora from C_2H_4 precursors to other carbon compounds as N is supplied by NO_3^- itself. In general nitrification enhances C_2H_4 catabolism as the enzyme involved, a monooxygenase, utilizes C_2H_4 as a co-substrate as discussed in Chapter 3. De heyder et al. (1997) have reported that C_2H_4 removal in a reactor packed with granular activated carbon biobed inoculated with a C_2H_4-degrading heterotrophic strain, *Mycobacterium* E3 was significantly enhanced by introducing nitrifying activity in the system. According to their study, nitrifying activity is associated with: (i) production of a soluble microbial product which can act as a co-substrate for heterotrophic microorganisms, and (ii) the co-oxidation of C_2H_4 (De heyder et al., 1997). However, further research is needed to understand the role of NO_3^- in C_2H_4 metabolism in soil.

Rigler and Zechmeister-Boltenstern (1996, 1998, 1999) conducted a series of studies to evaluate the effect of NH_4^+ and NO_3^- on C_2H_4 consumption/oxidation in two soils (acid spruce forest soil and deciduous forest soil). They found that with increasing NH_4^+ concentration, C_2H_4 uptake rates decreased in the case of the deciduous soil, but increased in the alpine spruce forest soil (Rigler and Zechmeister-Boltenstern, 1996). In another study, they observed that the addition of NH_4^+ as $(NH_4)_2SO_4$ had a promoting effect on C_2H_4 accumulation in the alpine spruce forest soil (Fig. 5.15) most likely due to an improved C:N ratio for more microbial activity or producing more substantial amino acids for C_2H_4 synthesis and possibly due to depletion of O_2 through nitrification and enhanced respiration (Rigler and Zechmeister-Boltenstern, 1998). Since the presence of acetylene did not affect the NH_4^+ stimulated C_2H_4 accumulation, it implied that NH_4^+ enhanced C_2H_4 accumulation by restricting C_2H_4 catabolism or uptake. In the deciduous soil, C_2H_4 uptake was inhibited by NH_4^+. This could be due to co-oxidation of C_2H_4 by methanotrophs. Since C_2H_4 oxidation (catabolism) is minor in the spruce forest soil compared to its production, C_2H_4 accumulated in this soil. Increased production of C_2H_4 was also noticed in the spruce forest soil in response to increased application rates of NO_3^- as KNO_3 up to 100 µg NO_3-N g^{-1} dry weight of soil, but remained the same with 500 µg NO_3-N g^{-1} dry weight of soil (Rigler and Zechmeister-Boltenstern, 1998). The presence of C_2H_2 did not show any clear affect on NO_3^--stimulated C_2H_4 accumulation in soil which may imply that the NO_3^- effect was due to less C_2H_4 catabolism/oxidation. Higher levels of 100-500 µg NO_3-N g^{-1} dry weight inhibited C_2H_4 oxidation (catabolism) completely and even led to the formation of stress C_2H_4 . Rigler and Zechmeister-Boltenstern (1998) concluded that NH_4^+ and CO_2 had a minor affect on C_2H_4 gross production but enhanced the net accumulation by decreasing C_2H_4 uptake or oxidation. Since in deciduous forest soils C_2H_4 uptake rates exceeded the production rates, no net accumulation was observed. Later, they confirmed that NH_4-N (100-500 µg g^{-1} dry weight of soil) inhibited C_2H_4 oxidation in deciduous forest soil, but promoted C_2H_4 oxidation in the acidic spruce soil (Rigler and Zechmeister-Boltenstern, 1999). Conversely, NO_3-N inhibited C_2H_4 oxidation in both soils, while at 10 µg g^{-1} dry weight soil either had no effect on the deciduous forest soil or increased C_2H_4 slightly in the spruce forest soil.

Zechmeister-Boltenstern and Smith (1998) examined C_2H_4 production in six soils and found a positive correlation between C_2H_4 production and total N content of soil (r = 0.96*, p = 0.002). Contrarily, no relationship could be found between ACC- and MET-

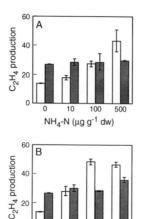

Fig. 5-15. Ethylene production without (☐) and with
(▨) acetylene in the alpine spruce forest soil as
affected by various ammonium (A) and nitrate (B)
concentrations. Production rates in pmol g^{-1} d.w. h^{-1},
error bars are standard deviation (n=4). Source: Rigler
and Zechmeister-Boltenstern (1998).

dependent C_2H_4 accumulation and the native soil N content (Frankenberger and Phelan, 1985b; Arshad and Frankenberger, 1991).

5.6. FERROUS IRON AS A STIMULATOR OF C_2H_4 PRODUCTION IN SOIL

Several studies have demonstrated that Fe(II) promotes abiotic C_2H_4 synthesis in soil. Smith et al. (1978) assessed the concentration-dependent influence of Fe(II) (100-10,000 mg kg^{-1} soil) on C_2H_4 production in soil and found that Fe(II) triggers an instantaneous release of C_2H_4, with an Fe(II) concentration of 1000 mg kg^{-1} soil being optimum for this reaction (Fig. 5.16). They concluded that Fe(II) promotes C_2H_4 production by a strictly chemical reaction, since this reaction occurred at an extremely high rate immediately after Fe(II) treatment excluding the possibility of involving any biological system. The comparative effectiveness of ferrous perchlorate in a 1:1 chloroform-methanol mixture in promoting C_2H_4 evolution in soil further supported that Fe(II) stimulates the release of C_2H_4 in soil nonbiologically. Arshad and Frankenberger (1991) also observed that Fe(II) at 100 mg kg^{-1} soil, or higher, promotes the abiotic production of C_2H_4 in soil (Table 5.10), confirming the findings of Smith et al. (1978). However, we also found that at low concentrations (0.5 mg kg^{-1} soil), Fe(II) stimulates the biological production of C_2H_4 in MET-amended soils. While comparing several soils for C_2H_4 production under anaerobic conditions, Zechmeister-Boltenstern and Nikodim (1999) found higher production of C_2H_4 in some soils compared to others and attributed these differences to soil humus, total N, texture, and Fe(II) content. Since the soils

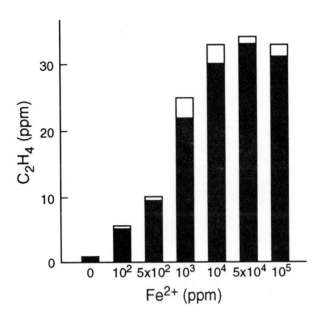

Fig. 5-16. Ethylene yield from soil, 3 h after treatment with various concentrations of Fe(II). Unshaded areas represent standard error. Source: Smith et al. (1978).

producing more C_2H_4 also contained higher amounts of Fe (II), they concluded that the higher rates of C_2H_4 in these soils may not only be due to microbial C_2H_4 formation, but also to chemical production. In another study, Rigler and Zechmeister-Boltenstern (1998) observed substantially more C_2H_4 accumulation in an alpine spruce soil which was rich in organic matter and contained over 9-fold Fe (510 mg kg^{-1} soil) compared to a deciduous forest soil which was poor in organic matter and Fe (57 mg kg^{-1} soil). They attributed more accumulation of C_2H_4 in the spruce forest soil because of Fe stimulated C_2H_4 production and the presence of high organic C (Rigler and Zechmeister-Boltenstern, 1998). Tang and Miller (1993) examined C_2H_4 accumulation in soil amended with poultry litter in three soils having Fe content ranging from 199 to 387 mg kg^{-1} soil. Incubation was carried out under waterlogged conditions. The soil with the highest Fe content produced the greatest amount of C_2H_4. However, they reported that Fe(II) may not be responsible for the observed differences in C_2H_4 production because under reduced conditions, all soils tend to have high Fe (II) content at the end of incubation.

Table 5.10. Effect of Fe(II) on C_2H_4 evolution from soils.

| Fe(II) | Hanford soil | | | | Garey soil | |
| | Unamended | | L-MET amended | | Unamended | |
(mg kg^{-1} soil)	Nonsterile	Sterile	Nonsterile	Sterile	Nonsterile	Sterile
			nmol of C_2H_4 kg^{-1} soil 7 d^{-1}			
0.0 (control)	35.4	19.7	836.2	100.4	241.1	77.0
0.5	35.8	19.0	1060.0	97.7	285.2	80.8
100.0	77.3	37.4	255.3	137.5	353.9	165.6
250.0	130.1	93.5	225.0	396.2	618.6	259.0
LSD ($p < 0.05$)	25.6	21.3	171.0	31.5	106.3	24.3

Source: Arshad and Frankenberger (1991).

The mechanism(s) of abiotic stimulation of C_2H_4 by Fe(II) is unclear. Smith et al. (1978) proposed that Fe(II) either has a direct effect on C_2H_4 precursors or simply establishes a favorable soil redox condition for C_2H_4 production. Hunt et al. (1982) suggested that Fe and Mn affect C_2H_4 synthesis by altering the redox potential of soil. However, Smith et al. (1978) found that a shift in E_h of -272 mV had virtually no effect on the amount or rate of Fe(II)-dependent C_2H_4 production (Fig. 5.17). Moreover, they reported that treating the soil with various reducing agents that mobilized Fe(II) resulted in enhanced C_2H_4 evolution, whereas, reducing agents that did not mobilize Fe(II) had no effect. This observation strongly implies that Fe(II) acts as a specific trigger for C_2H_4 evolution, and it appears to result from some direct action, rather than redox alterations. Arshad and Frankenberger (1991) also found that the addition of Fe(II) under completely aerobic conditions enhanced the nonbiological synthesis of C_2H_4, whereas, Fe(III) had a strongly negative effect on MET-derived C_2H_4 synthesis in soil. Smith et al. (1978) speculated that factors which affect Fe(II) mobility in soil may also affect the abiotic synthesis of C_2H_4 and under natural conditions, native Fe(II) may trigger C_2H_4 production. They discussed the role of Fe(II) in their proposed O_2-C_2H_4 cycle (see Section 5.2.2). Although they considered MET to be the major substrate of C_2H_4 in soil, Lynch and Harper (1980) reported that Fe(II) may have an important role in the final steps of the reaction (Fig. 5.18).

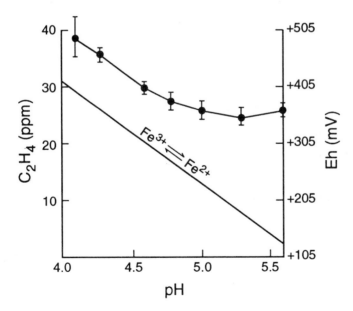

Fig. 5-17. Ethylene production in soil after 3 h of treatment with 10^4 ppm Fe(II), adjusted at various pH values. Source: Smith et al. (1978).

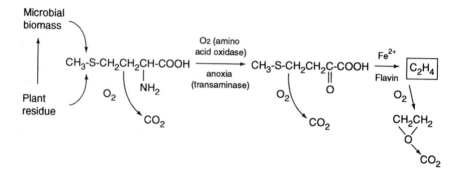

Fig. 5-18. Proposed role of Fe(II) on C_2H_4 production in soil. Source: Lynch and Harper (1980).

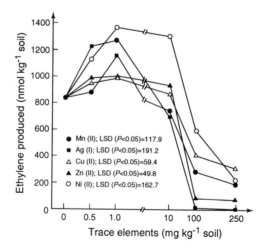

Fig. 5-19. Effect of Mn(II), Ag(I), Cu(II), Zn(II), and Ni(II) on L-methionine-dependent C_2H_4 accumulation in Hanford soil (1:1 soil:water suspension) incubated on a shaker (100 rpm) for 7 days at 30°C. Source: Arshad and Frankenberger (1991).

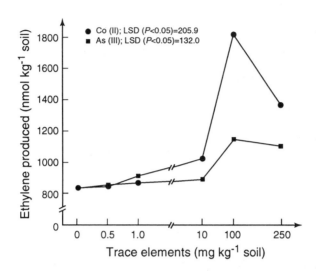

Fig. 5-20. Effect of Co(II) and As(III) on L-methionine-dependent C_2H_4 accumulation in Hanford soil (1:1 soil:water suspension) incubated on a shaker (100 rpm) for 7 days at 30°C. Source: Arshad and Frankenberger (1991).

Other studies have investigated the effect of other micronutrients on C_2H_4 accumulation in soil. Smith et al. (1978) did not find any increase in C_2H_4 production in response to amendments of soil with Co(II), Ni(II), Mn(II), or Cu(II) except Fe(II), revealing a specific role for Fe(II) in promoting C_2H_4 generation in soil. Smith (1976a) reported that Mn(II) strongly inhibited C_2H_4 production in soil, whereas, the addition of Co(II) significantly increased ACC-dependent C_2H_4 evolution (Frankenberger and Phelan, 1985a). Arshad and Frankenberger (1991) conducted a comprehensive study to assess the effects of 12 transition metals and metalloids on C_2H_4 generation in soil. The concentration-dependent effects of these trace elements on MET-derived C_2H_4 production and accumulation in soil varied considerably. Six trace elements [Ag(I), Cu(II), Fe(II), Mg(II), Zn(II), Al(III)] significantly stimulated C_2H_4 production when applied at low concentrations ($\leq$1.0 mg kg^{-1} soil), but inhibited the reaction at concentrations of 100 mg kg^{-1} soil or higher (Fig. 5.19). The addition of Ni(II) up to 10 mg kg^{-1} soil also significantly enhanced MET-dependent C_2H_4 production in soil. The most effective trace elements in promoting the production of C_2H_4 were Co(II) and As(III) when added at 100 mg kg^{-1} soil (Fig. 5.20). Ethylene generation was inhibited in the presence of Hg(II), Fe(II), and Mo(VI) at $\geq$10 mg kg^{-1} soil (Fig. 5.21). The influence of trace elements on C_2H_4 production in soil may be related to their effects on the redox potential, catalytic activity (cofactors vs. inhibitors) and toxicity to the microflora (Arshad and Frankenberger, 1991).

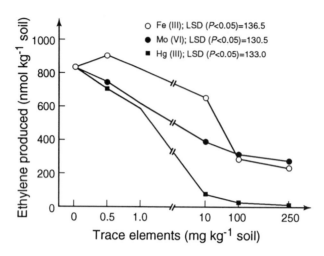

Fig. 5-21. Effect of Fe(III), Mo(VI), and Hg(II) on L-methionine-dependent C_2H_4 accumulation in Hanford soil (1:1 soil:water suspension) incubated on a shaker (100 rpm) for 7 days at 30°C. Source: Arshad and Frankenberger (1991).

5.7. PERSISTENCE OF ETHYLENE IN SOIL

Ecologically, the persistence/stability of C_2H_4 in soil is as important as its synthesis in soil, because plant roots exposed to C_2H_4 for longer periods may show a stronger physiological response. Ethylene in soil may be subjected to biological degradation, chemical decomposition, physical adsorption by soil components, or being a gas, may be diffused out into the atmosphere. All these processes could result in lowering the C_2H_4 level in soil below the physiologically active concentrations.

Several studies have demonstrated that C_2H_4 added to soil is subjected to degradation (Abeles et al., 1971; Smith et al., 1973; Cornforth, 1975; de Bont, 1976; Primrose, 1979; Arshad and Frankenberger, 1990c; Nohrstedt, 1983). de Bont (1976) observed a lag period of a few days before assimilation of added C_2H_4 by soil, indicating the involvement of biological activity (Fig. 5.22). Normally C_2H_4 oxidation by one C_2H_4 consumer, *Mycobacterium paraffinicum* is efficient enough to remove most of the C_2H_4 in soils (Abeles et al., 1992). Autoclaved soils do not absorb C_2H_4 (Abeles et al., 1971; Smith et al., 1973; Arshad and Frankenberger, 1990c). Soil sterilization by ethylene oxide also prevents C_2H_4 uptake (Abeles et al., 1971), confirming the premise that soil microbiota are involved in the removal of C_2H_4. Arshad and Frankenberger (1990c) found that C_2H_4 injected into the headspace of soil in a closed container was metabolized at a greater rate (9.8-fold) in nonsterile soil, compared to an autoclaved soil (Fig. 5.23). Our kinetic analysis revealed that C_2H_4 loss or degradation in a nonautoclaved soil under aerobic conditions follows a first-order reaction, with a rate constant (k) of 0.115 day^{-1} and a half-life ($t_{1/2}$) of 6.0 days (Arshad and Frankenberger, 1990c). By using autoclaved vs. non-autoclaved soil, Elsgaard (1996) confirmed the biological mediation of C_2H_4 consumption. Nikitin and Arakelyan (1979) reported that C_2H_4 could be used by soil bacteria as a source of energy and carbon. They demonstrated that C_2H_4 is consumed by several bacteria upon inoculating sterile soil and incubating in the presence of C_2H_4 in the gas mixture (Table 5.11). They found that *Pseudomonas fluorescens* was the most

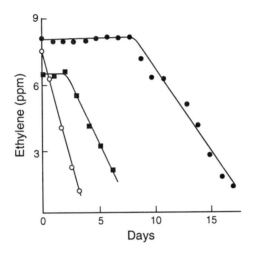

Fig. 5-22. Assimilation of C_2H_4 from the headspace of a clay (o) and two sandy (•,■) soils. Source: deBont (1976).

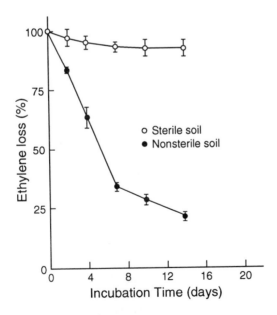

Fig. 5-23. Ethylene stability in sterilized vs. nonsterilized soil. Source: Arshad and Frankenberger (1990c).

Table 5.11. Utilization of C_2H_4 by bacteria in sterile soil in the presence of a gas mixture with C_2H_4 .

| Bacteria | Ethylene content | | | |
| | After 3 days | | After 8 days | |
	mg/10 cm^3	% of control	mg/10 cm^3	% of control
Sterile control	5.53	100.0	4.76	86
Bacillus polymyxa	5.54	100.0	2.25	47.6
Renobacter vacuolatum	4.55	82.5	1.17	24.7
Pseudomonas fluorescens	4.37	79.1	0.72	15.2

Source: Nikitin and Arakelyan (1979).

active utilizer of C_2H_4 . In non-inoculated sterile soils, no decomposition of C_2H_4 took place in 3 days of incubation, while in 8 days 14% disappeared (Nikitin and Arakelyan, 1979). This experiment further confirms the ability of soil microorganisms to catabolize C_2H_4 . Recently, Elsgaard and Andersen (1998) investigated C_2H_4 consumption by a peat soil either that was induced to microbial C_2H_4 consumption by exposing it to high levels of C_2H_4 for a long time or inoculated it with C_2H_4 –utilizing bacteria both in the presence and absence of *Begonia elatior*. They reported that both induced or inoculated soil showed high potential of C_2H_4 consumption (Fig. 5.24). A time course study of C_2H_4 consumption in the peat-soil, where the rate was increased by repeated additions of C_2H_4, indicated that C_2H_4 consumption was a microbial rather than a chemical process, and that proliferation of C_2H_4–oxidizing bacteria occurred. The specific rates of C_2H_4 consumption with the inoculated peat-soil were from 476 to 512 ng C_2H_4 g^{-1} dry weight soil and with induced peat soil were from 374 to 445 ng C_2H_4 g^{-1} dry weight soil. Degradation of a low C_2H_4 concentration (ca. 0.7 ppm) in the inoculated peat soil generally proceeded below the GC detection limit within less than 6 h, implying that nearly complete C_2H_4 removal in the peat soil inoculated with C_2H_4–oxidizing bacteria is possible (Elsgaard and Andersen, 1998). Frye et al. (1992) studied the removal of organic gases from air by a soil bed reactor in closed systems. They reported that the removal of C_2H_4 appears to be accelerated with prior exposure of soil to C_2H_4 (conditioning), suggesting that induction of microbial populations may be involved. Otani and Ae (1993) examined C_2H_4 decomposition under field conditions both in the rhizosphere and nonrhizosphere soil. They reported that C_2H_4 decomposition rates were faster in the rhizosphere soil (55 µl g^{-1} day^{-1}) than in the nonrhizosphere soil (3 µl g^{-1} day^{-1}), most likely due to higher populations of microorganisms in the former (Fig. 5.25). Cornforth (1975) and Sawada et al., (1985) also reported that the C_2H_4 decomposition capacity increased with increasing organic matter content. Zechmeister-Boltenstern and Smith (1998) examined C_2H_4 production and decomposition in six different soils. They reported that the amount of C_2H_4 decomposition seemed to be mainly influenced by land use and vegetation cover as they observed greater C_2H_4 decomposition in deciduous

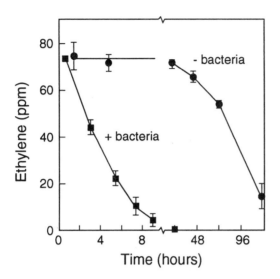

Fig. 5-24. Ethylene consumption by soil planted with *Begonia* and
inoculated with [■] or without (●) bacteria. Data represents the
mean ±SD of triplicate soil samples from plants that were processed
immediately after the addition of the bacterial suspension or water.
Note the scale break on the x-axis. Source: Elsgaard and Andersen
(1998).

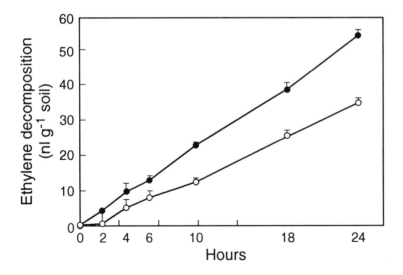

Fig. 5-25. Time course of C_2H_4 decomposition in rhizosphere soil
from a sorghum plot (●), and in non-rhizosphere soil from a fallow
plot (o). Bars represent SE, n=3. Source: Otani and Ae (1993).

woodland soils compared to an adjacent grassland or coniferous woodland soils. They found that C_2H_4 decomposition rates exceeded C_2H_4 production rates by factors from 10 to 100. However, the addition of MET and glucose inhibited C_2H_4 decomposition in soil. Zechmeister-Boltenstern and Nikodim (1999) reported that high rates of C_2H_4 decomposition in soils of deciduous forest sites were correlated with the amount of humus (r = 0.95) and nitrogen (r = 0.95) in soil, implying that soils rich in organic matter, such as deciduous and grassland soils, also contain a more active decomposing microflora than arable field soils. They reported that all soils immediately started to decompose C_2H_4, which indicates that the microflora capable of decomposing C_2H_4 had already existed in the soils.

Aerobic conditions favor C_2H_4 catabolism in soil. Bacteria capable of utilizing C_2H_4 as a sole carbon and energy source have been isolated from soil and most of these C_2H_4–utilizing bacteria oxidize C_2H_4 into ethylene oxide (see Section 3.7). Restriction in partial pressure of O_2 and elevation of $[CO_2]$ increases the persistence of C_2H_4 in soil (Abeles et al., 1971; Cornforth, 1975). Moreover, pure culture studies revealed that ammonia monooxygenase of *Nitrosomonas europae* which oxidizes NH_4^+ into NO_2^- (nitrification) also catalyzes the conversion of C_2H_4 into ethylene oxide (Hommes et al., 1998). However, the presence of soil modified the co-oxidation of C_2H_4 by this bacterium (Fig. 5.26). The rate of C_2H_4 uptake by soil decreased when O_2 concentrations

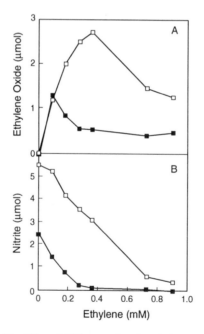

Fig. 5-26. Ethylene oxide (A) and NO_2^- production (B) vs. C_2H_4 concentration in the presence (■) or absence (□) of 0.5 g of soil with 10 mM NH_4^+. Samples were withdrawn from the headspace (for ethylene oxide determinations) or liquid (for nitrite determinations) after a 30-minute incubation with *Nitrosomonas europaea*. Source: Hommes et al. (1998).

fell below 2% (v/v) and was not measurable under anaerobic conditions (Abeles et al., 1971). An increase in CO_2 concentrations in the soil atmosphere reduced C_2H_4 oxidation and thus stimulated its accumulation in soil (Rigler and Zechmeister-Boltenstern, 1996, 1998). Zechmeister-Boltenstern and Nikodim (1999) also confirmed that C_2H_4 decomposition is an aerobic process as C_2H_4 degradation reached its maximum at a water potential of -30 kPa. They found that under aerobic conditions (-3 to -100 kPa), C_2H_4 decomposition rates exceeded production rates by a factor of 10 to 100, which implies that in most aerobic soils, C_2H_4 may not be detected. In another study, they evaluated the effect of CO_2 and mineral N [$(NH_4)_2SO_4$ and KNO_3] on the oxidation of C_2H_4 in Montana spruce forest soils and a mixed deciduous hardwood forest soil (Rigler and Zechmeister-Boltenstern, 1999). They observed increases in C_2H_4 uptake with elevated NH_4^+ concentrations in the acidic spruce forest soil and inhibition in the deciduous forest soil. Conversely, NO_3^- inhibited C_2H_4 oxidation in both soils. An enhancement of CO_2 in the soil air acts negatively upon the function of soil as a sink for volatile hydrocarbons (Rigler and Zechmeister-Boltenstern, 1999).

Soils also differ in their potentials to act as C_2H_4 sinks. Arshad and Frankenberger (1990c) reported that C_2H_4 was removed at an average rate of 10.3 nmol kg^{-1} soil over 14 days of incubation, which is considerably less than the average rates reported by others. Smith et al. (1973) observed that C_2H_4 decomposes at a rate of 140-970 nmol kg^{-1} day^{-1} over 15 days, whereas, Abeles et al. (1971) reported a rate of 13,000-14,000 nmol kg^{-1} day^{-1} over 5 days of incubation.

There is a strong possibility that some fraction of soil C_2H_4 is removed abiotically either through chemical decomposition or by physical adsorption or lost to the atmosphere through diffusion. These aspects have not been thoroughly investigated. Factors such as temperature, soil type, crop history and experimental pretreatments can influence the rate of C_2H_4 removal (de Bont, 1976; Abeles and Heggestad, 1973; Zechmeister-Boltenstern and Smith, 1998; Zechmeister-Boltenstern and Nikodim, 1999).

5.8. PLANT RESPONSES EVOKED BY ETHYLENE AND WATERLOGGING

Waterlogging of a soil has a major effect on natural vegetation and agricultural crops, limiting the growth of many vascular plants in humid regions (Jackson and Drew, 1984; Drew, 1991). Prolonged or intense periods of rainfall combined with poor soil drainage often causes inadequate soil oxygen supply and accumulation of other gases due to slow diffusion. As discussed earlier in Section 5.2.2, anaerobic and waterlogged soils are considered to be more favorable for C_2H_4 accumulation due to the enhanced stability and entrapment of C_2H_4 in water, leading to greater persistence and less degradation. Ethylene is also soluble in water (~140 ppm, 25°C), and hydrodiffusion of C_2H_4 is approximately 10,000 times less than its aerodiffusion. These two factors minimize C_2H_4 losses to the atmosphere from waterlogged soils. Moreover, under these conditions, aerobic respiration of plant roots and microorganisms very quickly reduces soil O_2 concentrations (Drew, 1992), which makes the conditions favorable for C_2H_4 accumulation in the soil atmosphere.

Plant physiologists have taken more interest than soil scientists in studying C_2H_4 formation in waterlogged soils because of the close similarities between C_2H_4-evoked plant responses and those caused by waterlogging, flooding, and submergence. Several studies conducted during the last two decades have revealed an C_2H_4-waterlogging interaction responsible for the physiological effects on plant growth and development (Jackson, 1985; Voesenek et al., 1992). Hypoxia may influence root and shoot hormone relations, such as C_2H_4 production which in turn affects plant responses to waterlogging (Jackson and Pearce, 1991). Kozlowski (1986) suggested that C_2H_4 together with other compounds appear to play an important causal role in the morphological adaptation of forest trees to soil anaerobiosis developed during flooding.

Because of C_2H_4 solubility in water and its greater persistence in waterlogged soils, exposure of plant roots to an exogenous source of C_2H_4 in soil may be sufficient to affect plant growth and development. However, in spite of numerous studies reporting similar effects of waterlogging and C_2H_4 treatment (Table 5.12), it is not clear what role C_2H_4 of soil origin plays in influencing plant growth. Endogenous levels of C_2H_4 and its precursor, ACC in plants increase dramatically after exposure to waterlogging, flooding, and submergence (Bradford and Yang, 1981; Hunt et al., 1981; Jackson et al., 1978; Jackson, 1982; Jackson and Campbell, 1976; Kawase, 1974, 1978; Jackson, 1985). Accumulation of C_2H_4 in plants is responsible for at least three major adaptive plant responses to flooding including: (i) shoot elongation enabling a totally submerged plant to emerge from the water surface (Musgrave et al., 1972; Cookson and Osborne, 1978; Voesenek and Blom, 1989); (ii) formation of aerenchyma which facilitates O_2 diffusion from shoots to submerged roots (Armstrong, 1979; Armstrong et al., 1994; Blom et al., 1990), and (iii) the formation of adventitious roots (Visser et al., 1996a). Recently, Blom (1999) critically reviewed the role of plant hormones in adaptation of plants to flooding stress. He suggested that C_2H_4 and other hormones are involved in the formation of flooding-induced adventitious roots. Plants growing under non-flooded conditions will also form adventitious roots upon application of C_2H_4. Yamamoto and Kozlowski (1985) reported that C_2H_4 was responsible for the initiation of adventitious roots. Drew et al. (1979) and Jackson et al. (1981) found that root formation was stimulated by C_2H_4 in maize and Bleecker et al. (1987) observed a similar response for deep water rice.

Ethylene concentrations in plant tissues can increase 500-fold in soil-flooded conditions (Visser et al., 1996b). It has not been confirmed whether or not exogenous soil C_2H_4 contributes to this elevated endogenous plant C_2H_4 pool. However, Jackson and Campbell (1975) have demonstrated uptake of labeled C_2H_4 from nutrient solution by tomato roots, being transported to the shoot where it exerts its physiological action. This observation was later confirmed by El-Beltagy et al. (1986) by exposing tomato plants to labeled C_2H_4. In another study, El-Beltagy et al. (1990) reported the uptake and movement of labeled C_2H_4 in waterlogged and control broad bean (*Vicia faba* L.) plants, and were of the view that soil-borne C_2H_4 does not significantly contribute to the observed increase in C_2H_4 levels of these plants. In waterlogged soils, the plant itself could contribute to the C_2H_4 pool in soil by releasing the precursor, ACC into its surrounding environment. It is highly likely that both soil C_2H_4 and increased endogenous C_2H_4 within the plant itself may contribute to the physiological response typical for plants growing in waterlogged soils. After observing increased C_2H_4 accumulation in soil and tobacco leaves in response to flooding, Hunt et al. (1981) proposed that soil- and plant-produced C_2H_4 may be important factors in reducing root permeability and increasing resistance to water uptake by tobacco. The use of C_2H_4 mutants, inhibitors of C_2H_4 biosynthesis in plants (such as AVG and AOA), and inhibitors of C_2H_4 action (such as Ag^+ and 2,5-norbornadiene) can be helpful in resolving

Table 5.12. Physiological responses of plants exposed to flooding/waterlogging/hypoxia stress and C_2H_4.

Plant	Stress	Ethylene source	Comments	Reference
Coleus blumei Benth.	Waterlogging	Ethrel	An increase in epinasty and decrease in water potential were observed after the first day of treatment with ethrel or waterlogging.	Pimenta et al. (1994)
Chickpeas	Waterlogging	Ethephon	Lower leaves became chlorotic, floral development ceased, and widespread leaf abscission occurred in waterlogged and ethephon treated plants.	Cowie et al. (1996)
Rumex palustris Sm.	Waterlogging	Ethylene gas	Adventitious root development at the base of the shoot is an important adaptation to flooded conditions. Decreases in endogenous auxins and C_2H_4 concentrations induced by application of inhibitors of either auxin transport or C_2H_4 biosynthesis reduced the number of adventitious roots formed by flooded plants, suggesting the involvement of these hormones in rooting process. The rooting response during flooding was preceded by an increase in endogenous C_2H_4 concentration in the root system. Exogenous application of C_2H_4 and/or auxin caused a very similar response in well aerated plants. The results suggested that the higher C_2H_4 concentrations in soil-flooded plants increased the activity of the root-forming tissues to endogenous IAA, thus initiating the formation of adventitious roots.	Visser et al. (1996b)
Wheat (two cultivars differing in waterlogging tolerance)	Hypoxia	C_2H_4 gas	Hypoxia induced root aerenchyma formation and enhanced root C_2H_4 production for both cultivars. Ethylene effects on root growth and aerenchyma formation were similar to those observed for hypoxia treatment.	Huang et al. (1997)

Table 5.12. (continued).

Rumex spp.	Waterlogging	C_2H_4 gas	Comparison of formation of adventitious roots in various *Rumex* spp. exposed to waterlogging and C_2H_4, and by using inhibitors of C_2H_4 biosynthesis, AVG or AOA, suggested that high C_2H_4 concentrations might be a prerequisite for the flooding-induced formation of adventitious roots in *Rumex* spp.	Visser et al (1996a)
Rumex spp.	Hypoxia	C_2H_4 gas	Ethylene had an even stronger effect on root elongation than hypoxia, indicated that C_2H_4 contributed strongly to the negative effects of flooding on root growth.	Visser et al (1997)
Rumex spp.	Submergence/ waterlogging	C_2H_4 gas	The growth responses under flooded conditions could be partially mimicked by exposing non-submerged plants to C_2H_4. The effects of C_2H_4, $AgNO_3$, and increased endogenous C_2H_4 concentrations during submergence suggested that C_2H_4 played a regulatory role in the growth responses of *Rumex* spp. under submerged conditions.	Voesenek et al. (1989)
Tomato	Flooding/ waterlogging	C_2H_4 gas	Soil flooding increased ACC oxidase activity in petioles of wild-type tomato plants within 6 to 12 hours. Flooding promoted epinastic curvature while exogenous C_2H_4 supplied to the well-drained plants also promoted epinastic curvature. Results suggested that increased ACC oxidase activity in shoots of flooded tomato plants raised C_2H_4 production to physiologically active levels.	English et al. (1995)
Corn	Flooding	C_2H_4 gas	The physiological responses to flooding were compared with those caused by exogenously applied C_2H_4 and it was concluded that increased concentrations of C_2H_4 during flooding contribute to growth alteration by accelerating the emergence of adventitious roots, inhibiting P accumulation in shoots, and by inhibiting plant growth.	Jackson et al. (1981)
Tomato and sunflower	Flooding		When plants were waterlogged, C_2H_4 concentrations increased in the waterlogged, exposed organs, which might serve as a sink for C_2H_4 generated by the plant and soil. In addition, increased synthesis of C_2H_4 in the tissues in response to stress was observed.	Kawase (1978)

Table 5.12. (continued).

Sunflower	Flooding	Ethephon	Damage symptoms caused by soil application of ethephon, such as reduced stem height, chlorophyll breakdown, epinasty of the second basal pairs of leaves, and hypocotyl hypertrophy, were almost identical with those caused by soil flooding. It was concluded that an increase in C_2H_4 concentration in flooded plants was largely although no exclusively, responsible for flooding damage symptoms.	Kawase (1974)
Tobacco	Flooding		Flooding promoted higher concentrations of C_2H_4 in soil and wilting developed progressively with increased level of C_2H_4 in soil. Leaf water potential, stem diameter, and relative water content, all decreased with increased exposure to C_2H_4 levels. It was suggested that soil and plant-produced C_2H_4 were responsible for reducing root permeability and increasing resistance to water uptake by tobacco.	Hunt et al. (1981)
Tomato	Flooding		C_2H_4 concentrations in soil might increase during waterlogging at levels that could allow sufficient movement into the shoot to stimulate epinasty. Treatment with substances that inhibited C_2H_4 action inhibited the epinastic response in waterlogged plants.	Jackson and Campbell (1976)
Barley	Flooding		Examination of waterlogged soils for C_2H_4 concentrations revealed that, in waterlogged soils, C_2H_4 could attain concentrations that were capable of inhibiting root extension.	Smith and Russell (1969)
Tomato	Flooding	C_2H_4 gas	Rapid movement to the shoots of [^{14}C]-labeled C_2H_4 applied to roots was demonstrated. It was proposed that increase in soil C_2H_4 and movement of the gas to the shoot system were factors contributing to the development of epinasty and other responses of the shoots to waterlogging of the plant.	Jackson and Campbell (1975)
Tomato	Flooding	Ethephon	Flooding reduced stem growth and leaf chlorophyll content of the lower leaves and promoted epinastic curvature of the leaf petiole and adventitious roots. Application of soil-drenched ethephon elicited a response similar to that of flooding.	Kuo and Chen (1980)

Table 5.12. (continued).

Tomato	Waterlogging	C_2H_4 gas	Waterlogging caused greater increases in C_2H_4 concentrations in plants grown in compost than those grown in gravel. The changes in plants grown in compost correlated with increased C_2H_4 concentrations in the soil. They also demonstrated the uptake of labeled C_2H_4 by roots and its subsequent transport to the shoot.	El-Beltagy (1986)
Taxodium distichum	Flooding		Flooding greatly stimulated C_2H_4 production in the submerged portions of the stem, which may be associated with hypertrophic stem growth. A possible role of auxin- C_2H_4 interaction was suggested.	Yamamoto (1992)

the issue of whether or not soil C_2H_4 has a significant effect on the growth and development of plants grown in waterlogged soils.

5.9. CONCLUDING REMARKS

Ethylene accumulation in soil is controlled by many factors including biological, chemical, and physical properties of soils; organic and inorganic amendments; and the stability/ persistance of C_2H_4 . However, the biochemistry of C_2H_4 production and decomposition in soil is not yet well understood and requires intensive work in the future. Moreover, there are little data regarding the physical adsorption/desorption of C_2H_4 by soil components and diffusion rates of C_2H_4 in soil. Since an extremely small amount of C_2H_4 in the root vicinity can be physiologically active to plants, it is highly likely that organic amendments which promote C_2H_4 accumulation in soil may also exert their physiological impact directly on the plant. Ethylene in the rhizosphere has been found to be involved in various physiological processes such as symbiosis (Chapter 6), pathogenesis (Chapter 7) and other stressed and nonstressed responses (Chapters 2, 4, and 8). Future research is needed on the various sources of C_2H_4 in soil and their agricultural applications.

5.10. REFERENCES

Abeles, F. B., and Heggestad, H. E., 1973, Ethylene: An urban air pollutant, *J. Air Pollut. Control Assoc.* **23**:517-521.

Abeles, F. B., Craker, L. E., Forrence, L. E., and Leather, G. R., 1971, Fate of air pollutants: Removal of ethylene, sulfur dioxide and nitrogen dioxide by soil, *Science* **173**:914-916.

Abeles, F. B., Morgan, P. W., and Saltveit, M. E., Jr., 1992, *Ethylene in Plant Biology*, 2nd Edition, Academic Press, Inc., San Diego, CA.

Adams, W. A., and Akhtar, N., 1994, The possible consequences for herbage growth of waterlogging compacted pasture soils, *Plant Soil* **162**:1-17.

Armstrong, W., 1979, Aeration in higher plants, *Adv. Botan. Res.* **7**:225-332.

Armstrong, W., Brandle, R., and Jackson, M. B., 1994, Mechanisms of flood tolerance in plants, *Acta Botanica Neexlandica* **43**:307-358.

Arshad, M., and Frankenberger, W. T., Jr., 1988, Influence of ethylene produced by soil microorganisms on etiolated pea seedlings, *Appl. Environ. Microbiol.* **54**:2728-2732.

Arshad, M., and Frankenberger, W.T., Jr., 1989, Biosynthesis of ethylene by *Acremonium falciforme*, *Soil Biol. Biochem.* **21**:633-638.

Arshad, M., and Frankenberger, W. T., Jr., 1990a, Response of *Zea mays* L. *and Lycopersicon esculentum* to the ethylene precursors, L-methionine and L-ethionine, applied to soil, *Plant Soil* **122**:219-227.

Arshad, M., and Frankenberger, W. T., Jr., 1990b, Ethylene accumulation in soil in response to organic amendments, *Soil Sci. Soc. Am. J.* **54**:1026-1031.

Arshad, M., and Frankenberger, W. T., Jr., 1990c, Production and stability of ethylene in soil, *Biol. Fert. Soils* **10**:29-34.

Arshad, M., and Frankenberger, W. T., Jr., 1991, Effects of soil properties and trace elements on ethylene production in soils, *Soil Sci.* **151**:377-386.

Arshad, M., and Frankenberger, W. T., Jr., 1992, Microbial biosynthesis of ethylene and its influence on plant growth, *Adv. Microbiol. Ecol.* **12**:69-111.

Babiker, H. M., and Pepper, I. L., 1984, Microbial production of thylene in desert soils, *Soil Biol. Biochem.* **16**:559-564.

Bibik, N. D., Letunova, S. V., Druchek, E. V., and Muromtsev. G. S., 1995, Effectiveness of a soil-acting ethylene producer in obtaining sanitized seed potatoes, *Russ. Agricul. Sci.* **9**:19-21.

Bleecker, A. B., Rose-John, S., and Kende, H., 1987, An evaluation of 2,5-norbornadiene as a reversible inhibitor of ethylene action in deepwater rice, *Plant Physiol.* **84**:395-398.

Blom, C. W. P. M., 1999, Adaptations to flooding stress: from plant community to molecule, *Plant Biol.* **1**:261-273.

Blom, C. W. P. M., Bögermann, G. M., Laan, P., van der Sman, A. J. M., Van der Steeg, H. M., and Voeseneki, L. A. C. J., 1990, Adaptations to flooding in plants from river areas, *Aquatic Bot.* **38**:29-47.

Bradford, K. J., and Yang, S. F., 1981, Physiological responses of plants to waterlogging, *Hortic. Sci.* **16**:25-30.

Burford, J. R., 1975, Ethylene in grassland soil treated with animal excreta, *J. Environ. Qual.* **4**:55-57.

Campbell, R. B., and Moreau, R. A., 1979, Ethylene in a compacted field soil and its effect on growth, tuber quality and yield of potatoes, *Am. Potato J.* **56**:199-210.

Chernys, J., and Kende, H., 1996, Ethylene biosynthesis in *Regnellidium diphyllum* and *Marsilea quadrifolia*, *Planta* **200**:113-118.

Conlin, T. S. S., and van den Driessche, R., 1996, Short-term effects of soil compaction on growth of *Pinus contorta* seedlings, *Can. J. For. Res.* **26**:727-739.

Considine, P. J., Flynn, N., and Patching, J. W., 1977, Ethylene production by soil microorganisms. *Appl. Environ. Microbiol.* **33**:977-979.

Cook. R. J., and Smith, A. M., 1977, Influence of water potential on production of ethylene in soil, *Can. J. Microbiol.* **23**:811-817.

Cookson, C., and Osborne, D. J., 1978, The stimulation of cell extension by ethylene and auxin in aquatic plants, *Planta* **144**:39-47.

Cornforth, I. S., 1975, The persistence of ethylene in aerobic soils, *Plant Soil* **42**:85-96.

Cowie, A. L., Jessop, R. S., and MacLeod, D. A., 1996, Effects of waterlogging on chickpeas. II. Possible causes of decreased tolerance of waterlogging at flowering, *Plant Soil* **183**:105-115.

de Bont, J. A. M., 1976, Oxidatioin of ethylene by soil bacteria, Antonie van Leeuwenhoek 42:59-71.

De heyder, B., Van Elst, T., Van Langgenhove, H., and Verstraete, W., 1997, Enhancement of ethene removal from waste gas by stimulating nitrification, *Biodegradation* **8**:21-30.

Dowdell, R. J., Smith, K. A., Crees, R., and Restall, S. W. F., 1972, Field studies of ethylene in the soil atmosphere--equipment and preliminary results, *Soil Biol. Biochem.* **4**:325-331.

Drew, M. C., 1991, Oxygen deficiency in the root environment and plant mineral nutrition, in: *Plant Life under Oxygen Deprivation*, M. B. Jackson et al., eds., Academic Publishing, The Hague, pp. 303-316.

Drew, M. C. , 1992, Soil aeration and plant root growth metabolism, *Soil Sci.* **154**:259-268.

Drew, M. C., Jackson, M. C., and Giffard, S. C., 1979, Ethylene-promoted adventitious rooting and development of cortical air spaces (aerenchyma) in roots may be adaptive resonses to flooding in *Zea mays* L., *Planta* **147**:83-88.

El-Beltagy, A. S., Madkour, M. A., and Hall, M. A., 1986, Uptake and movement of ethylene in tomatoes in relation to waterlogging, *Acta Horticul.* **190**:355-370.

Elsgaard, L., 1996, Ethylene degradation in peat-soil for horticultural practice, Abstract of the NATO Advanced Research Workshop on Biology and Biotechnology of the Plant Hormone Ethylene, 9-13 June, 1996. China, Greece, p. 97.

Elsgaard, L., and Andersen, L., 1998, Microbial ethylene consumption in peat-soil during ethylene exposure of *Begonia elatior*, *Plant Soil* **202**:231-239.

El-Sharouny, H. M., 1984, Screening of ethylene producing root infecting fungi in Egyptian soil, *Mycopathologia* **85**:13-15.

English, P. J., Lycett, G. W., Roberts, J. A., and Jackson, M. B., 1995, Increased 1-aminocyclopropane-1-carboxylic acid oxidase activity in shoots of flooded tomato plants raises ethylene production to physiologically active levels, *Plant Physiol.* **109**:1435-1440.

Frankenberger, W. T., Jr., and Arshad, M., 1995, Phytohormones in Soils: Microbial production and Functions, Marcel Dekker, Inc., New York.

Frankenberger, W. T., Jr., and Phelan, P. J., 1985a, Ethylene biosynthesis in soil. I. Method of assay in conversion of 1-aminocyclopropane-1-carboxylic acid to ethylene, *Soil Sci. Soc. Am. J.* **49**:1416-1422.

Frankenberger, W. T., Jr., and Phelan, P. J., 1985b, Ethylene biosynthesis in soil. II. Kinetics and thermodynamics in the conversion of 1-aminocyclopropane-1-carboxylic acid to ethylene, *Soil Sci. Soc. Am. J.* **49**:1422-1426.

Frye, R. J., Welsh, D., Berry, T. M., Stevenson, B. A., and McCallum, T., 1992, Removal of contaminant of organic gases from air in closed systems by soil, *Soil Biol. Biochem.* **24**:607-612.

Goodlass, G., and Smith, K. A., 1978a, Effects of pH, organic matter content and nitrate on the evolution of ethylene from soils, *Soil Biol. Biochem.* **10**:193-199.

Goodlass, G., and Smith, K. A., 1978b, Effects of organic amendments on evolution of ethylene and other hydrocarbons from soil, *Soil Biol. Biochem.* **10**:201-205.

Harvey, R. G., and Linscott, J. J., 1978, Ethyelene production in soil containing quackgrass rhizomes and other plant materials, *Soil Sci. Soc. Am. J.* **42**:721-724.

Hendrickson, O. Q., 1989, Implications of natural ethylene cycling processes for forest soil acetylene reduction assays, *Can. J. Microbiol.* **35**:713-718.

Hodges, C. F., and Campbell, D. A., 1998, Gaseous hydrocarbons associated with black layer induced by the interaction of cyanobacteria and *Desulfovibrio desulfuricans*, *Plant Soil* **205**:77-83.

Hommes, N. G., Russell, S. A., Bottomley, P. J., and Arp, D. J., 1998, Effects of soil on ammonia, ethylene, chloroethane, and 1,1,1-trichloroethane oxidation by *Nitrosomonas europaea*, *Appl. Environ. Microbiol.* **64**:1372-1378.

Huang, B., Johnson, J. W., Box, J. E., and NeSmith, D. S., 1997, Root characteristics and hormone activity of wheat in response to hypoxia and ethylene, *Crop Sci.* **37**:812-818.

Hunt, P. G., Campbell, R. B., and Moreau, R. A., 1980, Factors affecting ethylene accumulation in a Norfolk sandy loam soil, *Soil Sci.* **129**:22-27.

Hunt, P. G., Campbell, R. B., Sojka, R. E., and Parsons, J. E., 1981, Flooding-induced soil and plant ethylene accumulation and water status response of field-grown tobacco, *Plant Soil* **59**:427-439.

Hunt, P. G., Matheny, T. A., Campbell, R. B., and Parsons, J. E., 1982, Ethylene accumulation in southeastern coastal plain soils: Soil characteristics and oxidative-reductive involvement, *Commun. Soil Sci. Plant Anal.* **13**:267-278.

Hyodo, H., 1991, Stress/wound ethylene, in: *The Plant Hormone Ethylene*, A. K. Mattoo and J. C. Suttle, eds., CRC Press, Inc., Boca Raton, FL, pp. 43-63.

Ikeguchi, T., and Watanabe, I., 1991, Behavior of trace components in gases generated from municipal solid waste landfills, *Environ. Technol.* **12**:947-952.

Ishii, T., and Kadoya, K., 1984a, Ethylene evolution from organic materials applied to soil and its relation to the growth of grapevines, *J. Japan. Soc. Hort. Sci.* **53**:157-167.

Ishii, T., and Kadoya, K., 1984b, Growth of citrus trees as affected by ethylene evolved from organic materials applied to soil, *J. Japan. Soc. Hort. Sci.* **53**:320-330.

Ishii, T., and Kadoya, K., 1987, Lipids as ethylene-releasing substances in the soil, *Acta Hort.* **201**:69-76.

Ishii, T., Hashimoto, A., and Kadoya, K., 1983, Effects of soil ethylene on the growth of fruit trees, IV. Ethylene-releasing substances in organic materials, Abstr. Japan Soc. Hort. Sci., Autumn Meeting, pp. 58-59.

Jackson, M. B., 1982, Ethylene as a growth promoting hormone under flooded conditions, in: *Plant Growth Substances*, P. F. Wareing, ed., Academic Press, London, pp. 291-301.

Jackson, M. B., 1985., Ethylene and responses of plants to soil waterlogging and submergence, *Annu. Rev. Plant Physiol.* **36**:145-174.

Jackson, M. B., and Campbell, D. J., 1975, Movement of ethylene from roots to shoots, a factor in the responses of tomato plants to waterlogged soil conditions, *New Phytol.* **74**:397-406.

Jackson, M. B., and Campbell, D. J., 1976, Waterlogging and petiole epinasty in tomato: The role of ethylene and low oxygen, *New Phytol.* **76**:21-29.

Jackson, M. B., and Drew, M. C., 1984, Effects of flooding on growth and metabolism of herbaceous plants, in: *Flooding and Plant Growth*, T. T. Kozlowski, ed., Academic Press, New York, pp. 47-128.

Jackson, M. B., and Pearce, D. M. E., 1991, Hormones and morphological adaptation to aeration stress in rice, in: *Plant Life under Oxygen Deprivation*, M. B. Jackson et al., eds., Academic Publishing, The Hague, pp. 47-67.

Jackson, M. B., Gales, K., and Campbell, D. J., 1978, Effect of waterlogged soil conditions on the production of ethylene and on water relationships in tomato plants, *J. Exp. Bot.* **29**:183-193.

Jackson, M. B., Drew, M. C., and Giffard, S. C., 1981, Effects of applying ethylene to the root system of *Zea mays* on growth and nutrient concentration in relation to flooding tolerance, *Physiol. Plant.* **52**:23-28.

Kapulnik, E., Quick, J., and DeVay, J. E., 1985. Germination of propagules *of Verticillium dahliae* in soil treated with methionine and other substances affecting ethylene production, *Phytopathology* **75**:1348.

Kawase, M., 1974. Role of ethylene in induction of flooding damage in sunflower, *Physiol. Plant.* **31**:29-38.

Kawase, M., 1978, Anaerobic elevation of ethylene concentration in waterlogged plants, *Am. J. Bot.* **65**:736-740.

Kozlowski, T. T., 1986, Soil aeration and growth of forest trees (review article), *Scand. J. For. Res.* **1**:113-123.

Kuo, C. G., and Chen, B. W., 1980, Physiological responses of tomato cultivars to flooding, *J. Am. Soc. Hort. Sci.* **105**:751-755.

Lindberg, T., Granhall, U., and Berg, B., 1979, Ethylene formation in some coniferous forest soils, *Soil Biol. Biochem.* **11**:637-643.

Lynch, J. M., 1972, Identification of substrates and isolation of microorganisms responsible for ethylene production in soil, *Nature* **240**:45-46.

Lynch, J. M., 1975, Ethylene in soil. *Nature* **256**:576-577.

Lynch, J. M., 1983, Effects of antibiotics on ethylene production by soil microorganisms, *Plant Soil* **70**:415-420.

Lynch, J. M., and Harper, S. H. T., 1974a, Formation of ethylene by a soil fungus, *J. Gen. Microbiol.* **80**:187-195.

Lynch, J. M., and Harper, S. H. T., 1974b, Fungal growth rate and the formation of ethylene in soil, *J. Gen. Microbiol.* **85**:91-96.

Lynch, J. M., and Harper, S. H. T., 1980, Role of substrates and anoxia in the accumulation of soil ethylene, *Soil Biol. Biochem.* **12**:363-367.

Meek, B. D., Ehling, C. F., Stolzy, L. H., and Graham, L. E., 1983, Furrow and trickle irrigation effects of soil oxygen and ethylene and tomato yield, *Soil Sci. Soc. Am. J.* **47**:631-635.

Muromtsev, G. S., Karnenko, V. N., and Chernyaeva, I. I., 1988, Ethylene producing regulators of growth in plants, Inventor's Certificate No. 1372649555R, Byull. Izobret., No. 5.

Muromtsev, G. S., Letunova, S. V., Beresh, I. G., and Alekseeva, S. A., 1990, Soil ethylene as a plant growth regualtor and ways to intensify its formation in soil, *Biol. Bull. Acad. Sci. USSR* **16**:455-461.

Muromtsev, G. S., Krasinskaya, N. P., Letunova, S. V., and Beresh, I. G., 1991, Use of ethylene producing soil-acting preparation on citrus crops, *Soviet Agricul. Sci.* **2**:24-26.

Muromtsev, G. S., Letunova, S. V., Rentovich, L. N., Timpanova, Z. L., Gorbatenko, I. Y., Shapoval, O. A., Bibik, N. D., Stepanov, G. S., and Druchek, Y. V., 1993, Retprol – New ethylene-releasing preparation of soil activity, *Russ. Agricul. Sci.* **7**:19-26.

Muromtsev, G. S., Shapoval, O. A., Letunova, S. V., and Druchek, Y. V., 1995, Efficiency of new ethylene producing soil preparation Retprol on cucumber plants, *Selskokh. Biologia.* **5**:64-68.

Musgrave, A., Jackson, M. B., and Ling, E., 1972, *Callitriche* stem elongation is controlled by ethylene and gibberellin, *Nature New Biol.* **238**:93-96.

Nakano, R., and Kuwatsuka, S., 1979, Studies on formation and degradation of ethylene in flooded soils, IV. Ethylene and methane formation in soil incubated with plant materials, *J. Soc. Soil Manure, Japan* **50**:61-66.

Nakayama, M., and Ota, Y., 1980, Physiological action of ethylene in crop plants. V. Effects of water and compost on the ethylene production from soil, *Japan. J. Crop Sci.* **49**:359-365.

Nikitin, D. I., and Arakelyan, R. N., 1979, Utilization of ethylene by soil bacteria, *Biol. Bull. Acad. Sci. USSR* **6**:671-673.

Nohrstedt, H.-Ö., 1983, Natural formation of ethyhlene in forest soils and methods to correct results given by the acetylene-reduction assay, *Soil Biol. Biochem.* **15**:281-286.

Osborne, D. J., Walters, J., Milborrow, B. V., Norville, A., and Stange, L. M. C., 1996, Evidence for a non-ACC ethylene biosynthesis pathway in lower plants, *Phytochem.* **42**:51-60.

Otani, T., and Ae, N., 1993, Ethylene and carbon dioxide concentrations of soils as influenced by rhizosphere of crops under field and pot conditions, *Plant Soil* **150**:255-262.

Pazout, J., Wurst, M., and Vancura, V., 1981, Effect of aeration on ethylene production by soil bacteria and soil samples cultivated in a closed system, *Plant Soil* **62**:431-437.

Perret, P., and Koblet, W., 1984, Soil compaction induced iron-chlorosis in grape vineyards: Presumed involvement of exogenous soil ethylene, *J. Plant Nutrition* **7**:533-539.

Pimenta, J. A., Orsi, M. M., and Medri, M. E., 1994, Morphological and physiological aspects of *Coleus blumei* Benth. under waterlogged conditions and after application of ethrel and cobalt, *Rev. Barsil. Biol.* **53**:427-433.

Primrose, S. B., 1976, Ethylene-forming bacteria from soil and water, *J. Gen. Microbiol.* **97**:343-346.

Primrose, S. B., 1979, A review, ethylene and agriculture: The role of microbes, *J. Appl. Bacteriol.* **46**:1-25.

Rigler, E., and Zechmeister-Boltenstern, S., 1996, Einflub von CO_2 und stickstoff auf den microbiellen. Abbau von Methane und ethylene, mitteilungen der Deutschen, *Bodenkundichen Gesellschaft* **71**:205-208.

Rigler, E., and Zechmeister-Boltenstern, S., 1998, Influence of nitrogen and carbon dioxide on ethylene and methane production in two different soils, *Microbiol. Res.* **153**:227-237.

Rigler, E., and Zechmeister-Boltenstern, S., 1999, Oxidation of ethylene and methane in forest soils – effect of CO_2 and mineral nitrogen, *Geoderma* **90**:147-159.

Rovira, A. D., and Vendrell, M., 1972, Ethylene in sterilized soil. Its significance in studies of interactions between microorganisms and plants, *Soil Biol. Biochem.* **4**:63-69.

Sawada, S., Nakahata, K., and Totsuka, T., 1985, Fundamental studies on dynamics of ethylene in an ecosystem. III. Degradation capacity of atmospheric ethylene in soils taken from various vegetations, *Japanese J. Ecol.* **35**:453-459.

Sexstone, A. J., and Mains, C. N., 1990, Production of methane and ethylene in organic horizons of spruce forest soils, *Soil Biol. Biochem.* **22**:135-139.

Smith, A. M., 1973, Ethylene as a cause of soil fungistasis, *Nature* **246**:311-313.

Smith, A. M., 1975, Ethylene as a critical regulator of microbial activity in soil, Proc. First Intersect. *Congr. Int. Assoc. Microbiol. Soc. Tokyo* **2**:463-473.

Smith, A. M., 1976a, Ethylene in soil biology, *Annu. Rev. Phytopathol.* **14**:53-73.

Smith, A. M., 1976b, Ethylene production by bacteria in reduced microsites in soil and some implications to agriculture, *Soil Biol. Biochem.* **8**:293-298.

Smith, A. M., and Cook, R. J., 1974, Implications of ethylene production by bacteria for biological balance of soil, *Nature* **252**:703-705.

Smith, A. M., Milham, P. J., and Morrison, W. L., 1978, Plant diseases: Soil ethylene production specifically triggered by ferrous iron, in: *Microbial Ecology*, M. W. Loutit and J. A. R. Miles, eds., Springer-Verlag, New York, NY, pp. 329-336.

Smith, K. A., and Dowdell, R. J., 1974, Field studies of the soil atmosphere. I. Relationship between ethylene, oxygen, soil moisture content and temperature, *J. Soil Sci.* **25**:217-230.

Smith, K. A., and Restall, S. W. F., 1971, The occurrence of ethylene in anaerobic soil, *J. Soil Sci.* **22**:430-443.

Smith, K. A., and Russell, R. S., 1969, Occurrence of ethylene and its significance in anaerobic soil, *Nature* **222**:769-771.

Smith, K. A., Bremner, J. M., and Tabatabai, M. A., 1973, Sorption of gaseous atmospheric pollutants by soils, *Soil Sci.* **116**:313-319.

Sutherland, J. B., and Cook, R. J., 1980, Effects of chemical and heat treatments on ethylene production in soil, *Soil Biol. Biochem.* **12**:357-362.

Syslo, S. K., Myhre, D. L., and Biggs, R. H., 1990, Differences in ethylene production among three horizons of a Florida Spodosol, *Soil Sci. Soc. Am. J.* **54**:432-438.

Tang, T., and Miller, D. M., 1993, Ethylene production in anaerobically incubated soils amended with poultry litter, *Soil Sci.* **156**:186-192.

van Cleemput, O., El-Sebaay, A. S., and Baert, L., 1983, Evolution of gaseous hydrocarbons from soil, effect of moisture content and nitrate level. Soil Biol. Biochem. **15**:519-524.

Visser, E. J. W., Nabben, R. H. M., Blom, C. W. P., and Voesenek, L. A. C. J., 1997, Elongation by primary lateral roots and adventitious roots during conditions of hypoxia and high ethylene concentrations, *Plant, Cell Environ.* **20**:647-653.

Visser, E. J. W., Bögenman, G. M., Blom, C. W. P. M., and Voesenek, L. A. C. J., 1996a, Ethylene accumulation in waterlogged *Rumex* plants promotes formation of adventitious roots, *J. Exptl. Bot.* **47**:403-410.

Visser, E. J. W., Cohen, J. D., Barendse, G. W. M., Blom, C. W. P. M., and Voesenek, L. A. C. J., 1996b, An ethylene-mediated increase in sensitivity to auxin induces adventitious root formation in flooded *Rumex palustris* Sm., *Plant Physiol.* **112**:1687-1692.

Voesenek, L. A. C. J., and Blom, C. W. P. M., 1989, Growth responses of *Rumex* species in relation to submergence and ethylene, *Plant, Cell Environ.* **12**:433-439.

Voesenek, L. A. C. J., van der Sman, A. J. M., Harren, F. J. M., and Blom, C. W. P. M., 1992, An amalgamation between hormone physiology and plant ecology: A review on flooding resistance and ethylene, *J. Plant Growth Regul.* **11**:171-188.

Wainwright, M., and Kowalenko, C. G., 1977, Effects of pesticides, lime and other amendments on soil ethylene, *Plant Soil* **48**:253-258.

Yamamoto, F., 1992, Effects of depth of flooding on growth and anatomy of stems and knee roots of *Taxodium distichum*, *IAWA Bull.* **13**:93-104.

Yamamoto, F., and Kozlowski, T. T., 1985, Effects of flooding, tilting of stem and ethrel application on growth, stem anatomy, and ethylene production of *Acer platanoides,* Scand. J. Forest Res. 2:141-156.

Yoshida, T., and Suzuki, T., 1975, Formation and degradation of ethylene in submerged rice soils, Soil Sci. Plant. Nutr. (Tokyo) 21:129-135.

Zahir, Z. A., and Arshad, M., 1998, Response of *Brassica carinata* and *Lens culinaris* to ethylene precursors L-methionine and 1-aminocyclopropane-1-carboxylic acid, Soil Biol. Biochem. 30:2185-2188.

Zainol, E., Lal, R., Vantoai, T., and Fausey, N., 1991, Soil compaction and water table effects on soil aeration and corn growth in a greenhouse study, Soil Technol. 4:329-342.

Zechmeister-Boltenstern, S., and Nikodim, L., 1999, Effect of water tension on ethylene production and consumption in montane and lowland soils in Austria, European J. Soil Sci. 50:425-432.

Zechmeister-Boltenstern, S., and Smith, K. A., 1998, Ethylene production and decomposition in soils, Biol. Fertil. Soils 26:354-361.

6

ETHYLENE IN SYMBIOSIS

6.1. INTRODUCTION

The involvement of C_2H_4 in plant responses to a variety of biotic and abiotic stress is well known (Abeles et al., 1992). The biotic stress includes parasitic and non-parasitic plant-microbe interactions, which play an important role in plant growth and development (Arshad and Frankenberger, 1992, 1993, 1998; Frankenberger and Arshad, 1995; Boller, 1991; Abeles et al., 1992; Beyrle, 1995; Hirsch et al., 1997; Hirsch and Yang, 1994). The role of C_2H_4 in symbiotic associations is the major focus of this chapter.

6.2. SYMBIOTIC ASSOCIATIONS

Symbiotic associations primarily represent positive interactions between two partners for mutual benefit (symbiosis). Two well known plant-microbe symbiotic associations are nodulation and mycorrhizae formation. Lichens represent symbiosis between an alga and a fungus. The role of C_2H_4 in these three symbiotic associations will be critically discussed in the following sections.

6.3. NODULATION

The symbiotic relationship between N_2-fixing rhizobia and legume plants has been studied for over 100 years as a classic example of mutualistic association. This interaction between a microsymbiont (e.g., *Rhizobium*) and its host plant (e.g., legume) can result in the formation of unique and highly organized structures known as nodules. Compatible rhizobia trigger morphogenesis of a nodule organ and symbiotic N_2-fixation with legume host plants. This symbiosis is essential for the supply of atmospheric N to plants fixed by the rhizobia. It has been claimed that as much as 90% of the plant's requirements of N may come through this association (Drevon, 1983). Despite the beneficial aspects of this symbiosis, nodule formation is strictly controlled by the host

195

through an internal mechanism called autoregulation (Caetano-Anolles and Gresshoff, 1990) because it is energetically expensive to develop and maintain the nodules (Shantharam and Mattoo, 1997). This autoregulation involves early infection events that systemically suppress the development into nodules of subsequent infected cell-division foci (Caetano-Anolles and Gresshoff, 1990) resulting in restriction of nodulation to the root crown.

The formation of a N_2-fixing root nodule is a complex developmental event requiring the exquisite coordination of gene expression of two very distinct organisms, one a prokaryote and the other a eukaryote (Hirsch et al., 1997). This symbiosis is dependent on the specific recognition of signal molecules produced by each partner. It has been postulated that among other factors, plant hormones have on important regulatory role in the development and establishment of nodulation (Hirsch and Yang, 1994; Bladergroen and Spaink, 1998; Arshad and Frankenberger, 1998; Frankenberger and Arshad, 1995). Observations suggest that C_2H_4 is important in the development of the legume-*Rhizobium* association and may function as an autoregulator to control nodule formation and development (Ligero et al., 1991). Ethylene could be involved in several phases of symbiosis, including initial responses to bacterial *Nod* factors, nodule morphogenesis, regulation of host defense responses, and nodule senescence and abscission. Inhibition of normal longitudinal root growth and induction of root hair formation in response to inoculation closely resembles the effects of C_2H_4 on roots (Feldman, 1984; Goodlass and Smith, 1979; Goodwin and Mercer, 1983; Lieberman, 1979; Zaat et al., 1989). In the subsequent sections, the role of C_2H_4 as a plant hormone in the regulation of nodulation in legumes will be discussed comprehensively, taking into account the effects of exogenous C_2H_4 on nodulation and *Nod* factors, accelerated C_2H_4 production by infected roots, NO_3^-- and light-induced C_2H_4 production and its effects on nodulation, nodulation in mutants varying in C_2H_4 sensitivity, and the effect of C_2H_4 inhibitors (biosynthesis and action) on nodulation.

6.3.1. Effects of Exogenous Ethylene on Nodulation

The effects of exogenously applied C_2H_4 gas or C_2H_4 -releasing compounds such as 1-aminocyclopropane-1-carbolic acid (ACC) and 2-chloroethylphosphonic acid (ethephon/Ethrel) on nodule formation and development have been investigated by many scientists. These studies have revealed that exogenous C_2H_4 applied directly as a gas, or indirectly as ACC (an C_2H_4 precursor in higher plants) or ethephon/Ethrel (a nonenzymatically C_2H_4 releasing compound) acts as a strong inhibitor of nodulation (Drennan and Norton, 1972; Goodlass and Smith, 1979; Grobbelaar et al., 1971; Lee and LaRue, 1992a,c; Lee et al., 1993; van Workum et al., 1995; Schmidt et al., 1999; Fernandez-Lopez et al., 1998; Yuhashi et al., 2000; Duodu et al., 1999; Caba et al., 1999). Grobbelaar et al. (1971) demonstrated that nodulation decreased by 90% with bean explants treated with 0.4 ppm C_2H_4, and it was completely inhibited at higher concentrations of C_2H_4. Likewise, Goodlass and Smith (1979) found that the addition of 10 ppm C_2H_4 decreased nodule formation on pea by 70% and on clover by 44%. However, the rate of nitrogen fixation was not severely affected. Similarly, exogenous C_2H_4 application, even at a level of 0.01 ppm, inhibited nodulation of *Pisum sativum*, and nodulation was restored by treatment with 1 μM Ag^+ (Lee and LaRue, 1992a). In the presence of C_2H_4, the number of infected root hairs did not change; however, many infection threads were aborted and the epidermis or outer cortex and nodule primordia did not form. In another study, Lee and LaRue (1992c) found that nodule numbers were reduced to half in the presence of 0.07 μl L^{-1} (0.07 ppm) C$_2$H$_4$ applied continuously to the roots for 3 weeks. Inhibition was overcome by 1 μM Ag$^+$, by inhibiting C$_2$H$_4$ binding. Exogenous C$_2$H$_4$ also inhibited nodulation of sweet clover and pea mutants that were hypernodulating or had ineffective nodules. Lee and LaRue (1992c) observed that exogenous C$_2$H$_4$ did not decrease the number of infections per centimeter of lateral pea root, but nearly all of the infections were blocked when the infection thread was in the basal epidermal cell or in the outer cortical cells. Later, Lee et al. (1993) conducted studies on pea, sweet clover, and soybean inoculated with their respective *Rhizobium/Bradyrhizobium* microsymbiont to investigate effects of C$_2$H$_4$ and light on nodulation. They found that continuous application of C$_2$H$_4$ (up to 0.45 μl L^{-1}) decreased nodulation in pea and sweet clover, but not in soybean. However, exogenous C$_2$H$_4$ did not inhibit the very early stages of nodule development in pea, such as colonization of the rhizobia on the root hair or formation of an infection thread and its initial growth. The penetration of the infection thread from the epidermis into the outer cortex seems the most sensitive to C$_2$H$_4$ in pea. Ag$^+$ application reversed the negative effect of exogenous C$_2$H$_4$ on nodulation.

Application of Ethrel (2-chloroethylphosphonic acid) to the roots at concentrations >2 ppm severely inhibited nodulation in pea plants (Drennan and Norton, 1972). These investigators suggested that a partial effect of Ethrel on nodulation resulted from its inhibition of root growth. Nodulation was also inhibited on the treated, divided root system exposed to Ethrel, whereas, there was no inhibition on the neighboring untreated half (Drennan and Norton, 1972). Duodu et al. (1999) found that treatment with 100 μM ethephon resulted in a significant reduction in the number of mature nodules and in percentage of plants that developed mature nodules, demonstrating that nodule development on mung bean (*Vigna radiata*) is indeed sensitive to C$_2$H$_4$ (Fig. 6.1). Penmesta and Cook (1997) reported that exogenous application of ACC had a concentration-dependent negative effect on the nodule formation in *Medicago truncatula*; however, the time of application of ACC was also critical in inhibiting nodulation. They found that nodulation was suppressed when ACC was applied during the primary infection phase (24-48 hours), but not when ACC was applied after the appearance of the macroscopic nodule primordia (72 hours). Likewise, Schmidt et al. (1999) reported that ACC treatment also caused a decrease in the number of nodules formed on soybean wild type Hobbit 87 roots, but had no significant effect on the C$_2$H$_4$ insensitive mutant *etr1-1* line. However, they were of the opinion that control of nodule numbers is independent of C$_2$H$_4$ signaling and the effect of ACC on nodule number may be largely attributed to the stunted growth of Hobbit 87 roots. The effect of 1 μ*M* ACC on nodulation of *Macroptillium atropurpureum* inoculated with *Bradyrhizobium elkanii* was assessed by Yuhashi et al. (2000). It was observed that the number of nodules formed in the presence of ACC, 8 days after inoculation and later were significantly fewer than the number of nodules formed in the absence of ACC (Fig. 6.2). The addition of ACC also suppressed nodulation in the control and transgenic plants(Se 40) of *Medicago truncatula* confirming that C$_2$H$_4$ negatively regulates nodulation (Charon et al., 1999). A study on the C$_2$H$_4$ - mediated phenotype plasticity in root nodule development on *Sesbania rostrata* was conducted by Fernandez-Lopez et al. (1998). They found that in a Leonard jar assay

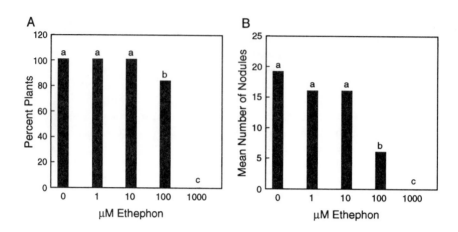

Fig. 6-1.   Sensitivity of nodulation to C_2H_4 in mung bean. Nodulation resopnse of mung bean plants treated with various concentrations of ethephon. Mature nodules were scored 21 days after inoculation with wild-type strain USDA61. Bars represent mean values of at least 12 plants. Bars with no letters in common are significantly different at the 0.05 confidence level by Fisher's protected least significant difference test. Source: Duodu et al. (1999).

Table 6.1.   Root nodulation assays on *Sesbania rostrata*.

Treatment	Tube assay		Leonard jar assay	
	Type	Number	Type	Number
Control	D	100.0 ± 11.6a	I	34.2 ± 2.1a
10 μM Ag₂SO₄	I	20.1 ± 2.0b	I	33.0 ± 4.7a
20 μM ACC	D	105.2 ± 10.3a	D	45.0 ± 6.6a
20 μM CEPA*	D	ND	D	44.8 ± 5.1a

Type (D, determinate; I, indeterminate) and number (average of three independent experiments with 10 plants each) of mature nodules developed on *S. rostrata* roots grown in tubes or in the vermiculite compartment of Leonard jars are tabulated. The same letter in a column indicates numbers that are significantly different (p > 0.05).
ND, not determined.
*, 2-chloroethylphosphonic acid (ethephon).
Source: Fernandez-Lobez et al. (1998).

system, in which normally indeterminate nodules are formed, addition of ACC or ethephon caused the formation of determinate nodules, while in a tube assay which normally favors the formation of determinate nodules, indeterminate nodules were formed in the presence of Ag^+ ions (Table 6.1), suggesting a major role of C_2H_4 in the switch of nodule type. Caba et al. (1999) conducted a comprehensive study on differential sensitivity of nodulation to C_2H_4 in soybean cv. Bragg and its supernodulating mutants. Both ACC and ethephon reduced nodule numbers per plant, nearly twofold more in wild type than in the supernodulating mutants; the recovery in response to treatment of Ag^+ was also more in the case of wild type, strongly suggesting the involvement of C_2H_4 in such an inhibition (Table 6.2). Interestingly, nodule growth and development were strongly enhanced by these C_2H_4 precursors as judged by nodule dry weight data per plant and acetylene reduction activity (ARA) in all the genotypes. This may imply that C_2H_4 precursors exert their effects at an early stage in nodule development and that prenodules, escaping this exposure, grow and develop faster achieving higher mass and greater ARA (Caba et al., 1999). The long term exposure of 1 μm ACC reduced nodule number in both genotypes, but the inhibition of Bragg was fourfold higher than that of the mutant. However, delayed application by six days at 100 μM ACC slightly decreased the nodule number in both genotypes implying that early stages in nodule development are more sensitive to C_2H_4 than the subsequent growth of prenodules (Caba et al., 1999).

6.3.2. Effects of Exogenous Ethylene on *Nod* Factor(s)

The rhizobia signals that initiate development of the nodule organ are specific lipo-chitin oligosaccharides called *Nod* factors (Lerouge et al., 1990; Carlson et al., 1995; Spaink, 1995). *Nod* factors alter the growth of two cell types in the roots as they induce root hair deformation by inducing tip growth in existing root hairs and furthermore, activate cortical cells to resume mitosis resulting in nodule primordia (Libbenga and Harkes, 1973; Lerouge et al., 1990; Spaink et al., 1991; Truchet et al., 1991). Ethylene is known to have an opposite effect on tip growth and cell division in roots of dicotyledonous plants. On one hand, C_2H_4 promotes tip growth, i.e., root hair formation (Tanimoto et al., 1995), indicating that it may act as a second messenger in *Nod* factor-induced hair deformation. On the other hand, C_2H_4 blocks cortical cell division (Grobbelaar et al., 1971; Goodlass and Smith, 1979; Lee and LaRue, 1992c). Some strains of *Rhizobium* induce the formation of thick, short roots (Tsr) in *Vicia sativa* in a manner that resembles the response to exogenous C_2H_4 (van Brussel et al., 1986), and this response is eliminated by AVG, an inhibitor of C_2H_4 synthesis (Zaat et al., 1989). Nodulation of *V. sativa* subsp. *nigra* by *R. leguminosarum* resulted in the development of thick short roots, (Zaat et al., 1989). This root phenotype, as well as root hair induction and root hair formation, are caused by a factor(s) produced by the bacterium in response to plant flavonoids. Zaat and co-workers (1989) reported that root growth inhibition and root hair induction, but not root hair formation, could be mimicked by an ethephon treatment. The addition of the C_2H_4 biosynthesis inhibitor, AVG to the co-cultures of *V. sativa* plus an inoculum (5×10^5 ml^{-1} of bacteria) suppressed the development of Tsr and restored nodulation to the normal pattern that was observed with very low populations of bacteria (0.5-5 ml^{-1} of bacteria). Similarly, van Spronsen et al. (1995) reported that the

Table 6.2. Growth, nodulation and acetylene reduction activity (ARA) in soybean cv. Bragg and its supernodulating mutant *nts* 382 inoculated with *Bradyrhizobium japonicum* USDA110 and treated with different concentrations of CEPA and Ag^+.

Genotype	CEPA** (μM)	Ag+ (μM)	Growth		Nodulation		
			Root (g.d.wt per plant)	Shoot	No. of nodules per plant	D.wt (mg per plant)	ARA* (μmol C_2H_4 g.d. wt^{-1} h^{-1})
Bragg	0	0	0.11	0.33	130	10.5	15.4
	1	0	0.13	0.33	109	11.7	26.6
	1	10	0.14	0.36	132	6.6	--
	10	0	0.13	0.35	75	12.7	39.0
	10	1	0.13	0.34	106	8.8	--
	10	10	0.13	0.35	119	7.1	--
	100	0	0.12	0.31	55	15.3	28.3
	100	1	0.13	0.32	96	12.6	--
	100	10	0.14	0.35	106	10.1	--
		$LSD_{0.05}$	0.02	0.03	8	1.7	4.3
nts 382	0	0	0.13	0.40	675	59.8	8.4
	10	0	0.15	0.35	563	73.5	12.2
	10	1	0.14	0.37	600	56.0	--
	100	0	0.14	0.35	454	72.0	16.0
	100	1	0.15	0.38	540	57.0	--
		$LSD_{0.05}$	0.02	0.04	50	10.6	3.0

*Because of the general negative effect of Ag^+ on nodule growth and activity, ARA values (nodule-dry-weight-based) are given for controls and ACC only-treated plants.

**2-Chloroethylphosphonic acid (ethephon).

Plants were grown in culture tubes with vermiculite and 1 mM NO_3^- solution. Plants were inoculated and treated 10 days after planting. Plants were harvested 18 days after inoculation, scored for nodulation, dried and weighed; means of 20 uniform plants per treatment are presented. ARA was determined in controls and CEPA only-treated plants with 7 replicates per treatment. Data were subjected to ANOVA and means were compared using the LSD test. (-), not determined.
Source: Caba et al. (1999)

development of the Tsr phenotype in *Vicia sativa* sp. *nigra* plants upon inoculation with nodule-inducing bacterium, *Rhizobium leguminosarum* biovar *viciae*, was correlated with the presence of *nod* (nodulation) genes in the rhizobia. *Nod* factors (lipochitin oligosaccharides), that are products of these *nod* genes, can induce the Tsr phenotype in the absence of rhizobia. The Tsr phenotype can be mimicked by addition of ethephon and inhibited by AVG (van Spronsen et al., 1995). This strongly suggests that the Tsr phenotype is caused by excessive C_2H_4 production. The C_2H_4 related changes mentioned above are also seen during infection thread formation, but only locally. Apparently, *Vicia* roots when grown under light overrespond to *Nod* factors leading to overproduction of C_2H_4 and to a non-local "ripening" process. These phenomena inhibit nodulation of the main root by preventing formation of pre-infected threads and by reducing formation of root nodule primordia. Local controlled production of C_2H_4 , as induced by *Nod* factors, may however, be an essential element of the nodulation process (van Spronsen et al., 1995). After studying the induction of chitinase activity in 161 cultivars of soybean, Xie et al. (1996) speculated that C_2H_4 -induced chitinases may play a role in nodule inhibition, while specific chitinases are able to cleave differentially substituted *Nod* factors of rhizobia and thereby determine their biological activities in the host plants (Hunter, 1993; Staehelin et al., 1994a,b).

Exopolysaccharides (EXO)- or lipopolysaccharide-defective (LPS)-mutants of *Rhizobium leguminosarum* bv. *viciae* are usually impaired in root nodule formation on their host plants (Perotto et al., 1994; van Workum et al., 1995). van Workum et al. (1995) found that *Vicia sativa* ssp. *nigra* (vetch) could be nodulated by such mutants if C_2H_4 production (resulting from rhizobial inoculation) by the host plant root was minimized. Under these circumstances, EXO mutants induced delayed formation of partially infected nodules. EXO mutants did not induce abnormally large amounts of C_2H_4 in host roots nor showed abnormal production of lipo-oligosaccharide *Nod* signals; thus, impaired nodulation could not be ascribed to the inability of the inoculum to produce EXO *Nod* signals. The nodulating ability of *R. leguminosarum* bv. *viciae* EXO mutants (deficient in exopolysaccharide synthesis) could be restored completely by coinoculation with a *Nod* $^-$ EXO$^+$ strain, indicating that impaired nodulation was indeed caused by the absence of exopolysaccharides. These findings led the authors to hypothesize that rhizobial *Nod* signals not only affect nodulation-related phenomena, but also induce C_2H_4 formation in the host plant roots. By influencing root cell growth, C_2H_4 inhibits proper root infection by rhizobia. In case of delayed nodulation due to exopolysaccharide deficiency, C_2H_4 formation precedes root infection and as a result nodulation is impaired (van Workum et al., 1995). By using AVG and Ag$^+$, Heidstra et al. (1997) demonstrated that C_2H_4 has no required positive role in the re-initiation of root hair tip growth induced by *Nod* factors, i.e., root hair deformation, whereas tip growth in epidermal cells leading to root hair formation in vetch is promoted by C_2H_4. They also reported that C_2H_4 is a negatively active factor controlling the position where nodule primordia are formed. They concluded that C_2H_4 is not involved in root hair deformation which excludes the possibility of its action during the early *Rhizobium*-host interaction.

One of the nodulation genes associated with the earliest phases of nodule organogenesis is *enod 40*. Charon et al. (1999) investigated the effect of alteration of *enod 40* on nodule development in transgenic *Medicato truncatula* and reported that *enod 40* action could be partially mimicked by treatment of the infected root with the C_2H_4 inhibitor, AVG. However, the interaction between *enod 40* and C_2H_4 may be much more complex.

6.3.3. Accelerated Ethylene Evolution from *Rhizobium*-Infected Roots

It has been generally observed that inoculated/nodulated legumes produce more C_2H_4 than non-inoculated/nodulated legumes (Ligero et al., 1986, 1987, 1991, 1999) and this inoculation stimulated C_2H_4 release (ISER) is most likely a plant response to the nodule bacteria (Zaat et al., 1989). Ligero and coworkers reported that inoculated roots of alfalfa produced more C_2H_4 than non-inoculated roots (Ligero et al., 1986, 1991; Caba et al., 1998). A similar kind of observation (ISER) was noted in many other plants (Zaat et al., 1989; Suganuma et al., 1995; van Spronsen et al., 1995). Suganuma et al. (1995) reported that production of C_2H_4 by soybean roots was stimulated by inoculation with *B. japonicum* and stimulation was maximal 3 days after inoculation. The rate of C_2H_4 production then fell to that in the roots of uninoculated plants. No ISER was detected upon inoculation with non-nodulating soybean mutants or after inoculation with a heterologous rhizobium, *Rhizobium leguminosarum* bv. *vicia*.

Likewise, Poveda et al. (1993) observed ISER by soybean roots within 2 hours after inoculation and higher intensity ISER in 8 mM NO_3^--fed roots than the control (1.5 mM NO_3^-, grown in vermiculite filled glass growth assembly covered with black paper). By day 4 after inoculation, C_2H_4 declined nearly to the respective original levels and then again increased (8 mM NO_3^- fed plants) until day 15 or day 20 (controls), where both treatments matched in C_2H_4 release. Over the periods of infection and nodule development, C_2H_4 evolution was significantly higher in 8 mM NO_3^--fed soybean plants, which further showed strong inhibition of nodulation (10 ± 2 nodules per plant vs. 60 ± 6 in the control plants) implying that endogenous C_2H_4 might be involved in the inhibitory effects of NO_3^- on nodulation (Poveda et al., 1993).

Liggero et al. (1999) compared NO_3^- and inoculation-induced C_2H_4 biosynthesis in soybean genotypes Bragg (wild type) and its supernodulating (*nts 382* and *nts 1007*) and *non*-nodulated (*nod 49* and *nod 139*) mutants. They found that regardless of the NO_3^- treatment, inoculation with *B. japonicum* significantly increased the root C_2H_4 evolution rate, reaching a plateau between 24 and 48 hours after inoculation. They suggested that this ISER response appears to be related to the infection process and nodule development, since treatment with Ag^+ at the moment of inoculation markedly increased nodule numbers of Bragg under both low and high NO_3^- concentrations. However, inoculation did not have any significant effect on ACC-oxidase activity, while this activity was strongly stimulated by NO_3^-. In another study, they compared C_2H_4 evolution activity in roots of soybean cv. Bragg (wild type) vs. the supernodulating mutants "*nts* 382 and *nts* 1007" after inoculation and treatment with ACC or ethephon (Caba et al., 1999). They observed that ISER was higher in case of Bragg compared to its mutants in the absence of ACC or ethephon (Table 6.3). However, at 100 μM ACC, C_2H_4 evolution was more in the case of mutants.

Hunter (1993) compared C_2H_4 formation and ACC concentrations in inoculated *B. japonicum* vs. uninoculated soybean plants grown hydroponically. He reported that the presence of nodules on the roots increased the amount of C_2H_4 produced in the soybean. Similarly, higher levels (>two-fold) of ACC were found in nodulated roots than in uninoculated roots, and [ACC] decreased in the roots when nodules were removed. The nodules contained the highest concentration of ACC (>four-fold), compared with the uninoculated roots. Nodules formed with an ineffective *B. japonicum* strain had less

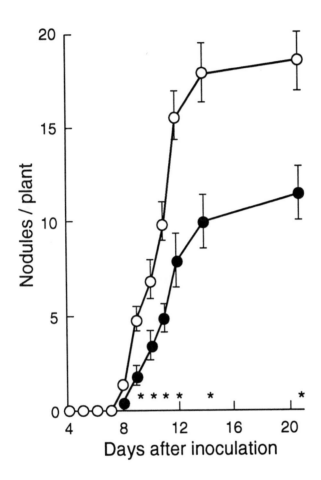

Fig. 6-2. Effect of ACC on nodulation of *Macroptillium atropupureum* cv. *Siratro* inoculated with *Bradyrhizobium elkanii* USDA94. Each point represents the mean number of nodules per plant. The bars indicate standard errors. Nine plants were treated and assessed for each data point. Symbols: O, no ACC; ●, 1 µM ACC. Asterisks indicate significant differences in mean nodule number between treatments at a confidence level of 0.05. Source: Yuhashi et al. (2000).

Table 6.3. Ethylene evolution and ACC oxidase activity in roots of soybean cv. Bragg and its supernodulating mutants *nts* 382 and *nts* 1007 after inoculation and treatment with different concentrations of ACC and CEPA.

Ethylene donor	Concen-tration (μM)	Genotype			
		Bragg	nts 382	nts 1007	$LSD_{0.05}$
(a) Ethylene evolution (nmol C_2H_4 g^{-1} d.wt.h^{-1})					
Control	0	2.39	1.97	1.78	0.21
ACC	10	5.96	4.63	4.92	1.17
	100	39.10	52.89	47.28	11.80
CEPA*	10	8.53	7.85	--	1.87
	100	38.15	49.86	--	7.64
	$LSD_{0.05}$	2.95	7.86	6.30	--
(b) ACC oxidase activity (nmol C_2H_4 g^{-1} d.wt.h^{-1})					
Control	0	52.7	46.3	41.9	7.34
ACC	10	126.5	68.7	73.2	17.2
	100	208.4	106.8	126.8	37.17
CEPA	10	99.7	79.5	--	22.75
	100	198.8	111.6	--	29.29
	$LSD_{0.05}$	37.95	18.65	23.4	--

Ten day-old seedlings were inoculated and treated with ACC and CEPA and 48 h later root C_2H_4 evolution and ACC oxidase activity of two sets of plants were measured. Means of 10 (C_2H_4 evolution) and eight (ACC oxidase activity) replicates per treatment are presented. Statistics and other details as in Table 6-2.
--, not determined.
*, 2-chloroethylphosphonic acid (ethephon).
Source: Caba et al. (1999).

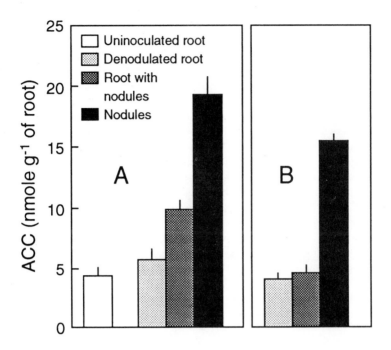

Fig. 6-3.   ACC contents of soybean root and nodule materials. (A) Data
for uninoculated plants and for plants that received the highly effective
wild-type inoculum. (B) Data for plants that received the ineffective
strain 5MT-7 inoculum. Plants were grown for 4 weeks in the growth
chamber. The values for roots and denodulated roots are averages ±
standard errors of nine measurements. In panel A the value for nodules is
the average ± standard error of seven measurements. In panel B, nodule
materials from nine plants were pooled to obtain enough mass to make
measurements: the value is the average ± standard error of two
measurements. Source: Hunter (1993).

ACC than an effective strain (Fig. 6.3). Hunter (1993) concluded that nearly all (>95%) of the C_2H_4 produced by nodules was of plant origin, with little produced by the bacteroids. However, the application of ethephon up to 10 μM did not affect the nodule number, although plant growth was significantly affected. Moreover, AVG did not stimulate nodulation in soybean grown with 0.8 and 2.5 mM NO_3^-, contrary to the other *Rhizobium*-legume systems. This may imply that the soybean-*B. japonicum* system is not as sensitive to C_2H_4 as the other *Rhizobium*-legume systems.   Hunter (1993) hypothesized that low and intermediate levels of C_2H_4 do not interfere with nodulation in soybean. Similarly, C_2H_4 production and ACC concentrations were also evaluated in inoculated vs. uninoculated bean (*Phaseolus vulgaris* L.) seedlings (Beard and Harrison, 1992). Preliminary results indicated that the pattern of C_2H_4 and ACC production in the various plant segments was similar for inoculated and uninoculated plants. However, the level of conjugated form of ACC, 1-(malonylamino)cyclopropane-1-carboxylic acid (MACC) was greatly increased in nodulated seedlings. Beard and Harrison (1992) reported the ability of *Rhizobium* spp. to produce C_2H_4 and detected ACC in the culture medium. Exogenous application of ACC resulted in a large burst of C_2H_4 in the rhizobial culture, indicating an active ACC oxidase system. Given these observations, Beard and Harrison (1992) suggested that nodulation may not change the overall C_2H_4 biosynthesis in bean seedlings, but may perhaps alter the level of ACC conjugation to MACC. It is not yet known whether rhizobial-infected nodules contribute C_2H_4 or ACC to the plant tissues. However, it is interesting that this *Rhizobium* spp. has the capacity for C_2H_4 production by the same biosynthetic pathway as found in higher plants (Beard and Harrison, 1992).

6.3.4. Nitrate and Light-Induced Ethylene Production and Nodulation

Both light and NO_3^- have been reported to increase C_2H_4 biosynthesis by roots and affect nodulation. Lee and LaRue (1992a) observed that exposure of *P. sativum* L. cv. Sparkle roots to light suppressed nodulation and induced an increase in C_2H_4 production by roots. Under dim light, treatment with Ag^+ increased the nodule numbers on the roots. Exposure of roots to NO_3^- also suppressed nodulation and induced an increase in C_2H_4 production by roots. However, induction of C_2H_4 production in roots by NO_3^- was less than that induced by dim light. They suggested that the inhibitory effects of light may occur via increased C_2H_4 production, but C_2H_4 does not mediate NO_3^- control of nodule numbers. Likewise, Lee et al. (1993) investigated the role of C_2H_4 in nodulation on pea, sweet clover and soybean. They found that dim as well as bright light stimulated C_2H_4 production and inhibited nodulation in *Pisum sativum* cv. Sparkle, suggesting the possible role of C_2H_4 in regulating nodule numbers. Ligero and co-workers also investigated the effects of NO_3^- and light on nodulation inhibition in alfalfa (Ligero et al., 1986, 1987, 1991; Poveda et al., 1993; Caba et al., 1998). It was demonstrated that NO_3^- and inoculation enhanced C_2H_4 evolution by alfalfa roots (Ligero et al., 1986, 1987) and that inhibition of C_2H_4 biosynthesis by treatment with AVG could eliminate the inhibitory effects of NO_3^- on nodulation (Ligero et al., 1991). Ligero et al. (1987) reported a positive correlation between NO_3^- concentrations and the quantity of C_2H_4 released from roots of alfalfa inoculated with *R. meliloti* in a mineral solution. They suggested that this may explain the inhibitory effects of combined nitrogen application on nodulation in legumes. In another study, Ligero et al. (1991) investigated the effects of various combinations and concentrations of AVG and NO_3^- on nodulation of alfalfa and observed that the effect of AVG was depressed with NO_3^- supply as AVG promoted as many as 4.3-, 2.4-, 2.3-, and 1.5-fold more nodules than infected plants growing in the presence of 20, 10, 5, and 2.5 mM NO_3^-, respectively by day 19 after inoculation. The concentration of AVG (10 and 20 µM), which induced more than two-fold the number of nodules per plant, also inhibited C_2H_4 biosynthesis by five- to ten-fold, 48 h after inoculation and AVG addition. These results reinforced their hypothesis that the NO_3^- effect on nodulation may be mediated through C_2H_4 production and C_2H_4 may be involved in the autoregulation of nodulation by plants. They concluded that: (i) NO_3^- inhibition of nodulation may be mediated by C_2H_4, (ii) this mechanism might be broadly distributed among legumes, and (iii) a connection between autoregulation and NO_3^- inhibition of nodulation exists as reported by Day et al. (1989) and Gresshoff (1993). Later, they found that high levels of NO_3^- (10 mol m^{-3}) applied upon planting inhibited nodulation to an equal extent (ca. 50%) in plants grown aeroponically or in darkened vermiculite filled tubes (Caba et al., 1998). In contrast, Ag^+ treatment increased nodule formation at all NO_3^- concentrations assayed under the two growth conditions with stimulation being higher in plants grown aeroponically. Nitrate induced inhibition of nodule formation after 24 h and was highest when applied at 10 mole m^{-3} and it paralleled an increase of ACC oxidase activity. Ethylene evolution rates markedly increased in inoculated and uninoculated roots treated with NO_3^- (Table 6.4). Caba et al. (1998) concluded that *Rhizobium* induces C_2H_4 biosynthesis which is further stimulated by NO_3^- and this surge of high amounts of C_2H_4 results in further restriction of infection or nodule formation. Similarly, Ligero et al. (1999) observed that an increase in NO_3^- concentration in the rooting medium strongly stimulated ACC oxidase activity in roots of soybean cultivar Bragg; however, inoculation had no effect on the enzyme activity. Moreover, NO_3^- strongly inhibited nodulation in a concentration-dependent manner, whereas Ag^+ enhanced nodule formation at all NO_3^- concentrations assayed with the effect (on a percentage basis) being proportional to the degree of NO_3^- inhibition. Since no negative effect of NO_3^- on root or shoot growth was observed, these effects appear to be specific to nodule formation(Table 6.5). Ruan and Peters (1992) also reported that NO_3^- suppressed nodulation in soybean.

Contrarily, Schmidt et al. (1999) reported that NO_3^- addition suppressed the nodule number in both C_2H_4 sensitive soybean wild type, Hobbit 87, and its insensitive mutant *etr 1-1* (T119N54) plants and that the presence of Ag^+ did not reverse this negative effect of NO_3^-. This led Schmidt et al. (1999) to suggest that any limitation of nodule formation by the host (soybean) in response to NO_3^- was independent of C_2H_4 .

6.3.5. Ethylene Sensitive and Insensitive Legume Mutants and Nodulation

The availability and development of legume mutants varying in their sensitivity and responsiveness against C_2H_4 or lacking autoregulation provides excellent tools to explore the role of C_2H_4 in nodulation. Few studies have demonstrated a defined role for C_2H_4 in nodule organogenesis. A mutant of *Medicago truncatula* exhibits an increase of more than an order of magnitude in the number of persistent rhizobial infections (Penmetsa and Cook, 1997). Physiological and genetic analyses indicate that this same mutation confers insensitivity to the plant hormone, C_2H_4 for multiple aspects of plant development,

Table 6.4. Ethylene evolution rates and ACC oxidase activity of inoculated and uninoculated alfalfa roots growing in vermiculite and treated with different levels of NO_3^-. Measurements were made at different times after inoculation.

NO_3^- (mol m^{-3})	Time of NO_3^- addition (DAP)	Inoculation (+/-)	Time after inoculation (h)	Ethylene evolution [nmol C_2H_4 (g DW)$^{-1}$ h^{-1}]	ACC oxidase activity
0	0	+	24	3.90	ND[a]
		-	48	3.64	32.3
		+	48	4.71	36.5
2	0	+	24	4.31	64.5
		-	48	3.82	49.7
		+	48	4.90	51.6
		+	72	5.0	32.5
10	10*	+	24	6.84	76.6
		-	48	5.46	61.2
		+	48	5.88	63.3
		+	72	6.01	41.0
			LSD$_{0.05}$	0.45	2.9

Values are means of 10 replications (ethylene evolution) or eight replications (ACC oxidase activity). Plants were grown and inoculated with *Rhizobium meliloti* GR4B. Ag+ was applied immediately after inoculation (10 days after planting). *NO_3^- supply increased from 2 mol m^{-3} to 10 mol m^{-3} 10 days after planting.
[a]Not determined.
Source: Caba et al. (1998).

including nodulation, implying that C_2H_4 is a component of the signaling pathway controlling rhizobial infection of legumes (Penmesta and Cook, 1997). Spaink (1997) also confirmed the insensitivity of this mutant of *M. truncatula* against ACC. They reported that in contrast to wild type plants, the formation of infection threads and nodule primordia in the *sickle* mutant is completely insensitive to high doses of externally added ACC. Likewise, Guinel and Sloetjes (2000) examined the role of C_2H_4 in nodulation by using a pleiotropic poorly nodulating mutant of *P. sativum*, R50 (*sym 16*). They found that C_2H_4 inhibitors (AVG and Ag$^+$) did not alter total number of infections, but at 28 days after inoculation, they all significantly increased nodulation in R50.

Ligero et al. (1999) studied C_2H_4 biosynthesis in inoculated roots of supernodulating (*nts 382* and *nts 1007*) and non-nodulating (*nod 49* and *nod 139*) mutants of soybean cv. Bragg. The supernodulating mutants had the least C_2H_4 production, on an average 25% and 40% less than the parental Bragg and *nod 139*, respectively. Similarly, ACC oxidase activity was also the lowest and highest in *nts 382* and *nod 139*, respectively. They suggested that the *nts* mutation not only altered C_2H_4 biosynthesis but might have also affected other physiological processes of the mutant plants including nodulation. In another study, they established a cause-effect relationship between C_2H_4 and nodulation by using hypernodulating mutants (*nts 382* and *nts 1007*) of soybean cv. Bragg (Caba et al., 1999). They observed that exogenous C_2H_4 suppressed nodulation more strongly in wild type 'Bragg' (approximately 2-fold) than its supernodulating mutants and

Table 6.5. Stimulation of nodule formation by silver thiosulfate in roots and effect on dry matter accumulation of soybean cv. Bragg plants grow in different concentrations of nitrate. Plants were grown in culture tubes with vermiculite and different Ag^+ treatments applied at inoculation. Nodule number and dry weight were recorded 18 days after inoculation.

	Ag^+	NO_3^- concentration (mmol L^{-1})			
	(μmol L^{-1})	0.5	1	5	8
Nodule number	0	129	135(+5)	51(-60)	6(-95)
per plant	1	--	158(+17)	80(+57)	15(+150)
	10	159(+23)	174(+29)	89(+75)	17(+173)
	$LSD_{0.05}$	13	15	9	4
Root growth	0	0.14	0.18	0.17	0.18
(g DW* per plant)	10	0.14	0.20	0.17	0.21
	$LSD_{0.05}$			0.02	
Shoot growth	0	0.35	0.48	0.58	0.81
(g DW per plant)	10	0.36	0.48	0.57	0.82
	$LSD_{0.05}$			0.05	

Values, mean of 20 plants, are representative of 3 experiments. (-), not determined. Data were subjected to analysis of variance (ANOVA2) and means were compared by the LSD test. Numbers in parentheses in the zero silver row are percentages relative to 0.5 mmol L^{-1} nitrate supply. Other values in parentheses show the percentage effect of applying silver ions.
*Dry weight
Source: Ligero et al. (1999).

application of Ag^+ reversed this nodulation suppressing effect of C_2H_4 more effectively in the wild type than its *nts* mutants. Furthermore, greater nodule formation in the *nts* 382 mutant than its parental wild type 'Bragg' was maintained even at lower (10^3 cfu) than normal (10^9 cfu) inoculum doses both in the presence or absence of 10 μM ACC. They suggested that sensitivity of nodulation to C_2H_4 might have been affected in supernodulating mutants since differences between the mutant and wild type in the triple response did not support differences in C_2H_4 perception on a whole plant basis (Caba et al., 1999). Xie et al. (1996) tested a total of 161 cultivars of soybean (*Glycine max* L. Merr.) for their responsiveness to C_2H_4 treatments, using senescence of the primary leaves and induction of chitinase activity in the roots as response markers. Cultivar "Gong jiao 6308-1" showed rapid chlorosis and higher chitinase induction upon treatments with C_2H_4 . The inducibility of chitinase by C_2H_4 increased with increasing age of the plant. In addition, it was found that upon repeated C_2H_4 treatments, nodule formation of cultivar "Gong jiao 6308-1" was completely blocked in the lower part of the root system when inoculated with *Bradyrhizobium japonicum*. In other cultivars, e.g., in "Bai tie jia qing", C_2H_4 treatments did not induce leaf senescence or induction of chitinase activity. Cultivar "Bar tie jia qing" also showed complete normal nodulation even after repeated C_2H_4 treatments. These results clearly demonstrate that different soybean cultivars show a wide variation and strongly differ in their C_2H_4 responsiveness and those sensitive to C_2H_4 did not nodulate or poorly nodulated upon exposure to C_2H_4 .

Contrarily, Schmidt et al. (1999) found no regulatory role of C_2H_4 in nodulation of soybean. They used soybean cv. Hobbit 87 and its mutants with decreased responsiveness to C_2H_4 and a mutant with defective regulation of nodule number. They reported that nodule number on C_2H_4 –sensitive mutants were similar to those on wild type plants. Moreover, hypernodulating mutants displayed C_2H_4 sensitivity similar to that of the wild type and NO_3^- - suppressed nodule numbers similarly in C_2H_4 insensitive and sensitive plants, as well as plants treated with Ag^+. They concluded that C_2H_4 has little role in regulating the number of nodules that form on soybeans (Schmidt et al., 1999).

6.3.6. Effects of Inhibitors of Ethylene Action and Ethylene Biosynthesis on Nodulation

Silver [Ag(I)] is a well established inhibitor of C_2H_4 action while AVG and AOA are well known inhibitors of C_2H_4 biosynthesis in higher plants (see Chapter 2). These inhibitors have been used by many workers to elucidate the regulatory role of C_2H_4 in nodule formation and development (Peters and Crist-Estes, 1989; Fearn and LaRue, 1991; Guinel and LaRue, 1991, 1992; Lee and LaRue, 1992a,b,c; Stokkermans et al., 1992; Caba et al., 1998, 1999; Ligero et al., 1991; Schmidt et al., 1999; Guinel and Sloetjes, 2000).

Ag^+ blocks binding of C_2H_4 to C_2H_4 receptors (Matoo and Suttle, 1991; Abeles et al., 1992), thus inhibiting C_2H_4 action. It has been extensively used in physiological studies to understand the role of C_2H_4 in nodulation (Schmidt et al., 1999; Caba et al., 1998, 1999; Ligero et al., 1999; Guinel and Sloetjes, 2000). Most of these studies indicate that application of Ag^+ restores partially or completely the NO_3^- or C_2H_4 suppressed nodulation. Caba et al. (1998) found that Ag^+ treatment increased nodule formation in alfalfa at all NO_3^- concentrations assayed under the two growth (aeroponic or darkened vermiculite) conditions, being more effective in plants grown aeroponically (Table 6.6). They observed that Ag^+ application before inoculation was effective in stimulation of nodulation, while application 48 h or later after inoculation did not have any significant effects.

Later, Ligero et al. (1999) confirmed the findings of Caba et al. (1998) and observed maximum stimulation of nodulation in soybean plants when Ag^+ was applied concurrent with *Bradyrhizobium* inoculation, whereas, later application of Ag^+ had no effect (Fig. 6.4). These studies may define a precise window for Ag^+ action on nodulation, presumably the first recognition of *Nod*-factors, or the induction of first cell division (Caba et al., 1998; Ligero et al., 1999). Markwei and LaRue (1997) reported that Ag^+ promoted nodule number in a pleiotropic mutant (E132) of *Pisum sativum* cv. Sparkle. Shirtliffe et al. (1996) found that exposure to Ag^+ (10 μM) for one day significantly increased nodulation in common bean cv. OAC Rico, but had no effect on its mutants R69 and R99.

Peters and Crist-Estes (1989) found that nodule formation by *R. meliloti* on *Medicago sativa* was stimulated two-fold when the C_2H_4 biosynthesis inhibitor, AVG was added along with the inoculum. Stimulation of nodule formation by AVG showed a similar concentration-dependent inhibiting effect on endogenous C_2H_4 biosynthesis,

Table 6.6. Effect of NO_3^- and Ag^+ on nodulation and growth of vermiculite-cultured alfalfa.

NO_3^- (mmol m^{-3})	Time of NO^{3-} addition (DAP)*	Ag^+ (mmol m^{-3})	Nodulation (nodules tube^{-1})	Plant growth (mg DW tube^{-1})**
0	0	0	53	5i
		10	70	53
2	0	0	66	91
		10	81	87
10	0	0	33	149
		10	45	145
10	9	0	27	126
		10	37	124
10	10	0	30	142
		10	40	141
10	11	0	43	139
		10	56	143
		LSD$_{0.05}$	4	10

Plants were grown and inoculated with *Rhizobium meliloti* GR4B. Nitrate was added as indicated and Ag^+ immediately after inoculation (10 DAP). Nodule number per tube (five plants) was recorded 22 days after inoculation for 20 culture tubes per treatment.
*DAP, days after planting.   **DW, dry weight.
Source: Caba et al. (1998).

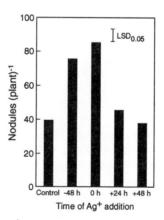

Fig. 6-4. Effect of time of Ag^+ addition, with respect to inoculation (0 h) on nodulation of soybean cv. Bragg grown with 5 mmol L^{-1} NO_3^-. Data, means of 8 replicates, were subjected to analysis ofvariace (ANOVA2) and are representative of three experiments. Source: Ligero et al. (1999).

suggesting that the primary action of AVG is in the inhibition of the endogenous C_2H_4 biosynthesis. These results imply that reduction in biosynthesis of endogenous C_2H_4 is associated with increased nodulation. However, on a per plant basis, the average rate of nitrogen fixation was unchanged by AVG treatment and was independent of nodule numbers. Fearn and LaRue (1991) found that treating the roots of poorly nodulating *sym5* mutants of *Pisum sativum* L. cv. Sparkle with AVG, Ag^+, and Co^{2+} increased nodulation. However, other *sym* lines did not respond to these treatments. Exposure of roots to Ag^+ significantly promoted the nodule numbers. Guinel and LaRue (1991) reported that the addition of Ag^+ increased cell division and nodule primordia in a poorly nodulating mutant E_2 of *P. sativum* L. cv. Sparkle. In another study, treatment of *P. sativum* L. cv. Sparkle with AVG and Ag^+ partly restored nodulation (Guinel and LaRue, 1992). Treatment with Ag^+ did not increase the number of infections, but approximately one-half of the infections went to completion (nodule formation). Very recently, Guinel and Sloetjes (2000) observed that AVG and Ag^+ restored the nodulation of the mutant to a number similar to those found in the wild type Sparkle pea. The treated plants had heavier nodules than the respective nontreated plants of R50 or Sparkle. Sections of the AVG-treated plants showed that the inner cortical cells that divide in preparation for the formation of the nodule primordium stage were sensitive to C_2H_4 (Guinel and Sloetjes, 2000). Similarly, the exogenous application of AVG improved nodulation in *Lotus japonicus* inoculated with *R. leguminosarum* (Bras et al., 2000).

On the other hand, a study conducted by Stokkermans et al. (1992) revealed that the use of AVG, Co^{2+}, and Ag^+ failed to restore nodulation in a soybean line carrying the Rj4 allele, which restricted nodulation upon inoculation with *B. japonicum* strain USDA61. Suganuma et al. (1995) reported that treatment of roots of wild-type soybean with AVG prevented the enhanced production of C_2H_4. However, the number of nodules formed on the roots was barely affected by the treatment with AVG. The results indicate that the production of C_2H_4 by soybean roots is transiently enhanced by infection with the homologous rhizobium. Schmidt et al. (1999) found that Ag^+ did not restore the NO_3^- suppressed nodulation in soybean wild type (Hobbit 87) as well as with the C_2H_4 insensitive mutant *etr 1-1* (T119N5U), which may imply that specifically in case of soybean, nodulation is independent of C_2H_4 signaling.

Rhizobitoxine, a structural analog of AVG, is synthesized by the legume microsymbiont *Bradyrhizobium elkanii* (Kuykendall et al., 1992; Minamisawa and Kume, 1987; Minamusawa, 1990; Minamisawa et al., 1990, 1997; Owens et al., 1972). By using labeled methionine, Owens et al. (1971) have demonstrated that rhizobitoxine inhibits production of C_2H_4. Compared with AVG, it is equally effective (Fig. 6.5) in inhibiting the activity of ACC synthase (Yasuta et al., 1999). Yasuta et al. (1999) also demonstrated the presence of rhizobitoxine in culture supernatant of *B. elkanii* USDA94.

In general, *B. elkanii* accumulates rhizobitoxine in cultures and in nodules, while *B. japonicum* does not (Devine et al., 1988; Kuykendall et al., 1992; Minamisawa, 1989; 1990). Since C_2H_4 has been implicated in regulation of nodulation, it was hypothesized that rhizobitoxine being an inhibitor of C_2H_4 synthesis, may play a positive role in nodule development (Duodu et al., 1999). Yuhashi et al. (2000) investigated the role of rhizobitoxine production on the nodulation and competitiveness of *B. elkanii* on *Macroptilium atropurpureum*. They used a *B. elkanii* strain that produces high levels of rhizobitoxine, *B. elkanii* USDA94 and an isogenic variant of rhizobitoxine-deficient mutant (RTS2). They found that inoculation with wild type USDA94 resulted in less

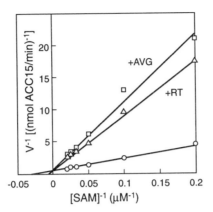

Fig. 6-5. Inhibition of ACC synthase activity by AVG and rhizobitoxine (RT). ACC synthase was incubated with SAM (5 to 50 μM)$^{-5}$ mM pyridoxal phosphate-0.1 M HEPES-KOH (pH 8.5) in the absence (○) or presence of 0.1 μM AVG (□) or 0.1 μM rhizobitoxine (△) for 15 min at 30°C. From the Lineweaver-Burk plot, the K_m for SAM was 29 μM, and the K_i values for AVG and RT were 0.019 and 0.025 μM, respectively. Source: Yasuta et al. (1999).

C_2H_4 production rates compared with RTS2 or the uninoculated control roots. Moreover, at 23 days after inoculation, C_2H_4 production by roots inoculated with USDA94 further decreased to nondetectable levels, while rates were maintained in the case of RTS2 inoculated roots. This may imply that more rhizobitoxine accumulated over time in the wild type. They also reported that there was no difference in the percentage of nodulated plants inoculated with rhizobitoxine producing wild type or its rhizobitoxine deficient mutant of *B. elkani*. However, the number of nodules were significantly different over time i.e, more in response to inoculation with wild type USDA94. Moreover, the competitiveness of *B. elkanii* for *M. atropurpureum* was also noted with the rhizobitoxine producing strains. Similarly, Duodu et al. (1999) screened host plants of *B. elkanii* for a differential nodulation response to the wild-type and rhizobitoxine mutant strains. In *Vigna radiata* (mung bean), the rhizobitoxine mutant strains induced many aborted nodules arrested at all stages of pre-emergent and post-emergent development and formed significantly fewer mature nodules than the wild type. Experiments revealed that nodulation of mung bean plants is sensitive to exogenous C_2H_4 , and that the C_2H_4 inhibitors, AVG and Co^{2+} were able to partially restore a wild-type nodulation pattern to the rhizobitoxine mutants. This is the first demonstration of a nodulation phenotype of rhizobitoxine mutants and suggests that rhizobitoxine plays a positive and necessary role in *Rhizobium*-legume symbiosis through its inhibition of C_2H_4 biosynthesis.

Stokkermans et al. (1992) used soybean lines that carry the Rj4 allele, which restricts nodulation by *B. japonicum* strain USDA61, but forms nodules normally with most other strains *of B. japonicum*. On cultivars carrying the Rj4 gene, USDA61 produces low levels of rhizobitoxine. Mutants of USDA61 that produce greater amounts of rhizobitoxine overcome the Rj4 host restriction and nodulate plants at a rate similar to nonrestricted strains of *B. japonicum*. However, nodulation assays, which included exogenous applications of rhizobitoxine, failed to overcome the nodulation restriction. They suggested that C_2H_4 does not mediate the Rj4 host restriction of soybean.

On the other hand, some reports have shown that there is not a significant difference in nodule number between plants inoculated with *B. elkanii* USDA61 and plants inoculated with rhizobitoxine-deficient mutants during nodulation of *Glycine max*, *Glycine soja*, *Vigna unguiculata*, and *Macroptilium atropurpureum* (Ruan and Peters, 1992; Xiong and Fuhrmann, 1996). By using mutants of *B. japonicum*, Ruan and Peters (1992) found that rhizobitoxine plays no apparent role in the nodulation of Rj4 soybean and did not overcome NO_3^- inhibition of nodule formulation, because nodule formulation induced by the wild-type (low rhizobitoxine producer) and induced by the mutant strains (high rhizobitoxine producer) were equally suppressed in the presence of NO_3^-. Although these findings do not seem to be consistent with the hypothesis that rhizobitoxine has a positive effect on nodulation, the inconsistency can be explained by differences in the C_2H_4 sensitivity of nodulation among leguminous species; nodulation of *G. max* is generally not sensitive to C_2H_4 (Hunter, 1993; Schmidt et al., 1999; Xie et al., 1996), while nodulation of *V. radiata* is sensitive (Duodu et al., 1999). The inconsistency could also result from differences in the abilities of the strains used in the experiments to produce rhizobitoxine; strain USDA61 is a weak producer of rhizobitoxine (Xiong and Fuhrmann, 1996).

Apart from *B. elkanii*, no other species of rhizobia are known to have mechanisms to limit plant C_2H_4 biosynthesis. Further research is needed to investigate alternative mechanisms responsible for suppressing C_2H_4 induction during the process of nodule formation.

6.3.7. Summary

As discussed above, the evidence suggesting the regulatory role of C_2H_4 in nodule organogenesis includes the following: (i) C_2H_4 effects on root growth, (ii) increased endogenous C_2H_4 synthesis upon infection by the microsymbiont, (iii) effect of C_2H_4 on *Nod* factors, (iv) inhibition of nodulation by exogenous application of C_2H_4, (v) restoration of nodulation in the presence of Ag^+ which inhibits C_2H_4 action, (vi) restoration of nodulation by AVG and AOA which are inhibitors of endogenous C_2H_4 biosynthesis, (vii) inhibition of nodulation by excessive amounts of NO_3^- which is known to stimulate endogenous synthesis of C_2H_4, (viii) elimination of NO_3^- and light effects on nodulation by Ag^+, AVG and AOA, and (ix) nodulation in mutants insensitive to C_2H_4 responsiveness even when exposed to exogenous C_2H_4. However, some legumes or their different cultivars like soybean do not show any response to C_2H_4, thus excluding or minimizing the role of C_2H_4 in regulation of nodulation in these legumes. Extensive work is still needed to precisely define the role of C_2H_4 in regulation of nodulation influencing the N-economy of plants.

Since all the phytohormones interact with each other and a physiological response is usually a result of a hormonal balance, it is very likely that the C_2H_4 effects in nodulation may be created due to changes in the balance of other hormones. Hirsch et al. (1997) and Bladergroen and Spaink (1998) have discussed the overall role of phytohormones in nodulation and in other plant-microbe symbiosis. Hirsch and Fang (1994) critically reviewed the relationship between PGRs and nodulation and based upon information regarding the presence of PGRs in nodules and the role on inhibitors on hormone transport in inducing nodule-like structures, they proposed a regulatory scheme as shown in Fig. 6.6. According to this proposed model, the exogenously produced PGRs by the microsymbiont as well as the endogenously produced PGRs by the host play critical roles in nodule organogenesis (Hirsch and Fang, 1994). Most studies reveal that C_2H_4 is a potent negative regulator of nodulation and inhibitors of endogenous C_2H_4 synthesis promotes nodule formation. Thus, any parameter that lowers the endogenous level of C_2H_4 may enhance nodulation. Glick and co-workers have demonstrated that *Pseudomonas putida* GR12-2 contains the enzyme, ACC deaminase (Jacobson et al., 1994; Glick et al., 1994a,b; Hall et al., 1996) which lowers C_2H_4 synthesis in roots upon inoculation (see Section 3.1). It would be interesting to test if this bacterium added with rhizobia promotes nodulation due to its ACC deaminase activity. Studies with inhibitors of C_2H_4 formation or action provide increasing evidence that C_2H_4 is somehow involved in the nodulation process. Future research will determine its specific role in regulation of this transformation.

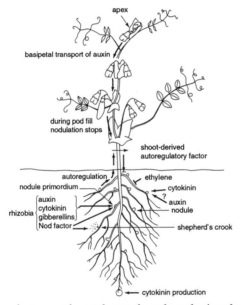

Fig. 6-6. Various plant growth regulators thought to be involved with nodulation. Cytokinin is produced in all the root tips, and auxin, produced in the shoot tip, is basipetally transported down the stem. Rhizobia synthesize *Nod* factor which affects both root hair deformation and cell divisios that foreshadow the nodule primordia. Although rhizobia also produce auxin, cytokinins, and gibberelins, it is not clear whether any of these are required for nodule primordia formation. Exogenous cytokinin can elicit nodule formation, and nodules contain high levels of cytokinin and auxin. Although auxin is involved in some way with nodulation, its role is as yet unclear. Ethylene and a shoot-derived autoregulatory factor inhibit nodule development. The stimulus to produce the shoot-derived factor is synthesized in the root and transferred to the shoot. During pod fill, nodulation stops. Source: Hirsch and Fang (1994).

Generally, it is considered that either C_2H_4 suppresses nodulation or has no apparent role in nodule formation. However, Arshad et al. (1993) reported that exogenous application of L-MET (a precursor of C_2H_4) to soil significantly influenced nodulation of *Albizia lebbeck* L. We found 1.88- and 4.48-fold increases in nodule numbers and nodule dry weight, respectively, in response to the application of 10^{-6} g kg^{-1} soil of L-MET, compared to the control (Table 6.7). This positive effect of L-MET was most likely due to microbially-derived C_2H_4 in extremely low concentrations within the rhizosphere and/or an interaction of L-MET-dependent microbially-derived C_2H_4 with other phytohormones.

Table 6.7.   Effect of L-methionine applied to soil on the nodulation of *Albizia lebbeck* L. weeks after treatment.

L-MET (g kg^{-1} soil)	Number of nodules $plant^{-1}$	Dry weight of nodules (g $plant^{-1}$)
Control	19.2	2.1
10^{-9}	21.8	3.2
10^{-8}	29.0	5.3**
10^{-7}	35.0	8.1**
10^{-6}	36.0*	9.4**
10^{-5}	32.0*	6.9**
10^{-4}	32.5	5.8**
10^{-3}	29.2	5.8**
10^{-2}	19.2	5.0**
10^{-1}	12.8	2.2

*, ** Means significantly different from control at $p \leq 0.05$ (*) and $p \leq 0.01$ (**) according to Duncan's test.
Source: Arshad et al. (1993).

6.4. MYCORRHIZAL ASSOCIATIONS

Mycorrhizal association represents another classic example of a symbiotic interaction between plant and microorganisms. It enhances the growth and survival of numerous plant species (Hayman, 1980) by acquiring more nutrition from a substantially greater soil volume than is typically explored by roots and supplies it to plants in exchange for fixed carbon (Harley and Smith, 1983). During mycorrhizal development, the plant and fungus undergo profound reorientation in their development due to the interaction between both the symbionts (Dexheimer and Pargney, 1991; Martin et al., 1997). This interaction is thought to be controlled by signal molecules exchanged very early between the symbionts, but so far very little is known about these molecules

(Ditengou and Lapeyrie, 2000). As with the *Rhizobium*-legume symbiosis, C_2H_4 along with other phytohormones may be important in regulating the establishment of mycorrhiza. Based upon the observation that mycorrhizal fungi increase C_2H_4 production in nonhost plants (rape, spinach, and lupin), but not in tomato, Vierheilig et al. (1994) suggested that C_2H_4 may help in distinguishing potential symbionts from potential pathogens. A possible role for C_2H_4 in ectomycorrhizal symbiosis is suggested by the observation that exogenous C_2H_4 and ectomycorrhizal colonization cause similar changes in mungo pine roots (Rupp and Mudge, 1985a,b). Low concentrations of C_2H_4 stimulated spore germination and hyphal growth of vesicular-arbuscular mycorrhiza (Ishii et al., 1996), and may therefore assist in the development of the association. Ethylene may have some role in mycorrhizal symbiosis as evident by the ability of the mycosymbiont to produce C_2H_4 *in vitro*, and some similarities between C_2H_4 –treated and mycorrhizal infected roots. Also, ACC oxidase activity of mycorrhizal roots and the effects of exogenous C_2H_4 or C_2H_4- releasing compounds on mycorrhiza development suggest a regulatory role of C_2H_4 in mycorrhizal symbiosis.

6.4.1. Ethylene Production by the Mycosymbiont

Numerous studies have demonstrated the ability of mycorrhizal fungi to produce C_2H_4 (Table 6.8). Graham and Linderman (1980) found that various ectomycorrhizal fungi produced C_2H_4 when grown in a medium containing DL-MET. Some of these required the renewal of fresh medium. Zhao et al. (1992) screened 21 ectomycorrhizal fungi for their commercial use against various environmental and ecological factors and found that many of these produced C_2H_4 . Similarly, Scagel and Linderman (1994) detected C_2H_4 by GC-MS produced by ectomycorrhizal fungi and grouped them as high, medium and low C_2H_4 producers. Livingston (1991) reported MET-dependent C_2H_4 production by *Laccaria lacata* and *L. bicolor*. Similarly, the ability of *L. lacata* to produce C_2H_4 was confirmed by DeVries et al. (1987).

Studies on the biochemistry of C_2H_4 biosynthesis by ectomycorrhizal fungi reveals that it is different from that of higher plants (Rupp et al., 1989b; Livingston, 1991). Rupp et al. (1989b) found that *L. laccata* and *Hebeloma crustiliniforme* produced C_2H_4 when grown in MET-amended media, but failed to yield C_2H_4 upon replacing MET with ACC. Moreover, AVG, an inhibitor of C_2H_4 biosynthesis in higher plants, did not inhibit MET-induced C_2H_4 production by these fungi. Livingston (1991) confirmed that the addition of MET promoted C_2H_4 generation by *L. laccata* and *L. bicolor,* but ACC had no promotional effect. It is certainly possible that these fungi could carry out C_2H_4 synthesis from MET via the KMBA pathway while others may be using any one of the possible pathways as discussed in Chapter 3; however, future research is needed to confirm this premise.

6.4.2. Ethylene Production by Mycorrhizal Roots

Unlike C_2H_4 production by nodulated vs. non-nodulated roots, $C_2H_{4\,p}$ production by mycorrhizal vs. non-mycorrizal plant roots has not been extensively studied (McArthur and Knowles, 1992; Rupp et al., 1989a; Vierheilig et al., 1994). Moreover, contrary to the consistent observations of accelerated C_2H_4 production by nodulated roots compared to non-nodulated roots, a decrease in C_2H_4 production by mycorrhizal roots compared to non-mycorrhizal roots has been noted in some studies. McArthur and Knowles (1992)

Table 6.8. Ethylene-producing mycorrhizal fungi.

Name	Reference
Amanita muscaria	Graham and Linderman (1980)
A. pantherina	Graham and Linderman (1980)
Calvatia fumosa	Graham and Linderman (1980)
Cenococcum geophilum	Graham and Linderman (1980)
C. graniforme	Strzelczyk et al. (1994)
Cortinarius elegantior	Graham and Linderman (1980)
Hebeloma crustuliniforme	Graham and Linderman (1980); Rupp et al. (1989b); Strzelczyk et al. (1994)
H. mesophaeum	Strzelczyk et al. (1994)
Laccaria bicolor	Livingston (1991)
L. laccata	Graham and Linderman (1980); DeVries et al. (1987); Rupp et al. (1989a,b); Livingston (1991); Scagel and Linderman (1998b)
Lycoperdon pyriforme	Graham and Linderman (1980)
Pisolithus tinctorius	Graham and Linderman (1980); Rupp et al. (1989a); DeVries et al. (1987)
Rhizopogon abietis	Graham and Linderman (1980)
R. ellenae	Graham and Linderman (1980)
R. sepelibilus	Graham and Linderman (1980)
R. subcaerulescens	Graham and Linderman (1980)
R. verisporus	Graham and Linderman (1980)
R. villosulus	Graham and Linderman (1980)
R. vinicolor	Scagel and Linderman (1998b)
R. vulgaris	Graham and Linderman (1980)
Suillus albidipes	Graham and Linderman (1980)
S. bovinus	Strzelczyk et al. (1994)
S. brevipes	Graham and Linderman (1980)
S. brunnescens	Graham and Linderman (1980)
S. lakei	Graham and Linderman (1980)
S. luteus	Scagel and Linderman (1998b)
S. ponderosus	Graham and Linderman (1980)
S. subolivacous	Graham and Linderman (1980)
S. tomentosus	Graham and Linderman (1980)
EEM fungus X	Strzelczyk et al. (1994)

studied the morphological and geochemical interaction between a vesicular-arbuscular (VAM) fungus, *Glomus fasiculatum* and potato under varying phosphorus levels. They observed that the capacity of excised roots from P-deficient plants to produce C_2H_4 in the presence or absence of exogenous ACC was markedly reduced by VAM infection (Fig. 6.7). ACC oxidase (ACC_{ox}) activity was localized to areas containing infected roots, as demonstrated in split-root studies. They detected a higher concentration of water-soluble phenolics in the leachate of VAM-infected roots than from non-mycorrhizal (NM) roots and found an inverse correlation between the rates of C_2H_4 formation and phenolic concentration in leachates from VAM roots. The application of this leachate from VAM infected roots suppressed the ACC-derived C_2H_4 by NM roots, suggesting that this inhibitor may be phenolic in nature. The *in vitro* ACC_{ox} activity was also greater in roots

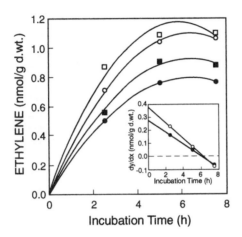

Fig. 6-7. Ethylene production (ACC_{ox} activity) by excised roots from 43-d-old potato plants grown with 0.5 mM (o,•) or 2.5 mM P (□,■) and inoculated with NM (o,□) or VAM (*Glomus fasciculatum*) (•, ■) clover inoculum. Roots were incubated in Mes (Tris) buffer (pH 6.5) at 23°C for 7.5 h. Inset: Change in C_2H_4 production rate for NM (o) and VAM (•) roots during incubation. Source: McArthur and Knowles (1992)

of plants grown on high levels of P compared with those grown on low levels, although the VAM infection partially counteracted the nutritional effects of P on ACC_{ox} activity. ACC_{ox} activity and extracellular peroxidase activity of roots increased linearly with increasing abiotic P supply, thus indicating a greater potential for resistance to VAM infection. These findings suggest that VAM fungi may alter the phenolic metabolism of roots so as to hinder C_2H_4 production and the root's ability to invoke a defense response. Raising the abiotic P supply to plants at least partially restores the capacity of roots to produce C_2H_4 and possibly increases the root's resistance to VAM infection. Besmer and Koide (1999) investigated the effect of mycorrhizal (*Glomus interaradices*) colonization and P on C_2H_4 production by two snapdragon (*Antirrhinum majus* L.) cultivars different in their C_2H_4 sensitivity. They observed that mycorrhizal colonization caused a reduction in flower C_2H_4 production by both the cultivars and that SN252 (the cultivar with greater sensitivity) had significantly lower rates of C_2H_4 production by flowers than MWI (less sensitive) (Table 6.9). They suggested that mycorrhizal colonization may be a viable alternative to C_2H_4 inhibitors such as silver thiosulfate. Application of P stimulated C_2H_4 production both in mycorrhizal and nonmycorrhizal plants (Besmer and Koide, 1999).

A tryptophan betain, hypaphorine (a major indole alkaloid) has been detected in *Pisolithus tinctorium* mycelium and exudates (Beguiristain et al., 1995). Berguiristain and Lapeyrie (1997) reported that infection with *P. tintorius* caused a 4-fold increase in the accumulation of hypaphorine within the host plant roots. Hypaphorine also counteracts with ACC when Eucalyptus (*Eucalyptus globulus*) seedlings are exposed to ACC. The effect of inoculation on the formation of hypocotyl apical roots was

Antagonized by hypaphorine application (Ditengou and Lapeyrie, 2000). This may imply that some mycorrhizal fungi produce compounds like hypaphorine which may somehow suppress C_2H_4 production by the host plant roots as the plant's defense against mycorrhizal infection gets weaker. Bragaloni and Rea (1996) also reported reduction of C_2H_4 production by plants inoculated with various (VAM) fungi. However, they hypothesized that the reduction in C_2H_4 was due to reduced nutritional stress because of mycorrhizal infection.

On the other hand, some studies have demonstrated more C_2H_4 production by the mycorrhizal infected plants. Rupp et al. (1989a) observed that mycorrhizal formation in *Pinus mungo* by *L. laccata* and *P. tinctorius* in a defined medium was associated with increases in C_2H_4 production four- and two-fold greater than the control (medium + fungus minus seedlings) by *L. laccata* and *P. tinctorius*, respectively, 5 weeks after the seedlings were inoculated. Dugassa et al. (1996) investigated the effects of arbuscular mycorrhizal (AM) symbiosis on the health of *Linus digitatissimum* L. infected by fungal pathogens. They detected increased concentrations of phytohormones including auxin- and gibberellin-like substances in shoots of AM plants along with increased production of C_2H_4 which might have a role in multiple mechanisms of underlying effects of AM on plant health and induction of tolerance against stresses.

Vierheilig et al. (1994) studied C_2H_4 biosynthesis and activities of chitinase and β-1,3-gluconase in the roots of host vs. non-host plants showed slightly enhanced C_2H_4 production and β-1,3-gluconase activity. They concluded that none of these reactions were involved in the host's inability to have VA mycorrhizal symbiosis. While investigating the effects of soil compaction on phosphorus uptake and growth of *Trifolium subterraneum* colonized by four species of VAM fungi, Nadian et al. (1998) speculated that the absence of any observable mycorrhizal growth in a highly compacted soil could be attributed to a significant decrease in the oxygen content of the soil atmosphere, change in soil pore size distribution and, presumably to C_2H_4 production.

6.4.3. Exogenous Ethylene and Mycorrhizae

Few studies have investigated the role of C_2H_4 on mycorrhizal symbiosis through exogenous application. Ishii et al. (1996) reported that exogenous application of C_2H_4 at concentrations ranging from 0.01 to 0.1 ppm stimulated spore germination and hyphal growth of VAM fungi, *Gigaspora ramisporophora* and *Glomus misceae*, whereas, concentrations at ≥ 0.2 ppm severely inhibited hyphal growth. Moreover, hyphal growth stimulated by 0.07 ppm C_2H_4 was severely retarded with an C_2H_4 absorbant. Similarly, VAM development in trifoliate orange trees was markedly enhanced upon exposure to 0.05 ppm C_2H_4, but was depressed at 1 ppm. At 0.05 ppm, tree growth was vigorous and leaf P content was high. This led Ishii et al. (1996) to propose that very low levels of C_2H_4 play an important role in the hyphal growth of VAM fungi and also in the infection and spread of mycorrhizae.

Investigations have been conducted to mimic the effect of mycorrhizal formation with C_2H_4 and other hormones by exogenous applications of phytohormones. The morphological changes characteristic of mycorrhizal infection include dichotomous branching of lateral roots, inhibition of root hair formation and enlargement of cortical cells. Similar to these effects, application of ethephon (≥ 50 μM) resulted in extensive

Table 6.9. Effect of mycorrhizal colonization and cultivar on various traits of snapdragon plants. Values shown are means ±se. Flower C_2H_4 production was measured 11 days after anthesis, and flower abscission was measured 48 h after a 1 μg ml⁻¹ C_2H_4 treatment. A two-factor analysis of variance was performed on all variables except root colonization and flower P concentration for which the least significant difference method was used to separate means. The same letter indicates there was no significant difference between means. Vegetative period was the period from sowing to harvest (when approximately 35% of flowers on a spike were open (M mycorrhizal, NM nonmycorrhizal, MWI Maryland White Improved [PanAmerican Seed]. SN252 experimental hybrid SN252 [PanAmerican Seed].

	Root colonization (%, n=10)	Vegetative period (days, n=30)	Shoot fresh weight (g, n=30)	Shoot height (cm, n=30)	Vase life (days, n=10)	Flowers per spike (n=10)	C_2H_4 (nl/g/h, n=10)	C_2H_4 treatment Abscised flowers (%, n=10)	Flower P concentration (%, n=5)
MWI M	71.2(4.6)a	118.3(0.5)	27.2(0.9)	82.3(1.5)	13.2(0.4)	14.6(0.5)	0.98(0.09)	43.8(5.1)	0.28(0.01)a
MWI NM	0.0	114.1(0.4)	24.2(0.6)	77.6(1.1)	12.3(0.3)	14.5(0.4)	2.47(0.42)	47.9(4.6)	Lost
SN252 M	72.0(3.1)a	120.3(0.6)	29.3(0.9)	92.7(1.3)	16.8(0.5)	14.2(0.6)	0.60(0.12)	72.6(3.8)	0.26(0.01)a
SN252 NM	0.0	116.0(0.4)	29.5(0.7)	92.6(1.3)	15.5(0.7)	14.3(0.5)	0.99(0.11)	69.8(3.9)	0.28(0.01)a
Analysis of variance									
Cultivar		0.0001	0.0840	0.0628	0.0001	0.5641	0.0001	0.0000	
M treatment		0.0001	0.0001	0.0001	0.0360	1.0000	0.0001	0.8856	
Interaction		0.9428	0.0379	0.0744	0.6986	0.8469	0.4380	0.4422	

Source: Besmer and Koide (1999).

dichotomous branching of mungo pine (*Pinus mungo*) roots, a characteristic of pine mycorrhizae (Rupp and Mudge, 1985a). Ethylene and auxin may be involved in the morphological changes associated with ectomycorrhizae formation. In another study, Rupp and Mudge (1985b) demonstrated that lateral roots of nonmycorrhizal root organ cultures grown in a defined medium underwent dichotomous branching in response to C_2H_4 released by ethephon, whereas root hair formation was inhibited. However, no effect on cortical cell dimension was observed. Later, Rupp et al. (1989a) observed that ethephon (100 μM) stimulated the dichotomous branching of roots inoculated with *P. tinctorius*, but it had no effect on those inoculated with *L. laccata*, or on noninoculated roots. Also, ethephon had no effect on the percentage of susceptible roots that became mycorrhizal with either fungus. Silver thiosulfate, an inhibitor of C_2H_4 action, decreased mycorrhizal formation by *L. laccata*, but had no significant effect on mycorrhizal formation by *P. tinctorius*. Rupp et al. (1989a) speculated that endogenous C_2H_4 may have influenced mycorrhizal formation and associated changes in root morphology. Morandi (1989) reported that applications of Ethrel (2-chloroethylphosphonic acid) generally reduced plant growth, except for mycorrhizal-infected soybeans, for which no significant effect was observed. However, VAM infection was significantly decreased by the Ethrel treatment. Stein and Fortin (1990) compared the pattern of root initiation on hypocotyl cuttings of *Latrix laricina* influenced by an ectomycorrhizal fungus, *L. bicolor*, with that induced by plant growth regulators. They found that a callus was not formed in both fungal-treated and ethephon (50, 75, 100 μM)-treated cuttings and a redistribution of roots along the hypocotyl was observed with 75 and 100 μM ethephon in a manner similar to that of the fungal treatment.

6.4.4. Relationship Between *in Vitro* Ethylene Production and Growth Response of Mycorrhizal Plants

Few attempts have been made by various workers to determine the relationship by the ability of mycosymbiont to produce C_2H_4 *in vitro* with the physiological responses of host plants to infection with these mycosymbionts. After confirming C_2H_4 production by various ectomycorrhizal fungi (see Section 6.4.1), Graham and Linderman (1980) investigated C_2H_4 production and mycorrhizal formation in Douglas fir roots. They observed that C_2H_4 was produced by aseptically grown Douglas fir seedlings inoculated with *Cenococcum geophilum*, *Hebeloma crustuliniforme*, and *Laccaria laccata*, and the appearance of C_2H_4 coincided with the formation of mycorrhizae. Lateral root formation was stimulated by inoculation with these three fungi, but was inhibited by *Pisolithus tinctorius*. *Fusarium*-inoculated seedlings produced C_2H_4 sooner and more than seedlings inoculated with mycorrhizal fungi, suggesting a possible differential role for C_2H_4 in symbiotic and pathogentic fungus host interactions (Graham and Linderman, 1980, 1981). Later, they reported that exogenously applied C_2H_4 did not have any effect on the development of mycorrhiza on Douglas fir inoculated with *Hebeloma crustuliniforme*, although root growth was modified by exposure to C_2H_4 (Graham and Linderman, 1981). After confirming the C_2H_4 -producing ability of *L. laccata* S238A, DeVries et al. (1987) noted an apparent correlation between C_2H_4 production and morphological effects, such as stimulation of lateral root formation by this fungus. However, he found a poor correlation between high C_2H_4 levels *in vivo* and low dichotomy suggesting that it might be due to inefficient mycorrhizal formation by *L. laccata* rather than a lack of C_2H_4 .

Similarly, they proposed that failure of the *P. tinctorius* isolates to stimulate lateral root formation could be related to the low levels of C_2H_4 produced by this fungus. Zhao et al. (1992) screened 21 ectomycorrhizal fungi for their commercial use against various environmental and ecological factors. They reported that, when the inoculum of *Suillus grevillei* (K1) Sing (capable of producing C_2H_4 and zeatin) was applied to *Pinus tabulaeformis* seedlings in the greenhouse, it accelerated the growth of seedlings as well as enhanced the biomass of trees.

Livingston (1991) conducted experiments to study the development of mycorrhizae on black spruce (*Picea mariana*) by two MET-dependent C_2H_4 producing ectomycorrhizal fungi, *Laccaria laccata* and *L. bicolor*. He found that, although all the laccarial isolates produced C_2H_4, this was not related to their ability to form ectomycorrhizae. However, Livingston (1991) was of the view that C_2H_4 production by the fungi in the presence of roots may have been derived differently from that produced in the absence of roots. Thus, the role and source of C_2H_4 production during ectomycorrhizal formation requires additional studies.

Recently, Scagel and Linderman (1998a) studied the relationship between *in vitro* indole-acetic acid (IAA) and C_2H_4 production capacity of various ectomycorrhizal fungi and conifer seedlings in symbiosis. By using HPLC and GC-MS, they quantified IAA and C_2H_4 production *in vitro* by these fungi and rated them as high, medium, and low producers of these phytohormones. They observed that inoculated Douglas fir, Engleman spruce, and lodge pole pine seedlings maintained under glasshouse conditions showed various morphological responses (seedling height, shoot weight, root collar diameter, number of lateral roots and root weight) which varied with tree species and mycorrhizal fungus combination. *In vitro* IAA production was significantly correlated to endogenous root IAA content and many morphological attributes of mycorrhizal Douglas fir seedlings; however, C_2H_4 production was poorly correlated to Douglas fir morphological responses, but positively correlated to root IAA for all the conifer species (Table 6.10), suggesting a relationship between auxin and C_2H_4 production by ectomycorrhizal fungi and changes in endogenous IAA content of roots that could affect growth responses of conifer seedlings (Scagel and Linderman, 1998a). In another study, they determined whether *in vitro* plant growth regulator production by mycorrhizal fungi correlated with conifer seedling growth and root IAA concentrations (Scagel and Linderman, 1998b). They used ectomycorrhizal fungi, having a high, moderate and low capacity to produce either IAA or C_2H_4 *in vitro* to inoculate Douglas fir (*Pseudotsuga menziesii* (Mirb.) Franco), lodgepole pine (Pinus *contorta* Dougl.) and ponderosa pine (*Pinus ponderosa* Dougl.). Data revealed that in seedlings transplanted to a nursery field, *in vitro* C_2H_4 producing capacity was highly correlated with more morphological features than *in vitro* IAA-producing capacity (Fig. 6.8). Both IAA- and C_2H_4 –producing capacity were significantly correlated with more morphological features in seedlings transplanted to a forest site than in seedlings transplanted to a nursery field. The highest percentage of colonization of plants was found in response to inocula capable of producing the highest amount of C_2H_4 in the case of seedlings transplanted to a nursery field (Table 6.11), whereas, seedlings transplanted to a clearcut site did not show this trend of colonization. They concluded that the growth-responses of conifer seedlings can be partially influenced by IAA and C_2H_4 produced by ectomycorrhizal fungal symbionts (Scagel and Linderman, 1998b).

The interaction between auxins and C_2H_4 is well established so it highly likely that the effect of C_2H_4 in mycorrhiza might be mediated by auxins. Recently, Beguiristain et

Table 6.10. Pearson Correlation Coefficients (p=0.05) between in vitro IAA or C_2H_4 production by mycorrhizal fungi (6 isolates/tree species) and inoculation responses to colonization on three conifer species four months after inoculation.

Characteristic	In vitro IAA production			In vitro C_2H_4 production		
	PSME*	PICO	PIEN	PSME	PICO	PIEN
Height growth						
2 months	ns	0.68	0.75	0.43	ns	0.73
4 months	0.51	ns	0.61	ns	0.51	ns
Relative growth rate	0.57	ns	ns	0.52	0.69	0.66
Diameter growth						
4 months	ns	9.67	ns	ns	ns	0.70
Relative growth rate	0.54	ns	ns	0.42	ns	ns
Dry weight						
Shoot dry weight	ns	ns	ns	ns	0.49	ns
Root dry weight	ns	0.44	ns	ns	0.49	0.81
Root growth and root IAA content						
Number primary lateral roots	0.81	ns	0.76	ns	ns	ns
Lateral roots per ml	0.79	0.56	ns	0.79	ns	ns
Primary laterals colonized	ns	0.58	ns	0.54	0.59	0.52
Root IAA content (conjugate)	ns	0.41	0.75	0.49	0.62	0.79

*Tree species inoculated PSME = *Psuedotsuga menziesii*, PICO = *Pinus contorta*, PIEN = *Picea englemanii*.
ns = No significant correlation between variables (p < 0.05).
Source: Scagel and Linderman (1998a).

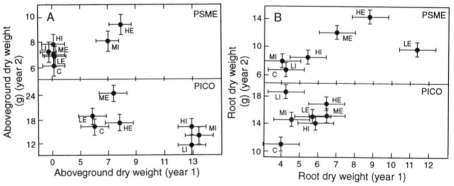

Fig. 6-8. Effects of mycorrhizal fungal isolates on aboveground dry weights (A) and root dry weights (B) of Douglas-fir (PSME) and lodgepole pine (PICO) seedlings after one growing season in the nursery (Year 1) and 6 months after being transplanted to a nursery field (Year 2). Inoculation treatments: C=control, HI=high IAA producing *Laccaria laccata* isolate; MI=moderate IAA producing *L. laccata* isolate; LI=low IAA producing *L. laccata* isolate; HE=high C_2H_4 producing *Rhizopogon vinicolor* (PSME) or *L. laccata* (PICO) isolate; ME=moderate C_2H_4 producing *R. vinicolor* (PSME) or *L. laccata* (PICO) isolate; and LE=low C_2H_4 producing *R. vinicolor* (PSME) or *L. laccata* (PICO) isolate. Bars represent Fischer's Protected LSD (P < 0.01). Source: Scagel and Linderman (1998b).

Table 6.11. Effects of mycorrhizal fungal isolates on morphology of Douglas-fir (PSME) and lodgepole pine (PICO) seedlings after one growing season in the nursery (Year 1) and 6 months after being transplanted to a nursery field (Year 2).

Tree species	Fungal isolate[1]	Height (cm)		Root collar diameter (mm)		Growth rate (% year^{-1})		Root:shoot ratio		Colonization (%)	
		Year 1	Year 2	Year 1	Year 2	Height	Diameter	Year 1	Year 2	Year 1	Year 2
Douglas fir	Control	23.15	33.70	2.76	5.41	37.55	67.30	1.52	0.98	22.3	8.3
	HI	24.16	39.83*[2]	2.94	6.02*	49.90	54.26	1.89	1.06	54.8*	41.8*
	MI	24.39	38.43*	2.81	5.69*	48.04	70.55	1.67	1.03	51.4*	26.0*
	LI	22.58	41.18*	2.86	5.60	60.00*	67.19	1.61	0.93	31.8	24.8*
	HE	26.35*	49.40*	3.15*	7.07*	62.85*	80.85*	1.79	1.52*	73.5*	52.5*
	ME	22.05	45.71*	3.34*	5.66*	72.90*	52.75	2.26*	1.69*	68.6*	31.9*
	LE	22.89	42.95*	3.01	5.90*	92.93*	67.30	3.39*	1.33	15.6	19.5
Lodgepole pine	Control	30.91	42.10	4.31	6.25	37.16	30.90	0.66	0.69	18.9	15.6
	HI	31.74	47.02*	4.46	7.39*	50.50*	39.30	0.44	0.98*	62.5*	53.8*
	MI	30.63	46.22*	4.30	7.64*	57.50*	41.14	0.33	1.11*	82.0*	61.9*
	LI	30.40	44.19	4.17	6.84	49.49	37.41	0.32	1.64*	52.3*	35.1*
	HE	31.56	58.39*	5.91*	8.07*	31.15	61.53*	0.86	0.99*	42.1*	39.0*
	ME	35.42*	60.74*	5.12*	8.00*	44.63	53.93	0.92*	0.60	27.4	19.4
	LE	37.84*	55.87*	5.27*	7.34*	33.13	38.97	1.02*	0.79	30.5	22.7

[1]Inoculation treatments: C=control; HI=high IAA producing *Laccaria laccata* isolate; MI=moderate IAA producing *L. laccata* isolate; LI = low IAA producing *L. laccata* isolate; HE=high C$_2$H$_4$ producing *Rhizopogon vinicolor* (PSME) or *L. laccata* isolate (PICO); ME=moderate C$_2$H$_4$ producing *R. vinicolor* (PSME) or *L. laccata* (PICO) isolate; and LE=low C$_2$H$_4$ producing *R. vinicolor* (PSME) or *L. laccata* isolate (PICO).

[2]An asterisk (*) indicates a value significantly greater than the control value ($p < 0.01$).

Source: Scagel and Linderman (1998b).

al. (1995) purified hypaphorine (the betane of tryptophan) from *Pisolithus tinctorius* mycelium and exudates, which counteracts the activity of IAA on *Eucalyptus* tap root elognation. The accumulation of hypaphorine increased fourfold soon after the hyphae came in contact with the plant host surface and had a prophagenetic effect by reducing root hair elongation (Beiguiristain and Lapeyrie, 1997). Hypaphorine regulates the expression of a plant symbiosis gene, the *EgHyper* gene, exhibiting a high degree of homology with plant-induced glutathion-S-transferase (Nehls et al., 1998). Ditengou and Lapeyrie (2000) hypothesized that fungal hypaphorine could regulate auxin activity during ectomycorrhizal symbiosis development. They conducted a comprehensive study to test this hypothesis by using *Eucalyptus globus* subsp. *bicostata* as a test plant. The data suggests that IAA and hypaphorine interact during the very early stages of IAA perception or the signal transduction pathway. Furthermore, while seedling treatment with ACC results in formation of a hypocotyl apical hook, hypaphorine application as well as root colonization by *P. tinctorius* (a hypaphorine-accumulating ectomycorrhizal fungus) stimulated hook opening. Hypaphorine counteraction with ACC is likely a consequence of hypaphorine interaction with IAA (Fig. 6.9). In most plant-microbe interactions studied, the interactions result in increased auxin synthesis or auxin accumulation in plant tissues. The *P. tinctorius*/eucalypt interaction is intriguing because in this interaction the VAM infection results in decreased auxin activity in the host plant. Hypaphorine might be the first specific IAA antagonist identified.

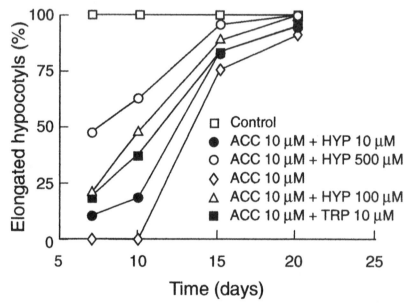

Fig. 6-9. Interaction between hypaphorine and 1-aminocyclopropane-1-carboxylic acid (ACC) controlling hypocotyl elongation. *Eucalyptus globulus* subsp. *bicostata* seedlings were cultivated over agar nutrient medium (control) supplemented with 10 μM ACC, a combination of ACC and hypaphorine (10, 100, 500 μM) or ACC and tryptophan (100 μM). The kinetics of the hypocotyl elongation were followed over 20 days. Percentages of elongated hypocotyls were recorded among at least 30 seedlings per treatment. Source: Ditengou and Lapeyrie (2000).

6.4.4. Summary

It is difficult to establish conclusions regarding the regulatory role of C_2H_4 in formation and development of mycorrhizae. It is likely that in addition to C_2H_4 other phytohormones might have a critical role in the establishment of mycorrhizal association as well as in the physiological responses of plants to this symbiosis. Studies with silver [Ag(I)] (inhibitor of C_2H_4 action) and with AVG or AOA (inhibitors of C_2H_4 biosynthesis) could be useful to elucidate the role of C_2H_4 in this symbiosis. Moreover, studies on the effects of coinoculation with C_2H_4-producing or C_2H_4-degrading microorganisms along with the mycorrhizal mycosymbiont on establishment of mycorrhizal associations would be of valuable interest and ecological significance.

6.5. LICHEN (AN ALGAL-FUNGAL SYMBIOSIS)

Lichen is a symbiotic relationship that consists of algal and fungal components that grow together to form a functional unit. In most lichens, the fungal component is ascomycetous and the algal component is either blue-green (cyanobacteria) or green algae. In morphogenesis, both bionts influence one another and together are responsible for a complex system of differentiation of thalli and apothocia (Jahns, 1988; Jahns and Ott, 1990; Ott, 1988). In symbiosis, the delicate balance between the partners is needed and phytohormones have been thought to play an important role in connection with control of the symbiosis in lichens (Remner et al., 1986; Epstein et al., 1986). Because of its gaseous nature, C_2H_4 could be of special interest for the development of lichens, particularly because of their anatomical and morphological structure, and especially in relation to their poorly developed system of transport in the thallus (Ott and Zwoch, 1992). Both algae and fungi are capable of producing C_2H_4 independently (Abeles et al., 1992; Lieberman, 1979; Arshad and Frankenberger, 1992, 1993); however, relatively less information is available on C_2H_4 production by their symbiotic unit, lichen.

6.5.1. Ethylene Production by Lichens

A number of lichens have been shown capable of producing C_2H_4 (Table 6.12). Epstein et al. (1986) reported that the lichen, *Ranalinan duriali* produces C_2H_4 and production was higher if collected from a polluted environment. Lurie and Garty (1991) observed that this lichen evolved C_2H_4 in a wetted state and various metals stimulated C_2H_4 release. Ott and Schieleit (1994) found a clear correlation between water content and C_2H_4 production by lichens (belonging to different genera, containing different photobionts and colonizing distinct environments), but there are intergeneric as well as intrageneric differences. Ott and Zwoch (1992) demonstrated species-specific variations in C_2H_4 production by lichens belonging to *Cetraria islandica* and *Cladonia arbuscula*.

Ott (1993) studied the effect of light on C_2H_4 production by different lichens. He found that C_2H_4 production in lichens is higher after incubation in light than in the dark. There was little difference in C_2H_4 release between *Cladonia* spp. *and Cetraria islandica* collected from a shady habitat upon light incubation, while intergeneric differences existed upon dark incubation. However, differences in C_2H_4 production were noted between ecotypes of *C. islandica*. It is very likely that in these organisms, fungal respiration appears to be increased in the light (Lawrey, 1984). Although a reduction of the CO_2 level during the assimilation of the photobiont certainly occurs, the respirating mycobiont continues to produce CO_2 which can even lead to an increasing internal CO_2 concentration that can influence C_2H_4 release by the lichen (Ott, 1993). Obviously, the photobiont has considerable direct or indirect influence on C_2H_4 production. Since the experiments conducted by Ott (1993) involved incubation in the darkness at relatively high temperatures, high CO_2 levels within the lichens are to be expected. Without complete knowledge of CO_2 concentrations inside the lichen thallus, the role of light and CO_2 on C_2H_4 production cannot be explained (Ott, 1993). Moreover, C_2H_4 biosynthesis by lichens or their bionts might be following a pathway different from that of higher plants, so the effect of various environmental factors on C_2H_4 release by lichens could be different.

Temporal and spatial fluctuations of C_2H_4 production by the lichen, *Romalina duriaei* were studied by Garty et al. (1993). They observed that lichens produced greater

Table 6.12. Ethylene-producing lichens.

Name of lichen	Reference
Alectoria sarmentosa	Garty et al. (1995)
Broyria fremontii	Garty et al. (1995)
Bryoria fuscescens	Garty et al. (1995)
Cetraria islandica	Ott and Schieleit (1994); Ott (1993); Ott and Zwoch (1992)
Cladina Stellans	Garty et al. (1995); Kauppi et al. (1998)
Cladonia arbuscula	Ott and Zwoch (1992); Ott (1993)
Cladonia cornuta	Ott and Zwoch (1992)
Cladonia pyxidata	Ott and Zwoch (1992)
Cladonia rangiferina	Ott and Zwoch (1992); Ott (1993); Schieleit and Ott (1996)
Cladonia sulfurina	Ott and Zwoch (1992)
Hypogymnia physodes	Garty et al. (1995)
Hypogymnia physodes	Garty et al. (1997b)
Peltigera aphthosa	Ott and Schieleit (1994)
Peltigera canina	Ott and Schieleit (1994)
Peltigera lactucifolia	Ott and Zwoch (1992)
Peltigera rufescans	Ott and Schieleit (1994)
Ramalina duriaei	Lurie and Garty (1991); Epstein et al. (1986); Garty et al. (1997a)
Usnea aurantiaco-atra	Ott and Zwoch (1992)
Usnea hirta	Garty et al. (1995); Garty et al. (1997c)

amounts of C_2H_4 in polluted industrial sites than control sites of unpolluted forest areas. Simiarly, it was noted that different lichens responded differently to S-containing chemicals, to acid rain or to heavy metal-polluted air, in terms of stress C_2H_4 production by these lichens. Garty et al. (1997a) compared stress C_2H_4 production by the lichen, *R. duriaei* collected from different habitats and found that high concentrations of various minerals (Cu, Mg, Fe, Ni, Pb, V, Cr, Mn, Sa, and S) in *R. duriaei* from natural reserves coincided with a high production rate of stress-C_2H_4. In another study, they observed a positive correlation between C_2H_4 production and Pb and Ca accumulation by the lichen, *Usnea hirta* (Garty et al., 1997b). They also found that Zn and Fe exhibited a significant positive correlation with the C_2H_4 concentration detected upon treatment of the lichen, *Hypogymnia physodes* with water or $NaHSO_3$ (Garty et al., 1997c). Similarly, in another study they investigated C_2H_4 production by the lichen, *Cladina stellaris* exposed to S and heavy metal-containing solutions under acid conditions (Kauppi et al., 1998). Exposure to KCl induced a higher rate of C_2H_4 production than the application of K_2SO_4 solutions. The application of Cu-containing solutions also enhanced the production of C_2H_4. The influence of Zn was smaller than that of Cu. Iron was the most effective heavy metal to promote the production of C_2H_4: very high C_2H_4 concentrations were detected upon the application of $FeCl_2$. Fe(II) proved more effective than Fe(III) in stimulating C_2H_4 production by *Cladina stellaris* (Table 6.13). Combined treatments of H_2SO_4 or H_2SO_4 + HNO_3 followed by either $FeCl_2$ or $FeSO_4$, yielded higher concentrations of C_2H_4 than the same treatments in reversed order. The role of Fe ions in the production of C_2H_4 could be linked with the indispensability of Fe ions for the biosynthesis of C_2H_4 via ACC (Kauppi et al., 1998). However, Fe has also been shown to stimulate C_2H_4 production via other pathways operative in microorganisms (see Chapter 3).

There is a correlation between C_2H_4 production by lichens and water content (Ott and Schieleit, 1994). In lichens with green algae, photosynthesis is only possible after uptake of water vapor, while lichens with cyanobacteria require liquid water (Lange et al., 1989). The phenomenon can be explained by the high diffusion resistance of the wall of cyanobacteria against water vapor. Thus, when water content is insufficient in lichens for photosynthesis and respiration, C_2H_4 will most likely not be synthesized.

6.5.2. Sources of Ethylene

Although several studies have demonstrated the ability of lichens to produce C_2H_4, it is not clear which of the symbiotic biont is more active in producing C_2H_4. Both the bionts have been found capable of producing C_2H_4 independently *in vitro*. This makes it difficult to assess the contribution of each partner. It is highly likely that both the bionts might contribute C_2H_4, but their relative contribution may be dependent to some extent on environmental factors. Schieleit and Ott (1996) reported that all bionts investigated showed C_2H_4 production activity and contained ACC. The level of ACC is nearly identical in the mycobiont and in the lichen. Ethylene production by the bionts was only enhanced after treatment with ACC, while KGA, a known precursor in ascomycetic fungi, was not effective. They concluded that both partners are involved in the production of C_2H_4. As both organisms are involved in the production of C_2H_4 in the lichen thallus, the organism specific effects can be superimposed upon one another and a mutual influence of the bionts on C_2H_4 production is probably regulated by the activity of enzymes which catalyzes various stages of the biosynthetic pathway. There were remarkable differences in C_2H_4 production by lichen thallus and the mycobiont (Schieleit and Ott, 1996).

On the other hand, studies do not indicate that the photobionts alone are responsible for C_2H_4 production. Huang and Chow (1984) pointed out that glucose and methionine are synthesized by cyanobacteria and released to the mycobiont. The physiological inactivity of the cyanobacterial photobiont in *Peltigera* can indirectly stop the production of C_2H_4 by the mycobiont, provided that the mycobiont contains insufficient reserves of these substances. Such an indirect influence seems probable as several ascomycetes are able to produce C_2H_4 (Fergus, 1954; Spalding and Lieberman, 1965; Chalutz et al., 1977; Thomas and Spencer, 1977). Green algae in lichens are physiologically active after

Table 6.13. Production of C_2H_4 by podetia of *Cladina stellaris* treated with Fe(II) and (III) -containing solutions and treatments with K_2SO_4 and KCl as controls to test the influence of anions.

		pH	n	$\bar{x}$ (nl g^{-1} h^{-1})	Duncan's test
H_2O		6.8	79	1.35	e
K_2SO_4	5.0 mM	3.5	10	1.35	e
KCl	5.0 mM	3.5	10	2.01	e
$FeSO_4$	0.5 mM	3.5	13	6.56	c
	5.0 mM	3.5	20	10.16	b
	10.0 mM	3.5	9	10.48	b
$FeCl_2$	0.5 mM	3.5	13	6.25	c
	5.0 mM	3.5	26	9.89	b
	10.0 mM	3.5	16	11.48	a
$Fe_2(SO_4)_3$	0.5 mM	3.5	21	4.75	d
$FeCl_3$	0.5 mM	3.5	21	4.69	d
$Fe(NO_3)_3$	0.5 mM	3.5	33	4.36	d

ANOVA
d.f. between groups   11
d.f. residual   259
F value   366.311
P value   0.001

The pH was adjusted by the acid corresponding to the anion of each salt. n, number of replicates; $\bar{x}$, mean values. Values of each vertical column followed by the same letter do not differ significantly at $p < 0.05$ using one-way ANOVA and Duncan's new multiple range test. d.f., degrees of freedom.
Source: Kauppi et al. (1998).

wetting with water vapor and can either produce C_2H_4 directly (Driessche et al., 1988) or can stimulate production by the mycobiont. This may explain the response of *C. islandica* to water content. Remner et al. (1986) investigated the possibility that a bacterial biont could possibly influence C_2H_4 production in lichens. However, after treatment with antibiotics (tetracycline) no change in C_2H_4 production was observed. An extract of the treated lichens was incubated on agar (Lurie-Broth-plates) and proved to be free of bacteria. Hence, they concluded that bacteria are not responsible for any measurable amount of C_2H_4 produced by the lichen (Remner et al., 1986).

6.5.3. Biochemistry of Ethylene Production by Lichens

Little efforts have been made to study the biochemistry of C_2H_4 synthesis by lichens. Lurie and Garty (1991) attempted to elucidate the biosynthetic pathway of C_2H_4 production by *R. duriaei*. They found that this lichen evolved C_2H_4 in a wetted state. Although addition of MET and ACC enhanced C_2H_4 synthesis, inhibitors of ACC $\rightarrow$ C_2H_4 biosynthesis (AOA, AVG, and Co^{2+}) and of C_2H_4 action (Ag^+) did not inhibit C_2H_4 production by *R. duriaei* (Table 6.14), implying that the C_2H_4 biosynthetic pathway in

Table 6.14. Effect of incubation time and C_2H_4 inhibitors, precursors and intermediates on C_2H_4 production by the lichen *Ramalina duriaei*.

	Ethylene production (nl g^{-1} h^{-1})	
Treatment	3 h	20 h
A		
Control	0.87b*	1.00b
1 mM ACC	0.99b	1.03b
1 mM ACC + 2 mM AOA	1.55a	1.35a
2 mM AOA	1.76a	1.46a
B		
Control	1.17b	1.19b
1 mM ACC	1.52a	1.48a
0.1 mM AVG	1.48a	1.40a
C		
Control	1.38d	
10 mM glutamic acid	2.52c	
10 mM methionine	2.97b	
10 mM methionine + 0.1 mM AVG	2.77c	
7 mM ACC	3.36a	
7 mM ACC + 0.1 mM AVG	3.01b	
D		
Control	1.05b	
20 mM CoCl₂	1.23ab	
2 mM AgS₂O₅	1.53a	

*Letters denote significant difference at p = 0.05 level according to Duncan's multiple-range test.
Source: Lurie and Garty (1991).

this lichen is different from that of higher plants. Moreover, the endogenous ACC content was similar whether the lichen was producing C_2H_4 at a basal rate, or during Fe(II) stimulation of C_2H_4 formation. Metal ions, including Pb(II), Zn(II), Mn(II), Cu(II), and Fe(II) (all added at 10 mM) stimulated C_2H_4 release (Table 6.15), with Fe(II) being the most effective. This stimulation was not inhibited by AVG, but free-radical

Table 6.15. Effect of metal ions on C_2H_4 production rates of *Ramalina duriaei*. Ethylene production was recorded in the presence of 20 mM solutions of various metal ions. The lichens were incubated for 30 min and then sealed for 3 h.

Solution (20 mM)	Ethylene (nl g^{-1} h^{-1})
Control	1.38e*
PbCl$_2$	3.12d
ZnCl$_2$	2.64d
MnCl$_2$	4.12c
CuCl$_2$	7.09b
FeCl$_2$	17.86a

*Letters denote significant difference at p = 0.05 level according to Duncan's multiple-range test.
Source: Lurie and Garty (1991).

scavengers, such as propylgallate and quercitin, had strong inhibitory effects. Malonicdialdehyde and aldehyde contents were higher in the presence of Fe(II). They suggested that the effect of Fe(II) and other metals might be similar to that observed in the stress situations in higher plants that lead to enhanced C_2H_4 evolution. Moreover, Fe is also known as a strong stimulator of C_2H_4 production by microorganisms which follow a different pathway than that of higher plants.

On the other hand, Ott and Zwoch (1992) observed a stimulation and increase in C_2H_4 release by various lichens in response to exogenous application of ACC. The presence of endogenous ACC was detected in all the lichen species tested and may have the same biosynthetic pathway as higher plants although concrete evidence is lacking. Similarly, studies conducted by Schieleit and Ott (1996) revealed that C_2H_4 production by the lichen, *Cladonia rangiferina* and its bionts (myco and photo) was only enhanced after treatment with ACC while KGA, a known C_2H_4 precursor in asomycetic fungi, had no effect (Table 6.16). They reported that all the bionts contain ACC and the level of ACC is nearly identical in the mycobiont and in the lichen thalli. The results presented here indicate that in the complex system of the lichen thallus both partners are involved in the production of C_2H_4 via the pathway identical to that of higher plants. ACC-dependent C_2H_4 production is not solely regulated by the concentration of ACC but is also ependent on the C_2H_4 forming enzyme (ACC-oxidase). Thus, in spite of the fact that both the lichen thallus and the mycobiont contained similar levels of ACC, the differences between C_2H_4 production in the lichen thallus and the mycobiont were

Table 6.16. Ethylene production of the thallus of *Cladonia rangiferina* and the isolated bionts after treatment with special C_2H_4 precursors.

Treatment	Lichen thallus x(SD) 3 h	24 h	Mycobiont x(SD) 3 h	24 h	Photobiont x(SD) 3 h	24 h
Control	2.10 (0.14)	0.40 (0.10)	n.d.	0.26 (0.02)	n.d.	0.17 (0.01)
ACC						
10 mM	**3.65 (0.18)**	**0.65 (0.11)**	n.d.	**0.33 (0.05)**	n.d.	n.d.
50 mM	**3.33 (0.37)**	**0.61 (0.07)**	n.d.	**0.52 (0.01)**	**5.83 (0.40)**	**1.03 (0.19)**
α-oxoglutarate						
10 mM	2.16 (0.12)	0.39 (0.03)		0.29 (0.01)	0	0
50 mM	1.78 (0.22)	0.37 (0.05)	n.d.	0.26 (0.03)	n.i.	n.i.

Ethylene production is given as nl/g/h; x = mean values; (SD) = standard deviation in parentheses; n.d. = not detectable; n.i. = not investigated. Bold figures are significant increases when compared with the water controls ($p = 0.05$).
Source: Schieleit and Ott (1996).

significant (Schieleit and Ott, 1996). The different consistency of the lichen thallus and the cultured mycobiont probably influences diffusion of the applied solutions which may be of relevance for C_2H_4 production. This could explain the observation that the ACC level in the lichen thallus and the isolated mycobiont are nearly identical while C_2H_4 production differs. Probably the two bionts in the lichen influence one another and the enhancement of C_2H_4 synthesis is caused by symbiotic interactions, whereas, the production of ACC by the mycobiont seems to be unaffected by symbiosis. Presumably, the ACC content measured in the lichen thallus is due to the mycobiont. The photobiont is able to produce ACC too, but the ratio of mycobiont/photobiont favors the mycobiont which represents the major part of the biomass. Schielit and Ott (1996) hypothesized that the synthetic pathway seems to be identical in all these organisms which belong to unrelated taxa. The exact nature of the interactive processes and the mutual influences between the fungus and algae are still unknown. Regulation may include participation of both partners as well as a restriction of process in one of them (Schieleist and Ott, 1996). As discussed in Chapter 3, it is worth noting that, in general, microorganisms cannot derive C_2H_4 from ACC, so future research with labelled substrates can resolve the issue that a mycobiont can generate C_2H_4 from ACC unequivocally.

6.5.4. Summary

The study conducted by Lurie and Garty (1991) reveals that the lichen, *Ramalina duriaei* produces C_2H_4 via a pathway different from that of higher plants, whereas, Schieleit and Ott (1996) and Ott and Zwoch (1992) provided convincing evidence that lichens they tested produced C_2H_4 by following the ACC pathway operative in higher plants. Schieleit and Ott (1996) also demonstrated that both myco- and photo-bionts contained ACC, whereas, studies with a majority of microorganisms have clearly indicated that microorganisms are in general incapable of deriving C_2H_4 from ACC (see Chapter 3). It is highly likely that the pathway of C_2H_4 biosynthesis in lichens varies with the nature and species of both symbiotic partners, i.e., myco- and photo-bionts.

6.6. CONCLUDING REMARKS

The plant-microbe interactions which result in symbiotic associations, no doubt play an important role in altering the growth and development of the host. For instance, nodulation is considered essential for nitrogen supply, whereas, mycorrhizae have been found to be beneficial for improving the supply of nutrients like phosphorus and other immobile trace micronutrients to the host. If the role of the plant hormone, C_2H_4 in these symbiotic associations is well characterized and understood, then these associations could be exploited more efficiently by plant scientists and soil microbiologists in favor of plant growth. Moreover, information on the role of C_2H_4 in these symbiotic associations could lead a molecular biologist to develop more efficient microsymbionts as well as more responsive host plants by using molecular and genetic engineering approaches for more successful plant-microbe symbiotic interactions. It is expected that future research in this discipline will provide new opportunities for improving food production through better control of these symbiotic associations.

6.7. REFERENCES

Abeles, F. B., Morgan, P. W., and Saltveit, M. E., Jr., 1992, *Ethylene in Plant Biology*, 2nd Ed. Academic Press, Inc., San Diego, CA.

Arshad, M., and Frankenberger, W. T., Jr., 1992, Microbial biosynthesis of ethylene and its influence on plant growth, *Adv. Microbiol. Ecol.* **12**:69-111.

Arshad, M., and Frankenberger, W. T., Jr., 1993, Microbial production of plant growth regulators, in: *Soil Microbial Ecology*, B. F. Metting, ed., Marcel Dekker, New York, NY, pp. 307-347.

Arshad, M., and Frankenberger, W. T., Jr., 1998, Plant growth-regulating substances in the rhizosphere: Microbial production and functions, D. L. Sparks, ed., *Adv. Agron.* **62**:45-151.

Arshad, M., Hussain, A., Javed, M., and Frankenberger, W. T., Jr., 1993, Effect of soil applied L-methionine on growth, nodulation and chemical composition of *Albizia lebbeck* L., *Plant Soil* **148**:129-135.

Beard, R., and Harrison, M. A., 1992, Effect of inoculation on ethylene production in beans, *Plant Physiol.* **99**:66.

Beguiristain, T., and Lapeyrie, F., 1997, Host plant stimulates hypaphorine accumulation in *Pisolithus tinctorius* hyphae during ectomycorrhizal infection while excreted fungus hypaphorine controls root hair development, *New Phytol.* **136**:525-532.

Beguiristain, T., Cote, R., Rubini, P., Jay-Allemand, C., and Lapeyrie, F., 1995, Hypaphorine accumulation in hyphae of the ectomycorrhizal fungus *Pisolithus tinctorius*, *Phytochem.* **40**:1089-1091.

Besmer, Y. L., and Koide, R. T., 1999, Effect of mycorrhizal colonization and phosphorus on ethylene production by snapdragon (*Antirrhinum majus* L.) flowers, *Mycorrhiza* **9**:161-166.

Beyrle, H., 1995, The role of phytohormones in the function and biology of mycorrhizae, in: *Mycorrhiza, Structure, Function, Molecular Biology and Biotechnology*, A. Varma and B. Hock, eds., Springer-Verlag, Berlin, Heidelberg, pp. 365-390.

Bladergroen, M. R., and Spaink, H. P., 1998, Genes and signal molecules involved in the rhizobia leguminoseae symbiosis, *Curr. Opinion Plant Biol.* **1**:353-359.

Boller, T., 1991, Ethylene in pathogenesis and disease resistance, in: *The Plant Hormone Ethylene*, A. K. Mattoo and J. C. Suttle, eds., CRC Press, Boca Raton, FL, pp. 293-314.

Bragaloni, M., and Rea, E., 1996, Impatto di endofiti micorrizici isolati da dune sabbiose su piante di interesse agrario, *Micologia Italiana* **25**:85-91.

Bras, C. P., Jorda, M. A., Wijfjes, A. H. M., Harteveld, M., Stuurman, N., Thomas-Oates, J. E., and Spaink, H. P., 2000, A *Lotus japonicus* nodulation system based on heterologous expression of the fucosyl transferase *NodZ* and the acetyl transferase *NolL* in *Rhizobium leguminosarum*, *Molecular Plant-Microbe Interact.* **13**:475-479.

Caba, J. M., Poveda, J. L., Gresshoff, P. M., and Ligero, F., 1999, Differential sensitivity of nodulation to ethylene in soybeans cv. Bragg and a supernodulating mutant, *New Phytol.* **142**:233-242.

Caba, J. M., Recalde, L., and Ligero, F., 1998, Nitrate-induced ethylene biosynthesis and the control of nodulation in alfalfa, *Plant, Cell Environ.* **21**:87-93.

Caetano-Anolles, G., and Gresshoff, P. M., 1990, Early induction of feedback regulatory responses governing nodulation in soybean, *Plant Sci.* **71**:69-81.

Carlson, R. W., Price, N. P. J., and Stacey, G., 1995, The biosynthesis of rhizobial lipo-oligosaccharide nodulation signal molecules, *Mol. Plant-Microbe Interact.* **7**:684-695.

Chalutz, E., Lieberman, M., and Sisler, H. D., 1977, Methionine-induced ethylene production by *Penicillium digitatum*, *Plant Physiol.* **60**:402-406.

Charon, C., Sousa, C., Crespi, M., and Kondorosi, A., 1999, Alteration of *enod40* expression modifies *Medicago truncatula* root nodule development induced by *Sinorhizobium meliloti*, *The Plant Cell* **11**:1953-1965.

Day, D. A., Carroll, B. J., Delves, A. C., and Gresshoff, P. M., 1989, Relationship between autoregulation and nitrate inhibition of nodulation in soybeans, *Physiol. Plant.* **75**:37-42.

Devine, T. E., Kuykendall, L. D., and O'Neill, J. J., 1988, DNA homology group and the identity of bradyrhizobial strains producing rhizobitoxine-induced foliar chlorosis on soybean, *Crop Sci.* **28**:938-941.

DeVries, H. E. II, Mudge, K. W., and Lardner, J. P., 1987, Ethylene production by several ectomycorrhizal fungi and effects on host root morphology, in: *Mycorrhizae in the Next Decade, Practical Applications and Research Priorities*, Proc. 7th North Amer. Conf. Mycorrhizae, May 3-8, 1987, Gainesville, FL, pp. 245.

Dexheimer, J., and Pargney, J. C., 1991, Comparative anatomy of the host-fungus interface in mycorrhizae, *Experientia* **47**:312-320.

Ditengou, F. A., and Lapeyrie, F., 2000, Hypaphorine from the ectomycorrhizal fungus *Pisolithus tinctorius* counteracts activities of indole-3-acetic acid and ethylene but not synthetic auxins in Eucalypt seedlings, *Mol. Plant-Microbe Interact.* **13**:151-158.

Drennan, D. S. H., and Norton, C., 1972, The effect of Ethrel on nodulation in *Pisum sativum* L., *Plant Soil* **36**:53-57.

Drevon, J. J., 1983, Various organisms that fix nitrogen, in: Technical Handbook on Symbiotic Nitrogen Fixation, Legume/Rhizobium, FAO, *The United Nations 1 Biol.* **2**:1-4.

Driessche, T. V., C., Kevers, C., and Collet, M., 1988, *Acetabularia mediterranea* and ethylene: Production in relation with development, circadian rhythms in emission and response to external application, *J. Plant Physiol.* **133**:635-639.

Dugassa, G. D., von Alten, A., and Schönbeck, F., 1996, Effects of arbuscular mycorrhiza (AM) on health of *Linum usitatissimum* L. infected by fungal pathogens, *Plant Soil* **185**:173-182.

Duodu, S., Bhuvaneswari, T. V., Stokkermans, T. J. W., and Peters, N. K., 1999, A positive role for rhizobitoxine in *Rhizobium*-legume symbiosis, *Molec. Plant-Microbe Inter.* **12**:1082-1089.

Epstein, E., Sagee, O., Cohen, J. D., and Garty, J., 1986, Endogenous auxin and ethylene in the lichen *Ramalina duriaei*, *Plant Physiol.* **82**:1122-1125.

Fearn, J. C., and LaRue, T. A., 1991, Ethylene inhibitors restore nodulation of sym 5 mutants of *Pisum sativum* L. cv. Sparkle, *Plant Physiol.* **96**:239-244.

Feldman, L. J., 1984, Regulation of root development, Annu. Rev. *Plant Physiol.* **35**:223-242.

Fergus, C. L., 1954, The production of ethylene by *Penicillium digitatum, Mycologia* **46**:543-555.

Fernandez-Lopez, M., Goormachtig, S., Gao, M., D'Haeze, W., Montagu, M. V., and Holsters, M., 1998, Ethylene-mediated phenotypic plasticity in root nodule development on *Sesbania rostrata, Proc. Natl. Acad. Sci. USA* **95**:12724-12728.

Frankenberger, W. T., Jr., and Arshad, M., 1995, *Phytohormones in Soils: Microbial Production and Functions*, Marcel Dekker, Inc., New York.

Garty, J., Kauppi, M., and Kauppi, A., 1995, Differential responses of certain lichen species to sulfur-containing solutions under acidic conditions as expressed by the production of stress-ethylene, *Environ. Res.* **69**:132-143.

Garty, J., Karary, Y., Harel, J., and Lurie, S., 1993, Temporal and spatial fluctuations of ethylene production and concentrations of sulfur, sodium, chlorine and iron on/in the thallus cortex in the lichen *Ramalina duriaei* (de not.) Bagl., *Environ. Exptl. Bot.* **33**:553-563.

Garty, J., Kloog, N., Wolfson, R., Cohen, Y., Karnieli, A., and Avni, A., 1997a, The influence of air pollution on the concentration of mineral elements, on the spectral reflectance response and on the production of stress-ethylene in the lichen *Ramalina duriaei, New Phytol.* **137**:587-597.

Garty, J., Kauppi, M., and Kauppi, A., 1997b, The influence of air pollution on the concentration of airborne elements and on the production of stress-ethylene in the lichen *Usnea hirta* (L.) Weber em. Mot. transplanted in urban sites in Oulu, N. Finland, *Arch. Environ. Contam. Toxicol.* **32**:285-290.

Garty, J., Kauppi, M., and Kauppi, A., 1997c, The production of stress ethylene relative to the concentration of heavy metals and other elements in the lichen *Hypogymnia physodes, Environ. Toxicol. Chem.* **16**:2404-2408.

Glick, B. R., Jacobson, C. B., Schwarze, M. M. K., and Pasternak, J. J., 1994a, Does the enzyme 1-aminocyclopropane-1-carboxylate deaminase play a role in plant growth-promotion by *Pseudomonas putida* GR12-2? in: *Improving Plant Productivity with Rhizosphere Bacteria*, M. H. Ryder, P. M. Stephens, and G. D. Bowen, eds., Commonwealth Scientific and Industrial Research Organization, Adelaide, Australia, pp. 150-152.

Glick, B. R., Jacobson, C. B., Schwarze, M. M. K., and Pasternak, J. J., 1994b, 1-Aminocyclopropane-1-carboxylic acid deaminase mutants of the growth-promoting rhizobacterium *Pseudomonas putida* GR12-2 do not stimulate root elongation, *Can. J. Microbiol.* **40**:911-915.

Goodlass, G., and Smith, K. A., 1979, Effects of ethylene on root extension and nodulation of pea (*Pisum sativum* L.) and white clover (*Trifolium repens* L.), *Plant Soil* **51**:387-395.

Goodwin, T. E., and Mercer, E. I., 1983, Introduction to Plant Biochemistry, 2nd Ed., Pergamon Press, Oxford, New York.

Graham, J. H., and Linderman, R. G., 1980, Ethylene production by ectomycorrhizal fungi, *Fusarium oxysporum* f. sp. *pini*, and by aseptically synthesized ectomycorrhizae and *Fusarium*-infected Douglas fir roots, *Can. J. Microbiol.* **26**:1340-1347.

Graham, J. H., and Linderman, R. G., 1981, Effect of ethylene on root growth, ectomycorrhiza formation, and *Fusarium* infection of Douglas fir, *Can. J. Bot.* **59**:149-155.

Gresshoff, P. M., 1993, Molecular genetic analysis of nodulation genes in soybean, *Plant Breeding Reviews* **11**:275-318.

Grobbelaar, N., Clarke, B., and Hough, M. C., 1971, The nodulation and nitrogen fixation of isolated roots of *Phaseolus vulgaris* L., *Plant Soil* (Spec. Vol.), pp. 215-223.

Guinel, F. C., and LaRue, T. A., 1991, Light microscopy study of nodule initiation in *Pisum sativum* L. cv. Sparkle and in its low-nodulating mutant E2 (sym 5), *Plant Physiol.* **97**:1206-1211.

Guinel, F. C., and LaRue, T. A., 1992, Ethylene inhibitors partly restore nodulation to pea mutants E107 (brz.), *Plant Physiol.* **99**:515-518.

Guinel, F. C., and Sloetjes, L. L., 2000, Ethylene is involved in the nodulation phenotype of *Pisum sativum* R50 (sym 16), a pleiotropic mutant that nodulates poorly and has pale green leaves, *J. Exptl. Bot.* **51**:885-894.

Hall, J. A., Peirson, D., Ghosh, S., and Glick, B. R., 1996, Root elongation in various agronomic crops by the plant growth promoting rhizobacterium *Pseudomonas putida* GR12-2, *Israel J. Plant Sci.* **44**:37-42.

Harley, J. L., and Smith, S. E., 1983, *Mycorrhizal Symbiosis*, Academic Press, London.

Hayman, D. S., 1980, Mycorrhiza and crop production, *Nature* (London):**287**:487-488.

Heidstra, R., Yang, W. C., Yalcin, Y., Peck, S., Emons, A., van Kammen, A., and Bisseling, T., 1997, Ethylene provides positional information on cortical cell division but is not involved in *Nod* factor-induced root hair tip growth in *Rhizobium*-legume interaction, *Development* **124**:1781-1787.

Hirsch, A. M., and Fang, Y., 1994, Plant hormones and nodulation: What's the connection? *Plant Mol. Biol.* **26**:5-9.

Hirsch, A. M., Fang, Y., Asad, S., and Kapulnik, Y., 1997, The role of phytohormones in plant-microbe symbiosis, *Plant Soil* **194**:171-184.

Huang, T. C., and Chow, T. J., 1984, Ethylene production by blue-green algae, *Bot. Bull. Acad. Sinica* **25**:81-86.

Hunter, W. J., 1993, Ethylene production by root nodules and effect of ethylene on nodulation in *Glycine max*, *Appl. Environ. Microbiol.* **59**:1947-1950.

Ishii, T., Shrestha, Y. H., Matsumoto, I., and Kadoya, K., 1996, Effect of ethylene on the growth of vesicular-arbuscular mycorrhizal fungi on the mycorrhizal formation of trifoliate orange roots, *J. Japan Soc. Hort. Sci.* 65:525-529.

Jacobson, C. B., Pasternak, J. J., and Glick, B. R., 1994, Partial purification and characterization of 1-aminocyclopropane-1-carboxylate deaminase from the plant growth promoting rhizobacterium *Pseudomonas putida* GR12-2, *Can. J. Microbiol.* **40**:1019-1025.

Jahns, H. M., 1988, The establishment, individuality and growth of lichen thalli, Bot. *J. Linnean Soc.* **96**:21-29.

Jahns, H. M., and Ott, S., 1990, Regulation of regenerative processes in lichens, *Bibliotheca Lichenologica* **38**:243-252.

Kauppi, M., Kauppi, A., and Garty, J., 1998, Ethylene produced by the lichen *Cladina stellaris* exposed to sulphur and heavy metal containing solutions under acidic conditons, *New Phytol.* **139**:537-547.

Kuykendall, L. D., Saxena, B., Devine, T. E., and Udell, S. E., 1992, Genetic diversity in *Bradyrhizobium japonicum* Jordan 1982 and a proposal for *Bradyrhizobium elkanii* sp. *nov.*, *Can. J. Microbiol.* **38**:501-505.

Lange, O. L., Bilger, W., Rimke, S., and Schreiber, U., 1989, Chlorophyll fluorescence of lichens containing green and blue-green algae during hydration by water vapor uptake and by addition of liquid water, *Bot. Acta* **102**:306-313.

Lawrey, J. D., 1984, *Biology of Lichenized Fungi*, Praeger, New York.

Lee, K. H., and LaRue, T. A., 1992a, Inhibition of nodulation of pea by ethylene, *Plant Physiol.* **99**(Suppl.):108.

Lee, K. H., and LaRue, T. A., 1992b, Ethylene as a possible mediator of light- and nitrate-induced inhibition of nodulation of *Pisum sativum* L. cv. Sparkle, *Plant Physiol.* **100**:1334-1338.

Lee, K. H., and LaRue, T. A., 1992c, Exogenous ethylene inhibits nodulation of *Pisum sativum* L. cv. Sparkle, *Plant Physiol.* **100**:1759-1763.

Lee, K. H., Fearn, J. C., Guinel, F. C., and LaRue, T. A., 1993, Ethylene and nodulation, in: *New Horizons in Nitrogen Fixation*, R. Palacios, J. Mora, and W. E. Newton, eds., Kluwer Academic Publishers, Dordrecht, Boston, London, pp. 303-308.

Lerouge, P., Roche, P., Faucher, C., Maillet, F., Truehet, G., Prome, J.-C., and Denarie, J., 1990, Symbiotic host-specificity of *Rhizobium meliloti* is determined by a sulphated and acylated glucosamine oligosaccharide signal, *Nature* **344**:781-784.

Libbenga, K. R., and Harkes, P. A. A., 1973, Initial proliferation of cortical cells in the formation of root nodules in *Pisum sativum* L., *Planta* **114**:17-28.

Lieberman, M., 1979, Biosynthesis and action of ethylene, Annu. Rev. *Plant Physiol.* **30**:533-591.

Ligero, F., Lluch, C., and Olivares, J., 1986, Evolution of ethylene from roots of *Medicago sativa* plants inoculated with *Rhizobium meliloti*, J. *Plant Physiol.* **125**:361-365.

Ligero, F., Lluch, C., and Olivares, J., 1987, Evolution of ethylene from roots and nodulation rate of alfalfa (*Medicago sativa* L.) plants inoculated with *Rhizobium meliloti* as affected by the presence of nitrate, *J. Plant Physiol.* **129**:461-467.

Ligero, F., Caba, J. M., Lluch, C., and Olivares, J., 1991, Nitrate inhibition of nodulation can be overcome by the ethylene inhibitor aminoethoxyvinylglycine, *Plant Physiol.* **97**:1221-1225.

Ligero, F., Poveda, J. L., Gresshoff, P. M., and Caba, J. M., 1999, Nitrate and inoculation enhanced ethylene biosynthesis in soybean roots as a possible mediator of nodulation control, *J. Plant. Physiol.* **154**:482-488.

Livingston, W. H., 1991, Effect of methionine and 1-aminocyclopropane-1-carboxylic acid on ethylene production by *Laccaria bicolor* and L. *laccata*, Mycologia 83:236-241.

Lurie, S., and Garty, J., 1991, Ethylene production by the lichen *Ramalina duriaei*, *Ann. Bot.* **68**:317-319.

Markwei, C. M., and LaRue, T. A., 1997, Phenotypic characterization of *sym21*, a gene conditioning shoot-controlled inhibition of nodulation in *Pisum sativum* cv. Sparkle, *Physiol. Plant.* **100**:927-932.

Martin, F., Lapeyrie, F., and Tagu, D., 1997, Altered gene expression during ectomycorrhizal development, in: *The Mycota, Vol., V*, A. G. Carroll and P. Tudzynski, eds., Springer-Verlag, Berlin, pp. 223-242.

Mattoo, A. K., and Suttle, J. C., eds., 1991, *The Plant Hormone Ethylene*, CRC Press, Boca Raton.

McArthur, D. A. J., and Knowles, N. R., 1992, Resistance response of potato to vesicular-arbuscular mycorrhizal fungi under varying abiotic phosphorus levels, *Plant Physiol.* **100**:341-351.

Minamisawa, K., 1989, Comparison of extracellular polysaccharide composition, rhizobitoxine production, and hydrogenase phenotype among various strains of *Bradyrhizobium japonicum*, *Plant Cell Physiol.* **30**:877-884.

Minamisawa, K., 1990, Division of rhizobitoxine-producing and hydrogen-uptake positive strains of *Bradyrhizobium japonicum* by *nifDKE* sequence divergence, *Plant Cell Physiol.* **31**:81-89.

Minamisawa, K., and Kume, N., 1987, Determination of rhizobitoxine and dihydrorhizobitoxine in soybean plants by amino acid analyzer, *Soil Sci. Plant Nutr.* **33**:645-649.

Minamisawa, K., Fukai, K., and Asami, T., 1990, Rhizobitoxine inhibition of hydrogenase synthesis in free-living *Bradyrhizobium japonicum*, *J. Bacteriol.* **172**:4505-4509.

Minamisawa, K., Onodera, S., Tanimura, Y., Kobayashi, N., Yuhashi, K.-I., and Kubota, M., 1997, Preferential nodulation of *Glycine max*, *Glycine soja*, and *Macroptilium atropurpureum* by two *Bradyrhizobium* species *japonicum* and *elkanii*, *FEMS Microbiol. Ecol.* **24**:49-56.

Morandi, D., 1989, Effect of xenobiotics on endomycorrhizal infection and isoflavonoid accumulation in soybean roots, *Plant Physiol. Biochem.* **27**:697-701.

Nadian, H., Smith, S. E., Alston, A. M., Murray, R. S., and Siebbert, B. D., 1998, Effect of soil compaction on phosphorus uptake and growth of *Trifolium subterraneum* colonized by four species of vesicular-arbuscular mycorrhizal fungi, *New Phytol.* **140**:155-165.

Nehls, U., Beguiristain, T., Ditengou, F. A., Lapeyrie, F., and Martin, F., 1998, The expression of a symbiosis-regulated gene in eucalypt roots is regulated by auxins and hypaphorine, the tryptophan betaine of the ectomycorrhizal basidiomycete *Pisolithus tinctorius*, *Planta* **207**:296-302.

Ott, S., 1988, Photosymbiodemes and their development in *Peltigera venosa*, *Lichenologist* **20**:361-368.

Ott, S., 1993, The influence of light on the ethylene production by lichens, *Bibliotheca Lichenologica* **53**:185-190.

Ott, S., and Schieleit, P., 1994, Influence of exogenous factors on the ethylene production by lichens, I. Influence of water content and water status conditions on ethylene production, *Symbiosis* **16**:187-201.

Ott, S., and Zwoch, I., 1992, Ethylene production by lichens, *Lichenologist* **24**:73-80.

Owens, L. D., Liebermann, M., and Kunishi, A., 1971, Inhibition of ethylene production by rhizobitoxine, *Plant Physiol.* **48**:1-4.

Owens, L. D., Thompson, J. F., Pitcher, R. G., and Williams, T., 1972, Structure of rhizobitoxine, an antimetabolic enol-ether-acid from *Rhizobium japonicum*, *J. Chem. Soc. Chem. Commun.* **1972**: 714.

Penmetsa, R. V., and Cook, D. R., 1997, A legume ethylene-insensitive mutant hyperinfected by its rhizobial symbiont, *Science* **275**:527-530.

Perotto, S., Brewin, N. J., and Kannenberg, E. L., 1994, Cytological evidence for a host defense response that reduces cell and tissue invasion in pea nodules by lipopolysaccharide-defective mutants of *Rhizobium leguminosarum* strain 3841, *Molec. Plant-Microbe. Interac.* 7:99-112.

Peters, N. K., and Crist-Estes, D. K., 1989, Nodule formlation is stimulated by the ethylene inhibitor aminoethoxyvinylglycine, *Plant Physiol.* **91**:690-693.

Poveda, J. L., Caba, J. M., Lluch, C., and Ligero, F., 1993, Nitrate, ethylene and nodulation, *Plant Physiol.* **102**:176 (Abstract 1010).

Remner, S. B., Ahmadjian, V., and Livdahl, T. P., 1986, Effects of IAA (indole-3-acetic acid) and kinetin (6-furfuryl-amino-purine) on the synthetic lichen *Cladonia cristatella* and its isolated symbionts, *Lichen Physiol. Biochem.* 1:1-25.

Ruan, X., and Peters, N. K., 1992, Isolation and characterization of rhizobitoxine mutants of *Bradyrhizobium japonicum*, *J. Bacteriol.* **174**:3467-3473.

Rupp, L. A., and Mudge, K. W., 1985a, Is ethylene involved in ectomycorrhizae formation on mungo pine, in: *Proc. 6th North Amer. Conf. on Mycorrhizae*, June 25-29, 1984, Bend, OR, pp. 355.

Rupp, L. A., and Mudge, K. W., 1985b, Ethephon and auxin induce mycorrhiza-like changes in the morphology of root organ cultures of mugo pine, *Physiol. Plant.* **64**:316-322.

Rupp, L. A., Mudge, K. W., and Negm, F. B., 1989a, Involvement of ethylene in ectomycorrhiza formation and dichotomous branching of roots of mugo pine seedlings, *Can. J. Bot.* **67**:477-482.

Rupp, L. A., DeVries, H. E., II, and Mudge, K. W., 1989b, Effect of aminocyclopropane carboxylic acid and aminoethoxyvinylglycine on ethylene production by ectomycorrhizal fungi, *Can. J. Bot.* **67**:483-485.

Scagel, C. F., and Linderman, R. G., 1994, Increases in endogenous IAA content of conifer roots mediated by mycorrhizal fungi and exogenously applied plant growth regulators, Plant Physiol. 105(Suppl):143 (Abstract 781).

Scagel, C. F., and Linderman, R. G., 1998a, Relationships between differential *in vitro* indole-acetic acid or ethylene production capacity by ectomycorrhizal fungi and conifer seedling responses in symbiosis, *Symbiosis* **24**:13-34.

Scagel, C. F., and Linderman, R. G., 1998b, Influence of ectomycorrhizal fungal inoculation on growth and root IAA concentrations of transplanted conifers, *Tree Physiol.* **18**:739-747.

Schieleit, P., and Ott, S., 1996, Ethylene production and 1-aminocyclopropane-1-carboxylic acid content of lichen bionts, *Symbiosis* **21**:223-231.

Schmidt, J. S., Harper, J. E., Hoffman, T. K., and Bent, A. F., 1999, Regulation of soybean nodulation independent of ethylene signalling, *Plant Physiol.* **119**:951-959.

Shantharam, S., and Mattoo, A. K., 1997, Enhancing biological nitrogen fixation: An appraisal of current and alternative technologies for N input into plants, *Plant Soil* **194**:205-216.

Shirtliffe, S. J., Vessey, J. K., Buttery, B. R., and Park, S. J., 1996, Comparison of growth and N accumulation of common bean (*Phaseolus vulgaris* L.) cv. OAC Rico and its two nodulation mutants, R69 and R99, *Can. J. Plant. Sci.* **76**:73-83.

Spaink, H. P., 1995, The molecular basis of infection and nodulation by rhizobia: The ins and outs of sympathogenesis, Annu. Rev. *Phytopathol.* **33**:345-368.

Spaink, H. P., 1997, Ethylene as a regulator of *Rhizobium* infection, *Trends Plant Sci.* **2**:203-204.

Spaink, H. P., Sheeley, D. M., van Brussel, A. A. N., Glushka, J., York, W. S., Tak, T., Geiger, O., Kennedy, E. P., Reinhold, V. N., and Lughtenberg, B. J. J., 1991, A novel highly unsaturated fatty acid moeity of lipo-oligosaccharide signals determines host specificity of *Rhizobium*, *Nature* **354**:125-130.

Spalding, D. H., and Lieberman, M., 1965, Factors affecting the production of ethylene by *Penicillium digitatum*, *Plant Physiol.* **40**:645-648.

Stein, A., and Fortin, J. A., 1990, Pattern of root initiation by an ectomycorrhizal fungus on hypocotyl cuttings of *Larix laricina*, *Can. J. Bot.* **68**:492-498.

Staehelin, C., Grando, J., Muller, J., Wiemken, A., Mellor, R. B., Felix, G., Regenass, M., Broughton, W. J., and Boller, T., 1994a, Perception of *Rhizobium* nodulation factors by tomato cells and inactivation by root chitinases, *Proc. Natl. Acad. Sci. USA* **91**:2196-2200.

Staehelin, C., Schultze, M., Kondorosi, E., Kondorosi, A., Mellor, R. B., and Boller, T., 1994b, Structural modifications in *Rhizobium meliloti Nod* factors influence their stability against hydrolysis by chitinases, *Plant J.* **5**:319-330.

Stokkermans, T. J. W., Sanjuan, J., Ruan, X., Stacey, G., and Peters, N. K., 1992, *Bradyrhizobium japonicum* rhizobitoxine mutants with altered host-range on Rj4 soybean, *Plant Physiol.* **99**:110.

Strzelczyk, E., Kampert, M., and Pachlewski, R., 1994, The influence of pH and temperature on ethylene production by mycorrhizal fungi of pine, *Mycorrhiza* **4**:193-196.

Suganuma, N., Yamauchi, H., and Yamamoto, K., 1995, Enhanced production of ethylene by soybean roots after inoculation with *Bradyrhizobium japonicum*, *Plant Sci.* **111**:163-168.

Tanimoto, M., Roberts, K., and Dolan, L., 1995, Ethylene is a positive regulator of root hair development in *Arabidopsis thaliana*, *Plant J.* **8**:943-948.

Thomas, K. C., and Spencer, M., 1977, L-Methionine as an ethylene precursor in *Saccharomyces cerevisiae*, *Can. J. Microbiol.* **23**:1669-1674.

Truchet, G., Roche, P., Lerouge, P., Vasse, J., Camut, S., de Billy, F., Prome, J.-C., and Denarie, J., 1991, Sulphated lipo-oligosaccharide signals of *Rhizobium meliloti* elicit root nodule organogenesis in alfalfa, *Nature* **351**:670-673.

van Brussel, A. A. N., Zaat, S. A. J., Cremers, H. C. J. C., Wijffelman, C. A., Pees, E., Tak, T., and Lugtenberg, B. J. J., 1986, Role of plant root exudate and *sym* plasmid-localized nodulation genes in the synthesis by *Rhizobium leguminosarum* of Tsr factor, which causes thick and short roots on common Vetch., *J. Bacteriol.* **165**:517-522.

van Spronsen, P. C., van Brussel, A. A. N., and Kijne, J. W., 1995, *Nod* factors produced by *Rhizobium leguminosarum* biovar. *viciae* induce ethylene-related changes in root cortical cells of *Vicia sativa* ssp. *nigra*, *European J. Cell Biol.* **68**:463-469.

van Workum, W. A. T., van Brussel, A. A. N., Tak, T., Wijffelman, C. A., and Kijne, J. W., 1995, Ethylene prevents nodulation of *Vicia sativa* ssp. *nigra* by exopolysaccharide-deficient mutants of *Rhizobium leguminosarum* bv. *viciae, Molec. Plant-Microbe Interact.* **8**:278-285.

Vierheilig, H., Alt, M., Mohr, U., Boller, T., and Wiemken, A., 1994, Ethylene biosynthesis and activities of chitinase and β-1,3-glucanase in the roots of host and non-host plants of vesicular-arbuscular mycorrhizal fungi after inoculation with *Glomus mosseae, J. Plant. Physiol.* **143**:337-343.

Xie, Z.-P., Staehelin, C., Wiemken, A., and Boller, T., 1996, Ethylene responsiveness of soybean cultivars characterized by leaf senescence, chitinase induction and nodulation, *J. Plant Physiol.* **149**:690-694.

Xiong, K., and Fuhrmann, J. J., 1996, Soybean response to nodulation by wild-type and an isogenic *Bradyrhizobium elkanii* mutant lacking rhizobitoxine production, *Crop. Sci.* **36**:1267-1271.

Yasuta, T., Satoh, S., and Minamisawa, K., 1999, New assay for rhizobitoxine based on inhibition of 1-aminocyclopropane-1-carboxylate synthase, *Appl. Environ. Microbiol.* **65**:849-852.

Yuhashi, K.-I., Ichikawa, N., Ezura, H., Akao, S., Minakawa, Y., Nukui, N., Yasuta, T., and Minamisawa, K., 2000, Rhizobitoxine production by *Bradyrhizobium elkanii* enhances nodulation and competitiveness on *Macroptilium atropurpureum, Appl. Environ. Microbiol.* **66**:2658-2663.

Zaat, S. A. J., van Brussel, A. A. N., Tak, T., Lugtenberg, B. J. J., and Kijne, J. W., 1989, The ethylene-inhibitor aminoethoxyvinylglycine restores normal nodulation by *Rhizobium letguminosarum* biovar. *viciae* on *Vicia sativa* ssp. *nigra* by suppressing the "thick and short roots" phenotype, *Planta* **177**:141-150.

Zhao, Z., Wang, X., and Guo, X., 1992, Selection of fungi for the production of ectomycorrhizal fungus inoculum, *Acta Microbiol. Sin.* **32**:227-232.

7

ETHYLENE IN PATHOGENESIS

7.1. INFECTION: A BIOTIC STRESS

Stress C_2H_4 represents collectively the accelerated C_2H_4 production in plants induced by various abiotic (wounding, physical load, chilling temperatures, waterlogging, and exposure to chemicals) and biotic (disease and insect damage) factors. Plant C_2H_4 synthesis is often significantly increased during infection by pathogens and can also be induced by treatment with pathogen-derived elicitors (Boller, 1991; Pegg, 1976b; Frankenberger and Arshad, 1995). It has been proposed that C_2H_4 acts as a messenger during plant-microbe interactions. This accelerated stress C_2H_4 during pathogenesis may be a stimulus for defense responses that lead to resistance or conversely, it may play a role in disease symptom development and in the weakening of endogenous resistance (Ben-David et al., 1986; Boller, 1991; Pegg, 1976b; Stall and Hall, 1984; Yang and Hoffman, 1984; Abeles et al., 1992; Lund et al., 1998). By using various mutants of soybean altered in C_2H_4 sensitivity and a number of pathogens (virulent and avirulent), Hoffman et al. (1999) concluded that the reduced C_2H_4 sensitivity could be beneficial against some pathogens but deletrious to resistance against other pathogens. This chapter deals mainly with infection-induced C_2H_4 production and its possible role in disease or resistance development in the infected hosts. Excellent reviews related to this subject are published elsewhere (Abeles et al., 1992; Boller, 1982, 1991, 1990; Hislop et al., 1973b; Archer and Hislop, 1975; Pegg, 1976b; Frankenberger and Arshad, 1995).

7.2. ENHANCED ETHYLENE PRODUCTION DURING PATHOGENESIS

Production of C_2H_4 at an accelerated rate at various stages of pathogenesis is commonly observed (Table 7.1) and is considered an early biochemical event in many plant-pathogen interactions (Boller, 1982, 1990; 1991; Yang and Pratt, 1978; Yang and Hoffman, 1984; Archer and Hislop, 1975; Pegg, 1976b). All known types of diseases show this enhanced C_2H_4 response caused by fungi, bacteria, viruses and nematodes. Since the first report by Ross and Willliamson (1951) on virus-induced C_2H_4 production in many host-virus interactions, several studies have shown that viral infection resulting

Table 7.1. Ethylene production in response to host-pathogen infection.

Pathogen	Host	Specific response to infection	Reference
Agrobacterium sp.	Carrot discs	Infection resulted in production of elevated levels of C_2H_4.	Goodman et al. (1986)
A. tumefaciens	*Nicotiana* and *Lycopersicon*	Synthesis of ACC and conversion of ACC to C_2H_4 were influenced by crown gall transformation.	Miller and Pengelly (1984)
A. tumefaciens	Tomato	Treatment with $CoCl_2$ completely inhibited crown gall tumor formation, which might indicate that accumulation of ACC prevented this transformation. Co^{2+} inhibited conversion of ACC to C_2H_4.	Davis et al. (1992)
A. tumefaciens and *A. rhizogenes*	Carrot discs	Inoculation with *A. rhizogenes* (root inducing bacteria with Ri plasmid) produced about twice as much C_2H_4 as those inoculated with *A. tumefaciens* (tumor-inducing bacteria with Ti plasmid). Both strains of *Agrobacterium* caused carrot discs to produce more C_2H_4 than controls. Strains of these bacteria devoid of the Ri of Ti plasmid produced neither roots nor tumors nor C_2H_4 above the controls.	Canfield and Moore (1983)
Alternaria and *Glocosporium*	Navel orange	C_2H_4 and fungal infection were associated with blossom end yellowing of navel orange.	Southwick et al. (1982)
Bipolaris sorokiniana	*Poa pratensis* leaves	Enhanced C_2H_4 production was responsible for much of the chlorophyll loss.	Hodges and Coleman (1984)
Botrytis cinerea	leaves of tomato, pepper, bean, and cucumber	Infection resulted in much higher C_2H_4 production by the leaves than noninfected leaves. Maximum C_2H_4 production was observed in leaves showing mild symptoms of the disease. Exogenous supply of C_2H_4 induced 75-350% more necrosis than the infection application AVG; AOA and Ag+ reduced the development of grey mold disease significantly.	Elad (1990)

Table 7.1. (continued).

Ceratocystis fimbriata	Sweet potato	Vigorous production of C_2H_4 took place in sweet potato root tissue with infection. However, the contribution of C_2H_4 by the fungus to that of the total amount produced was small and was produced by a pathway different from that of the plant.	Hirano et al. (1991)
Cercospora arachidicola	Peanut leaves	Increased C_2H_4 production coincided with disease symptoms (i.e., defoliation and abscission).	Ketring and Melouk (1982)
Colletotrichum lagenarium	Melon seedlings	C_2H_4 and cell wall hydroxyproline-rich glycoprotein biosynthesis were greatly enhanced.	Toppan et al. (1982)
C. lagenarium	Melon tissue	Infection led to early stimulation of C_2H_4.	Toppan and Esquerre-Tugaye (1984)
C. lagenarium	Bean	Addition of ACC, chitin oligosaccharides, or an elicitor derived from the pathogen, to a transformed protoplast resulted in rapid and marked increase in the expression of the chimeric gene.	Roby et al. (1991)
C. lagenarium	Melon hypocotyl	Infection triggered C_2H_4 production, which might be involved in hydroxyproline-rich glycoprotein (HRGP) synthesis.	Roby et al. (1985)
C. lagenarium	Melon leaves or seedlings	Increased chitinase activity with a simultaneous increase in C_2H_4 production occurred as a result of infection.	Roby et al. (1986)
Erwinnia carotovora	Green bell pepper	C_2H_4 production by infected green bell pepper was higher than controls and correlated with the extent of lesion development.	Ibe and Gorgan (1985)
Fusarium spp.	Tomato	Increased C_2H_4 production reached a peak within 9-10 days, which was coincident with marked foliar wilting and basal leaf abscission.	Gentile and Matta (1975)
Fusarium spp. pathogenic to tulips	Tulip bulb	Produced 2000 times more C_2H_4 than other nonpathogenic *Fusarium*.	Swart and Kamerbeek (1976)

Table 7.1. (continued).

F. oxysporum f. sp. *tulipae*	Tulip bulb	Infection increased C_2H_4 production; caused gummosis and bud necrosis.	deMunk (1971)
F. solani f. sp. *phaseoli* (nonpathogenic) f. sp. *pisi* (pathogenic)	Pea pods	Fungal infection strongly increased C_2H_4 production, but C_2H_4 and fungal infection were independent signals for induction of chitinase and β-1,3-glucanase.	Mauch et al. (1984)
Mycospharella citri	Rough lemon and grapefruit leaves	Infection induced C_2H_4 production and caused leaf chlorosis and abscission.	Graham et al. (1984)
Penicillium digitatum	Grapefruit	Fungal invasion was associated with increases in both ACC and C_2H_4 production, but the ability of the plant tissue to convert ACC to C_2H_4 decreased with the development of infection.	Achilea et al. (1985a)
P. italicum	Orange fruit	Fungal inoculation increased C_2H_4 production. Treatment with C_2H_4 (1000 μl L^{-1}) and inoculation enhanced respiration rates of fruit.	El-Kazzaz et al. (1983c)
Peronospora tabacina Adam.	Tobacco	Infection increased the level of C_2H_4, causing growth retardation and accumulation of scopoletin in the upper stem.	Reuveni and Cohen (1978)
Phytophthora citrophthora	Citrus	C_2H_4 production by the infected stem tissues was a direct factor influencing duct development.	Gedalovich and Fahn (1985)
P. megasperma	Parsley cells	Peak of ACC-synthase activity preceded maximal PAL activity.	Chappell et al. (1984)
P. megasperma f. sp. *glycinea*	Soybean	Infected seedling roots resulted in compatible and incompatible interactions displayed by an increase in C_2H_4 biosynthesis. In the compatible interactions, the rate of C_2H_4 biosynthesis started to increase about 6 h after infection and reached 10- to 15-fold induction 11-12 h after inoculation. In the incompatible interaction, C_2H_4 production was stimulated 5- to 10-fold as early as 3 h after infection and reached its maximum (50-fold induction) in 6 h after infection. In soybean roots, an early burst of C_2H_4 biosynthesis was a characteristic symptom of the incompatible reaction.	Reinhardt et al. (1991)

Table 7.1. (continued).

P. megasperma f. sp. *glycinea*	Tomato	Treatment with elicitors prepared from this pathogen stimulated C_2H_4 biosynthesis 10- and 20-fold in exponentially growing cells and more than 100-fold in stationary cells. Activity of both ACC synthase and EFE strongly increased in response to elicitor treatment.	Felix et al. (1991)
P. megasperma var. *sojae*	Soybean cotyledons	Infection led to increased C_2H_4 formation, phenylalanine ammonia lyase (PAL) activity, and glyceollin accumulation.	Paradies et al. (1980)
P. parasitica var. *nicotianae*	Tobacco	Treatment with an elicitor (prepared from the pathogen) caused an increase in C_2H_4 synthesis and proteinase inhibitor production in tobacco cells. However, C_2H_4 was not a signal for elicitation of protein inhibitors.	Rickauer et al. (1992)
P. parasitica var. *nicotianae*	Tobacco	Tobacco cell cultures treated with fungal elicitors responded by an early stimulation of C_2H_4 and lipoxygenase activity.	Rickauer et al. (1990)
Phytophthora infestans	Potato tuber disc	A burst of C_2H_4 production was observed within 6 h after infection with an incompatible race of the pathogen but not with compatible races. Later additional C_2H_4 was synthesized coincidentally with symptoms appearance in both compatible and incompatible combinations.	Gwinn et al. (1989)
P. infestans	Tomato leaves	Levels of ACC synthesis, ACC content, EFE activity and C_2H_4 production all increased following infection suggested that normal C_2H_4 pathway was functioning.	Spanu and Boller (1989a,b)
Piricularia oryzae	Rice	Considerable increase in C_2H_4 evolution, creating stunting of blast.	Kozaka and Teraoka (1978)
Pseudomonas syringae pv. *atropurpurea*	Tobacco	The rate of C_2H_4 released from the leaves was proportional to the concentration of coronatine (a toxin produced by the pathogen) applied to the leaf surface. ACC accumulated in the coronatine-treated tissue. Coronatine induced the synthesis of C_2H_4 from MET.	Kenyon and Turner (1992)

Table 7.1. (continued).

Rhizopus stolonifer	Nonripening tomato mutants	Infection markedly stimulated C_2H_4 production followed by accelerated climacteric-like patterns of respiration.	Barkai-Golan and Kopeliovitch (1983)
Septoria nodorum	Wheat (40 cultivars)	A close correlation between high rates of C_2H_4 production and increased susceptibility was observed.	Wendland and Hoffman (1988)
Trichoderma viride	*Nicotiana tabacum* cv. Xanthi leaves	Xylanase from the fungus increased C_2H_4 biosynthesis in the treated leaves which was correlated to the accumulation of ACC synthase and ACC oxidase transcript.	Avni et al. (1994)
Trichoderma viride	Tobacco	Treatment with a fungal elicitor (C_2H_4-inducing xylanase) enhanced C_2H_4 production and tissue necrosis in whole plants at sites far away from the point of treatment when applied through a cut petiole.	Bailey et al. (1991)
Uromyces phaseoli	Bean leaves, detached	C_2H_4 production increased at the time the injection pegs penetrated the stomata.	Montalbini and Elstner (1977)
U. phaseoli	Bean hypocotyl segment, detached	C_2H_4 production increased at the time the infection pegs penetrated the stomata. A second peak in C_2H_4 production occurred during the expression of necrotic spots. ACC synthase activity increased ACC level in tissue and might represent an increase in EFE activity.	Paradies et al. (1979)
Ustilago maydis	Maize seedlings	Reduction in elongation of leaves and shoots, increase in basal diameter, and a decrease in weight were preceded by a 4-fold increase in C_2H_4 synthesis.	Andrews et al. (1981)
Verticillium dahliae Kleb.	Cotton	Infection accelerated C_2H_4 production in defoliating plants; however, the rate of C_2H_4 production was not related to the relative virulence of various pathotypes in cotton.	Tzeng and DeVay (1985)

Table 7.1. (continued).

Organism	Plant/tissue	Description	Reference
Xanthomonas compestris	Citrus leaves	Infected leaves with a virulent strain of the bacteria showed three peaks of C_2H_4 production. The first peak occurred when plants were treated with virulent or avirulent strains, heat-killed cells, and extracellular polysaccharides or oligosaccharides. The last peak of C_2H_4 production was associated with the growth of bacteria in the leaves.	Goto and Hyodo (1985)
X. campestris	Pepper	Infected plants produced more C_2H_4 than the healthy control and there was a direct correlation between C_2H_4 production in diseased plants, the number of bacteria in the tissues and disease development.	Ben-David et al. (1986)
X. campestris pv. *citri*	Swingle citrumelo	Inoculation with the pathogen induced significant C_2H_4 production by the leaves, whereas the nonpathogen did not. Leaves inoculated with the pathogen had higher endogneous [ACC] than the leaves treated with AVG, or streptomycin before inoculation. All inhibitors of C_2H_4 biosynthesis were effective in suppression of C_2H_4 production from pathogen-inoculated leaves. These inhibitors also suppressed the rate of disease progression. Pathogen-induced C_2H_4 biosynthesis was closely associated with disease progression. *De novo* induction and activation of ACC synthase is the key step in the induction of C_2H_4 production. It was suggested that the induction of C_2H_4 production as a result of pathogen invasion might be a plant defense response, since it led to the abscission of leaves, which helped retard the spread of the pathogen into the plant system.	Dutta and Biggs (1991)
X. campestris pv. *vesicatoria*	Pepper leaves	A chlorotic zone surrounding a necrotic lesion of a bacterial spot was associated with increased C_2H_4 synthesis in diseased leaves.	Stall and Hall (1984)
X. citri	Citrus leaves	Development of disease symptoms was related directly to an increase in C_2H_4 production. Rate of defoliation was related directly to rate of C_2H_4 production.	Goto et al. (1980)

Source: Frankenberger and Arshad (1995).

in local lesions induce stress C_2H_4 in various plants including bean (Nakagaki et al., 1970), tobacco (De Laat and van Loon, 1982, 1983a,b; Roggero and Pennazio, 1988), cucumber (Marco et al., 1976), tomato and *Gynura aurantiaca* leaves (Belles and Conejero, 1989), tomato cell suspension culture (Belles et al., 1989), and *Tetragonia expansa* (Gaborjanyi et al., 1971). Increased C_2H_4 production is often associated with the beginning of tissue necrosis. Ethylene production in tobacco plants infected with tobacco mosaic virus (TMV) has been studied extensively (Balazs et al., 1969; Nakagaki et al., 1970; De Laat et al., 1981; De Laat and van Loon, 1982, 1983a,b; Roggero and Pennazio, 1988). These studies demonstrated that tobacco cultivars, which carry the *N* genes and react hypersensitively to the virus, show a surge of C_2H_4 production about 48 h after infection, just before the first symptoms of hypersensitive response are seen (De Laat et al., 1981; De Laat and van Loon, 1982). De Laat et al. (1981) clearly demonstrated that C_2H_4 in tobacco plants reacting hypersensitively to TMV was derived from methionine (MET) and that 1-aminocyclopropane-1-carboxylic acid (ACC) accumulated at the time of increased C_2H_4 production.

Similarly, increased C_2H_4 release from plants or plant tissues is observed with bacterial infection which is mostly correlated with lesion development. Studies with green bell peppers, other peppers, cauliflower, citrus leaves, carrot disks and banana infected with *Erwinia carotivora*, *Xanthomonas campestris*, *Erwinia carotovora*, *Xanthomonas campestris*, *Agribacterium tumefaciens/rhizogenes* and *Pseudomonas solanacearum*, respectively, showed accelerated C_2H_4 production as a result of plant-pathogen interactions (Ben-David et al., 1986; Ibe and Gorgan, 1985; Canfield and Moore, 1983; Lund and Mapson, 1970; Goto and Hyodo, 1985; Freebairn and Buddenhagen, 1964).

Like viral and bacterial infections, the fungal-infected plant tissues also show peak production of C_2H_4. The rate of this enhanced C_2H_4 production is mostly correlated with the amount of tissue damage (Hebard and Shain, 1988; Ketring and Melouk, 1982; Spanu and Boller, 1989b). Infection enhanced (stress) C_2H_4 was observed in beet leaves infected with *Cercospora beticola* (Koch et al., 1980), castor bean leaves infected with *Fusarium oxysporum* (VanderMolen et al., 1983), bark plugs of American and Chinese chestnut infected with *Endothia parasitica* (Hebard and Shain, 1988), and rose and carnation flowers infected with *Botrytis cinerea* (Elad, 1988). Stress C_2H_4 was also found to be associated with rots in banana, mango and apple fruits infected with *Diplodia natalensis*, *Trichothecum roseum*, *Penicillium expansum*, *Colletotrichum gloeosporoides*, *Botrytis cinerea*, *Pestalotia* sp., and *Alternaria citri* (Schiffmann-Nadel et al., 1985). Very recently, Wachter et al. (1999) reported 140 times greater C_2H_4 production by *Agrobacterium tumefaciens*-induced stem tumors of *Ricinus communis* compared with non-tumorized control stems. Accumulation of ACC preceded C_2H_4 emission with a maximum 2 weeks after tumor induction.

Moreover, the treatment of plant tissues with cell-free compounds of pathogens, so called elicitors, also results in production of stress C_2H_4 . The elicitors obtained from various pathogens stimulated C_2H_4 production in elicitor treated tissues of parsley, melon, tomato, and tobacco cells (Felix et al., 1991; Rickauer et al., 1990, 1992; Paradies et al., 1979; Chappell et al., 1984).

Hodges and Coleman (1984) suggested that C_2H_4 may function late in the pathogenesis of host (*Poa pratensis*) - pathogen (*Bipolaris sorokiniana*) interactions and is responsible for much of the chlorophyll loss. Later, Hodges and Campbell (1993) reported that C_2H_4 surge in *Poa pratensis* leaves infected with *Bipolaris sorokiniana* could be more effectively blocked by the inhibition of S-adenosylmethionine (SAM) conversion into ACC (i.e., ACC synthase inhibitors) than by inhibitors that block the conversion of ACC to C_2H_4 (i.e., ACC oxidase inhibitors). This may indirectly imply that infection could lead to greater formation of ACC from SAM which may be a regulating step in accelerated C_2H_4 synthesis in pathogenesis. The first peak in increased C_2H_4 production from barley leaves inoculated with conidia of barley powdery mildew (*Erysiphe graminis*) coincided with the penetration of the fungus through epidermal cells (Portmann and Elstner, 1983).   Ethylene production from detached bean leaves (Montalbini and Elstner, 1977) and hypocotyl segments (Paradies et al., 1979) infected with *Uromyces phaseoli* increased at the time the infection pegs penetrated the stomate. A second peak in C_2H_4 production occurred during the expression of necrotic spots. The first rise in C_2H_4 production seems to be a common denominator for fungal-induced stress and is not associated with increased ACC synthase activity or increased ACC levels in the tissue, but may represent an increase in C_2H_4 –forming enzyme (EFE) activity (Paradies et al., 1979), which is contrary to the findings of Hodges and Campbell (1993). It is most likely that the C_2H_4 surge during pathogenesis might be regulated at different levels of the ACC biosynthetic pathway.

Flaishman and Kolattukudy (1994) observed that C_2H_4 stimulated germination and appressorium formation in *Colletotrichum* sp. penetrating climacteric fruit but not in other *Colletotrichum* strains. *Colletotrichum gloeosporioides* spores formed multiple appressoria on normally ripening C_2H_4 -producing tomato, whereas on nonproducing C_2H_4 transgenic tomato and orange fruits, appressorium formation required exogenous C_2H_4. Taken together, Flaishman and Kolattukudy (1994) suggested that these fungi must have evolved to develop a mechanism to use the host's ripening hormone (C_2H_4) as a signal to differentiate into multiple infection structure and thus control the infection process.

7.3. SOURCE OF ACCELERATED ETHYLENE EVOLUTION DURING PATHOGENESIS

There is no doubt about the plant's (host) origin of accelerated C_2H_4 in the case of viral infection or the case of plant tissue treated by elicitors; however, the source of C_2H_4 (pathogen or host) of accelerated C_2H_4 production during pathogenesis is subject to controversy because many of the bacterial and fungal pathogens have been demonstrated as capable of synthesizing C_2H_4 *in vitro*. Most of these pathogens derive C_2H_4 from MET which also serves as a sole precursor of C_2H_4 in higher plants. This further complicates the source with respect to differentiating the accelerated C_2H_4 during pathogenesis. It is now well established that the majority of bacteria and fungi use MET as a substrate of C_2H_4 but follow a pathway of C_2H_4 biosynthesis different from that of higher plants (see Chapter 3). However, some pathogens produce C_2H_4 independently of MET and use α-ketoglutarate (KGA) as an immediate precursor of C_2H_4.

Many strains of *Pseudomonas solanacearum* causing rot of banana fruit were long considered to be the only bacteria to produce substantial amounts of C_2H_4 independent of MET (Freebairn and Buddenhagen, 1964). A strain of *Pseudomonas syringae* pv. *phaseolicola*, pathogenic to the Japanese weed, *Puerari kuzdu* was found to synthesize C_2H_4 from KGA (Goto and Hyodo, 1987; Goto et al., 1985; and also see Section 3.4.2).

Penicillium digitatum, a common mould on citrus fruits, has been studied extensively and found capable of producing large amounts of C_2H_4 *in vitro* in the presence of MET (Chalutz et al., 1977; Chou and Yang, 1973; Fukuda et al., 1986). However, in the absence of MET, it produces C_2H_4 most likely from glutamic acid or KGA (Chalutz et al., 1977; Chou and Yang, 1973; Fukuda et al., 1986; see also Section 3.4.1).

Similarly, strains of *Fusarium oxysporum* f.sp. *tulipae*, a tulip pathogen produce large amounts of C_2H_4 in liquid media (Swart and Kamerbeek, 1976; 1977; Hottiger and Boller, 1991). Studies on C_2H_4 production by 19 various species of *Fusarium* indicate that C_2H_4 production is a constant trait of the isolates of *Fusarium oxysporum* from tulip (Swart and Kamerbeek, 1976). Abnormally high levels of C_2H_4 production by *F. oxysporum* f. sp. *tulipae* have also been reported *in vitro* (deMunk and De Rooy, 1971), 4,000 times more than other fungi tested. Ethylene production was reported in the fungus, *Bipolaris sorokiniana* when fed MET (Coleman and Hodges, 1986). Blastomycosis pathogens including *Blastomyces dermatitidis, B. brassiliensis, Histoplasma capsulatum, Candida vartiovaari, Trichosporum cutaneum,* and *Mucor hiemalis* have also been reported to produce C_2H_4 *in vitro* (Bernheim, 1942; Nickerson, 1948; Smith and Restall, 1971). Ethylene production by *Asperigillus* and a *Mucor* sp. has been reported by Dasilva et al. (1974). Sparophytes and facultative parasites also produce C_2H_4, although its significance, if any, in the growth and metabolism of the microorganism is not clear (see Chapter 4). All of these studies reveal that many pathogens are capable of producing C_2H_4 *in vitro* particularly in the presence of substrates such as MET or KGA.

7.3.1. Contribution by the Pathogen

Contradicting reports have appeared in the literature regarding the contribution of pathogens in stress C_2H_4 released during pathogenic infections. Some studies have demonstrated a significant contribution by the pathogen in C_2H_4 released during pathogenesis. Achilea et al. (1985a) indicated that a relatively low rate of C_2H_4 production in infected grapefruit was mostly from the fruit tissues, whereas a later and higher rate of C_2H_4 synthesis originated mostly from the fungus, *P. digitatum*. Achilea et al. (1985b) used radio-labeled precursors to study the sources of C_2H_4 in *P. digitatum*-infected grapefruit and found that C_2H_4 produced by healthy portions of the fruit originated from MET; whereas, C_2H_4 produced by the infected fruit was derived mostly from glutamic acid. Furthermore, C_2H_4 production from the healthy fruit was markedly enhanced by ACC and, to a lesser extent, by $CuSO_4$ (an inducer of stress C_2H_4), but inhibited by aminovinylglycine (AVG). In contrast, production of C_2H_4 by the *P. digitatum*-infected peel was not affected by ACC, but was markedly inhibited by $CuSO_4$, and to a lesser extent by AVG. From these observations, they suggested that C_2H_4 production in the healthy fruit was of plant origin, whereas markedly enhanced production of C_2H_4 by *P. digitatum*-infected fruit was mostly or entirely of fungal origin. Hodges and Coleman (1984) suggested that the pathogen may stimulate C_2H_4 production from the root tissue early in pathogenesis, whereas later during the infection process, the pathogen may also directly contribute C_2H_4 and ACC. Coleman and Hodges (1986) found that the addition of ACC to the leaf blade infusion media of *Poa pratensis* resulted in low C_2H_4 production compared with that produced when infected by the pathogen, *Bipolaris sorokiniana*, with the addition of MET. They concluded that the pathogen did not convert ACC to C_2H_4 and that the C_2H_4 was produced by more than one pathway. In

another study, Coleman and Hodges (1987) reported that in response to the interaction of
P. pratensis with *B. sorokiniana*, two peaks of C_2H_4 production were observed,
accompanied by substantial but erratic increases in ACC. Because it is suspected that
this pathogen also produces ACC, it remains unclear whether C_2H_4 comes from the plant
or from the fungus. In the case of *Pueraria lobata* infected with *P. syringae* pv.
phaseolicola, it is likely that the bacterium contributes the largest portion of C_2H_4 formed
in the diseased tissue (Goto et al., 1985). The plant's contribution to C_2H_4 production was
negligible in contrast to the C_2H_4 production of citrus leaves infected with *Xanthomonas
citri* (Goto et al., 1980) and of tomato plants inoculated with *Ps. solanacearum* (Lu et al.,
1989; Pegg and Cronshaw, 1976b). Recently, Weingart and Völksch (1997)
demonstrated unequivocally the contribution of a pathogen in accelerated C_2H_4
production by infected plant tissue (Fig. 7.1). They found that untreated leaf discs of
soybean and bean produced almost no C_2H_4 , while leaves inoculated with *Ps. syringae*
pv. *phaseolicola* 6/0 (which did not produce detectable amounts of C_2H_4 *in vitro*)
produced C_2H_4 at a very low rate (about 100 pl h^{-1} cm^{-1}) throughout the experimental
period (Fig. 7.C1). However, in soybean and bean leaves inoculated with *Ps. syringae*
pv. *glycineae* (8.83) and *Ps. syringae* pv. *phaseolicola* (KZ2w) isolated from kudzu
(potent producers of C_2H_4 *in vitro*), respectively, C_2H_4 production began after inoculation
at the beginning of the multiplication of the bacterial strains and continued to increase
until the bacteria entered a stationary phase (Fig. 7.A1 and 7.B1). The kinetics of C_2H_4
production in plants were similar with the kinetics *in vitro*. Typical disease symptoms
developed in bean leaves after 3 to 4 days. The phytotoxic coronatine released by the
pathogen also enhanced C_2H_4 production in plant tissues which was inhibited by AVG,
implying that the coronatine-stimulated C_2H_4 was of plant origin. In contrast, the C_2H_4
produced by inoculated soybean and bean plants was not inhibited by AVG, further
confirming that the C_2H_4 in soybean and bean leaves infected with C_2H_4 -producing *P.
syringae* pv. *glycinea* or *phaseolicola* was mostly of bacterial (pathogen) origin
(Weingart and Völksch, 1997)

7.3.2. Contribution by the Host

Contrary to the observations summarized in the previous section (Section 7.3.1),
several studies have shown indirectly that increased C_2H_4 release from infected tissues
may be a plant metabolic product stimulated by infection stress. According to Boller
(1991), most of the C_2H_4 synthesized upon bacterial or fungal diseases is generally of
plant origin. Induction of C_2H_4 biosynthesis in response to elicitor treatment (Chappell et
al., 1984; Toppon and Esquerre-Tugaye, 1984) and by fungal enzymes (Fuchs and
Anderson, 1987; Fuchs et al., 1989) supports the premise that accelerated C_2H_4 release
during pathogenesis is most likely a plant metabolite. Similarly, accelerated C_2H_4
production in case of virus infection is completely of plant origin since viruses are unable
to synthesize C_2H_4 *in vitro*. Lund and Mapson (1970) observed a marked increase in the
rate of C_2H_4 production by cauliflower florets infected by *Erwinia carotovora* (soft rot
bacterium), and these increases were correlated directly with bacterial numbers, while the
bacterium did not produce C_2H_4 *in vitro*. They were of the view that, although bacterial
infection resulted in enhanced C_2H_4 production, the C_2H_4 was most likely a plant
metabolite, since the pure culture of *E. carotovora* failed to produce C_2H_4 when
cultivated on either pectate medium or cauliflower extract supplemented with MET. The

effect was apparently due to production of specific enzymes by the microorganisms because C_2H_4 inducing activity was heat labile and the specific enzymes involved were

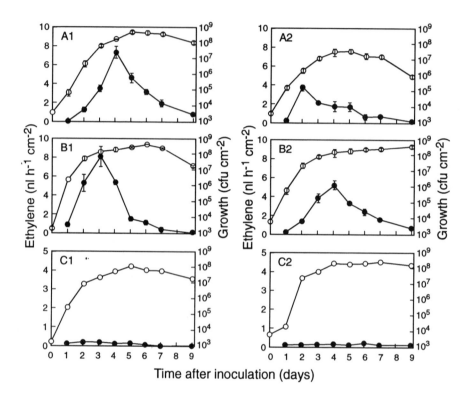

Fig. 7-1. Growth kinetics (o) and C_2H_4 production (●) by *P. syringae* pv. *glycinea* 8/83 (A), *P. syringae* pv. *phaseolicola* KZ2w from kudzu (B), and *P. syringae* pv. *phaseolicola* 6/0 from bean (C) in soybean (1) and bean (2) leaves. The data are the means and standard errors of three independent experiments. Source: Weingart and Völksch (1997).

shown to be pectate lyase and polygalacturonase (Lund and Mapson, 1970; Lund, 1973). A cell-free culture fluid of this bacterium containing pectolytic enzymes caused marked C_2H_4 production in cauliflower florets (Lund, 1973). Ethylene appeared to be derived from MET, indicating that the 'normal' biosynthetic pathway of C_2H_4 was induced. After comparing the C_2H_4-producing capacity of *Ps. solanacearum* in culture with that produced by infected banana cells, Pegg and Cronshaw (1976b) were of the view that C_2H_4 contributed by the pathogen was negligible in banana wilt. Toppan et al. (1982) revealed that enhanced C_2H_4 in melon infected by *Colletotrichum lagenarium* was suppressed by specific inhibitors of the plant C_2H_4 biosynthetic pathway, namely L-canaline and AVG, implying that C_2H_4 biosynthesis was a plant metabolic product. Similarly, the possibility of greater contribution by *Ceratocystis fimbriata* was precluded because the fungus itself produced only a small amount of C_2H_4 compared to that found in infected tissue (Stahmann et al., 1966; Imaseki et al., 1968; Chalutz and DeVay, 1969; Hirano et al., 1991). Hirano et al. (1991) concluded that the ratio of C_2H_4 contribution by *Ceratocystis fimbriata* to the total C_2H_4 production by the diseased tissues of sweet potato was extremely small, and the C_2H_4 synthetic pathway of the fungus involved L-MET, but not ACC. By using [U-C^{14}]MET, Kenyon and Turner (1992) demonstrated that the C_2H_4 released by coronatine (a phytotoxin produced by the pathogen, *Ps. syringae* pv. *savastanoi*)-treated tobacco leaves was exclusively from MET origin. This may imply that the pathogen accelerates C_2H_4 release from the host. Similarly, several studies have shown that high rates of C_2H_4 production triggered by infection were suppressed by AVG in pea pods infected by *Fusarium solani* f. sp. *pisi* (Mauch et al., 1984); parsley cells infected by *Phytophthora megasperma* (Chappell et al., 1984); melon seedlings treated with *Colletotrichum lagenarium* (Roby et al., 1985, 1986); and swingle citrumelo leaves infected by *Xanthomonas campestris* pv. *citri* (Dutta and Biggs, 1991). These studies further support the hypothesis that C_2H_4 production in the infected host is a plant metabolite, since AVG inhibits the conversion of SAM to ACC only in plant systems. Moreover, it has been indirectly shown that an increase in C_2H_4 production in viroid (Belles et al., 1989), viral (De Laat and van Loon, 1982), fungal (Elad and Volpin, 1988), and bacterial (David et al., 1986) infections, as well as that from phytotoxin action is related to induction of ACC synthase (Ferguson and Mitchell, 1985). Spanu and Boller (1989b) investigated the pathway of C_2H_4 biosynthesis in tomato leaves infected by *Phytophthora infestans*. Infected leaves reacted hypersensitively and started to produce large amounts of C_2H_4 just before necrotic lesions became visible. They measured ACC synthase *in vitro* and concluded that induction of C_2H_4 formation in infected tissues required induction of ACC synthase. Levels of ACC synthase, ACC content, EFE activity, and C_2H_4 production all increased following infection of tomato leaves with *Phytophthora infestans* (Spanu and Boller, 1989a,b). Since no pathogen has been reported capable of using SAM as a substrate (ACC synthase converts SAM to ACC), this may indicate that enhanced C_2H_4 production during pathogenesis is primarily of host origin.

Some studies have demonstrated that C_2H_4 released during pathogensis might be originated from the host via an MET-ACC-independent pathway. Kato and Uritani (1972) reported poor incorporation of radioactive MET into C_2H_4, suggesting that MET was not associated with the main C_2H_4 biosynthetic pathway in infected tissue. Evidence that the massive C_2H_4 synthesis occurring in infected tissue was inhibited very little, if at all, by AVG (an inhibitor of ACC synthase) and that externally applied ACC stimulated C_2H_4 production to a minor extent (Hyodo and Uritani, 1984), supports the premise that C_2H_4 production in infected tissues is independent of the ACC pathway. Hirano et al. (1991) further showed that Co^{2+}, an inhibitor of ACC oxidase, failed to inhibit C_2H_4 production in the presence of ACC in infected tissue. Taken together, these data indicate the involvement of a pathway differing from the MET-ACC pathway in the infected tissue. Very recently, Okumura et al. (1999) investigated the mechanism of C_2H_4 synthesis in the sweet potato root tissue infected by *Ceratocystis fimbriata*. Incorporation of L-[$^{14}C(U)$]MET into C_2H_4, 24 h after inoculation occurred at a low rate. When ACC was supplied externally, the rate of C_2H_4 production was not enhanced. Moreover, the activity of ACC oxidase was extremely low when compared with the C_2H_4 production rate. Okumura et al. (1999) hypothesized that the predominant C_2H_4 generated in the site adjacent to the invaded region of sweet potato root tissue may originate from a pathway independent of ACC.

7.4. ROLE OF ETHYLENE IN DISEASE DEVELOPMENT

The relationship between pathogen-induced C_2H_4 and disease development in the host is very complex and it is difficult to assess the role of C_2H_4 as a causal agent in disease development. Pathogenesis is further complicated by C_2H_4 interactions with other plant hormones, particularly with auxin; the effects of C_2H_4 on the physiology and metabolism of plants in the absence of obvious growth effects, and secretions of toxins by the parasite inducing C_2H_4 synthesis by the host. It is not possible to be conclusive about any correlation between virulence of the pathogen, C_2H_4 production and disease development. However, some "cause and effect" relationships between the disordered growth of the host and C_2H_4 production in response to pathogen infection have been reported in the literature. The effect of C_2H_4 on disease development is variable. Ethylene produced by the pathogen or host, or applied externally can increase, have no effect, or decrease disease development (Table 7.2). By using various mutants of soybean differing in C_2H_4 sensitivity and pathogens varying in virulence, Hoffman et al. (1999) observed that reduction of C_2H_4 sensitivity had a neutral or beneficial effect on the plant repsonses to some pathogens, but a detrimental effect on others. Furthermore, C_2H_4 sensitivity of the host had differential effects on gene-for-gene resistance against strains of the same pathogen species (Hoffman et al., 1999).

Recently, Song and Zheng (1998) reported a role of C_2H_4 in disease development (*Fusarium* wilt) in cotton caused by *Fusarium oxysporum* f. sp. *vasinfectum*. They found that pretreatment with trifluralin inhibited C_2H_4 production by the infected tissues of cotton plant which was positively related to the reduction of disease incidence and disease severity. The spraying with ethephon had reverse effects. Moreover, foliar application of an inhibitor of C_2H_4 biosynthesis (2,4-dinitrophenol) or C_2H_4 action (Ag^+) after infection with the pathogen resulted in the reduction of disease incidence and disease severity. These observations suggest a correlation between inhibition of C_2H_4 production and trifluralin-induced resistance of cotton seedlings against *F. oxysporum* f. sp. *vasinfectum* (Song and Zheng, 1998).

Studies have been conducted to understand the role of accelerated C_2H_4 production in disease development by (i) exogenous applications of C_2H_4 or C_2H_4–releasing compounds, (ii) comparing the effects of C_2H_4 production by various strains of pathogens differing in their virulence, (iii) using hosts differing in their sensitivity to C_2H_4 and (iv) studying the pathogen induced changes in the host at the molecular level. These aspects are discussed in the following sections.

Table 7.2. The effect of ethephon on resistance of various host plants to different pathogens.

Plant	Pathogen Disease development
Tomato	*Fusarium oxysporum* Increased
Tomato	*Verticillium albo-atrum* Increased
Tomato	*Botrytis cinerea* Increased
Tomato	*Myzus persicae* Increased
Barley	*Helminthosporium sativum* Increased
Barley	*Puccinia hordei* Decreased
Barley	*Erysiphe graminis* Decreased
Barley	*Rhopalosiphum padi* Increased
Wheat	*Helminthosporium sativum* Increased
Wheat	*Erysiphe graminis* Decreased
Wheat	*Macrosiphum avenae* Increased
Bean	*Colletotrichum lindemuthianum* Increased
Bean	*Pseudomonas phaseolicola* Increased
Bean	*Xanthomonas phaseoli* Increased
Bean	*Uromyces phaseoli* Decreased
Cucumber	*Erysiphe cichoracearum* Decreased
Tobacco	TMV Decreased
Datura	TMV Decreased

Source: Dehne and Spengler (1982).

7.4.1. Effect of Exogenous Ethylene

Scientists have evaluated the effect of C_2H_4 on disease development through its exogenous applications. It has been reported that exogenous C_2H_4 promotes the development of disease by stimulating growth of the pathogen or by increasing the development of disease symptoms. Ethylene treatments increased disease development in barley (Dehne et al., 1981), citrus (Barmore and Brown, 1985; Brooks, 1944; Grierson and Newhall, 1955), Douglas fir seedlings (Graham and Linderman, 1981), gladiolus corms (Halevy et al., 1970), grapefruit (Hatton and Cubbedge, 1981), leaves of geranium and *Ruscus hypoglossum* (Elad and Volpin, 1988), rose and carnation flowers (Elad, 1988), strawberries (El-Kazzaz et al., 1983a), tomatoes (Barkai-Golan and Lavy-Meir, 1989; Geeson et al., 1986), watermelon fruit (Elkashif et al., 1989), and wheat plants (Daly et al., 1970). In the tulip bulb, exogenously applied C_2H_4 caused disappearance of an antibiotic substance, tulipalin (Beijersbergen and Bergman, 1973). It was proposed that the tulip pathogen, *F. oxysporum* f. sp. *tulipae* produces C_2H_4 as a pathogenicity factor, eliminating resistance factor (such as tulipalin) of the host plant (Swart and Kamerbeek, 1977; Beijersbergen and Bergman, 1973). Four species of ornamental pot plants (*Campanula isophylla*, *Chrysanthemum morifolium*, *Pelargonium zonale*, and *Saintpaulia ionantha*) exhibited increased microbial attack after C_2H_4 treatment (Woltering, 1987). Ethylene treatment of citrus fruit significantly enhanced the incidence of stem-end rot caused by *Diplodia natalensis* and stimulated disease development (Barmore and Brown, 1985; Grierson and Newhall, 1955; McCormack, 1971). Brown and Lee (1993) studied the interaction of C_2H_4 with citrus stem-end rot caused by *D. natalensis* and suggested that one role of C_2H_4 could be that of stimulating more rapid growth of the fungus, *D. natalensis* and invasion of tissue in spite of the presence of threshold inhibitory levels of scoparone. Elad (1990) reported that application of Ag^+, AOA and AVG significantly reduced the development of grey mold (*Botrytis cinerea*) disease in bean, tomato, pepper and cucumber. In another study, Elad (1993) investigated the ability of several compounds to reduce development of grey mold (*Botrytis cinerea*) on rose, tomato, pepper, eggplant, French bean, and *Sencio* sp. Removal of C_2H_4 from the surroundings of rose flowers and leaves of tomato and pepper resulted in lower grey mold development. Similarly, inhibitors (2,5-norbornadiene, Co^{2+}, AOA, uncomplexed 2,4-dinitrophenol and radical scavenger) of C_2H_4 activity also controlled the disease on various crops differentially (Table 7.3), suggesting an important role of C_2H_4 in the development of disease (Elad, 1993). Elad and Evensen (1995) have critically discussed the interaction between C_2H_4 and host susceptibility to *B. cinerea*.

Some studies have demonstrated that C_2H_4 treatment of plants profoundly affects the growth of parasites within the hosts, the development of disease, and manifestations of the symptoms characteristic of those induced by the parasites. For instance, cotton plants exposed to C_2H_4 (7.1 x 10^{-8} M) developed both epinasty and defoliation, the characteristics of *Verticillium albo-atrum* infection (Wiese and DeVay, 1970). Goto et al. (1980) found that exposure of immature citrus leaves to 3 x 10^{-3} M C_2H_4 caused the falling of young and mature leaves after 9 and 16 days of post-treatment, respectively, similar to the infection caused by *Xanthomonas citri*. Ethrel application to tobacco leaves produces necrotic spots resembling virus-induced local lesions (van Loon, 1977). Brown and Barmore (1977) reported the time-dependent dual effect of C_2H_4 treatment on disease development in Robinson tangerines. Exposure to C_2H_4 3 days before inoculation reduced disease severity, whereas treatment immediately after inoculation enhanced the disease by inducing the development of infectious hyphae that penetrated the head tissue. All of these studies suggest a direct or indirect role for C_2H_4 in pathogenicity.

7.4.2. Ethylene Production and Virulence

Efforts have been made to investigate if there is any correlation between C_2H_4 production and the virulence of the pathogen. Wiese and DeVay (1970) showed that a

Table 7.3. Effect of 2,5-norbornadiene (NBD), cobalt sulfate, 2,4-dinitrophenol (DNP), salicylic acid and benzyladenine on the incidence of grey mould on rose flowers and leaves of various hosts.

Host plant	NBD			CoSO$_4$			DNP		Salicylic acid			Benzyladenine	
	0.1	1.0	10.0	0.1	1.0	10.0	0.1	1.0	0.1	1.0	10.0	0.1	1.0
Rose	15[a]	10	62*	45	50	85*	0	10	15	16	4	16	39*
Pepper	61*	69*	89*	74*	63*	91*	61*	70*	44*	20	12	15	10
Tomato	43	48	20	38	45	25	71*	37	28*	60*	7	42*	37*
Eggplant	85*	71*	65*	62*	60*	52*	11	38	0	8	34	0	-5
Bean	17	37	100*		5	0	42	56*	--	--	--	--	--
Senecio sp.	65*	45	62*	94*	47	85*	0	56*	28	4	5	82*	99*

[a]Disease reduction was calculated according to the formula (a–b) x 100/a, where a=disease severity of nontreated control and b=disease severity of leaves or flowers treated with certain dose of a compound.
*=Significant reduction of disease severity according to Duncan's Multiple Range Test (p<0.05); -- = not determined.
Source: Elad (1993).

strain of *Verticillum dahliae* (T-9), causing defoliation in cotton, produced twice as much C_2H_4 *in vitro*, compared with a pathogenic nondefoliating strain (SS4). Similarly, Tzeng and DeVay (1985) demonstrated enhanced production of C_2H_4 from cotton plants inoculated with a defoliating pathotype of *V. dahliae* compared to a non-defoliating one. Prince et al. (1988) found that both C_2H_4 -producing and non-C_2H_4 -producing strains of *Penicillium* grew well on agar, but C_2H_4 producing strains appeared to be more virulent on tulip bulbs, particularly in competition studies with the non-C_2H_4 -producing strains. This study clearly indicates that C_2H_4 production provides a selective advantage for a tulip bulb pathogen. Lu et al. (1989) screened 12 strains of *Pseudomonas solanacearum* for their *in vitro* and *in vivo* C_2H_4 production and their virulence on tomato plants. They observed that C_2H_4 production by the tested bacteria coincided well with the active growth of the culture and the loss of virulence was generally accompanied with reduction of C_2H_4 synthesizing activity. The tomato plant inoculated with virulent strain PS61 showed more C_2H_4 production which coincided well with the development of wilt symptom compared to an avirulent one. Moreover, the ACC content of the tissue inoculated with the virulent strain was much higher than that inoculated with the avirulent strain. Lu et al. (1989) suggested that C_2H_4 production as a plant pathogen interaction might have some role in pathogenesis. Vanzyl and Wingfield (1998) measured C_2H_4 in the bark of two *Eucalyptus* clones that were artificially inoculated with virulent and hypovirulent isolates of *Cryphonectria cubensis*. Trees inoculated with the hypovirulent isolate produced extremely low amounts of C_2H_4. No significant differences were found in C_2H_4 production by a susceptible clone (ZG 14) and a disease-tolerant clone (TAG 5) inoculated with a hypovirulent isolate. However, clone ZG 14 produced significantly more C_2H_4 than TAG 5 when inoculated with the virulent isolate of *C. cubensis*. This supports the view that trees more susceptible to fungal attack tend to produce greater amounts of C_2H_4 than disease-tolerant clones. There was a correlation established between C_2H_4 production after infection with *C. cubensis* and relative susceptibility or resistance of *Eucalyptus* clones to infection by the pathogen. The amount of C_2H_4 produced in response to *C. cubensis* infection may thus be a important and rapid means of distinguishing a susceptible response from resistance which may be a valuable factor in breeding for resistance to this pathogen (Vanzyl and Wingfield, 1998).

In bacterial wilt of banana caused by *Ps. solanacearum*, C_2H_4 is most likely the primary factor of this disease (Freebairn and Buddenhagen, 1964). Bonn et al. (1972) reported that bacterial C_2H_4 production in infected banana was proportional to the virulence of the strain, but Pegg and Cronshaw (1976b) could not support this claim. Chalutz (1979) demonstrated that a non-C_2H_4 -producing mutant of *P. digitatum*, causing rot of orange and lemon fruit, was equally as pathogenic as the wild-type (C_2H_4 -producing), suggesting little or no role for C_2H_4 in pathogenicity. Michniewicz et al. (1983) also observed no correlation between C_2H_4 producting ability of the *Fusarium culmorum* strains and their pathogenicity to wheat seedlings. It is highly likely that under certain conditions, microbial generated C_2H_4 acts as a *prima facie* cause of disease symptoms, whereas in others, it is only a secondary symptom.

7.4.3. Ethylene Production and Development of Disease Symptoms

Ethylene may play an important role in symptom development in the infected tissues, an aspect of disease susceptibility. Usually the triggered C_2H_4 production in response to the host-parasite interaction coincides with or precedes the development of disease symptoms and the subsequent physiological changes in plants or C_2H_4 outburst occurs at later stages of disease development (Table 7.1). Because C_2H_4 production in plant tissues in response to numerous environmental and developmental cues is known to cause necroses (Abeles et al., 1992), host-derived C_2H_4 has been hypothesized to be an important signal for disease development. Chlorosis, senescence, and abscission are well-known responses of plants to C_2H_4 , and in some cases, a clear correlation between C_2H_4 production and pathogen-induced tissue damage has been observed (Ben-David et al., 1986; Boller, 1991; Pegg, 1976b; Stall and Hall, 1984; Yang and Hoffman, 1984). Likewise, studies have shown a correlation between the timing of increased C_2H_4 evolution in response to pathogen infections and the development of chlorotic and wilting symptoms in many plant species (Goto et al., 1980; Pegg, 1981; Stall and Hall, 1984; Ben-David et al., 1986; Boller, 1991; Gentile and Matta, 1975; Pegg and Cronshaw, 1976a,b; Elad, 1990). Many symptoms of fusarial wilt of tomato are those of advanced senescence, suggesting that the disease is a result of C_2H_4 action (Dimond and Waggoner, 1953). However, peak C_2H_4 production coincides with foliar chlorosis only in older leaves of tomato and often precedes the appearance of symptoms in young leaves by several days (Gentile and Matta, 1975; Pegg and Cronshaw, 1976a). Smith et al. (1964) reported that in fusarial disease of tulips, C_2H_4 functions directly as a gaseous, stunting agent and distorts flowers.

The accumulation and increased activities of the C_2H_4 biosynthetic enzymes, ACC synthase and ACC oxidase have been localized in chlorotic tissue directly surrounding primary lesions in tobacco leaves in response to TMV (De Laat and van Loon, 1983a) and *Phytophthora infestans* (Spanu and Boller, 1989b) infections. This may suggest that host C_2H_4 production may be a response to cell death that occurs during primary lesion formation. From the observation that chlorosis spreading beyond the fungal stroma can be well advanced before abscission, Williamson (1950) attributed abscission in rose leaves infected with *Diplocarpon rosae* to C_2H_4 action. Goto et al. (1980) reported that inoculation with *X. compestri* caused a sharp increase in infected citrus leaves and C_2H_4 production was directly related to disease symptom development. Maximum C_2H_4 production coincided with severity of the disease. The rate of defoliation was directly related to the rate of C_2H_4 production and leaves inoculated with higher inoculum doses produced C_2H_4 earlier than those with low inoculum doses. Likewise, Resende et al. (1996) reported that a defoliating isolate of *V. dahliae* induced accumulation of C_2H_4 in newly developed cocoa leaves where the first symptoms generally appeared and C_2H_4 was responsible for the accelerated senescence and defoliation as demonstrated by reversal with the application of silver thiosulfate, an inhibitor of C_2H_4 action. Increased symptoms and C_2H_4 production occurred in the upper leaves and coincided with more intensive colonization of this part of the plant by the pathogen (Resende et al., 1996).

Recently, Ohtsubo et al. (1999) showed the direct involvement of C_2H_4 in the formation of necrotic lesions and in the induction of a basic pathogenesis-related (PR) protein gene in tobacco. Increase in C_2H_4 production as well as accumulation of ACC oxidase gene transcript preceded the lesion appearance in TMV-infected leaves in an N gene-dependent synchronous lesion formation system. Inhibitors of C_2H_4 biosynthesis or action significantly suppressed both lesion formation and the basic PR gene expression. Induction of these genes was enhanced in ACC-treated leaves especially in ACC oxidase-overexpressing transgenic tobacco plants. Further, Ohtsubo et al. (1999) found that C_2H_4 production during a hypersensitive reaction was restricted at the level of ACC oxidase activity. Similarly, Wachter et al. (1999) investigated the role of C_2H_4 in *Agrobacterium tumefaciens*-induced stem tumors of *Ricinus communis*. They observed a substantial release of C_2H_4 from the tumorized stems compared to the non-tumorized stems and accumulation of ACC preceeded C_2H_4 emission by tumorized stems with concomitant changes in the xylem. Wachter et al. (1999) concluded that, in addition to cytokinin and auxins, C_2H_4 plays an important role in the differentiation of *A. tumefaciens* induced tumors (Fig. 7.2).

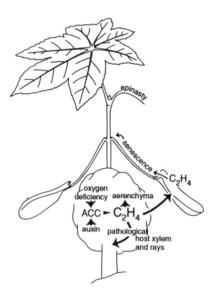

Fig. 7-2. Summarizing diagram of C_2H_4 relations of a stem tumor of *Ricinus communis*. Both auxin and oxygen deficiency induce ACC accumulation in the tumor. ACC is converted to C_2H_4. Ethylene induces the development of aerenchyma in the tumor, thus improving oxygen supply; C_2H_4 -induced anatomical changes of the tumor adjacent host xylem enhance water and nutrient supply to the tumor; the emitted C_2H_4 causes leaf senescence to provide assimilates from the leaves for the tumor as a metabolic sink. Source: Wachter et al. (1999).

7.4.4. Ethylene and Toxins Produced by Pathogens

In many cases, production of toxins by a pathogen is considered responsible for the stress C_2H_4 which subsequently promotes disease development. Fusicoccin, a major phytotoxic metabolite produced by *Fusicoccum amygdali*, which is responsible for many of the pathological symptoms induced by the fungus on peach and almond trees (Ballio et al., 1964), significantly increases C_2H_4 production in detached rice leaves. This stimulatory effect is due to enhancement in ACC conversion to C_2H_4 (Chen and Kao, 1993). Similarly, exogenous application of coronatin, a phytotoxin produced by *Ps.*

syringae (Mitchell, 1982), increased C_2H_4 release from tobacco leaves (Kenyon and Turner, 1992) and in bean leaves (Weingart and Völksch, 1997). Völksch and Weingart (1997) compared various strains of *Ps. syringae* pv. *glycinea* and *Ps. syringae* pv. *phaseolicola* isolated from kudzu (*Pueraria lobata*) and bean (*Phaseolus vulgaris*). They divided these strains into three groups and reported that the *Ps. phaseolicola* strains which are strongly pathogenic to kudzu, bean and soybean also possess the ability to produce phaseolotoxin in addition to C_2H_4 , compared to weak pathogenic pathovars. These studies imply that increased C_2H_4 evolution from the infected tissue may be the indirect effect of the phytotoxins released by the pathogen.

7.4.5. Ethylene Mutants

Controversy remains as to whether the C_2H_4 produced by the host is a signal that acts to amplify cell death during the susceptible response or whether increased C_2H_4 synthesis is simply correlated with advanced disease symptoms and has no biological significance to the process. Because C_2H_4 appears to increase disease severity, insensitivity to C_2H_4 may increase plant tolerance to plant pathogens. Early evidence for this premise has come from studies on pepper with enhanced sensitivity to C_2H_4 that exhibited more rapid development of chlorosis in response to *X. campestris* pv. *vesicola*, the causal agent of caterial spot disease (Stall and Hall, 1984). Investigations with C_2H_4 insensitive mutants of *Arabidopsis* revealed enhanced disease tolerance to *Ps. syringae* pv. *tomato* or pv. *madelicola* and *Xanthomonas campestris* pv. *campestris* pathogens (Bent et al., 1992). Bent et al. (1992) were the first to use a genetic approach to determine whether C_2H_4 perception plays a causal role in disease development. They observed that a mutant, *ein2-1*, impaired in a signal transduction component of the C_2H_4 response (McGrath and Ecker, 1998), showed less macroscopically visible chlorosis and less chlorophyll degradation compared with wild-type *Arapidopsis* plants. However, when the bacteria multiplying in *ein2-1* and wild-type plants were counted, no significant difference was found. It therefore appears that C_2H_4 does not play a role in actual resistance to these bacteria but, rather, in the development of pathogen-induced chlorosis symptoms, suggesting a role for C_2H_4 in disease development during the susceptible response (Bent et al., 1992). Later, Lund et al. (1998) investigated the role for C_2H_4 perception in disease development by using the solanaceous species tomato. They hypothesized that if C_2H_4 is an important host-derived signal that promotes disease development in tomato, then mutant lines impaired in C_2H_4 perception should exhibit reduced disease symptoms when infected by virulent pathogens (Lund et al., 1998). An C_2H_4 –insensitive mutant of tomato, *Never ripe (Nr)*, impaired in C_2H_4 receptors was used by Lund et al. (1998). They observed that the mutant, *Nr* exhibited a significant reduction in disease symptoms in comparison to the wild type after inoculation of both genotypes with virulent bacterial (*X. campestris* pv. *vesicatoria* and *Ps. syringae* pv. *tomato*) and fungal *(Fusarium oxysporum* f. sp. *lycopersici)* pathogens. Bacterial spot disease symptoms were also reduced in tomato genotypes impaired in C_2H_4 synthesis (ACC deaminase) and perception, thereby confirming a reducing effect for C_2H_4 insensitivity on foliar disease development. The reduction in foliar disease symptoms in *Nr* plants was a specific effect of C_2H_4 insensitivity and was not due to reduction in bacterial populations or decreased C_2H_4 synthesis (Lund et al., 1998). Similarly, Wendland and Hoffman (1988) observed a correlation between high rates of C_2H_4 production and increased susceptibility to infection of 40 wheat cultivars and 21 breeding lines of *Septoria nodorum*.

7.5. ROLE OF ETHYLENE IN DISEASE RESISTANCE

There is some evidence suggesting that C_2H_4 has a role in disease resistance. Studies have shown that treatment of plants with exogenous C_2H_4 enhanced the resistance to microorganisms and, conversely, treatment with C_2H_4 inhibitors adversely affected the resistance level of various hosts in a number of plant-pathogen interactions (Esquerre-Tugaye et al., 1979; El-Kazzaz et al., 1983a; Marte et al., 1993). However, for many plant-pathogen combinations, pretreatment with C_2H_4 either had no effect on resistance or actually diminished the resistance level (El-Kazzaz et al., 1983b,c; Brown and Lee, 1993; van Loon and Pennings, 1993). These contradictory results have made the role of C_2H_4 in host defense a controversial subject.

Plants react to pathogen attack by an array of biochemical changes that are thought to represent a defense mechanism against pathogens and C_2H_4 has been implicated in the induction process (Hislop et al., 1973a,b; Archer and Hislop, 1975; Pegg, 1976b; Boller, 1982; Bell, 1981; Ecker and Davis, 1987; van Loon, 1985). Studies have shown that a strong and early induction of C_2H_4 biosynthesis is correlated with the resistance against a pathogen (Montalbini and Elstner, 1977; Boller, 1990). Ethylene is thought to enhance plant resistance against pathogen infection by activating the activities of pathogenesis-related (PR) proteins and enzymes such as chitinase, peroxidase, phenylalanine ammonia-lyase, choline synthase and vacuolar hydrolases, polyphenol oxidase and hydroxyproline-rich cell wall proteins, and/or by accumulation of defense-related compounds such as phenolics, lignins and suberins to suppress multiplication of pathogens in the infected and adjacent tissues (Pritchard and Ross, 1975; Boller, 1991; Vera et al., 1993; Eker and Davis, 1987; van Loon, 1983; Smith-Becker et al., 1998; Broglie et al., 1986; Chappell et al., 1984; Ecker and Davis, 1987; Mauch et al., 1984; Toppan et al., 1982; Deikman, 1997). The premise that C_2H_4 is somehow involved in resistance of plants against infection is further supported by the reports that exogenous application of C_2H_4 to plants activates the genes encoding antimicrobial pathogenesis-related (PR) proteins (Boller et al., 1983; Mauch and Staehelin, 1989; Memelink et al., 1990; Eyal et al., 1992; Beffa et al., 1995; Penninckx et al., 1996; Knoester et al., 1999), or cell wall-strengthening hydroxyproline-rich glycoproteins (Esquerre-Tugaye et al., 1979; Ecker and Davis, 1987; Tagu et al., 1992), or enzymes involved in the synthesis of phenylpropanoids (Ecker and Davis, 1987). Synthesis of these proteins is also observed after pathogen infection or treatment with pathogen-derived elicitors, and roles for these PR proteins in defense have been postulated by various scientists (Dixon and Harrison, 1990; Dixon and Lamb, 1990; Lamb et al., 1989). del Campillo and Lewis (1992) proposed that PR-related proteins were either involved in a cell separation process itself or protected the plant from pathogenic attack.

Treatment of plants with C_2H_4 (C_2H_4 gas or Ethrel) reduces the frequency of several diseases (Stahmann et al., 1966; Lockhart et al., 1968; Hislop et al., 1973b; Ross and Pritchard, 1972; Brown, 1978; El-Kazzaz et al., 1983b; Stermer and Hammerschmidt, 1987). Based upon the observations that plants treated with C_2H_4 gas or Ethrel showed diminished infection or disease symptoms, C_2H_4-induced changes in enzymes that are associated presumably with resistance, and C_2H_4 stimulated biosynthesis of known anti-fungal chemicals, Pegg (1981) supported the hypothesis that C_2H_4 induces disease resistance in host plants.

A number of workers have used *Arabidopsis* mutants for understanding the role of C_2H_4 in developing resistance against pathogenic infection (Penninckx et al., 1998; Lawton et al., 1994; Thomma et al., 1999; Lee et al., 2000). Inoculation of leaves of wild-type plants with an avirulent *Ps. syringae* pv. *tomato* strain was found to trigger a systemic defense response that protected the leaves against subsequent inoculation with either virulent strains of *Ps. parasitica* or *Ps. syringae* pv. *tomato* (Lawton et al., 1995; Pieterse et al., 1998). This systemic response was equally effective in the C_2H_4 – insensitive *etr1-1* mutant (Lawton et al., 1995; Pieterse et al., 1998). On the other hand, Pieterse et al. (1998) observed that inoculating *Arabidopsis* roots with a nonpathogenic root-colonizing strain of *Ps. fluorescence* conferred systemic resistance in wild-type plants but not *etr1-1* mutants to subsequent inoculation of the leaves with a virulent *Ps. syringae* pv. *tomato* strain. Therefore, a systemic resistance response triggered by leaf inoculation with an avirulent bacterium appears to be C_2H_4 –independent, while that induced by inoculating roots with a nonpathogenic bacterium is C_2H_4 –dependent (Pieterse et al., 1998). Root colonization of *Arabidopsis thaliana* by the nonpathogenic, rhizosphere-colonizing, biocontrol bacterium *Ps. flurescens* WCS417r has been shown to elicit induced systemic resistance (ISR) against *Ps. syringae* pv. *tomato* (Pst) (Knoester et al., 1999). Knoester et al. (1999) tested several C_2H_4 –response mutants and showed essentially normal symptoms of Pst infection. ISR was abolished in the C_2H_4 –insensitive mutants, indicating that the expression of ISR requires the complete signal-transduction pathway of C_2H_4 known so far. The induction of ISR by WCS417r was not accompanied by increased C_2H_4 production in roots or leaves, nor by an increase in the expression of the genes encoding the C_2H_4 biosynthetic enzymes, ACC synthase and ACC oxidase. The mutant, displaying C_2H_4 insensitivity in the roots only, did not express ISR upon application of WCS417r to the roots, but did exhibit ISR when the inducing bacteria had infiltrated into the leaves. These results demonstrate that, for the induction of ISR, C_2H_4 responsiveness is required at the site of application of inducing rhizobacteria.

Thomma et al. (1999) observed that inoculation of wild-type *Arabidopsis* plants with the fungus, *Alternaria brassicicola* results in systemtic induction of genes encoding a plant defensin (*PDF1.2*), a basic chitinase (*PR-3*), and an acidic hevein-like protein (*PR-4*). Pathogen-induced induction of these three genes is almost completely abolished in the C_2H_4 –insensitive *Arabidopsis* mutant *ein2-1*. This indicates that a functional C_2H_4 signal transduction component (*EIN2*) is requried in this response. The *ein2-1* mutants were found to be markedly more susceptible than wild-type plants to infection by two different strains of the gray mold fungus *Botrytis cinerea*. In contrast, no increased fungal colonization of *ein2-1* mutants was observed after challenge with avirulent strains of either *Peronospora parasitica* or *A. brassicicola*. The data support the conclusion that C_2H_4 –controlled responses play a role in resistance of *Arabidopsis* to some but not all types of pathogens. Penninckx et al. (1998) investigated the interactions between the C_2H_4 and jasmonate signal pathways for the induction of a defense response. Inoculation of wild-type *Arabidopsis* plants with the fungus, *A. brassicicola* led to a marked increase in production of jasmonic acid, and this response was not blocked in the *ein2-1* mutants. Likewise, *A. brassicicola* infection caused stimulated emission of C_2H_4 in both wild-type plants and in *coi1-1* mutants. However, treatment of either *ein2-1* or *coi1-1* mutants with methyl jasmonate or C_2H_4 did not induce *PDF1.2*, as it did in wild-type plants. Penninckx et al. (1998) concluded from these experiments that both the C_2H_4 and jasmonate signaling pathways need to be triggered concomitantly, and not sequentially, to activate *PDF1.2* upon pathogen infection. Similarly, Lee et al. (2000) hypothesized that both C_2H_4 and slicylic acid may be involved in the signal transduction pathway of pepper defense – or pathogenesis-related plant responses.   It is most likely that C_2H_4 , jasmonate and salicylic acid may function coordinatedly in the induction of systemic resistance. Shetty and Kumar (1999) have also comprehensively discussed the roles of C_2H_4, jasmonate and salicylic acid in signalling of plants during induction of resistance against pathogens.

There is little direct evidence that C_2H_4 plays a causal role in resistance. Many defense proteins are inducible by C_2H_4 -independent pathways (Dixon and Harrison, 1990; Dixon and Lamb, 1990; Lamb et al., 1989) and treatment of plants with inhibitors of C_2H_4 biosynthesis does not necessarily alter the macroscopic responses of plants to pathogens or block synthesis of phytoalexins (Nemestothy and Guest, 1990; Paradies et al., 1980).   Thus, C_2H_4 production may correlate with the induction of resistance responses but may not necessarily cause resistance.

7.5.1. Induction of Pathogenesis-Related (PR) Proteins and Enzymes

Of the many apparent plant defense responses to invasion by various pathogens, one of the most studied is the synthesis of a group of host-encoded proteins referred to as pathogenesis-related (PR) proteins. Ethylene biosynthesis is a characteristic response of plant cells to certain microbial proteins (Anderson et al., 1993, 1997; Bailey et al., 1997a,b; Felix et al., 1991; Fuchs et al., 1989; Hammond-Kosack et al., 1996; Yu, 1995). Pathogenesis-related proteins were first discovered as polypeptides that accumulated in genotypes of tobacco that responded hypersensitively to infection with TMV (Gianinazzi et al., 1970; van Loon and van Kammen, 1970). Since then, many PR proteins have been described as occurring in a wide variety of plant species following infection with various pathogens (Cutt and Klessig, 1992; Strauch et al., 1990; Kim and Hwang, 1994; Lee and Hwang, 1996; van Loon and van Strien, 1999). The PR proteins are not only induced by infections (Vera and Conejero, 1989a; Christ and Mosionger, 1989), but also by ethephon in tobacco (van Loon, 1977, 1983) and in tomato leaves (Vera and Conejero, 1989a,b; Christ and Mosionger, 1989).   Since C_2H_4 is produced in large amounts in the hypersensitive response of tobacco, van Loon (1983) proposed that C_2H_4 is a natural signal for the induction of PRs. Coupe et al. (1997) used *Sambucus nigra* as a model system for investigating temporal and spatial expression of mRNA encoding PR proteins during C_2H_4 promoted abisssion. They observed that PR –mRNAs were not expressed in freshly excised plant material but accumulated primarily in the abscission zone tissue after 18 hours of exposure to C_2H_4 at a time when abscission of the leaflet explants had reached 70%. They concluded that these genes encode PR-proteins that may function to protect the exposed fracture surfaces from pathogenic attack. However, in pepper leaves, PR proteins appear to be induced by TMV both in the presence and absence of induced C_2H_4 synthesis (Tobias et al., 1989).

The PR proteins and genes that encode these proteins have been categorized into five major groups. Three of the PR protein groups, PR-1, PR-3 (chitinases), and PR-5 (osmotins), have now been reported to contain gene members that can convey increased resistance to phytopathogenic fungi when overexpressed in transgenic plants (Broglie et al., 1991; Alexander et al., 1993; Liu et al., 1994). Increasing evidence is now revealing that many members of the PR protein superfamily have antifungal activity in *in vitro* assays (Bol et al., 1990). Bol et al. (1996) have critically reviewed the regulation of the expression of plant defense genes induced as a hypersensitive response of a host to pathogenic infection. Kitajima and Sato (1999) have discussed the molecular mechanisms of gene expression and protein function of PR proteins. Most of the studies on PR proteins have focussed on induction of chitinases, β-1,2-glucanases, phenylalamine ammonia-lyase, and xylanases which will be discussed in detail in the following sections.

7.5.1.1. Chitinases and β-1,3-Glucanases

Since chitin and β-1,3-glucans are important elements of fungal cell walls, the two enzymes, chitinase and β-1,3-glucanase, that hydrolyze them have been hypothesized to function in the defense of plants especially against fungal pathogens (Boller, 1982, 1985, 1987, 1988; Boller and Vogeli, 1984; Abeles and Forrence, 1970; Abeles et al., 1971; Mauch and Staehelin, 1989; Pegg, 1977; van Loon, 1989; Bol et al., 1990; Linthorst, 1991). These "antifungal hydrolases" are induced by several biotic (e.g., infection with pathogens) stress situations (Bowles, 1990).

An increase in chitinase and β-1,3-glucanase in response to C_2H_4 treatment (Pegg, 1976a,b; Abeles et al., 1971; Boller et al., 1983; Broglie et al., 1986; Vogeli et al., 1988; Ishige et al., 1991) and in response to pathogen attack (Nichols et al. 1980; Pegg and Young, 1982; Roby et al., 1989; Roby and Esquerre-Tugaye, 1987) may also induce disease resistance, resulting in lysis of the pathogen (Abeles et al., 1971; Wargo, 1975; Hadwiger and Beckman, 1980). Three lines of indirect evidence support this hypothesis (Pegg, 1977; Boller, 1985; 1987) including (i) chitinase and β-1,3-glucanase are strongly induced in the course of pathogenesis (Pegg and Young, 1981; Mauch et al., 1984, 1988a,b; Yoshikawa et al., 1993) and upon treatments with elicitors (Mauch et al., 1984; Chappell et al., 1984; Kombrink and Hahlbrock, 1986); (ii) chitinase and β-1,3-glucanase degrade isolated fungal cell walls and has been shown to inhibit the growth of some fast-growing fungi due to their antifungal activity (Schlumbaum et al., 1986; Roberts and Selitrennikoff, 1988; Broekaert et al., 1988; Young and Pegg, 1982); and (iii) some pathogenic fungi possess proteins that inhibit plant β-1,3-glucanase (Albersheim and Valent, 1974). In addition, many other fungi which are not inhibited by chitinase alone, are strongly inhibited by the combination of chitinase and β-1,3-glucanase (Mauch et al., 1988b). Taken together, these data indicate that the two hydrolases have strong antifungal properties (Boller, 1991).

In bean and tobacco, the best-studied models, the main C_2H_4-induced chitinases and β-1,3-glucanases are basic homologs of the corresponding PR proteins (Carr and Klessig, 1989; Boller, 1988). Studies with *Sambuscus nigra* plants have revealed that during C_2H_4-stimulated leaflet abscission, the activities of both polygalacturonases and β-1,4-glucanase are increased (Taylor et al., 1993; Webb et al., 1993) and that this correlates with an increase in the expression of genes encoding these enzymes (Taylor et al., 1994; Roberts et al., 1997). Broglie et al. (1986; 1989) isolated cDNA as well as genomic clones for chitinase from bean and tomato and obtained evidence that C_2H_4 regulated the level of chitinase mRNA. They introduced one of the bean chitinase genes, the CHB5 gene into tobacco and found that C_2H_4 strongly induced the bean gene in the transgenic tobacco plants. Comparison of the induction kinetics of chitinase and β-1,3-glucanase in C_2H_4-treated bean leaves demonstrated a closely parallel increase of the enzymatic activities as well as of the levels of translatable mRNA (Vögeli et al., 1988). Upon withdrawal of C_2H_4, a similar parallel decay of mRNA activities was observed with a half-life of about 10 h, whereas enzyme activities continued to increase for about 20 h and then reached a plateau (Vögeli et al., 1988). These data indicate that chitinase and β-1,3-glucanase are coordinately regulated at the level of mRNA.

The chitinase and β,1-3-glucanase were induced and accumulated in the pepper leaves or stems by *Pseudomonas solanacearum*, *Xanthomonas compestris* pv. *vesicatoria* or *Phytophthora capsici* infections, more distinctly in the incompatible than compatible host-parasite interactions (Kim and Hwang, 1994; Lee et al., 1994; Lee and Hwang, 1996). Evidence to support the role of pepper chitinases in defense against pathogenic fungi was based upon the observations that chitinases purified from pepper stems have antifungal activity *in vitro* (Kim and Hwang, 1996), which can be synergistically enhanced by β-1,3-gluconases (Kim and Hwang, 1997). In addition, chitinases and β-1,3-glucanases can also be activated upon exposure to certain elicitors (Mauch et al., 1984; Kombrink and Hahlbrock, 1986; Ebel and Scheel, 1991; Kirsch et al., 1993). Popp et al. (1997) measured constitutive and elicitor-induced levels of C_2H_4 production, and chitinase and glucanase activities in cells of loblolly pine (*Pinus taeda* L.). Increased C_2H_4 production was induced similarly by a live fungus (*Ophiostoma minus* Hudgc. H. P. Sydow) and chitosan as an elicitor. Culture age affected the constitutive level of all defense responses and had a pronounced effect on the ability of the cells to produce C_2H_4 as well as cellular and secreted chitinase and glucanase (Popp et al., 1997).

It has been suggested that in heterotrophic suspension-cultured sunflower cells, the elicitor-induced activity of chitinase and β-1,3-glucanase is antagonistically regulated by C_2H_4 and ACC (Siefert et al., 1994). Experiments with metabolic products, in combination with inhibitors of C_2H_4 biosynthesis (such as the synthetic growth retardant BAS 111W, $CoCl_2$, salicylic acid, and AVG) have demonstrated that the induction process is triggered by an increase in the internal ACC/C_2H_4 ratio. ACC appears to function as a promoting factor, whereas C_2H_4 particularly at higher rates of formation might be inhibitory to elicitor induced enzymatic activity (Siefert et al., 1994). In another study, Siefert and Grossmann (1997) further investigated the regulation of chitinase and β-1,3-glucanase activity induced in the reaction of suspension-cultured sunflower cells to the fungal Pmg elicitor (a fungal elicitor prepared from *Phytophthora megasperma* f. sp. *glycinea*) by internal C_2H_4 and ACC levels. They found that within 5h of elicitor application, the production of C_2H_4 and particularly ACC had been transiently stimulated, depending on the elicitor concentration. The elicitor dose-dependent stimulation in the intracellular chitinase activity showed a continuous increase during the initial 24h of elicitor treatment, while β-1,3-glucanase activity was stimulated only after a lag phase of 5h after treatment. Treatment with 5 μg ml^{-1} elicitor stimulated C_2H_4 and ACC levels 1.6-fold and 4-fold, relative to the control, respectively. The molar ratio of ACC to C_2H_4 changed from approximately 3:1 in controls to 9:1 in treated cells which later on increased to 56:1 in elicitor-treated cells. Treatment with AVG decreased the elicitor-induced enzyme activities and the levels of both C_2H_4 and ACC. Elicitor effects on chitinase and β-1,3-glucanase activities could be fully restored by exogenous application of ACC. Concomitantly, C_2H_4 formation increased slightly, whereas intracellular ACC accumulated considerably, resulting in an increased ACC/C_2H_4 ratio. Neither treatments with ACC alone, which simultaneously increased internal ACC and C_2H_4 levels, nor treatments with AVG alone, which simultaneously reduced ACC and C_2H_4 levels, could stimulate chitinase or β-1,3-glucanase activities in the cells. It was suggested that ACC functions as a promoting factor in the induction of chitianse and β-1,3-glucanase activity triggered by the Pmg elicitor and also reverses the inhibitory influence of C_2H_4 (Siefert and Grossmann, 1997).

Contrarily, Hedrick et al. (1988) found that in bean suspension cultures, the chitinase gene was induced within 15 min by a fungal elicitor. Since induction of C_2H_4 biosynthesis requires at least 15 min in the fastest responding system, chitinase gene activation by the elicitor is almost certainly not caused by C_2H_4 in this system. Mauch et al. (1984) have studied induction of chitinase and β-1,3-glucanase by infection or by elicitor treatment in the absence or presence of AVG. AVG reduced C_2H_4 production below the level of uninfected controls but did not interfere with the induction of chitinase and β-1,3-glucanase in response to infection or elicitors. Ethylene also induces chitinase activity in cucumber leaves, but there appeared to be no ACC transport out of the infected leaves (Metraux and Boller, 1986) and no enhanced C_2H_4 production in the noninfected leaves, indicating that C_2H_4 did not contribute to systemic induction of chitinase in cucumber (Boller, 1991). Endogenous C_2H_4 production appeared to be necessary for the induction of the β-1,3-glucanase enzyme in media without auxin or cytokinin, since cobalt and norbornadiene (inhibitors of C_2H_4 biosynthesis and action, respectively) inhibited the induction of β-1,3-glucanase (Felix and Meins, 1987).

7.5.1.2. Phenylalanine Ammonia-Lyase

The key enzymes of flavonoid biosynthesis, phenylalanine ammonia-lyase (PAL), 4-coumarate:CoA ligase (4-CL), and chalcone synthase (CHS) are known to be involved in defense against parasitic infection (Bell, 1981; Ecker and Davis, 1987; Lawton and Lamb, 1987), since they are strongly induced by pathogen elicitors. In addition, other products of PAL and 4-CL, including polyphenols and lignin, are thought to play a role in the defense response. Ethylene produced by pathogen infection is considered a signal molecule that codes for PAL (Rickey and Belknap, 1991).

Several studies have examined the possible involvement of C_2H_4 in the induction of PAL. PAL has been found to be induced by C_2H_4 in a number of tissues, including citrus fruit peel (Riov et al., 1969; Lisker et al., 1983), pea (Hyodo and Yang, 1971), carrot (Chalutz, 1973) and Swede roots (Rhodes and Wooltorton, 1978). In carrot roots, RNA blot hybridization analysis revealed marked increases in mRNAs of PAL, 4-CL, and CHS in response to elicitors (Ecker and Davis, 1987). Ethephon and ACC were found to induce PAL in rice leaves (Haga et al., 1988). In lettuce, C_2H_4 causes a metabolic disorder called russet spotting, characterized by brown, lignified flecks and an induction of PAL activity (Hyodo et al., 1978; Ke and Saltveit, 1988). Induction of PAL activity resembles a hypersensitive reaction to infection and may be considered a defense reaction. In parsley cells, in which PAL can be induced by an elicitor but not by exogenous C_2H_4, AVG partially reduced elicitor-stimulated C_2H_4 production and elicitor-induced PAL activity, suggesting a link between induction of ACC synthase and PAL (Chappell et al., 1984). ACC but not C_2H_4 reversed the effect of AVG on PAL activity. Apparently the functioning of the ACC pathway is important for PAL induction in parsley cells. Subsequently, experiments with elicitor-treated bean leaves yielded similar results (Hughes and Dickerson, 1989), indicating that even though C_2H_4 does not induce PAL, it enhances induction by elicitors.

In grapefruit flavedo, where PAL can readily be induced by exogenous C_2H_4 (Riov et al., 1969; Lisker et al., 1983), the C_2H_4-producing pathogen *Penicillium digitatum* did not cause induction of PAL. Interestingly, infection even suppressed induction of PAL by exogenous C_2H_4, indicating that the fungus is able to prevent the defense response of the host (Lisker et al., 1983). It should be noted that PAL is frequently not induced by C_2H_4 in plant cells when PAL is strongly induced by pathogensis or elicitors (Chappell et al., 1984).

7.5.1.3. Ethylene Biosynthesis-Inducing Xylanase

The C_2H_4 biosynthesis-inducing xylanase (EIX) is known to be a potent elicitor of C_2H_4 biosynthesis and other responses when applied to plant tissue. EIX was isolated from filtrates of xylan-induced *Trichoderma viride* liquid cultures (Dean and Anderson, 1991) and similar xylanases have been identified in xylan-induced culture filtrates of other plant pathogenic fungi (Dean et al., 1989). EIX elicits enhanced C_2H_4 biosynthesis, altered membrane permeability, and necrosis when applied to leaves of *Nicotiana tabacum* L. cv. *Xanthi* (Bailey et al., 1990, 1991). Xylanase stimulates C_2H_4 synthesis in *N. tabacum* cv. *Xanthi* leaves, which is accompanied by the enhanced accumulation of both ACC synthase and ACC oxidase transcripts (Avni et al., 1994). Ethylene treatment accelerated the levels of C_2H_4 and necrosis produced in tobacco leaves in response to a xylanase elicitor (Bailey et al., 1995). Ethylene-pretreated leaves subsequently treated with xylanase had greatly elevated ACC synthase transcript levels compared to levels in xylanase treated control (air-pretreated) leaves. Bailey et al. (1995) concluded that continual presence of C_2H_4 is required for maintenance of xylanase-induced effects (necrosis and induction of ACC synthase transcript) and that the timing of the induction of C_2H_4 effects and subsequent loss of these effects are closely coordinated whether at the molecular or whole tissue level. Xylanase elicits numerous defense responses when applied to tobacco *Xanthi* leaves (Anderson et al., 1993), including the synthesis of a PR-protein (Lotan and Fluhr, 1990), tissue necrosis, and electrolyte leakage (Bailey et al., 1990). Other fungal endoxylanases elicit defense responses, such as a phytoalexin product (Farmer and Helgeson, 1987), lipid oxidation (Ishii, 1988), vascular galling (VanderMolen et al., 1983), and cell death (Bucheli et al., 1990) of various plant tissues. Most of the responses to EIX identified in *Xanthi* tobacco are characteristics of plant responses observed in cases of other elicitor-treated tissues (Toppan and Esquerre-Tugaye, 1984; DeWit et al., 1985; Kurosaki et al., 1987; Daniel et al., 1990; Keen et al., 1990; Blein et al., 1991; Conrath et al., 1991), as well as of plant resistance to infection by pathogenic organisms (DeWit et al., 1985; Atkinson et al., 1990). Induction of defense responses by EIX is controlled by a single dominant gene in both tobacco and tomato (Bailey et al., 1993).

It is most likely that xylanase-induced C_2H_4 biosynthesis might have a role in defense responses to pathogenic infection.

7.5.1.4. Others

Other enzymes such as peroxidases, catalases, and phenol oxidases are also considered to be involved in resistance mechanisms against pathogenic infection (Fehrmann and Dimond, 1967; Macko et al., 1968; Benedict, 1972). The increase in the enzymes is usually accompanied by accelerated production of C_2H_4. Lovrekovich et al.

(1968) found that infection of tobacco with heat-killed cells of *Pseudomonas tabaci* induced a defense reaction, protecting against subsequent inoculation with live cells, while showing a substantial increase in peroxidase. Lipoxygenase has been proposed to have a regulatory role in a defense response by means of the product it forms from polyunsaturated fatty acids (Preisig and Kuc, 1987). Tobacco cell cultures treated with an elicitor of *Phytophthora parasitica* var. *nicotianae* responded by an early stimulation of C_2H_4 and lipoxygenase activity (Rickauer et al., 1990, 1992).

Another group of defense-related proteins has been identified to be thionins, which are a class of basic and cystein-rich low molecular weight proteins found in a variety of plants (Bohlmann, 1994; Bohlmann and Apel, 1991). The involvement of C_2H_4, salicylic acid and methyl jasmonate in the signal transduction pathway of defense- or pathogenesis-related plant responses for thionin synthases in pepper plants has been demonstrated (Lee et al., 2000). Lee et al. (2000) found that treatment of pepper plants with ethephon strongly induced the expression of genes encoding thionin in pepper cells.

Nep1 is one of the primary polypeptides found in culture filtrates of several *Fusarium* species (Bailey et al., 1997b) and necrosis is one of the visible responses of leaves to Nep1 (Bailey, 1995), the other response is the production of C_2H_4 (Jennings et al., 2000). Jennings et al. (2000) reported that rates of C_2H_4 production by isolated leaves of various plants in response to Nep1 treatment were correlated to the concentration of the protein, Nep1 (Jennings et al., 2000).

Recently, Kim and Hwang (2000) isolated a new basic PR-1 protein from a cDNA library of pepper leaves infected with *Xanthommas compestris* pv. *vesicatoria*, which was intrinsically associated with the induction of C_2H_4 biosynthesis. Treatment with AVG strongly suppressed the induction of PR-1-mRNA expression in the bacteria-infected leaves, suggesting that C_2H_4 may function as a strong signal elicitor in the activation of PR-1 genes, eventually mediating a plant defense response (Kim and Hwang, 2000).

7.5.2. Induction of Phytoalexins and Antibiotics

The accumulation of antibiotic secondary compounds, so-called phytoalexins during pathogenesis is also considered as one of the defense reactions of the host against pathogens. Accumulation of phytoalexins can often be induced not only by living pathogens but also by elicitors. Several studies have demonstrated that exogenous C_2H_4 can induce phytoalexin production in carrot (Chalutz et al., 1969; Coxon et al., 1983; Jaworski and Kuc, 1974; Hoffman and Heale, 1987). In carrot discs infected with *Botrytis cinerea*, treatment with AVG suppressed phytoalexin accumulation and reduced resistance, indicating that endogenous stress C_2H_4 was involved in phytoalexin accumulation (Hoffmann et al., 1988). However, Paradies et al. (1980) showed that AVG did not reduce the phytoalexin response of infected or elicitor-treated soybean plants, implying that C_2H_4 was an indicator but not an inducer of phytoalexin biosynthesis in soybean. It is very likely that when C_2H_4 is not an inducer of phytoalexin production in plant tissues, C_2H_4 will enhance the plant response to phytoalexins (Boller, 1991). In many cases, although phytoalexins have been found to accumulate in response to C_2H_4, the amounts of phytoalexins accumulated are often much smaller than that observed in infected or elicitor-treated plants (Chalutz and Stahmann, 1969; Uegaki et al., 1980). However, the induction of the phytoalexin, 6-methoxymellein in carrot is strongly stimulated by exposure to C_2H_4 (Chalutz et al., 1969; Coxon et al., 1983; Jaworski and Kuc, 1974; Hoffman and Heale, 1987). Similarly, C_2H_4 stimulates the antifungal agent,

6-methoxy-8-hydroxy-3,4-dihydroisocoumarin (MMHD) in carrot, with enhanced production being proportional to C_2H_4 exposure time (Carlton et al., 1961; Chalutz et al., 1969). Chalutz and Stahmann (1969) showed that pisatin may be induced by C_2H_4 treatment. In addition, C_2H_4 exposure increases the antifungal agents, phytuberin, phytuberol, rishitin, and lubimin in treated tissues of potato (Henfling et al., 1978; Alves et al., 1979). Evidence suggests an important role for C_2H_4 in disease resistance; however, timing of exposure in relation to infection is critical (Pegg, 1981). Potato tuber slices pretreated with ethephon (Henfling et al., 1978) or C_2H_4 (Alves et al., 1979) formed phytoalexins more rapidly than control slices when treated with an elicitor. Thus, C_2H_4 appeared to sensitize the tissue to react more rapidly to fungal elicitors; however, Boller (1982) reported that C_2H_4 has a sensitizing effect only in tubers stored for long periods of time (>4 months); tubers freshly harvested or stored for two months are less sensitive to elicitors when treated with C_2H_4 (Boller, 1982). Among the plant genes activated in the course of a pathogen attack are those that code for the hydroxyproline-rich glycoproteins (HPRGP), also called extensins. Although accumulation of HPRGP is often considered a defense reaction, their function in defense is not clear (Cassab and Varner, 1988). HPRGP are also induced in response to treatment with exogenous C_2H_4 (Esquerre-Tugaye et al., 1979), a process shown to involve gene activation in carrot discs (Ecker and Davis, 1987). Stermer and Hammerschmidt (1987) found that cell walls of heat-shocked cucumber seedlings became more resistant to degradation, an effect that did not appear to depend on lignification. However, C_2H_4 biosynthesis and extensin accumulation were induced in parallel.

In conifers, the development of a monoterpene-soaked lesion is thought to be a defensive response that inhibits the colonization and spread of *Ophiostoma* fungi vectored by bark beetles (Berryman, 1972). Artificial inoculation with different species of *Ophiostoma* fungi has indicated that the size and monoterpene concentration of lesions increases with increasing fungal virulence (Paine, 1984; Cook and Hain, 1985, 1986; Paine and Stephen, 1987; Owen et al., 1987; Harrington, 1993; Popp et al., 1995a), suggesting the presence of a regulatory mechanism. Exogenous application of ethephon is also known to increase the concentration of terpenoid compounds in pines (Peters and Roberts, 1977; Peters et al., 1978), implying that C_2H_4 regulates the production of monoterpenes in pine stem lesions following invasion by bark beetle vectored fungi (Popp et al., 1995a). Popp et al. (1995b) determined the rate of C_2H_4 production, monoterpene concentration, and size of induced lesions following inoculation of stems of slash pine and loblolly pine with *Ophiostoma* fungi. Fungal inoculation significantly increased the rate of C_2H_4 production compared with that of stems inoculated with sterile water and greater rates of production were associated with greater fungal virulence. Changes in monoterpene concentration generally mimicked changes in the rate of C_2H_4 production, i.e., high monoterpene concentrations were associated with high rates of C_2H_4 production (Popp et al., 1995b).

7.5.3. Induction of Structural and Developmental Changes

As discussed in Chapters 1 and 2, C_2H_4 affects plant structure and development in a variety of ways. Many of these well-known C_2H_4 effects, for example, epinasty or stunted growth, also frequently occur as symptoms of diseases. These C_2H_4 –induced symptoms may or may not have any direct significance in the plant's defense against pathogens. Several workers have suggested that induction of C_2H_4 production as a result of pathogen invasion may be a plant defense response, since it leads to the premature abscission of the tissue, which helps retard the spread of the pathogen into the plant system (Tzeng and DeVay, 1985; Ketring and Malouk, 1982; Goto et al., 1980; Ben-David et al., 1986; Martin et al., 1988; Boller, 1991). Some pathogens are capable of moving through the plant vascular system to induce disease symptoms in infected tissues, whereas some plant responses to infection such as hypersensitive necrosis are believed to be defense responses used by plants to limit progression of the pathogen into healthy tissues (Hahn et al., 1989). Exogenously applied C_2H_4 provokes premature abscission of leaves, fruits, or flowers in many model systems (Sexton and Roberts, 1989; Osborne, 1989). Similarly, in citrus leaves infected with *Xanthomonas citri* (Goto et al., 1980) and in pepper leaves infected with *X. campestris* (Ben-David et al., 1986), C_2H_4 production was correlated with abscission. In peanut leaflets infected with *Cercospora arachidicola*, abscission was correlated with the ability of plant genotypes to produce pathogen-induced C_2H_4 (Ketring and Melouk, 1982). Tzeng and DeVay (1985) found that in cotton plants infected with various strains of *Verticillium dahliae*, abscission was related to C_2H_4 production in response to the pathogens (Goto et al., 1980; Ben-David et al., 1986; Ketring and Malouk, 1982; Tzeng and DeVay, 1985; Martin et al., 1988). Induction of cellulase plays a key role in C_2H_4-induced abscission. In the abscission zones, antifungal hydrolases like chitinase are induced in addition to cellulase (Gomez et al., 1987). One possible explanation for this effect is that the scars remaining after abscission need special biochemical protection against pathogens.

Localized lignification and apposition of new cell wall material (Bell, 1981) are typical defense responses against various pathogens (Bell, 1981; Vance et al., 1980). These structural reinforcements of the wall may hinder penetration by fungi as well as spread of viruses (Bell, 1981; Vance et al., 1980). Lignification may be stimulated by C_2H_4 in certain tissues, e.g., in Swede roots (Rhodes and Wooltorton, 1973, 1978). Induction of PAL (Ecker and Davis, 1987; Stahmann et al., 1966; Ke and Saltveit, 1988) and of peroxidase may be important for enhanced lignification. The lignification response observed within 3 to 4 h of peeling did not occur in the presence of Ag^+, Co^{2+}, or AVG (Geballe and Galston, 1983). The inhibitory effect of AVG on the development of resistance was reversed by the addition of ACC (Geballe and Galston, 1982). These results indicate that stress C_2H_4 is necessary for lignification and the development of resistance to cellulase in response to wounding (Vance et al., 1980).

Formation of vascular gels or "gums" may be a local response to infection. Functionally, gummosis resembles abscission, isolating infection sites and preventing the invasion of pathogens into the plant (Olien and Bukovac, 1982). Similarly, involvement of pathogen induced C_2H_4 biosynthesis in the formation of vascular gels has been demonstrated. Exogenously-applied C_2H_4 promotes gum formation in the gum ducts of some woody plants, including *Prunus* species (Olien and Bukovac, 1982) and induces the formation of vascular gels in castor bean (VanderMollen et al., 1983). It has been also proposed that C_2H_4 may have a role in another defense response, i.e., compartmentalization (Shain and Miller, 1988).

7.5.4. Summary

Ethylene may be involved not only in the induction of defense reactions, but it may also increase disease severity in certain plant-pathogen interactions. In fact, some of the C_2H_4 effects summarized in the foregoing discussion as "defense reactions" are frequently considered as grave symptoms of disease. Recently, however, conclusive evidence has been presented that C_2H_4 is indeed involved in host resistance, albeit only to a particular class of pathogens and not to others, thus reconciling previous conflicting data (Knoester et al., 1998; Hoffman et al., 1999; Lund et al., 1998). Knoester et al. (1998) made use of transgenic tobacco plants transformed with a dominant-negative mutant allele of the *Arabidopsis* C_2H_4 receptor gene ETR1. The transgenic plants with a disrupted C_2H_4 response were more susceptible than wild-type plants to normally nonpathogenic soil-borne *Pythium* spp., whereas their level of resistance to TMV was unaffected. Hoffman et al. (1999) found that some soybean mutants with reduced C_2H_4 sensitivity had a tendency toward more severe symptoms compared with wild-type plants when challenged with virulent strains of the fungi, *Septoria glycines* and *Rhizoctonia solani* and some but not all avirulent strains of *Phytophthora sojae*. On the other hand, some of the C_2H_4 –insensitive soybean mutants showed less-severe chlorotic symptoms relative to their wild-type parents upon inoculation with virulent strains of *Pseudomonas syringae* pv. *glycinea*. Less-severe chlorosis was also observed in the C_2H_4 –insensitive *Never ripe* tomato strain compared with wild type plants when inoculated with either *Xanthomonas campestris* pv. *vesicatoria* or *Pseudomonoas syringae* pv. *tomato*. In addition, the *Never ripe* tomato mutants also showed less-severe wilting symptoms upon challenge with the fungal vascular pathogen, *Fusarium oxysporum* f. sp. *lycopersici* (Lund et al., 1998). Ethylene is known to promote events such as chlorophyll degradation (Stall and Hall, 1984) and xylem occlusion (VanderMolen et al., 1983), which are positively correlated with severity of disease symptoms such as chlorosis and wilting, respectively.

Exogenously applied C_2H_4 promotes disease severity of grey mold, *Botrytis cinerea* on rose and carnation flowers (Elad, 1988) and on detached leaves of tomato, pepper, bean, and cucumber (Elad, 1990). Prevention of synthesis or action of endogenous C_2H_4 by spraying tissues with AVG or silver thiosulfate significantly decreased severity (Elad, 1988, 1990). Similarly, C_2H_4 treatment has been reported to stimulate disease development in green tangerines infected with *Colletotrichum gloeosporiodes* (Brown and Barmore, 1977) and strawberries exposed to *B. cinerea* (El-Kazzaz et al., 1983a). In conclusion, it appears that C_2H_4 controls both disease resistance responses and symtpoms expression; therefore, this hormone can influence specific plant-pathogen interactions in different ways, depending on the offensive strategies of the pathogen, the efficacy of the defense genes, and the nature of the physiological reactions that are triggered by the pathogen. This may imply that under certain conditions by regulating C_2H_4 synthesis during pathogenesis, damage caused by the disease could be minimized.

7.6. SOIL ETHYLENE AS A WEEDICIDE

Several species of parasitic weeds are considered extremely damaging to host crops. For instance, *Striga* species (*S. asiatica*, *S. aspera*, *S. hermonthica* and *S. gesnerioides*) may cause 100% yield losses in cereals and 83% in cowpea (Lagoke et al., 1991; Aggarwal and Ouedrago, 1989). It is estimated that, on an average, 40% loss in crop production in Africa is due to *Striga* spp. and this accounts for an annual loss in agricutlural revenue of $7 billion (Robson and Broad, 1989). Among several methods suggested for control of these parasites in agricultural systems, the most effective of these is soil injection of C_2H_4 gas as an eradication program for *S. asiatica* as demonstrated in two southeastern U.S. states (Eplee, 1975; Sand and Manley, 1990).

A control program in North Carolina for eradication of *Striga* based on the injection of C_2H_4 into the soil was instituted (Eplee, 1975) and several thousand hectares were treated (Egley, 1982). The objective was to cause suicidal germination of *Striga* seed in the absence of obligate host roots. Three annual applications of C_2H_4 in sequence along with control of escaped plants achieved eradication in the treated areas. Partial eradication has also been achieved at various sites in Africa with the same treatments (Bebawi and Eplee, 1986).

As discussed in Chapters 3 and 4, *Pseudomonas syringae* pathovar *glycinea* synthesizes relatively large amounts of C_2H_4. Very recently, Berner et al. (1999) used these C_2H_4 –producing bacteria for stimulation of *Striga* spp. seed germination. In a laboratory study, they tested various strains of *Ps. syringae* pv. *glycinea* (Psg strains) for their efficacy in stimulating germination of seeds of *Striga* spp. For comparison, they also included C_2H_4 gas, a root exudate (known to stimulate germination) and a synthetic stimulant for germination of *Striga* spp. They found that Psg strains were consistently better stimulators of germination of the parasite seeds than C_2H_4 gas, or the root exudate (Table 7.4). The Psg strains were also as effective in germinating *S. aspera* and *S. gesnerioides* as the synthetic stimulant. They concluded that C_2H_4 –producing bacteria may provide a practical means as a soil amendment of biological control of *Striga* spp. in Africa and other locations (Berner et al., 1999).

7.7. CONCLUDING REMARKS

Studies have provided evidence supporting the involvement of C_2H_4 in both disease development and disease resistance, most likely depending upon the sensitivity of plant tissues to C_2H_4, biochemical changes occurring in the infected tissues and the genetic and physiological responses activated by C_2H_4 or pathogenic infection. This information can be used in breeding of crops which discourage infection by regulating C_2H_4 synthesis. For instance, biotechnologists have successfully developed transgenic plants by using ACC deaminase genes (see Section 8-3) which have reduced the ability of synthesizing C_2H_4. Moreover, microorganisms with ACC-deaminase activity (see Section 3.3.1) can be used as biocontrol agents in cases where C_2H_4 enhances the disease severity. The presence of these microflora in the rhizoplane can lead to lowering the ACC levels of the host plant roots, thus contributing to decreasing C_2H_4 generation. Buchenauer (1998) has highlighted the role of rhizobacteria with ACC-deaminase activity in the biological control of soil-borne diseases. On the other hand, where C_2H_4 production is in favor of defense against infection, the plants with greater capacity of C_2H_4 production could be constructed by using bacterial C_2H_4-forming enzyme genes (see Section 8.3.3). Similarly, transgenic plants varying in sensitivity against C_2H_4 have been developed (Hoffman et al., 1999; Konoester et al., 1998; Bent et al., 1992). These developments can provide opportunities of exploiting the endogenous C_2H_4 potential of the host to produce stress C_2H_4 in favor of controlling parasitic infections and ultimately decreasing the damage/loss caused by plant diseases.

Table 7.4. Germination of *Striga* spp. seeds stimulated by strains of *Pseudomonas syringae* pv. *glycinea* (Psg), ethylene gas, a synthetic germination stimulant (GR24), and roots of a cowpea (*Vigna unguiculata*) cultivar.

(% Seed germination)*

Striga species	Psg strains			Ethylene (2.5 mg)	GR24 (10mg/liter)	Cowpea roots (1g)	Mean
	16/83	19/84	8/83				
S. aspera	10.3	10.0	10.1	7.1	12.8	8.1	9.7
S. gesnerioides	14.2	14.5	13.1	14.1	13.4	13.8	13.8
S. hermonthica	14.6	14.8	13.6	2.0	44.1	7.2	16.0
Mean	13.1	13.1	12.3	7.7	23.4	9.7	13.2

*Least squares means of actual percentage of germination that was adjusted, through analysis of covariance, using viability of seeds of the three species in the three experiments as a coavriate in the analysis.
Source: Berner et al. (1999).

7.8. REFERENCES

Abeles, F. B., and Forrence, L. E., 1970, Temporal and hormonal control of β-1,3-glucanase in *Phaseolus vulgaris* L., *Plant Physiol.* **45**:395-400.

Abeles, F. B., Morgan, P. W., and Saltveit, M. E., Jr., 1992, *Ethylene in Plant Biology*, Academic Press, San Diego, CA, pp. 414.

Abeles, F. B., Bosshart, R. P., Forrence, L. E., and Habig, W. H., 1971, Preparation and purification of glucanase and chitinase from bean leaves, *Plant Physiol.* **47**:129-134.

Achilea, O., Chalutz, E., Fuchs, Y., and Rot, I., 1985a, Ethylene biosynthesis and related physiological changes in *Penicillium digitatum*-infected grapefruit (*Citrus paradisi*), *Physiol. Plant Pathol.* **26**:125-134.

Achilea, O., Fuchs, Y., Chalutz, E., and Rot, I., 1985b, The contribution of host and pathogen to ethylene biosynthesis in *Penicillium digitatum*-infected citrus fruit, *Physiol. Plant Pathol.* **27**:55-63.

Aggarwal, V. D., and Ouedrago, J. T., 1989, Estimation of cowpea yield loss from *Striga* infestation, *Trop. Agric.* **66**:91-92.

Albersheim, P. and Valent, B. S., 1974, Host-pathogen interactions. VII. Plant pathogens secrete proteins which inhibit enzymes of the host capable of attacking the pathogen, *Plant Physiol.* **53**:684-687.

Alexander, C., Goodman, R. M., Gui-Rella, M., Glascock, C., Weymann, K., Friedrich, L., Maddox, D., Ahi-Goy, P., Luntz, T., Ward, E., and Ryals, J., 1993, Increased tolerance to two oomycete pathogens in transgenic tobacco expressing pathogenesis-related protein, *Proc. Natl. Acad. Sci. USA* **90**:7327-7331.

Alves, L. M., Heisler, E. G., Kissinger, J. C., Patterson, J. M., III, and Kalan, E. B., 1979, Effects of controlled atmospheres in production of sesquiterpenoid stress metabolites by white potato tuber, *Plant Physiol.* **63**:359-362.

Anderson, J. D., Bailey, B. A., Taylor, R., Sharon, A,. Avni, A., Mattoo, A. K., and Fuchs, Y., 1993, Fungal xylanase elicits ethylene biosynthesis and other defense responses in tobacco, in: *Cellular and Molecular Aspects of the Plant Hormone Ethylene*, J. G. Pech, Latche, A., and Balague, C., eds., Kluwer Academic Publ., Dordrecht, The Netherlands, pp. 197-204.

Anderson, J. D., Cardinale, F. C., Jennings, J. C., Norman, H. A., Avni, A., Hanania, U., and Bailey, B. A., 1997, Involvement of ethylene in protein elicitor-induced plant responses, in: *Biology and Biotechnology of the Plant Hormone Ethylene*, A. K. Kanellis et al., eds., Kluwer Academic Publishers, Dordrecht, pp. 267-274.

Andrews, E., Hanowski, C., and Beiderbeck, R., 1981, Wachstumsveränderungen an Maiskeimlingen nach Befall mit haploiden Linien des Maisbrandes (*Ustilago maydis*), *Phytopath. Z.* **102**:10-20.

Archer, S. A., and Hislop, E. C., 1975, Ethylene in host-pathogen relationships, *Ann. Appl. Biol.* **81**:121-126.

Atkinson, M. M., Keppler, I. D., Orlandi, E. W., Baker, C. J., and Mischke, C. F., 1990, Involvement of plasma membrane calcium influx in bacterial induction of the K^+/H^+ and hypersensitive responses in tobacco, *Plant Physiol.* **92**:215-221.

Avni, A., Bailey, B. A., Mattoo, A. K., and Anderson, J. D., 1994, Induction of ethylene biosynthesis in *Nicotiana tabacum* by a *Trichoderma viride* xylanase is correlated to the accumulation of 1-aminocyclopropane-1-carboxylic acid (ACC) synthase and ACC oxidase transcripts, *Plant Physiol.* **106**:1049-1055.

Bailey, B. A., 1995, Purification of a protein from culture filtrates of *Fusarium oxysporum* that induces ethylene and necrosis in leaves of *Erythroxylum coca*, *Phytopathology* **85**:1250-1255.

Bailey, B. A., Dean, J. F. D., and Anderson, J. D., 1990, An ethylene biosynthesis-inducing endoxylanase elicits electrolyte leakage and necrosis in *Nicotiana tabacum* cv. *xanthi* leaves, *Plant Physiol.* **94**:1849-1854.

Bailey, B. A., Taylor, R., Dean, J. F. D., and Anderson, J. D., 1991, Ethylene biosynthesis-inducing endoxylanase is translocated through the xylem of *Nicotiana tabacum* cv. *xanthi* plants, *Plant Physiol.* **97**:1181-1186.

Bailey, B. A., Korcak, R. F., and Anderson, J. D., 1993, Sensitivity to an ethylene biosynthesis-inducing endoxylanase in *Nicotiana tabacum* L. cv. *xanthi* is controlled by a single dominant gene, *Plant Physiol.* **101**:1081-1088.

Bailey, B. A., Avni, A., and Anderson, J. D., 1995, The influence of ethylene and tissue age on the sensitivity of Xanthi tobacco leaves to a *Trichoderma viride* xylanase, *Plant Cell Physiol.* **36**:1669-1676.

Bailey, B. A., Jennings, J. C., and Anderson, J. D., 1997a, Sensitivity of *Erythroxylum coca* var. *coca* to ethylene and fungal proteins, *Weed Sci.* **45**:716-721.

Bailey, B. A., Jennings, J. C., and Anderson, J. D., 1997b, The 24 kDa protein from *Fusarium oxysporum* f. sp. *erythroxyli*: Occurrence in related fungi and the effect of growth medium on its production, *Can. J. Microbiol.* **43**:45-55.

Balazs, E., Gaborjanyi, R., Toth, A., and Kiraly, Z., 1969, Ethylene production in xanthi tobacco after systemic and local virus infections, *Acta Phytopathol. Acad. Sci. Hung.* **4**:355-358.

Ballio, A., Chain, E. B., De Leo, P., Erlanger, B. F., Mauri, M., and Tonolo, A., 1964, Fusicoccin, a new wilting toxin produced by *Fusicoccum amygdali* Del., *Nature* **203**:297.

Barkai-Golan, R., and Kopeliovitch, E., 1983, Induced ethylene evolution and climacteric-like respiration in *Rhizopus*-infected *rin* and *nor* tomato mutants, *Physiol. Plant Pathol.* **22**:357-362.

Barkai-Golan, R., and Lavy-Meir, G., 1989, Effects of ethylene on the susceptibility to *Botrytis cinerea* infection of different tomato genotypes, *Ann. Appl. Biol.* **114**:391-396.

Barmore, C. R., and Brown, G. E., 1985, Influence of ethylene on increased susceptibility of oranges to *Diploidia natalensis*, Plant Dis. **69**:228-230.

Bebawi, F. F., and Eplee, R. E., 1986, Efficacy of ethylene as a germination stimulant of *Striga hermonthica* seed, *Weed Sci.* **34**:694-698.

Beffa, R., Szell, M., Meuwly, P., Pay, A., Vogeli-Lange, R., Metraux, J.-P., Neuhaus, G., Meins, F., and Nagy, F., 1995, Cholera toxin elevates pathogen resistance and induces pathogenesis-related gene expression in tobacco, *EMBO J* **14**:5753-5761.

Beijersbergen, J. C. M., and Bergman, B. H. H., 1973, The influence of ethylene on the possible resistance mechanism of the tulip (*Tulipa* spp.) against *Fusarium oxysporum, Acta Bot. Neerl.* **22**:172.

Bell, A.A., 1981, Biochemical mechanisms of disease resistance, *Annu. Rev. Plant Physiol.* **32**:21-81.

Belles, J. M., and Conejero, V., 1989, Evolution of ethylene production, ACC and conjugated ACC levels accompanying symptoms development in tomato and *Gynjura aurantiaca* DC leaves infected with Citrus exocortis viroid (CEV), *J. Phytopathol.* **127**:81-85.

Belles, J. M., Granell, A., Duran-Vila, N., and Conejero, V., 1989, ACC synthesis as the activated step responsible for the rise of ethylene production accompanying citrus exocortis viroid infection in tomato plants, *Phytopath.* **125**:198-208.

Benedict, W. G. , 1972, Influence of light on peroxidase activity associated with resistance of tomato cultivars to *Septoria lycopersici, Can. J. Bot.* **50**:1931-1936.

Ben-David, A., Bashan, Y., and Okon, Y., 1986, Ethylene production in *pepper (Capsicum annuum)* leaves infected with *Xanthomonas campestris* pv. *vesicatoria, Physiol. Mol. Plant Pathol.* **29**:305-316.

Bent, A. F., Innes, R. W., Ecker, J. R., and Staskawicz, B. J., 1992, Disease development in ethylene-insensitive *Arabidopsis thaliana* infected with virulent and avirulent *Pseudomonas* and *Xanthomonas* pathogens, *Mol. Plant-Microbe Interactions* **5**:372-378.

Berner, D. K., Schaad, N. W., and Völksch, B., 1999, Use of ethylene-producing bacteria for stimulation of *Striga* spp. seed germination, *Biol. Control* **15**:274-282.

Bernheim, F., 1942, The effect of various substances on the oxygen uptake of *Blastomyces dermatiditis, J. Bacteriol.* **44**:533-539.

Berryman, A. A., 1972, Resistance of conifers to invasion by bark beetle fungus associatons, *BioSci.* **22**:598-602.

Blein, J. P., Milat, M. L., and Ricci, P., 1991, Responses of cultured tobacco cells to cryptogein, a proteinaceous elicitor from *Phytophthora cryptogea, Plant Physiol.* **95**:486-491.

Bohlmann, H., 1994, The role of thionins in plant protection, *Critical Rev. Plant Sci.* **13**:1-16.

Bohlmann, H., and Apel, K., 1991, Thionins, *Annu. Rev. Plant Physiol. Plant Molecu. Biol.* **42**:227-240.

Bol, J. F., Buchel, A. S., Knoester, M., Baladin, T., Van Loon, L. C., and Linthorst, H. J. M., 1996, Regulation of the expression of plant defence genes, *Plant Growth Regula.* **18**:87-91.

Bol, J. F., Linthorst, H. J. M., and Cornelissen, B. J. C., 1990, Plant pathogenesis-related proteins induced by virus infection, *Annu. Rev. Phytopathol.* **28**:113-138.

Boller, T., 1982, Ethylene-induced biochemical defenses against pathogens, in; *Plant Growth Substances*, P. F. Wareing, ed., Academic Press, London, pp. 303-312.

Boller, T., 1985, Induction of hydrolases as a defense reaction against pathogens, in: *Cellular and Molecular Biology of Plant* Stress, J. L. Key and T. Kosuge, eds., Alan R. Liss, Inc., New York, pp. 247-262.

Boller, T., 1987, Hydrolytic enzymes in plant idsease resistance, in: *Plant-Microbe Interactions*, Vol. 2, T. Kosuge and Nester, E. W. Nester, eds., Macmillan, New York, pp. 385-413.

Boller, T., 1988, Ethylene and the regulation of antifungal hydrolases in plants, in: *Oxford Surveys of Plant Molecular and Cell Bio*logy, Vol. 5, B. J. Miflin, ed., Oxford University Press, Oxford, pp. 145-174.

Boller, T. , 1990, Ethylene and plant-pathogen interactions, in: *Polyamines and Ethylene: Biochemistry, Physiology, and Interactions*, H. E. Flores, Arteca, R. N., and Shannon, J. C., eds., Am. Soc. Plant Physiologists, Rockville, MD, pp. 138-145.

Boller, T., 1991, Ethylene in pathogenesis and disease resistance, in: *The Plant Hormone Ethylene*, A. K. Mattoo and Suttle, J. C., eds., CRC Press, Boca Raton, FL, pp. 293-314.

Boller, T., and Vogeli, U., 1984, Vacuolar localization of ethylene-induced chitinase in bean leaves, *Plant Physiol.* **74**:442-444.

Boller, T., Gehri, A., Mauch, F., and Vogeli, U., 1983, Chitinase in bean leaves: Induction by ethylene, purification, properties, and possible function, *Planta* **157**:22-31.

Bonn, W. G., Selqueira, L., and Upper, C. D., 1972, Determination of the rate of ethylene production by *Pseudomonas solanacearum*, *Proc. Can. Phytopathol. Soc. No. 39.*

Bowles, D. J., 1990, Defense-related proteins in higher plants, *Annu. Rev. Biochem.* **59**:873-907.

Broekaert, W. F., van Parijs, J., Allen, A. K., and Peumans, W. J., 1988, Comparison of some molecular, enzymatic and antifungal properties of chitinases from thorn-apple, tobacco and wheat, *Physiol. Mol. Plant Pathol.* **33**:319-331.

Broglie, K. E., Gaynor, J. J., and Broglie, R. M., 1986, Ethylene-regulated gene expression: Molecular cloning of the genes encoding an endochitinase from *Phaseolus vulgaris*, *Proc. Natl. Acad. Sci. USA* **83**:6820-6824.

Broglie, K. E., Biddle, P., Cressman, R., and Broglie, R. M., 1989, Functional analysis of DNA sequences responsible for ethylene regulation of a bean chitinase gene in transgenic tobacco, *Plant Cell.* **1**:599-607.

Broglie, K., Chet, I., Hollidat, M., Cressman, R., BiddleP., Knowlton, S., Mauvais, C. J., and Broglie, R., 1991, Transgenic plants with enhanced resistance to the fungal pathogen *Rhizoctonia solani*, *Science* **254**:1194-1197.

Brooks, C., 1944, Stem end rot of oranges and factors affecting its control, *J. Agric. Res.* **68**:363-381.

Brown, G. E., 1978, Hypersensitive response of orange-colored robinson tangerines to *Colletotrichum gloeosporioides* after ethylene treatments, *Phytopathol.* **68**:700-706

Brown, E. G., and Barmore, C. R., 1977, The effect of ethylene on susceptibility of Robinson tangerines to anthracnose, *Phytopathol.* **67**:120-123.

Brown, G. E., and Lee, H. S., 1993, Interactions of ethylene with citrus stem-end rot caused by *Diplodia natalensis*, *Phytopathol.* **83**:1204-1208.

Bucheli, P., Doares, S. H., Albersheim, P., and Darvill, A., 1990, Host-pathogen interactions XXXVI. Partial purification and characterization of heat-labile molecules secreted by the rice blast pathogen that solubilize plant cell wall fragments that kill plant cells, *Physiol. Mol. Plant Pathol.* **36**:159-173.

Buchenauer, H., 1998, Biological control of soil-borne diseases by rhizobacteria, *J. Plant Diseases and Protection* **105**:329-348.

Canfield, M. L., and Moore, L. W., 1983, Production of ethylene by *Daucus carota* inoculated with *Agrobacterium tumefaciens* and *Agrobacterium rhizogenes*, *Z. Pflenzenphysiol.* **112**:471.

Carlton, B. C., Peterson, C. E., and Tolbert, N. E., 1961, Effects of ethylene and oxygen on production of a bitter compound by carrot roots, *Plant Physiol.* **36**:550-557.

Carr, J. P., and Klessig, D. F., 1989, The pathogenesis-related proteins of plants, in: *Genetic Engineering, Principles and Methods*, Vol. II, J. K. Satlow, ed., Plenum Press, New York, pp. 65-109.

Cassab, G. I., and Varner, J. E., 1988, Cell wall proteins, *Annu. Rev. Plant Physiol. Plant Mol. Biol.* **39**:321-353.

Chalutz, E., 1973, Ethylene-induced phenylalanine ammonia-lyase activity in carrot roots, *Plant Physiol.* **51**:1033-1036.

Chalutz, E., 1979, No role for ethylene in the pathogenicity of *Penicillium digitatum*, *Physiol. Plant Pathol.* **14**:259-262.

Chalutz, E., and DeVay, J. E., 1969, Production of ethylene *in vitro* and *in vivo* by *Ceratocystis fimbriata* in relation to disease development, *Phytopathol.* **59**:750-755.

Chalutz, E., and Stahmann, M. A., 1969, Induction of pisatin by ethylene, *Phytopathol.* **59**:1972-1973.

Chalutz, E., DeVay, J. E., and Maxie, E. C.,1969, Ethylene-induced isocoumarin formation in carrot root tissue, *Plant Physiol.* **44**:235-241.

Chalutz, E., Lieberman, M., and Sisler, H. D., 1977, Methionine-induced ethylene production by *Penicillium digitatum*, *Plant Physiol.* **60**:402-406.

Chappell, J., Hahlbrock, K., and Boller, T., 1984, Rapid induction of ethylene biosynthesis in cultured parsley cells by fungal elicitor and its relationship to the induction of phenylalanine ammonia-lyase, *Planta* **161**:475-480.

Chen, C. T., and Kao, C. H., 1993, Characteristics of fusicoccin-induced production of ethylene in detached rice leaves, *Plant Physiol. Biochem.* **31**:121-124.

Chou, T. W., and Yang, S. F., 1973, The biogenesis of ethylene in *Penicillium digitatum*, *Arch. Biochem. Biophys.* **157**:73-82.

Christ, U., and Mosionger, E., 1989, Pathogenesis-related proteins of tomato. I. Induction by *Phytophthora infestans* and other biotic and abiotic inducers and correlations with resistance, *Physiol. Mol. Plant Pathol.* **35**:53-65.

Coleman, L. W., and Hodges, C. F., 1986, The effect of methionine on ethylene and 1-aminocyclopropane-1-carboxylic acid production by *Bipolaris sorokiniana*, *Phytopathol.* **76**:851-855.

Coleman, L. W., and Hodges, C. F., 1987, Ethylene biosynthesis in *Poa pratensis* leaves in response to injury or infection by *Biopolaris sorokiniana*, *Phytopathol.* **77**:1280-1283.

Conrath, U., Jeblick, W., and Kauss, H., 1991, The protein kinase inhibitor, K-252a, decreases elicitor-induced Ca^{2+} uptake and K^+ release, and increases coumarin synthesis in parsley cells, *FEBS* **279**:141-144.

Cook, S. P., and Hain, F. P., 1985, Qualitative examination of the hypersensitive response of loblolly pine, *Pinus taeda* L., inoculated with two fungal associates of the southern pine beetle, *Dendroctonus frontalis* Zimmermann (*Coleoptera*: *Scolytidae*), *Environ. Entomol.* **14**:396-400.

Cook, S. P., and Hain, F. P., 1986, Defensive mechanisms of loblolly and shortleaf pine against attack by southern pine beetle, *Dendroctonus frontalis* Zimmermann and its fungal associate, *Ceratocystis minor* (Hedgecock) Hunt., *J. Chem. Ecol.* **12**:1397-1406.

Coupe, S. A., Taylor, J. E., and Roberts, J. A., 1997, Temporal and spatial expression of mRNAs encoding pathogenesis-related proteins during ethylene-promoted leaflet abscission in *Sambucus nigra*, *Plant, Cell Environ.* **20**:1517-1524.

Coxon, D. T., Curtis, R. F., Price, K. R., and Levett, G., 1983, Abnormal metabolites produced by *Daucus carota* roots stored under conditions of stress, *Phytochemistry* **12**:1881-1885.

Cutt, J. R., and Klessig, D. F., 1992, Pathogenesis-related proteins, in: *Genes Involved in Plant Defense*, T. Boller and Meins, F., eds., Springer-Verlag, New York, pp. 209-243.

Daly, J. W., Seevers, P. M., and Ludden, P., 1970, Studies on wheat stem rust resistance controlled at the Sr6 locus, III. Ethylene and disease reaction, *Phytopathol.* **60**:1648-1652.

Daniel, S., Tiemann, K., Wittkampf, U., Bless, W., Hinderer, W., and Barz, W., 1990, Elicitor-induced metabolic changes in cell cultures of chickpea (*Cicer arietinum* L.) cultivars resistant and susceptible to *Ascochyta rabiei*, *Planta* **182**:270-278.

Dasilva, E. J., Henriksson, E., and Henriksson, L. E., 1974, Ethylene production by fungi, *Plant Sci. Lett.* **2**:63-66.

David, A. B., Bashan, Y., and Okon, Y., 1986, Ethylene production in pepper *(Capsicum annuum)* leaves infected with *Xanthomonas campestris* pv. *vesicatoria*, *Physiol. Mol. Plant Pathol.* **29**:305-316.

Davis, M. E., Miller, A. R., and Lineberger, R. D., 1992, Studies on the effects of ethylene on transformation of tomato cotyledons (*Lycopersicon esculentum* Mill.) by *Agrobacterium tumefaciens*, *J. Plant Physiol.* **139**:309-312.

Dean, J. F. D., and Anderson, J. D., 1991, Ethylene biosynthesis-inducing xylanase. II. Purification and physical characterization of the enzyme produced by *Trichoderma viride*, *Plant Physiol.* **95**:316-323.

Dean, J. F. D., Gamble, H. R., and Anderson, J. D., 1989, The ethylene biosynthesis-inducing xylanase: Its induction in *Trichoderma viride* and certain plant pathogens, *Phytopathol.* **79**:1071-1078.

Dehne, H.-W., and Spengler, G., 1982, Untersuchungen zum einfluß von ethephon auf pflanzenkrankheiten, *Phytopathol. Z.* **104**:27-38.

Dehne, H.-W., Blankenagel, R., and Schonbeck, F., 1981, Influence of ethylene-releasing substances on the occurrences of *Helminthosporium sativum* on winter barley and on the yield under practical conditions, *Z. Pflanzen. Pflanzenshutz.* **88**:206-209.

Deikman, J., 1997, Molecular mechanisms of ethylene regulation of gene transcription, *Physiol. Plant.* **100**:561-566.

del Campillo, E., and Lewis, L. N., 1992, Occurrence of 9.5 cellulase and other hydrolases in flower reproductive organs undergoing major cell wall disruption, *Plant Physiol.* **99**:1015-1020.

De Laat, A. M. M., van Loon, L. C., and Vonk, C. R., 1981, Regulation of ethylene biosynthesis in virus-infected tobacco leaves, I. Determination of the role of methionine as the precursor of ethylene, *Plant Physiol.* **68**:256-260.

De Laat, A. M. M., and van Loon, L. C., 1982, Regulation of ethylene biosynthsis in virus-infected tobacco leaves, *Plant Physiol.* **69**:240-245.

De Laat, A. M. M. and van Loon, L. C., 1983a, The relationship between stimulated ethylene production and symptom expression in virus-infected tobacco leaves, *Physiol. Plant Pathol.* **22**:261-273.

De Laat, A. M. M., and van Loon, L. C., 1983b, Effects of temperature, light and leaf age on ethylene production and symptoms expression in virus-infected tobacco leaves, *Physiol. Plant Pathol.* **22**:275-283.

deMunk, W. J., 1971, Bud necrosis, a storage disease of tulips. II. Analysis of disease-promoting storage conditions, *Neth. J. Plant Pathol.* **77**:177-186.

deMunk, W. J., and De Rooy, M., 1971, The influence of ethylene on the development of 5°C pre-cooled "Apeldoorn" tulips during forcing, *HortSci.* **6**:40-41.

DeWit, P. J. G. M., Hofman, A. E., Velthuis, G. C. M., and Kuc, J. A., 1985, Isolation and characterization of an elicitor of necrosis isolated from intercellular fluids of compatible interactions of *Cladosporium fulvum* (Syn. *Fulvia fulva*) and tomato, *Plant Physiol.* **77**:642-647.

Dimond, A. E., and Waggoner, P. E., 1953, The cause of epinastic symptoms in *Fusarium* wilt of tomatoes, *Phytopathol.* **43**:663-669.

Dixon, R. A., and Harrison, M. J., 1990, Activation, structure and organization of genes involved in microbial defense in plants, *Adv. Genet.* **28**:165-234.

Dixon, R. A., and Lamb, C. J., 1990, Molecular communication in interactions between plants and microbial pathogens, *Annu. Rev. Plant Physiol. Plant Mol. Biol.* **41**:339-367.

Dutta, S., and Biggs, R. H., 1991, Regulation of ethylene biosynthesis in citrus leaves infected with *Xanthomonas campestris* pv. *citri*, *Physiol. Plant.* **82**:225-230.

Ebel, J., and Scheel, D., 1991, Elicitor recognition and signal transduction, in: *Plant Gene Research: Genes Involved in Plant Defense*, T. Boller, and Meins, F., eds., Springer-Verlag, Berlin, pp. 183-205.

Ecker, J. R., and Davis, R. W., 1987, Plant defense genes are regulated by ethylene, *Proc. Natl. Acad. Sci. USA* **84**:5202-5206.

Egley, G. H., 1982, Ethylene stimulation of weed seed germination, *Agric. and Forest Bull.* **5**:13-18. (Univ. Alberta).

Elad, Y., 1988, Involvement of ethylene in the disease caused by *Botrytis cinerea* on rose and carnation flowers and the possibility of control, *Ann. Appl. Biol.* **113**:589-598.

Elad, Y., 1990, Production of ethylene by tissues of tomato, pepper, French-bean and cucumber in response to infection by *Botrytis cinerea*, *Physiol. Mol. Plant Pathol.* **36**:277-287.

Elad, Y., 1993, Regulators of ethylene biosynthesis or activity as a tool for reducing susceptibility of host plant tissues to infection by *Botrytis cinerea*, *Neth. J. Plant Pathol.* **99**:105-113.

Elad, Y., and Evensen, K., 1995, Physiological aspects of resistance to *Botrytis cinerea*, *Phytopathol.* **85**:637-643.

Elad, Y., and Volpin, H., 1988, The involvement of ethylene and calcium in gray mold of pelargonium, ruscus, and rose plants, *Phytoparasitica* **16**:119-131.

El-Kazzaz, M. K., Sommer, N. F., and Fortlage, R. J., 1983a, Effect of different atmospheres on postharvest decay and quality of fresh strawberries, *Phytopathol.* **73**:282-285.

El-Kazzaz, M. K., Sommer, N. F., and Kader, A. A., 1983b, Ethylene effects on *in vitro* and *in vivo* growth of certain postharvest fruit-infecting fungi, *Phytopathol.* **73**:998-1001.

El-Kazzaz, M. K., Chordas, A., and Kader, A. A., 1983c, Physiological and compositional changes in orange fruit in relation to modification of their susceptibility to *Penicillium italicum* by ethylene treatments, *J. Am. Soc. Hortic. Sci.* **108**:618-622.

Elkashif, M. E., Huber, D. J., and Brecht, J. K., 1989, Respiration and ethylene production in harvested watermelon fruit: Evidence for nonclimacteric respiratory behavior, *J. Amer. Soc. Hort. Sci.* **114**:81-85.

Eplee, R. E., 1975, Ethylene, a witchweed seed germination stimulant, *Weed Sci.* **23**:433-436.

Esquerre-Tugaye, M.-T., Lafitte, C., Mazau, D., Toppan, A., and Touze, A., 1979, Cell surfaces in plant-microorganism interactions. II. Evidence for the accumulation of hydroxyproline-rich glycoproteins in the cell wall of diseased plants as a defense mechanism, *Plant Physiol.* **64**:320-326.

Eyal, Y., Sagele, O., and Fluhr, R., 1992, Dark-induced accumulation of a basic pathogenesis-related (PR-1) transcript and a light requirement for its induction by ethylene, *Plant Mol. Biol.* **19**:589-599.

Farmer, E. E., and Helgeson, J. P., 1987, An extracellular protein from *Phytophthora parasitica* var. *nicotianae* is associated with stress metabolite accumulation in tobacco callus, *Plant Physiol.* **85**:733-740.

Fehrmann, H., and Dimond, A. E., 1967, Peroxidase activity and *Phytophthora* resistance in different organs of the potato plant, *Phytopathol.* **57**:68-72.

Felix, G., and Meins, F., 1987, Ethylene regulation of β-1,3-glucanase in tobacco, *Planta* **172**:386-392.

Felix, G., Grosskopf, D. G., Regenass, M., Basse, C. W., and Boller, T., 1991, Elicitor-induced ethylene biosynthesis in tomato cells: Characterization and use as a bioassay for elicitor action, *Plant Physiol.* **97**:19-25.

Ferguson, I. B., and Mitchell, R. E., 1985, Stimulation of ethylene production in bean leaf discs by the *Pseudomonas* phytotoxin coronatine, *Plant Physiol.* **77**:969-973.

Flaishman, M. A., and Kolattukudy, P. E., 1994, Timing of fungal invasion using host's ripening hormone as a signal, *Proc. Natl. Acad. Sci. USA* **91**:6579-6583.

Frankenberger, W. T., Jr., and Arshad, M., 1995, Phytohormones in Soils: Microbial Production and Function, Marcel Dekker, Inc., New York.

Freebairn, H. T., and Buddenhagen, I. W., 1964, Ethylene production by *Pseudomonas solanacearum*, *Nature* **202**:313-314.

Fuchs, Y., and Anderson, J. D., 1987, Purification and characterization of ethylene inducing proteins from cellulysin, *Plant Physiol.* **84**:732-736.

Fuchs, Y., Saxena, A., Gamble, H. R., and Anderson, J. D., 1989, Ethylene biosynthesis-inducing protein from cellulysin is an endoxylanase, *Plant Physiol.* **89**:138-143.

Fukuda, H., Fujii, T., and Ogawa, T., 1986, Preparation of a cell-free ethylene-forming system from *Penicillium digitatum*, *Agric. Biol. Chem.* **50**:977-981.

Gaborjanyi, R., Balazs, E., and Kiraly, E., 1971, Ethylene production, tissue sensescence and local virus infections, *Acta Phytopathol. Acad. Sci. Hung.* **6**:51-55.

Geballe, G. T., and Galston, A. W., 1982, Ethylene as an effector of wound-induced resistance to cellulase in oat leaves, *Plant Physiol.* **70**:788-790

Geballe, G. T., and Galston, A. W., 1983, Wound-induced lignin formation and resistance to cellulase in oat leaves, *Phytopathol.* **73**:619-629.

Gedalovich, E., and Fahn, A., 1985, Ethylene and gum duct formation in citrus, *Ann. Bot.* **56**:571-577.

Geeson, J. D., Browne, K. M., and Guaraldi, F., 1986, The effects of ethylene concentration in controlled atmosphere storage of tomatoes, *Ann. Appl. Biol.* **108**:605-610.

Gentile, I. A., and Matta, A., 1975, Production of and some effects of ethylene in relation to *Fusarium* wilt of tomato, *Physiol. Plant Pathol.* **5**:27-37.

Gianinazzi, S., Martin, C., and Vallee, J. C., 1970, Hypersensibilite aux virus, temperature et proteines solubles chez le *Nicotiana xanthi* n.c. Apparition de nouvelles macromolecules lors de la repression de la synthese virale, *C. R. Acad. Sci. Paris D* **270**:2283-2286.

Gomez, Lim, M. A., Kelly, P., Sexton, R., and Trewavas, A. J., 1987, Identification of chitinase mRNA in abscission zones from bean (*Phaseolus vulgaris* Red Kidney) during ethylene-induced abscission, *Plant Cell Environ.* **10**:741-746.

Goodman, T. C., Montoya, A. L., Williams, S., and Chilton, M. D., 1986, Sustained ethylene production in *Agrobacterium*-transformed carrot disks caused by expression of the T-DNA *tms* gene products, *J. Bacteriol.* **167**:387-388.

Goto, M., and Hyodo, H., 1985, Role of extracellular polysaccharides of *Xanthomonas campestris* pv. *citri* in the early stage of infection, *Ann. Phytopathol. Soc. Jpn.* **51**:22-31.

Goto, M., and Hyodo, H., 1987, Ethylene production by cell-free extract of the Kudzu strain of *Pseudomonas syringae* pv. *phaseolicola*, *Plant Cell Physiol.* **28**:405-414.

Goto, M., Yaguchi, Y., and Hyodo, H., 1980, Ethylene production in citrus leaves infected with *Xanthomonas citri* and its relation to defoliation, *Physiol. Plant Pathol.* **16**:343-350.

Goto, M., Ishida, Y., Takikawa, Y., and Hyodo, H., 1985, Ethylene production by the Kudzu strain of *Pseudomonas syringae* pv. *phaseolicola* causing halo blight in *Pueraria lobata* (Wild) Ohwi., *Plant Cell Physiol.* **26**:141-150.

Graham, J. H., and Linderman, R. G., 1981, Effect of ethylene on root growth, ectomycorrhiza formation, and *Fusarium* infection of Douglas-fir, *Can. J. Bot.* **59**:149-155.

Graham, J. H., Whiteside, J. O., and Barmore, C. R., 1984, Ethylene production by *Mycosphaerellay citri* and greasy spot-infected citrus leaves, *Phytopathol.* **74**:817.

Grierson, W., and Newhall, W. F., 1955, Tolerance to ethylene of various types of citrus fruits, *Proc. Am. Soc. Hortic. Sci.* **65**:244-250.

Gwinn, K. D., Stelzig, D. A., and Bhatia, S. K., 1989, Differential ethylene production by potato tuber inoculated with a compatible or an incompatible race of *Phytophthora infestans*, *Amer. Potato J.* **66**:417-423.

Hadwiger, L. A., and Beckman, J. M., 1980, Chitosan as a component of pea-*Fusarium solari* interactions, *Plant Physiol.* **66**:205-211.

Haga, M., Haruyama, T., Kano, H., Sekizawa, Y., Urushikazi, S., and Matsumoto, K., 1988, Dependence on ethylene of the induction of phenylalanine ammonia-lyase activity in rice leaf infected with blast fungus, *Agric. Biol. Chem.* **52**:943-950.

Hahn, M. G., Bucheli, P., Cervone, F., Doares, S. H., O'Neill, R. A., Darvill, A., and Albersheim, P., 1989, Roles of cell wall constituents in plant-pathogen interactions, in: *Plant-Microbe Interactions, Molecular and Genetic Perspectives*, Vol. 3, T. Kosuge and E. W., Nester, eds., McGraw-Hill, New York, pp. 131-181.

Halevy, A. H., Shilo, R., and Simchon, S., 1970, Effect of 2-chloroethylphosphonic acid (ethrel) on health, dormancy, and flower and corm yield of gladioli, *J. Hortic. Sci.* **45**:427-434.

Hammond-Kosack, K. E., Silverman, P., Raskin, I., and Jones, J. D. G., 1996, Race-specific elicitors of *Cladosporium fulvum* induce changes in cell morphology and the synthesis of ethylene and salicylic acid in tomato plants carrying the corresponding *Cf* disease resistance gene, *Plant Physiol.* **110**:1381-1394.

Harrington, T. C., 1993, Plant diseases caused by *Ophiostoma* and *Leptographium*, in: *Cerotycystis and Ophiostoma: Taxonomy, Ecology and Pathology*, M. J. Wingfield, Seifert, K. A., and Webber, J. F., eds., APS Press, St. Paul, MN, pp. 120-132.

Hatton, T. T., and Cubbedge, R. H., 1981, Effects of ethylene on chilling injury and subsequent decay of conditioned early Marsh grapefruit during low-temperature storage, *HortSci.* **16**:783-784.

Hebard, F. V., and Shain, L., 1988, Effects of virulent and hypovirulent *Endothia parasitica* and their metabolites on ethylene production by bark of American and Chinese chestnut and scarlet oak, *Phytopathol.* **78**:841-845.

Hedrick, S. A., Bell, J. N., Boller, T., and Lamb, C. J., 1988, Chitinase cDNA cloning and mRNA induction by fungal elicitor, wounding, and infection, *Plant Physiol.* **86**:182-186.

Henfling, J. W. D. M., Lisker, N., and Kuc, J., 1978, Effect of ethylene on phytuberin and phytuberol accumulation in potato tuber slices, *Phytopathol.* **68**:857-862.

Hirano, T., Uritani, I., and Hyodo, H., 1991, Further investigation on ethylene production in sweet potato root tissue infected by *Ceratocystis fimbriata*, *Nippon Shokuhin Kogyo Gakhaishi* **38**:249-259.

Hislop, E. C., Archer, S. A., and Hoad, G. V., 1973a, Ethylene production by healthy and S*clerotinia fructigena*-infected apple peel, *Phytochem.* **12**:2081-2086.

Hislop, E. C., Hoad, G. V., and Archer, S. A., 1973b, The involvement of ethylene in plant diseases, in: *Fungal Pathogenecity and the Plant's Response*, W. Byrde and Cutting, C. V., eds., Academic Press, New York, pp. 87-117.

Hodges, C. F., and Campbell, D. A., 1993, Inhibition of ethylene biosynthesis and endogenous ethylene and chlorophyll content of *Poa pratensis* leaves infected by *Bipolaris sorokiniana*, *J. Plant Physiol.* **142**:699-703.

Hodges, C. F., and Coleman, L. W., 1984, Ethylene-induced chlorosis in the pathogenesis of *Bipolaris sorokiniana* leaf spot of *Poa pratensis*, *Plant Physiol.* **75**:462-465.

Hoffman, R. M., and Heale, J. B., 1987, 6-Methoxymellein accumulation and induced resistance to *Botrytis cinerea* Pers. ex Pers. in carrot slices treated with phytotoxic agents and ethylene, *Physiol. Mol. Plant Pathol.* **30**:67-75.

Hoffman, R., Roebroeck, E.,, and Heale, J. B., 1988, Effects of ethylene biosynthesis in carrot root slices on 6-methoxymellein accumulation and resistance to *Botrytis cinerea*, *Physiol. Plant.* **73**:71-76.

Hoffman, T., Schmidt,, J. C., Zheng, X., and Bent, A. F., 1999, Isolation of ethylene-insensitive soybean mutants that are altered in pathogen susceptibility and gene-for-gene disease resistance, *Plant Physiol.* **119**:935-949.

Hottiger, T., and Boller, T., 1991, Ethylene biosynthesis in *Fusarium oxysporum* f. sp. *tulipae* proceeds from glutamate/2-oxoglutarate and requires oxygen and ferrous ions, *Arch. Microbiol.* **157**:18-22.

Hughes, R. K., and Dickerson, A. G., 1989, The effect of ethylene on phenylalanine ammonia lyase (PAL) induction by a fungal elicitor in *Phaseolus vulgaris*, *Physiol. Mol. Plant Pathol.* **34**:361-378.

Hyodo, H., and Uritani, I., 1984, Ethylene production in sweet potato root tissue infected by *Ceratocyctis fimbriata*, *Plant Cell Physiol.* **25**:1147-1152.

Hyodo, H., and Yang, S. F., 1971, Ethylene-enhanced synthesis of phenylalanine ammonia-lyase in pea seedlings, *Plant Physiol.* **47**:765-770.

Hyodo, H., Kuroda, H., and Yang, S. F., 1978, Induction of phenylalanine ammonia-lyase and increase in phenolics in lettuce leaves in relation to the development of russet spotting caused by ethylene, *Plant Physiol.* **62**:31-35.

Ibe, S. N., and Gorgan, R. C., 1985, Effect of chilling, carbon dioxide, and ethylene inhibitors on ethylene production and infestation of green bell peppers *Capsicum annuum* L. by *Erwinia carotovora* susp. *carotovora*, *Biol. Afric.* **2**:24-32.

Imaseki, H., Teranishi, T., and Uritani, I., 1968, Production of ethylene by sweet potato roots infected by black rot fungus, *Plant Cell Physiol.* **9**:769-781.

Ishige, F., Yamazaki, K., Mori, H., and Imaseki, H., 1991, The effects of ethylene on the coordinated synthesis of multiple proteins: Accumulation of an acidic chitinase and a basic glycoprotein induced by ethylene in leaves of azuki bean, *Vigna angularis*, *Plant Cell Physiol.* **32**:681-690.

Ishii, S., 1988, Factors influencing protoplast viability of suspension-cultured rice cells during isolation process, *Plant Physiol.* **88**:26-29.

Jaworski, J. G., and Kuc, J., 1974, Effect of ethrel and *Ceratocystis fimbriata* on the synthesis of fatty acids and 6-methoxy mellein in carrot root, *Plant Physiol.* **53**:331-336.

Jennings, J. C., Apel-Birkhold, P. C., Bailey, B. A., and Anderson,, J. D., 2000, Induction of ethylene biosynthesis and necrosis in weed leaves by a *Fusarium oxysporum* protein, *Weed Sci.* **48**:7-14.

Kato, Y., and Uritani, I., 1972, Ethylene biosynthesis in diseased sweet potato root tissue with special reference to the methionine system, *Agric. Biol. Chem.* **36**:2601-2604.

Ke, D., and Saltveit, M. E., 1988, Plant hormone interaction and phenolic metabolism in the regulation of russet spotting in iceberg lettuce, *Plant Physiol.* **88**:1136-1140.

Keen, N. T., Tamaki, D., Kobayashi, D., Gerhold, D., Stlayton, M., Shen, H., Gold, S., Lorang, J., Thordal-Christensen, H., Dahlbeck, D., and Staskawicz, B., 1990, Bacteria expressing avirulence gene D produce a specific elicitor of the soybean hypersensitive reaction, *Mol. Plant Microbe. Interact.* **3**:112-121.

Kenyon, J. S., and Turner, J. G., 1992, The stimulation of ethylene synthesis *in Nicotiana tabacum* leaves by the phytotoxin coronatine, *Plant Physiol.* **100**:219-224.

Ketring, D. L., and Melouk, H. A., 1982, Ethylene production and leaflet abscission of three peanut genotypes infected with *Cercospora arachidicola* Hori., *Plant Physiol.* **69**:789-792.

Kim, Y. J., and Hwang, B. K., 1994, Differential accumulation of β-1,3-glucanase and chitinase isoforms in pepper stems infected by compatible isolates of *Pseudomonas solanacearum*, Physiol. Mol. *Plant Pathol.* **45**:195-209.

Kim, Y. J., and Hwang, B. K., 1996, Purification, N-terminal amino acid sequencing and antifungal activity of chitinases from pepper stems treated with mercuric chloride, *Physiol. Mol. Plant Pathol.* **48**:417-432.

Kim, Y. J., and Hwang, B. K., 1997, Isolation of a basic 34-kilodalton β-1,3-glucanase with inhibitory activity against *Phytophthora capsici* from pepper stems, *Physiol. Mol. Plant Pathol.* **50**:103-115.

Kim, Y. J., and Hwang, B. K., 2000, Pepper gene encoding a basic pathogenesis-related 1 protein is pathogen and ethylene inducible, Physiol. Plant. 108:51-60.

Kirsch, C., Hahlbrock, K., and Kombrink, E., 1993, Purification and characterization of extracellular, acidic chitinase isoenzymes from elicitor-stimulated parsley cells, *Eur. J. Biochem.* **213**:419-425.

Kitajima, S., and Sato, F., 1999, Plant pathogenesis-related proteins: Molecular mechanisms of gene expression and protein function, *J. Biochem.* **125**:1-8.

Knoester, M., van Loon, L. C., van den Heuvel, J., Hennig, J., Bol, J. F., and Linthorst, H. J. M., 1998, Ethylene-insensitive tobacco lacks non-host resistance against soil-borne fungi, *Proc. Natl. Acad. Sci. USA* **95**:1933-1937.

Knoester, M., Pieterse, C. M. J., Bol, J. F., van Loon, L. C., 1999, Systemic resistance in *Arabidopsis* induced by rhizobacteria requires ethylene-dependent signaling at the site of application, *Molec. Plant. Micro. Interact.* **12**:720-727.

Koch, F., Baur, M., Burba, M., and Elstner, E. F., 1980, Ethylene formation by *Beta vulgaris* leaves during systemic (beet mosaic virus and beet mild yellowing virus, BMV and BMYV) or necrotic (*Cercospora beticola* Sacc.) disease, *Phytopathol. Z.* **98**:40-46.

Kombrink, E., and Hahlbrock, K., 1986, Responses of cultured parsley cells to elicitors from phytopathogenic fungi. Timing and dose dependency of elicitor-induced reactions, *Plant Physiol.* **81**:216-221.

Kozaka, T., and Teraoka, T., 1978, Ethylene evolution by rice plants infected with *Pyricularia oryzae* Cav. in relation to stunting of diseased plants, *Ann. Phytopathol. Soc. Jpn.* **43**:549-556.

Kurosaki, F., Tsurusawa, Y., and Nishi, A., 1987, Breakdown of phosphatidylinositol during the elicitation of phytoalexin production in cultured carrot cells, *Plant Physiol.* **85**:601-604.

Lagoke, S. T. O., Parkinson, V., and Agunbiade, R. M., 1991, Parasitic weeds and control methods in Africa, in: *Combating Striga in Africa*, S. K. Kim, ed., Proc. Intl. Workshop Organized by IITA, ICRISAT, and IDRC, 22-24, August 1988: IITA, Ibadan, pp. 3-14.

Lamb, C. J., Lawton, M. A., Dron, M., and Dixon, R. A., 1989, Signals and transduction mechanisms for activation of plant defenses against microbial attack, *Cell* **56**:215-224.

Lawton, M. A., and Lamb, C. J., 1987, Transcriptional activation of plant defense genes by fungal elicitor, wounding and infection, *Mol. Cell. Biol.* **7**:335-341.

Lawton, K. A., Potter, S. L., Uknes, S., and Ryals,, J., 1994, Acquired resistance signal transduction in *Arabidopsis* is ethylene independent, *Plant Cell* **6**:581-588.

Lawton, K. A., Weymann, K., Friedrich, L., Vernooij, B., Uknes, S., and Ryals,, J., 1995, Systemic acquired resistance in *Arabidopsis* requires salicylic acid but not ethylene, *Mol. Plant-Microbe Interact.* **8**:863-870.

Lee, S. C., Hong, J. K., Kim, Y. J., and Hwang, B. K., 2000, Pepper gene encoding thionine is differentially induced by pathogens, ethylene and methyljasmonate, *Physiol. Molec. Plant Pathol.* **56**:207-216.

Lee, Y. K., and Hwang, B. K., 1996, Differential induction and accumulation of β-1,3-glucanase and chitinase isoforms in the intercellular space and leaf tissues of pepper by *Xanthomonas compestris* pv. *vesicatoria* infection, *J. Phytopathol.* **144**:79-87.

Lee, Y. K., Kim, Y. J., and Hwang, B. K., 1994, Bacterial multiplications and electrophoretic patterns of soluble proteins in compatible and incompatible interactions of pepper leaves with *Xanthomonas compestris* pv. *vesicatoria*, *Korean J. Plant Pathol.* **40**:305-313.

Linthorst, H. J. M., 1991, Pathogenesis-related proteins of plants, *Plant Sci.* **10**:123-150.

Lisker, N., Cohen, L., Chalutz, E., and Fuchs, Y., 1983, Fungal infection suppress ethylene-induced phenylalanine ammonia-lyase in grapefruits, *Physiol. Plant Pathol.* **22**:331-338.

Liu, D., Raghothama, K. G., Hasegawa, P. M., and Bressan, R. A., 1994, Resistance to the pathogen *Phytophthora infestans* in transgenic potato plants that over-express osmotin, *Proc. Natl. Acad. Sci. USA* **91**:1888-1892.

Lockhart, C. L., Forsyth, F. R., and Eaves, C. A., 1968, Effect of ethylene on development of *Gloeosporium album* in apple and on growth of the fungus in culture, *Can. J. Plant Sci.* **48**:557-559.

Lotan, T., and Fluhr, R., 1990, Xylanase, a novel elicitor of pathogenesis-related proteins in tobacco, uses a nonethylene pathway for induction, *Plant Physiol.* **93**:811-817.

Lovrekovich, L., Lovrekovich, H., and Stahmann, M. A., 1968, The importance of peroxidase in the wildfire disease, *Phytopathol.* **58**:193-198.

Lu, S.-F., Tzeng, D. D., and Hsu, S.-T., 1989, *In vitro* and *in vivo* ethylene production in relation to pathogenesis of *Pseudomonas solanacearum* (Smith) on *Lycopersicon esculentum* M., *Plant Protection Bull.* **31**:60-70.

Lund, B. M., 1973, The effect of certain bacteria on ethylene production by plant tissue, in: *Fungal Pathogenicity and the Plant's Response*, W. Byrde and Cutting, C. V., eds., Academic Press, New York, pp. 69.

Lund, B. M., and Mapson, L. W., 1970, Stimulation of *Erwinia carotovora* of the synthesis of ethylene in cauliflower tissue, *Biochem. J.* **119**:251-263.

Lund, S. T., Stall, R. E., and Klee, H. J., 1998, Ethylene regulates the susceptible response to pathogen infection in tomato, *Plant Cell* **10**:371-381.

Macko, V., Woodbury, W., and Stahmann, M. A., 1968, The effect of peroxidase on the germination and growth of mycelium of *Puccinia graminis* f. sp. *tritici*, *Phytopathol.* **58**:1250-1254.

Marco, S., Levy, D., and Aharoni, N., 1976, Involvement of ethylene in the suppression of hypocotyl elongation in CMV-infected cucumbers, *Physiol. Plant. Pathol.* **8**:1-17.

Marte, M., Buonaurio, R., and Dellatorre, G., 1993, Induction of systemic resistance to tobacco powdery mildew by tobacco mosaic virus, tobacco necrosis virus or ethephon, *J. Phytopathol.* **138**:137-144.

Martin, W. R., Morgan, P. W., Sterling, W. L., and Kenerley, C. M., 1988, Cotton fleahopper and associated microorganisms as components in the production of stress ethylene by cotton, *Plant Physiol.* **87**:280-285.

Mauch, F., and Staehelin, L. A., 1989, Functional implications of the subcellular localization of ethylene-induced chitinase and β-1,3-glucanase in bean leaves, *Plant Cell.* **1**:447-457.

Mauch, F., Hadwiger, L. A., and Boller, T., 1984, Ethylene: Symptom, not signal for the induction of chitinase and β-1,3-glucanase in pea pods by pathogens and elicitors, *Plant Physiol.* **76**:607-611.

Mauch, F., Hadwiger, L. A., and Boller, T., 1988a, Antifungal hydrolases in pea tissue. I. Purification and characterization of two chitinases and two β-1,3-glucanases differentially regulated during development and in response to fungal infection, *Plant Physiol.* **87**:325-333.

Mauch, F., Mauch-Mani, B., and Boller,, T., 1988b, Antifungal hydrolases in pea tissue. II. Inhibition of fungal growth by combinations of chitinase and β-1,3-glucanase, *Plant Physiol.* **88**:936-942.

McCormack, A. A., 1971, Effect of ethylene degreening on decay of Florida citrus fruit, *Proc. Fla. State Hortic. Soc.* **84**:270-272.

McGrath, R. B., and Ecker, J. R., 1998, Ethylene signalling in *Arabidopsis*: Events from the membrane to the nucleus, *Plant Physiol. Biochem.* **36**:103-113.

Memelink, J., Linthorst, H. J. M., Schilperoort, R. A., and Hoge, J. H. C., 1990, Tobacco genes encoding acidic and basic isoforms of pathogenesis-related proteins display different expression patterns, *Plant Mol. Biol.* **14**:119-126.

Metraux, J. P., and Boller, T., 1986, Local and systemic induction of chitinase in cucumber plants in response to viral, bacterial and fungal infections, *Physiol. Mol. Plant Pathol.* **28**:161-169.

Michniewicz, M., Czerwinska, E., Rozej, B., and Bobkiewicz, W., 1983, Control of growth and development of isolates of *Fusarium culmorum* (W.G.Sm) Sacc. of different pathogenicity to wheat seedlings by plant growth regulators, II. Ethylene, *Acta Physiol. Plant.* **5**:189-198.

Miller, A. R., and Pengelly, W. L., 1984, Ethylene production by shoot-forming and unorganized crown-gall tumor tissues of *Nicotiana* and *Lycopersicon* cultured *in vitro*, *Planta* **161**:418-424.

Mitchell, R. E., 1982, Coronatine production by some phytopathogenic pseudomonads, *Physiol. Plant Pathol.* **20**:83-89.

Montalbini, P., and Elstner, E. F., 1977, Ethylene evolution by rust-infected, detached bean (*Phaseolus vulgaris* L.) leaves susceptible and hypersensitive to *Uromyces phaseoli* (Pers.) Wint., *Planta* **135**:301-306.

Nakagaki, Y., Hirai, T., and Stahmann, M. A., 1970. Ethylene production by detached leaves infected with tobacco mosaic virus, *Virology* **40**:1-9.

Nemestothy, G. S., and Guest, D. I., 1990, Phytoalexin accumulation, phenylalanine ammonia lyase activity and ethylene biosynthesis in fosetyl-Al treated resistant susceptible tobacco cultivars infected with *Phytophthora nicotianae* var. *nicotianae*, *Physioil. Mol. Plant Pathol.* **37**:207-219.

Nichols, E. J., Beckman, J. M., and Hadwiger, L. A., 1980, Glycosidic enzyme activity in pea tissue and pea-*Fusarium solani* interactions, *Plant Physiol.* **66**:199-204.

Nickerson, W. J., 1948, Ethylene as a metabolic product of the pathogenic fungus, *Blastomyces dermatitidis*, *Arch. Biochem.* **17**:225-233.

Ohtsubo, N., Mitsuhara, I., Koga, M., Seo, S., and Ohashi, Y., 1999, Ethylene promotes the necrotic lesion formation and basic PR gene expression in TMV-infected tobacco, *Plant Cell Physiol.* **40**:808-817.

Okumura, K., Hyodo, H., Kato, M., Ikoma, Y., and Yano, M., 1999, Ethylene biosynthesis in sweet potato root tissue infected by black rot fungus (*Ceratocystis fimbriata*), *Postharvest Biol. Technol.* **17**:117-125.

Olien, W. C., and Bukovac, M. J., 1982, Ethephon-induced gummosis in sour cherry, *Plant Physiol.* **70**:547-555.

Osborne, D. J., 1989, Abscission, Crit. Rev. Plant Sci. 8:103-129.

Owen, D. R., Lindhal, K. Q., Jr., Wood, D. L., and Parmeter, J. R., Jr., 1987, Pathogenicity of fungi isolated from *Dendroctonus valens*, *D. brevicomis*, and *D. ponderosae* to pine seedlings, *Phytopathol.* **77**:631-636.

Paine, T. D., 1984, Seasonal response of ponderosa pine to inoculation of the mycangial fungi from the western pine beetle, *Can. J. Bot.* **62**:551-555.

Paine, T. D., and Stephen, F. M., 1987, Fungi associated with the southern pine beetle: Avoidance of the induced response in loblolly pine, *Oecologia* **74**:377-379.

Paradies, I., Humme, B., Hoppe, H. H., Heitefuss, R., and Elstner, E. F., 1979, Induction of ethylene formation in bean (*Phaseolus vulgaris*) hypocotyl segments by preparations isolated from germ tube cell walls of *Uromyces phaseoli*, *Planta* **146**:193-197.

Paradies, I., Konze, J. R., Elstner, E. F., and Paxton, J., 1980, Ethylene: Indicator but not inducer of phytoalexin synthesis in soybean, *Plant Physiol.* **66**:1106-1109.

Pegg, G. F., 1976a, The response of ethylene-treated tomato plants to infection by *Verticillium albo-atrum*, *Physiol. Plant Pathol.* **9**:215-226.

Pegg, G. F., 1976b, The involvement of ethylene in plant pathogenesis, in: *Encyclopedia of Plant Pathology, New Series,* R. Heltfuss and Williams, P. H., eds.,Springer-Verlag, Heidelberg, pp. 582-591.

Pegg, G. F., 1977, Glucanohydrolases of higher plants: A possible defense mechanism against parasitic fungi, in: *Cell Wall Biochemistry Related to Specificity in Host-Plant Pathogen Interactions*, B. Solheim and Raa, J., eds., Universitatsforlaget, Tromso, pp. 305-345.

Pegg, G. F., 1981, The involvement of growth regulators in the diseased plant, in: *Effects of Disease on the Physiology of the Growing Plant*, P. G. Ayres, ed., Soc. Exp. Biol. Semin. Ser. 11, Cambridge Univ. Press, Cambridge, pp. 149-177.

Pegg, G. F., and Cronshaw, D. K., 1976a, Ethylene production in tomato plants infected with *Verticillium albo-atrum, Physiol. Plant Pathol.* **8**:279-295.

Pegg, G. F., and Cronshaw, D. K., 1976b, The relationship of *in vitro* to *in vivo* ethylene production in *Pseudomonas solanacearum* infection of tomato, *Physiol. Plant Pathol.* **9**:145-154.

Pegg, G. F., and Young, D. H., 1981, Changes in glycosidase activity and their relationship to fungal colonization during infection of tomato by *Verticillium albo-atrum, Physiol. Plant Pathol.* **19**:371-382.

Pegg, G. F., and Young, D. H., 1982, Purification and characterization of chitinase enzymes from healthy and *Verticillium albo-atrum*-infected tomato plants and from *V. albo-atrum, Physiol. Plant Pathol.* **21**:389-409.

Penninckx, I. A. M. A., Eggermont, K., Terras, F. R. G., Thomma, B. P. H. J., De Samblanx, G. W., Buchala, A., Metraux, J.-P., Manners, J. M., and Broekaert, W. F., 1996, Pathogen-induced systemic activation of a plant defensin gene in *Arabidopsis* follows a salicylic acid-independent pathway, *Plant Cell* 8:2309-2323.

Penninckx, I. A. M. A., Thomma, B. P. H. J., Buchala, A., Metraux, J.-P., and Broekaert, W. F., 1998, Concomitant activation of jasmonate and ethylene response pathways is required for induction of a plant defensin gene in *Arabidopsis, Plant Cell* **10**:2103-2113.

Peters, W. J., and Roberts, D. R., 1977, Ethrel bipyridilium synergism in slash pine, in: *Proc. of Lightwood Res. Coord. Council*, Atlantic Beach, FL, pp. 78-83.

Peters, W. J., Roberts, D. R., and Munson, J. W., 1978, Ethrel-diquat-paraquat interactions in lightwood formation, in: *Proc. of Lightwood Res. Coord. Council*, Atlantic Beach, FL, pp. 31-39.

Pieterse, C. M. J., van Wees, S. C. M., van Pelt, J. A., Knoester, M., Laan, R., Gerrits, H., Weisbeek, P. J., and van Loon, L. C., 1998, A novel signaling pathway controlling induced systemic resistance in *Arabidopsis, Plant Cell* **10**:1571-1580.

Popp, M. P., Johnson, J. D., and Lesney, M. S., 1995a, Response of slash pine to inoculation with bark beetle vectored fungi, *Tree Physiol.* **15**:619-623.

Popp, M. P., Johnson, J. D., and Lesney, M. S., 1995b, Changes in ethylene production and monoterpene concentration in slash pine and loblolly pine following inoculation with bark beetle vector fungi, *Tree Physiol.* **15**:807-812.

Popp, M. P., Lesney, M. S., and Davis, J. M., 1997, Defense responses elicited in pine cell suspension cultures, *Plant, Cell, Tissue, Organ, Culture* **47**:199-206.

Portmann, A., and Elstner, E. F., 1983, Ethylene formation in leaves of different barley cultivars during primary infection by *Erysiphe graminis* f. sp. *hordei*, *Z. Pflanzen. Pflanzenshutz*. **90**:634-640.

Preisig, C. L., and Kuc, J. A., 1987, Inhibition by salicylhydroxamic acid, BW755C, eicosatetraynoic acid, and disulfiram of hypersensitive resistance elicited by arachidonic acid or poly-L-lysine in potato tuber, *Plant Physiol.* **84**:891-894.

Pritchard, D. W., and Ross, A. F., 1975, The relationship of ethylene to formation of tobacco mosaic: Virus lesions in hypersensitive responding tobacco leaves with and without induced resistance, *Virology* **64**:295-307.

Prince, A., Stephens, C. T., and Herner, R. C., 1988, Pathogenicity, fungicide resistance and ethylene production of *Penicillium* spp. isolated from tulip bulbs, *Phytopathol.* **78**:682-686.

Reinhardt, D., Wiemken, A., and Boller, T., 1991, Induction of ethylene biosynthesis in compatible and incompatible interactions of soybean roots with *Phytophthora megasperma* f. sp. *glycinea* and its relation to phytoalexin accumulation, *J. Plant Physiol.* **138**:394-399.

Reuveni, M., and Cohen, Y., 1978, Growth retardation and changes in phenolic compounds, with special reference to scopoletin, in mildewed and ethylene-treated tobacco plants, *Physiol. Plant Pathol.* **12**:179-189.

Resende, M. L. V., Mepsted, R., Flood, J., and Cooper, R. M., 1996, Water relations and ethylene production as related to symptom expression in cocoa seedlings infected with defoliating and non-defoliating isolates of *Verticillium dahliae*, *Plant Pathol.* **45**:964-972.

Rhodes, J. M., and Wooltorton,, L. S. C., 1978, The biosynthesis of phenolic compounds in wounded plant storage tissue, in: *Biochemistry of Wounded Plant Tissues*, G. Kahl, ed., Walter de Gruyter, Berlin, pp. 243-286.

Rhodes, M. J. C., and Wooltorton, L. S. C., 1973, Stimulation of phenolic acid and lignin biosynthesis in swede root tissue by ethylene, *Phytochemistry* **12**:107-118.

Rickauer, M., Bottin, A., and Esquerre-Tugaye, M.-T., 1992, Regulation of proteinase inhibitor production in tobacco cells by fungal elicitors, hormonal factors and jasmonate, *Plant Physiol. Biochem.* **30**:579-584.

Rickauer, M., Fournier, J., Pouenat, M. L., Berthalon, E., Bottin, A., and Esquerre-Tugaye,, M.-T., 1990, Early changes in ethylene synthesis and lipoxygenase activity during defense induction in tobacco cells, *Plant Physiol. Biochem.* **28**:647-653.

Rickey, T. M., and Belknap, W. R., 1991, Comaprison of the expression of several stress-responsive genes in potato tubers, *Plant Molecu. Biol.* **16**:1009-1018.

Riov, J., Monselise, S. P., and Kahan, R. S., 1969, Ethylene-controlled induction of phenylalanine ammonia lyase in citrus fruit peel, *Plant Physiol.* **44**:631-635.

Roberts, J. A., Coupe, S. A., Whitelaw, C. A., and Taylor, J. E., 1997, Spatial and temporal expression of abscission-related genes during ethylene-promoted organ shedding, in: *Biology and Biotechnology* of *the Plant Hormone Ethylene*, A. Kanellis, Chang, C., Kende, H., and Grierson, D., eds., NATO Series, Kluwer Academic Press, Dordrecht, pp. 185-190.

Roberts, W. K., and Selitrennikoff, C. P., 1988, Plant and bacterial chitinases differ in antifungal activity, *J. Gen. Microbiol.* **134**:169-176.

Robson, T. O., and Broad, H. R., 1989, *Striga*: Improved Management in Africa, *FAO Plant Production and Protection Paper 96*, Rome, pp. 205.

Roby, D., Toppan, A., and Esquerre-Tugaye,, M.-T., 1985, Cell surfaces in plant-microorganism interactions, V. Elicitors of fungal and of plant origin trigger the synthesis of ethylene and of cell wall hydroxyproline-rich glycoprotein in plants, *Plant Physiol.* **77**:700-704.

Roby, D., Toppan, A., and Esquerre-Tugaye, M.-T., 1986, Cell surfaces in plant-microorganism interactions, VI. Elicitors of ethylene from *Colletotrichum legenarium* trigger chitinase activity in melon plants, *Plant Physiol.* **81**:228-233.

Roby, D., and Esquerre-Tugaye, M.-T., 1987, Induction of chitinases and of translatable mRNA for the enzymes in melon plants infected with *Colletotrichum legenarium*, *Plant Sci.* **52**:175-185.

Roby, D., Toppan, A., and Esquerre-Tugaye, M.-T., 1989, Systematic induction of chitinase activity and resistance in melon plants upon fungal infection or elicitor treatment, *Physiol. Mol. Plant Pathol.* **33**:409-417.

Roby, D., Broglie, K., Gaynor, J., and Broglie, R., 1991, Regulation of a chitinase gene promotor by ethylene and elicitors in bean protoplasts, *Plant Physiol.* **97**:433-439.

Roggero, P., and Pennazio, S., 1988, Biochemical changes during the nectrotic systemic infection of tobacco plants by potato virus Y, necrotic strain, *Physiol. Molecu. Plant Pathol.* **32**:105-113.

Ross, A. F., and Pritchard, D. W., 1972, Local and systemic effects of ethylene on tobacco mosaic virus lesions in tobacco, *Phytopathol.* **62**:786-788.

Ross, A. F., and Williamson, C. E., 1951, Physiologically active emanations from virus-infected plants, *Phytopathol.* **41**:431-438.

Sand, P. F., and Manley, J. D., 1990, The witchweed eradication program survey, regulatory and control, in: *Witchweed Research and Control in the United States*. P. F. Sand, Eplee, R. E., and Westbrooks, R. G., eds., Weed Sci. Soc. Am., Champaign, pp. 144-150.

Schiffmann-Nadel, M., Michaely, H., Zauberman, G., and Chet, I., 1985, Physiological changes occurring in picked climacteric fruit infected with different pathogenic fungi, *Phytopathol. Z.* **113**:227-284.

Schlumbaum, A., Mauch, F., Vogeli, U., and Boller, T., 1986, Plant chitinases are potent inhibitors of fungal growth, Nature 324:365-367.

Sexton, R., and Roberts, J. A., 1989, Cell biology of abscission, *Annu. Rev. Plant Physiol.* **33**:133-162.

Shain, L., and Miller, J. B., 1988, Ethylene production by excised sapwood of clonal eastern cottonwood and the compartmentalization and closure of seasonal woods, *Phytopathol.* **78**:1261.

Shetty, H. S., and Kumar, V. U., 1999, Signaling in plants during induction of resistance against pathogens, *Curr. Sci.* **76**:640-646.

Siefert, F., and Grossmann, K., 1997, Induction of chitinase and β-1,3-glucanase activity in sunflower suspension cells in response to an elicitor from *Phytophthora megasperma* f.sp. *glycinea* (Pmg). Evidence for regulation by ethylene and 1-aminocyclopropane-1-carboxylic acid (ACC), *J. Exptl. Bot.* **48**:2023-2029.

Siefert, F., Langebartels, C., Boller, T., and Grossmann, K., 1994, Are ethylene and 1-aminocyclopropane-1-carboxylic acid involved in the induction of chitinase and β-1,3-glucanase activity in sunflower cell-suspension cultures? *Planta* **192**:431-440.

Smith, K. A., and Restall, S. W. F., 1971, The occurrence of ethylene in anaerobic soil, *J. Soil Sci.* **22**:430-443.

Smith, W. H., Meigh, D. F., and Parker, J. C., 1964, Effect of damage and fungal infection on the production of ethylene by carnations, *Nature* **204**:92-93.

Smith-Becker, J., Marois, E., Huguet, E. J., Midland, S. L., Sims, J. J., and Keen, N. T., 1998, Accumulation of salicylic acid and 4-hydroxybenzoic acid in phloem fluids of cucumber during systemic acquired resistance is preceded by a transient increase in phenylalanine ammonia-lyase activity in petioles and stems, *Plant Physiol.* **116**:231-238.

Song, F.-M., and Zheng, Z., 1998, The correlation between inhibition of ethylene production and trifluralin-induced resistance of cotton seedlings against *Fusarium oxysporum* f. sp. *vasinfectum*, *Acta Phytophysiol. Sinica.* **24**:111-118.

Southwick, S. M., Davies, F. S., El-Gholl, N. E., and Schoulties, C. L., 1982, Ethylene, fungi and summer fruit drop of navel orange, *J. Am. Soc. Hortic. Sci.* **107**:800-804.

Spanu, P., and Boller, T., 1989a, Ethylene biosynthesis in tomato infected by *Phytophthora infestans,* in: *Biochemical and Biophysiological Aspects of Ethylene Production in Lower and Higher Plants,* H. Clijsters, ed., Kluwer Academic Publishers, pp. 255-260.

Spanu, P., and Boller, T., 1989b, Ethylene biosynthesis in tomato plants infected by *Phytophthora infestans, J. Plant. Physiol.* **134**:533-537.

Stahmann, M. A., Clare, B. G., and Woodbury, W., 1966, Increased disease resistance and enzyme activity induced by ethylene and ethylene production by black rot infected sweet potato tissue, *Plant Physiol.* **41**:1505-1512.

Stall, R. E., and Hall, C. B., 1984, Chlorosis and ethylene production in pepper leaves infected by *Xanthomonas campestris* pv. *vesicatoria, Phytopathol.* **74**:373-375.

Stermer, B. A., and Hammerschmidt, R., 1987, Association of heat shock induced resistance to disease with increased accumulation of insoluble extensin and ethylene synthesis, Phyhsiol. Mol. Plant *Pathol.* **31**:453-461.

Strauch, J. C., Dow, J. M., Miligan, D. E., Parra, R., and Daniels, M. J., 1990, Induction of hydrolytic enzymes in *Brassica campestris* in response to pathovars of *Xanthomonas campestris, Plant Physiol.* **93**:238-243.

Swart, A., and Kamerbeek, G. A., 1976, Different ethylene production *in vitro* by several species and formae speciales of *Fusarium, Neth. J. Plant Pathol.* **82**:81-84.

Swart, A., and Kamerbeek, G. A., 1977, Ethylene production and mycelium growth of the tulip strain of *Fusarium oxysporum* as influenced by shaking of and oxygen supply to the culture medium, *Physiol. Plant.* **39**:38-44.

Tagu, D., Walker, N., Ruiz-Avila, L., Burgess, S., Martinez-Izquierdo, J. A., Leguay, J. J., Netter, P., and Puigdomenech, P., 1992, Regulation of the maize HRGP gene expression by ethylene and wounding mRNA accumulation and qualitative expression analysis of the promoter by microprojectile bombardment, *Plant Mol. Biol.* **20**:529-538.

Taylor, J. E., Coupe, S. A., Picton, S., and Roberts, J. A., 1994, Characterization and accumulation pattern of an mRNA encoding an abscission-related β-1, 4-glucanase from leaflets of *Sambucus nigra, Plant Molec. Biol.* **24**:961-964.

Taylor, J. E., Webb, S. T. J., Coupe, S. A., Tucker, G. A., and Roberts, J. A., 1993, Changes in polygalacturonase activity and solubility of polyuronides during ethylene-stimulated leaf abscission in *Sambucus nigra, J. Exptl. Bot.* **44**:93-98.

Thomma, B. P. H. J., Eggermont, K., Tierens, K. F. M.-J., and Broekaert, W. F., 1999, Requirement of functional ethylene-insensitive 2 gene for efficient resistance of *Arabidopsis* to infection by *Botrytis cinerea, Plant Physiol.* **121**:1093-1101.

Tobias, L., Fraser, R. S. S., and Gerwitz, A., 1989, The gene-for-gene relationship between *Capsicum annuum* L. and tobacco mosaic virus effects on virus multiplication, ethylene, asynthesis and accumulation of pathogenesis-related proteins, *Physiol. Mol. Plant Pathol.* **35**:271-286.

Toppan, A., Roby, D., and Esquerre-Tugaye, M.-T., 1982, Cell surfaces in plant-microorganism interactions, III. *In vivo* effect of ethylene on hydroxyproline-rich glycoprotein accumulation in the cell wall of diseased plants, *Plant Physiol.* **70**:82-86.

Toppan, A., and Esquerre-Tugaye, M.-T., 1984, Cell surfaces in plant-microorganism interactions. IV. Fungal glycopeptides which elicit the synthesis of ethylene in plants, *Plant Physiol.* **75**:1133-1138.

Tzeng, D. D., and DeVay, J. E., 1985, Physiological responses of *Gossypium hirsutum* L. to infection by defoliating and non-defoliating pathotypes of *Verticillium dahliae* Kleb., *Physiol. Plant Pathol.* **26**:57-72.

Uegaki, R., Fujimori, T., Kaneko, H., Kubo, S., and Kato, K., 1980, Phytuberin and phytuberol, sesquiterpenes from *Nicotiana tabacum* treated with ethrel, *Phytochemistry* **19**:1543-1544.

Vance, C. P., Kirk, T. K., and Sherwood, R. T., 1980, Lignification as a mechanism of disease resistance, *Annu. Rev. Phytopathol.* **18**:259-288.

van Loon, L. C., 1977, Induction by 2-chloroethylphosphonic acid of viral-like lesions, associated proteins and systemic resistance in tobacco, *Virology* **80**:417-420.

van Loon, L. C., 1983, The induction of pathogenesis-related proteins by pathogens and specific chemicals, Neth. J. Plant Pathol. 89:265-273.

van Loon, L. C., 1985, Pathogenesis-related proteins, *Plant Mol. Biol.* **4**:111-116.

van Loon, L. C., 1989, Stress proteins in infected plants, in: *Plant-Microbe Interactions, Vol. 3*, T. Kosuge and E. W. Nester, eds., McGraw-Hill, New York, pp. 198-237.

van Loon, L. C. and van Kammen, A., 1970, Polyacrylamide disc electrophoresis of the soluble leaf proteins from *Nicotiana tabacum* var. 'Samsun' and 'Samsun NN', II. Changes in protein constitution after infection with tobacco mosaic virus, *Virology* **40**:199-211.

van Loon, L. C., and Pennings, G. G. H., 1993, Involvement of ethylene in the induction of systemic acquired resistance in tobacco, in: *Mechanisms of Plant Defense Responses,* B. Fritig, and Legrand, M., eds., Kluwer Academic Publishers, Dordrecht, pp. 156-159.

van Loon, L. C., and Van Strien, E. A., 1999, The familities of pathogenesis-related proteins, their activities, and comparative analysis of PR-1 type proteins, *Physiol. Mol. Plant Pathol.* **55**:85-97.

VanderMolen, G. E., Labavitch, J. M., Strand, L. L., and DeVay, J. E., 1983, Pathogen-induced vascular gels: Ethylene as a host intermediate, *Physiol. Plant* **59**:573-580.

Vanzyl, L. M., and Wingfield, M. J., 1998, Ethylene production by *Eucalyptus* clones in response to infection by hypovirulent and virulent isolates of *Cryphonectria cubensis*, *South Africa J. Sci.* **94**:193-194.

Vera, P., and Conejero, V., 1989a, The induction and accumulation of the pathogenesis-related P69 proteinase in tomato during citrus exocortis viroid infection and in response to chemical treatments, *Physiol. Mol. Plant Pathol.* **34**:323-334.

Vera, P., and Conejero, V., 1989b, Effect of ethephon on protein degradation and the accumulation of pathogenesis-related (PR) proteins in tomato leaf discs, *Plant Physiol.* **92**:227-233.

Vera, P., Tornero, P., and Conejero, V., 1993, Cloning and expression analysis of a viroid-induced peroxidase from tomato plants, *Mol. Plant Microbe Interact.* **6**:790-794.

Vögeli, U., Meins, F., Jr. and Boller, T., 1988, Coordinated regulation of chitinase and β-1,3-glucanase in bean leaves, *Planta* **174**:364-372.

Völksch, B., and Weingart, H., 1997, Comparison of ethylene-producing *Pseudomonas syringae* strains isolated from kudzu (*Pueraria lobata*) with *Pseudomonas syringae* pv. *phaseolicola* and *Pseudomonas syringae* pv. *glycinea*, *European J. Plant Pathol.* **103**:795-802.

Wächter, R., Fischer, K., Gabler, R., Kuhnemann, F., Urban, W., Bogemann, G. M., Voesenek, L. A. C. J., Blom, C.W.P.M., and Ullrich, C. I., 1999, Ethylene production and ACC-accumulation in *Agrobacterium tumefaciens*-induced plant tumours and their impact on tumour and host stem structure and function, *Plant, Cell and Environ.* **22**:1263-1273.

Wargo, P. M., 1975, Lysis of the cell wall of *Armillaria mellea* by enzymes from forest trees, Physiol. Plant Pathol. 5:99-105.

Webb, S.T.J., Taylor, J. E., Coupe, S. A., Ferrarese, L., and Roberts, J. A., 1993, Purification of β-1,4-glucanase from ethylene treated leaflet abscission zones of *Sambucus nigra*, *Plant, Cell and Environ.* **16**:329-333.

Weingart, H., and Völksch, B., 1997, Ethylene production by *Pseudomonas syringae* pathovars *in vitro* and *in planta*, *Appl. Environ. Microbiol.* **63**:156-161.

Wendland, M., and Hoffman, G. M., 1988, Differenzierung der quantitative Resistenz von Weizensorten und-linien gegen *Septoria nodorum* auf der Grundlage der postinfectionellen Ethylenbildung, *Z. Pflanzen. Pflanzenshutz.* **95**:113-123.

Wiese, M. V., and DeVay, J. E., 1970, Growth regulator changes in cotton associated with defoliation caused by *Verticillium albo-atrum*, *Plant Physiol.* **45**:304-309.

Williamson, C. E., 1950, Ethylene, a metabolic product of diseased or injured plants, *Phytopathol.* **40**:205-208.

Woltering, E. J., 1987, Effects of ethylene on ornamental pot plants: A classification, *Scientia Hortic.* **31**:283-294.

Yang, S. F., and Hoffman, N. E., 1984, Ethylene biosynthesis and its regulation in higher plants, *Annu. Rev. Plant Physiol.* **35**:155-189.

Yang, S. F., and Pratt, H. K., 1978, The physiology of ethylene in wounded plant tissues, in: *Biochemistry of Wounded Plant Tissues*, G. Kahl, ed., Walter de Gruyter, Berlin, pp. 595-622.

Yoshikawa, M., Yamaoka, N., and Takeuchi, Y., 1993, Elicitors: Their significance and primary modes of action in the induction of plant defense reactions *Plant Cell Physiol.* **34**:1163-1173.

Young, D. H., and Pegg, G. F., 1982, The action of tomato and *Verticillium albo-atrum* glycosidases on the hyphal wall of *Verticillium albo-atrum*, Physiol. Plant Pathol. 21:411-423.

Yu, L. M., 1995, Elicitins from *Phytophthora* and basic resistance in tobacco, *Proc. Natl. Acad. Sci. USA* **92**:4088-4094.

8

ETHYLENE IN AGRICULTURE: SYNTHETIC AND NATURAL SOURCES AND APPLICATIONS

8.1. INTRODUCTION

Ethylene as well as other plant growth regulators (PGRs) are important chemicals in agricultural production. Plant growth regulators are now used worldwide on a diversity of crops each year (Thomas, 1982). The plant hormone, C_2H_4 strongly influences nearly every development stage in plant growth, from germination to fruit ripening and senescence. Moreover, its critical role in post-harvest physiology of agricultural products has also been well documented. Obviously, a compound with so many different effects may be useful in many ways to modify plant growth and development as required by growers. However, many factors including its gaseous nature and some negative effects on plant growth, restrict the extensive practical usefulness of C_2H_4. Furthermore, the consistency of results observed under controlled experimental conditions may not always be achieved under conditions of practical applications (i.e., under natural field conditions).

It is well accepted that the application of C_2H_4 at the appropriate time and dose can be a vital chemical to promote agricultural production by altering the growth pattern of treated plants and crops. For instance, C_2H_4 can substantially reduce the cost of production by reducing the fruit removal force which can save the crop from damage during mechanical harvest. Additionally, C_2H_4 can be used for manipulation of the harvest date (by accelerating maturity) in a wide range of crops. In the case of cereals, C_2H_4 can promote the yield by reducing lodging. It also plays an important role in quality and shelf life of agricultural produce. Most applications of C_2H_4 or C_2H_4 – releasing compounds are confined to high-value horticultural crops rather than field crops, although there are several exceptions.

According to Abeles et al. (1992), there are two ways of identifying the uses for C_2H_4 in agriculture. The first is to use the known effects of C_2H_4 as a "shopping list". The second is to select a specific crop and identify C_2H_4 -induced effects which might be useful. Lürssen (1991) has summarized the physiological effects of C_2H_4 useful in agriculture and horticulture particularly its role in abscission, germination, sprouting, apical dominance and dormancy, female flower induction and flowering, growth inhibition, male sterility and ripening. Similarly, Abeles et al. (1992) also discussed the potential applications of C_2H_4 in agriculture. However, these authors put the major focus on synthetic C_2H_4 gas or C_2H_4 –releasing compounds and neglected natural sources of C_2H_4, i.e., soil microorganisms, which can have a significant agronomic impact.

As discussed in Chapter 1, C_2H_4 is a plant hormone and an atmospheric air pollutant. The agricultural sources of C_2H_4 may have a direct or indirect impact on crop productivity with respect to quality and yield of produce. These agricultural sources include C_2H_4 production by higher and lower plants, biotic and abiotic production of C_2H_4 in soil in response to amendments, and commercially available C_2H_4 gas and C_2H_4 – releasing compounds. The non-agricultural sources of C_2H_4 include fires, automobiles and various industry operations which contribute C_2H_4 as a pollutant in the environment. This chapter will focus on agricultural sources of C_2H_4 for the betterment of agriculture production.

8.2. ETHYLENE SOURCES AND THEIR APPLICATIONS IN AGRICULTURE

The phytohormone, C_2H_4 is available commercially as a gas, as well as C_2H_4 – releasing compounds such as ethephon, ethrel and others. The concentrations of C_2H_4 in the rhizosphere can be increased through various amendments such as Retprol (a calcium carbide-based commercial product), methionine, ethionine, and other organic compounds. The possibility of utilizing these various sources of C_2H_4 in favor of crop production will be discussed in the following sections.

8.2.1. Ethylene Gas

As discussed in Chapter 1, C_2H_4 may be regarded as the first PGR used in agriculture, because it was unknowingly employed in the form of fumes. Its nature as an active ingredient in fumes was not known at that time, nor was its role as a plant hormone established. Long before C_2H_4 was recognized as an active agent, emanations from ripe fruit or smoke were used to hasten fruit ripening (Knight and Crocker, 1913; Chace and Sorber, 1936; Cousins, 1910; Miller, 1947) or to promote flowering (Rodriguez, 1932; Gonzalez, 1924; Traub et al., 1940). The first intentional use of C_2H_4 in agriculture was to degreen citrus fruits (Denny, 1923, 1924). This was followed by its use to blanch celery (Harvey, 1925) and ripen fruit (Harvey, 1928). Similarly, C_2H_4 was also used unknowingly to ripen bananas in East Africa. When C_2H_4 produced by kerosene stoves was shown to be the active agent for degreening citrus, these stoves were also used to ripen mature-green bananas (Miller, 1947) and subsequently C_2H_4 was also used to ripen bananas in the United States, Great Britain, and Australia (Simmonds, 1959). Early workers recommended C_2H_4 for ripening of tomatoes, pineapples, cantaloupes, dates, jujubes, persimmons, pears, mangoes, pomegranates, peppers, avocados, honeydew melons, apples, plums, papayas, cherimoyas, plantains, chicory, and endive (Chace and Church, 1927; Chace and Sorber, 1936; Harvey, 1928; Hills and Haywood, 1946; Miller, 1947; Rosa, 1925).

Recognition of the date of maturity is important in assuring high-quality fruit. The ability of exogenous C_2H_4 to induce the accumulation of endogenously produced C_2H_4 (Liu, 1978; Dilley and Dilley, 1985) has been used to predict the autogenous C_2H_4 climacteric a week in advance and allows preclimacteric apples to be harvested and held in storage successfully (Dilley, 1989; Dilley et al., 1989). A formula based on 1-aminocyclopropane-1-carboxylic acid (ACC) content of the fruit, C_2H_4 concentration, and exposure time needed to ripen bananas has also been published (Inaba and Nakamura, 1988). Currently, in the United States, C_2H_4 -enhanced ripening is used in the marketing of tomatoes, bananas, mangoes, and honeydew melons (Sherman, 1985). Similarly, guides on the use of C_2H_4 to ripen tomatoes (Gull, 1981; Sherman and Gull, 1981) and honeydew melons (Kasmire et al., 1970) are available. Ethylene sources for degreening include catalytic gas generators, compressed C_2H_4, and compressed C_2H_4 diluted with an inert gas such as "banana gas" (Sherman, 1985) applied by "shot", "trickle", and "flow-through" methods.

Exogenous C_2H_4 has been used to promote flowering of geophytes. Ethylene accelerates the flowering of iris bulbs (Stuart et al., 1966), promotes flowering of iris, freesia and narcissus geophytes (De Munk and Duineveld, 1986; Schipper, 1982; Uyemura and Imanishi, 1983) and induces flowering in pineapple (Rodriguez, 1932). Ethylene has also been used to promote curing of tobacco leaves after harvest (Crocker, 1948) and to defoliate nursery stock after digging (Milbrath et al., 1940). As indicated by Abeles et al., (1992) and Lürssen (1991), C_2H_4 is still used for these purposes either in the form of a gas or as ethephon.

As discussed in Chapter 7 (Section 7.6), *Striga hermonthica* and *Striga asiatica* (witchweed) are phanerogamic parasites of sorghum and millet. Ethylene was found to be a germination stimulant for *Striga* (Egley and Dale, 1970; Eplee, 1975). A control program in North Carolina based on the injection of C_2H_4 into the soil was applied (Eplee, 1975) and several thousand hectares were treated (Egley, 1982). The objective of this program was to cause suicidal germination of the *Striga* seed in the absence of obligate host roots. Three annual applications of C_2H_4 in sequence along with control of escaped *Striga* plants achieved eradication in treated areas. Partial eradication has been achieved at sites in Africa with the same treatments (Bebawi and Eplee, 1986).

The presence of C_2H_4 in storage areas has been found detrimental for the market life of produce (Knee et al., 1985; Wills et al., 1989). Exogenous application of C_2H_4 to stored produce resulted in a significant decrease in the market life of lettuce (Kim and Wills, 1995), strawberries (Wills and Kim, 1995), green beans (Wills and Kim, 1996), the Asian leafy vegetables, bok choi, choi sum, and gai lai (Wills and Wong, 1996), Chinese cabbage and orange (Wills et al., 1999), and kiwi fruit (Agar et al., 1999). On the other hand, the removal of C_2H_4 around the stored produce gave a significant increase in its market life. Similarly, use of low levels of methylcyclopropene gas, as an inhibitor of C_2H_4 action, was found effective in extending the post-harvest life of strawberries exposed to C_2H_4 (Ku et al., 1999). This C_2H_4 antagonistic gas also protects cut flowers against C_2H_4 induced undesired changes (Serek et al., 1995; Sisler et al., 1996; Macnish et al., 1999).

8.2.2. Ethylene-Releasing Compounds

Ethylene is commercially available as a gas, but its early use as a PGR was limited to greenhouse and fruit storage facilities and for practical purposes its application in the field was not undertaken on a large scale because of its gaseous nature which makes it difficult to handle in agriculture. The major breakthrough in commercial use of C_2H_4 came at the end of the 1960s when an C_2H_4-releasing compound in liquid form was introduced under the trade name of "ethephon" (Sterry, 1969; Cooke and Randall, 1968).

Today, this compound is still one of the major PGRs on the market with many different uses. Later, several other compounds with similar modes of action were introduced (Draber, 1977). Two other compounds, etacelasil (2-chloroethyl-tris-ethoxymethoxy silane) and silaid (2-chloroethyl-bis-phenylmethoxy silane) decompose to C_2H_4, but much more rapidly than ethephon, and are less sensitive to changes in pH. These C_2H_4-releasing compounds may have different spectra in terms of usefulness in agriculture. The plant response(s) to these compounds depend upon: (i) the penetration of the C_2H_4-releasing compound into the plant and possibly into different cell compartments, (ii) the speed and duration of release, (iii) the mode of C_2H_4 generation, and (iv) the sensitivity of the plants due to their developmental stage and the plant species. The total amount of C_2H_4 released seems to play a minor role (Lürssen, 1982; Konze and Lürssen, 1983; Lürssen and Konze, 1985).

Very recently, the All-Russian Scientific Research Institute of Agricultural Biotechnology and St. Petersburg Institute of Basic Chemical Industry jointly developed a new C_2H_4-releasing preparation under the trade name of "Retprol" which is discussed in detail in Section 8.2.4. Another plant growth regulator, 2-hdyroxyethylhydrazine, also yields typical C_2H_4 reactions in plants. However, this compound does not release C_2H_4 in aqueous solutions, but may be converted enzymatically to C_2H_4 within the plant (Block and Young, 1971; Dollwet and Kumamoto, 1972). Methionine (a physiological precursor of C_2H_4 in higher plants and microorganisms) and ACC (an intermediate of C_2H_4 biosynthesis in higher plants), and its derivatives were found to exert C_2H_4-like reactions in plants in a manner suitable for agricultural uses (Schröder and Lürssen, 1978; Arshad and Frankenberger, 1990a; Zahir and Arshad, 1998; also see Section 8.4.5).

Of all the commercially prepared C_2H_4-releasing compounds, ethephon (also prepared under the trade name ethrel, Amchem 66-329, and CEPA) has the broadest spectrum of usefulness in agriculture.

8.2.3. Ethephon and Its Applications

Ethephon (2-chloroethanephosphonic acid) was first synthesized by the Russian workers, Kabachnik and Rosiiskaya (1946) and later by the GAF Corporation (Wayne, New Jersey) and it was tested for growth regulator activity by Amchem Products, Inc. (Cooke and Randall, 1968; Morgan, 1982). The appearance of treated plants and the chemistry of ethephon suggested that it was converted into C_2H_4. In 1967, Amchem Products, Inc. released it for large-scale testing (Anonymous, 1967) and upon appearance of reports regarding its physiological effects (Edgerton and Blanpied, 1968; Morgan, 1969; Warner and Leopold, 1969; Yang, 1969) its commercial preparation was acted upon rapidly (DeWilde, 1971).

The hydrolysis of ethephon at pH 5 and above yields chloride, phosphate, and C_2H_4 as final products (Maynard and Swan, 1963). The chemical reaction involves a nucleophilic attack on the phosphate dianion by water yielding C_2H_4 (Maynard and Swan, 1963; Yang, 1969), along with the elimination of chlorine and the formation of phosphate as shown below:

$$Cl - CH_2 - CH_2 - PO_3H_2 + OH^- \rightarrow CH_2 = CH_2 + Cl^- + H_3PO_4$$
$$\text{(ethephon)} \qquad\qquad\qquad \text{(ethylene)}$$

The absorption and breakdown of ethephon varies from plant to plant within plant tissues. The breakdown of ethephon to C_2H_4 occurs primarily on the leaf surface (Beaudry and Kays, 1988). Lavee and Martin (1974) observed that most of the [^{14}C]-ethephon applied to peach fruit was broken down and only a small portion was absorbed. However, Bukovac et al. (1971) found rapid uptake of ethephon and later conversion to C_2H_4 in cherry. Interestingly, Van Andel and Verkerke (1978) found that *Poa pratensis* and *Avena sativa* shoots treated with ethephon were still releasing C_2H_4 after 25 days. In a study with squash, cucumber, and tomato a 50% conversion of [^{14}C]-ethephon to C_2H_4 occurred in 7 days (Yamaguchi et al., 1971). Because ethephon participates in a wide variety of biological reactions, its mode of application is extremely important. Furthermore, factors such as plant growth stage, plant stress status, plant foliage spray coverage, ethephon rates and environmental conditions determine the response obtained. The effectiveness of ethephon varies with the dose level and the date of application (Gianfagna et al., 1986; Irving, 1987).

Ethephon, commonly known as "liquid ethylene", is registered in the United States for many preharvest and harvest-aid processes as summarized in Table 8.1 (Abeles et al., 1992). It has been reported that over 50% of the ethephon used in the United States is applied to field crops such as wheat, barley, and cotton (Morgan, 1986). Ethephon is also used in large amounts in tropical regions in the production of coffee, pineapple, rubber and sugarcane (Table 8.2). Murray et al. (1995) identified five major agricultural applications of ethephon: (i) increase boll opening in cotton, (ii) enhance pistallate flower induction in hybrid squash seed, (iii) accelerate flower maturity in processing tomatoes, (iv) enhance hull splitting in walnuts and (v) reduce lodging in wheat. Reports describing new uses for ethephon are continually appearing in the scientific literature.

Ethephon has many applications in agriculture. It has been used to break the dormancy of peanut (Ketring, 1977), geranium (Rogers, 1987), and witchweed seeds (Egley and Dale, 1970). Similarly, Banno et al. (1986) found that ethephon applied to pear trees in the summer after cessation of shoot growth increased the number of flowers three-fold the following year. Increased frost protection by delaying flowering is another use of ethephon. Delay of flowering in sweet cherry and plum was observed following ethephon application during the previous fall (Dennis, 1976). However, deleterious side effects of ethephon were also observed. Ethephon hastens the development of female flowers in *Cucurbis* (Iwahori et al., 1969; McMurray and Miller, 1968; Robinson et al., 1968; Rudich et al., 1969) and *Cannabis sativa* (Mohan and Jaiswal, 1970). Ethephon also promotes the female flowers of zucchini, pumpkin, muskmelon, and squash. However, no increase in pistillate flowers on watermelon was observed (Lower and Miller, 1969; Rudich et al., 1969). Ethephon has been used to induce male sterility in small flowers (Rowell and Miller, 1971; Fairey and Stoskopf, 1975) (Table 8.1). Thankur and Rao (1988) indicated that ethephon induced male sterility in pearl millet; however, female fertility in a male-sterile line was not affected by ethephon. Ethephon has also been used as a harvest aid in crops (citrus, cranberries, plums, apples, tomatoes, and peppers) in which the development of yellow or red pigments is important (Abeles et al., 1992).

Application of ethephon to increase stem stiffness and reduce lodging in cereals is an established practice in the United States (Table 8.1) and Europe (Table 8.2). The variable effects of ethephon on growth and yield parameters of various cereals are comprehensively summarized in Table 8.3. The response of cereal crops to a given

Table 8.1. Registered uses for ethephon in the United States.[a]

Purpose	Crop	Product
Preharvest		
Promote lateral branching	Azalea, geranium	Florel®
Increase bud hardiness and delay bloom	Sweet cherry	Ethrel®
Initiate uniform flowering	Pineapple, bromeliad	Ethrel®, Florel®
Promote flowering on young trees	Apple	Ethrel®
Modify flowering and sex expression, early fruit set	Cucumber, squash, pumpkin	Florel®
Thin flowers and promote return bloom	Apple	Ethrel®
Prevent preharvest fruit drop and promote fruit color	Apple	Ethrel®, Fruitone N[c]
Prevent lodging, shorten stems	Wheat, barley	Cerone®, Florel®
Harvest aid		
Hasten yellowing and reduce curing time	Tobacco	Ethrel®
Promote and/or hasten uniform ripening, color development, and maturity	Cherry, table grape, raisin grape, pepper, blackberry, tomato, apple, boysenberry, pineapple	Ethrel®. Florel®
Promote fruit abscission and nut hull dehiscence	Cherry, apple, black-berry, cantaloupe, walnut	Ethrel®, Florel®
Promote mature fruit dehiscence and enhance defoliation	Cotton	Prep®
Promote defoliation, remove shoots	Rose, tallhedge buckthorn, apple nursery stock, dwarf and leafy mistletoe	Florel®

[a]Information taken from specimen labels in product label guide, Rhone-Poulenc Agricultural Company, 1990, for Cerone®, Ethrel®, Prep®, and Florel®, which are all registered trade names.
Source: Abeles et al. (1992)

ethephon treatment is highly dependent on management practices, particularly number and time of irrigations (d'Andria et al., 1997; Kasele et al., 1994). Ethephon has been extensively tested on barley (Caldwell et al., 1988; Dahnous et al., 1982; Simmons et al., 1988; Hill et al., 1982; Green et al., 1988; Pearson et al., 1989; Tindall et al., 1989; Bridger et al., 1995; Foster and Taylor, 1993; Webster and Jackson, 1993; Ma et al., 1992). It reduced the plant height and increased yield when lodging occurred in the untreated control. Ramos et al. (1989) observed that an increase in yield in spring barley was due to an increase in the number of ears per plant. Lodging in maize occurs principally with sweet corn and in high-density stands of hybrid maize with favorable moisture and fertilizer. Post anthesis application of ethephon to sweet corn prevented lodging and increased the yield in some cases (Gaska and Oplinger, 1988a; Bratsch and Mack, 1990; Cox and Andrade, 1988; Kasele et al., 1994; d'Andria et al., 1997). However, yield reductions were noted on a dry year and with lodging-resistant cultivars.

Table 8.2. Worldwide uses of ethephon under brand names Cerone, Ethrel, Floral, and Prep.[a,b]

Area/Country	Apples/pears	Cereals	Cherries	Citrus	Coffee	Cotton	Cucumbers/squash	Grapes	Maize	Olive	Ornamentals	Peaches/plums	Peppers	Pineapple	Rice	Rubber	Small fruit	Sugarcane	Tobacco	Tomato
AFRICA/MIDDLE EAST																				
Israel	x							x		x										
Saudi Arabia	x	x						x					x	x				x	x	x
South Africa	x	x	x	x				x	x			x						x	x	
Zimbabwe	x				x				x					x				x	x	x
CENTRAL/SOUTH AMERICA																				
Argentina	x	x		x	x	x						x	x					x	x	x
Brazil					x							x	x	x				x		x
Chile	x															x			x	x
Costa Rica				x	x	x									x	x		x	x	
Dominican Republic						x									x					
Ecuador	x				x				x				x	x	x			x	x	x
El Salvador					x								x	x		x		x	x	x
Guatemala				x	x								x	x		x		x	x	x
Honduras				x	x								x	x		x		x	x	x
Mexico	x	x			x			x					x	x	x				x	x
Nicaragua				x	x								x	x		x		x	x	x
Panama					x								x	x		x		x	x	x
EUROPE																				
Austria											x								x	
Belgium		x	x								x						x			x
Czechoslovakia			x										x				x		x	x
Denmark	x	x	x										x				x			
France	x	x		x					x		x	x					x			
Germany	x	x	x								x						x			x

Table 8.2. (continued)

CROP USES[c]

Area/Country	Apples/pears	Cereals	Cherries	Citrus	Coffee	Cotton	Cucumbers/squash	Grapes	Maize	Olive	Ornamentals	Peaches/plums	Peppers	Pineapple	Rice	Rubber	Small fruit	Sugarcane	Tobacco	Tomato
Greece				x			x	x		x		x	x							x
Hungary		x																		
Italy	o						o	o		o		o	o						o	o
Netherlands	x										x									x
Norway	x											x					x			x
Poland			x														x			x
Romania			x				x					x	x							x
Spain	x			o						x		x	x							x
Sweden	x		x								x						x			
Switzerland		x	x								x									
UK	x	x									x						x			x
Yugoslavia										x		x								x
PACIFIC																				
Australia	x	x		x		x		x				x		x				x		x
Indonesia	x		x		x									x		x			x	
Japan	x	x	x	x								x		x		x				x
Malaysia			x											x		x				
New Zealand	x							x				x					x			x
Philippines														x		x				
Taiwan								x				x		x				x	x	
Thailand					x									x		x		x	x	

x = use approved or registered, o = use application filed Information on uses in Libera, Israel, Nigeria, West Africa, Zaire, Indonesia, Malaysia, New Zealand, Philippines, Taiwan, Thailand, Yugoslavia, and Chile.

[a] Harvest aid uses include promoting and concentrating maturity and color, reducing fruit removal force, degreening citrus Other uses include promoting early flowering on young apple trees, thinning fruit on apples and peaches, delay flowering in cherry and peach, defoliating apple and peach nursery stock.

[c] Less common registered uses include bananas in Ecuador and Indonesia, flax in Belgium, France and Netherlands; gums and resins in Brazil and Ecuador; macadamia nuts in Australia, melons in Italy and Romania, onions in the Netherlands; persimmon in Japan; oil seed rape in Denmark and Germany; soybeans in Ecuador; and sunflowers in France.

Source Abeles et al (1992)

Table 8.3. Effects of ethephon on various growth processes of cereals.

Plant	Concentration applied	Physiological responses	Reference
Barley	0.28-0.84 kg ha^{-1}	Treatment significantly reduced lodging and increased the 1000-grain weight, but increases in yields were not significant.	Foy (1983); Foy and Witt (1984)
Barley	0.3-0.6 kg ha^{-1}	Injection of ethephon into the base promoted tillering and reduced elongation of the upper stem. Foliar treatment, which reduced plant height, also increased post-treated tillering.	Foster et al. (1992)
Barley	50 mM	Both water shortage and C_2H_4 treatment reduced ear fertility and tillering when applied at joints and pre-anthesis.	Bergner and Teichmann (1993)
Corn	0.56-2.24 kg ha^{-1}	Foliar treatment caused a reduction in plant height, leaf area, leaf efficiency, and yield.	Earley and Slife (1969)
Corn	0.14-0.42 kg ha^{-1}	High rates of treatment reduced lodging by 87% and increased brace root development by 50%.	Gaska and Oplinger (1987)
Corn	0.14-0.56 kg ha^{-1}	Increasing application rates of treatment generally decreased yields. High application rates decreased lodging by 63%.	Gaska and Oplinger (1988b)
Corn	0.14-0.84 kg ha^{-1}	Treatment had a highly significant effect on grain yield, plant height, ear height, and brace root development. Yield of one hybrid variety generally decreased with increasing ethephon rates, whereas, the lowest rate (140 g ha^{-1}) increased the yield of another hybrid by as much as 700 kg ha^{-1}.	Langan and Oplinger (1987)
Corn	0.28-0.56 kg ha^{-1}	Foliar treatment reduced plant height by 20-24%, with subsequent reduction in lodging.	Sagaral and Parrish (1989)

Table 8.3 (continued)

Crop	Rate	Description	Reference
Corn	1.168 L ha^{-1}	When applied foliarly in combination with Triggrr (a cytokinin-containing material, 0.58 µL ha^{-1}), there were 2093 and 1110 kg ha^{-1} greater yields during the first and second year, respectively, over the respective controls. Lodging was also reduced markedly.	Sagaral and Parrish (1990)
Corn	0.28–0.84 kg ha^{-1}	Data of two years of field trials revealed that foliar application of ethephon reduced plant height and leaf area index by 10–40% relative to control. Ethephon application either had no effect or decreased yields in 1985 under all irrigation and plant density treatments, because of a lack of significant drought stress. However, when drought occurred in 1990, ethephon application decreased yields at low plant densities, but enhanced yields at high densities with a maximum of 37% yield increase for the intermediate ethephon rate implying that ethephon can improve resistance to drought in corn.	Kasele et al. (1994)
Corn	0.28–0.84 kg a.i. ha^{-1}	Increasing ethephon rates caused a significant reduction in leaf area index, plant height, and dry matter accumulation, resulting in reduction in water consumption. In the rainfall control, increasing the ethephon rate resulted in a linear increase in yield with maximum increase of about 200% (1.6 t ha^{-1}) and 75% (1.7 t ha^{-1}) in 1993 and 1994, respectively. By contrast, treatments irrigated four times caused a linear decrease of about 13% in both years. In another irrigation treatment trial, ethephon had no effect on yield.	d'Andria et al. (1997)
Rice	0.24–0.36 kg ha^{-1}	Foliar treatment increased yield by 10–45%.	Rao and Fritz (1987)
Spring Barley	0.48 kg a.i. ha^{-1}	Treatment increased tiller-derived spikes per square meter up to 265% compared with the control. Application reduced plant height and lodging, but decreased yield by 27% because of reduction in number of grains per spike, the main culm or spike number, or weight per grain.	Ma and Smith (1992)

Table 8.3 (continued)

Crop	Rate	Description	Reference
Spring barley	0.1–0.5 kg a.i. ha⁻¹	Ethephon increased the yield of Leger barley up to 15% in one experiment under favorable conditions, but during the next wet years, yield in one experiment was reduced by 11% at the higher ethephon dose.	Bridger et al. (1995)
Spring barley	0.48 kg a.i. ha⁻¹	Ethephon consistently reduced plant height and reduced lodging. Ethephon increased yield of Leger by 26% and decreased yield of Argle by 19%.	Bulman and Smith (1993)
Spring barley	0.1–0.5 kg a.i. ha⁻¹	Plant height and lodging were reduced; however, differences in cultivars response were observed. Ethephon was an effective anti-lodging agent only under moderate lodging conditions. It is not likely to increase yield under conditions of intense irrigation and high fertility.	Foster and Taylor (1993)
Spring wheat	0.56 kg a.i. ha⁻¹	Ethephon did not provide complete lodging control, but reduced plant height, delayed the onset of severe lodging until later stages of grain-fill, and increased grain yield by 5–21% depending on cultivars and lodging severity.	Webster and Jackson (1993)
Sweet corn	0.28–0.56 kg a.i. ha⁻¹	Ethephon reduced plant height by 12 to 26%. Effects of ethephon on husked yield varied from an 8% increase in yield to an 18% decrease, depending on rate, timing and season.	Bratsch and Mack (1990)
Wheat	0.28–0.56 kg ha⁻¹	Treatment significantly reduced lodging, but increases in yields were not significant.	Foy (1983); Foy and Witt (1984)
Wheat	0.42 kg ha⁻¹	Treatment reduced stem height significantly, without significant effect on yields.	Parrish et al. (1985)
Wheat	0.28–0.56 kg ha⁻¹	Foliar treatments either had no effect or reduced the yield of winter wheat by 6%, and decreased lodging and plant height slightly.	Nafziger et al. (1986)

Table 8.3 (continued)

Wheat	0.22 kg a.i. ha^{-1}	Treatment reduced lodging and plant height, and increased grain yield by an average of 280 kg ha^{-1} (6.4% increase). Grain yield increases were obviously only in cases where lodging was decreased.	Wiersma et al. (1986)
Wheat	1000-10,000 ppm	Applications to wheat decreased both leaf and stem elongation and promoted tillering.	Poovaiah and Leopold (1973)
Wheat, barley and rye	0.28-0.56 kg ha^{-1}	Foliar treatment progressively reduced plant height as applied concentrations increased, resulting in a consistent and total prevention of lodging. There were significant increases in mechanically harvested grain yields in the treated plots compared with the control.	Schwartz et al. (1982)
Wheat, barley, and triticales	0.28-0.85 kg ha^{-1}	Treatment reduced elongation of tall cereals and increased the harvestable yield by reducing lodging.	Dahnous et al. (1982)
Wheat and oat	0.28-2.24 kg ha^{-1}	Foliar application at low rates (up to 0.56 kg ha^{-1}) effectively reduced plant height and lodging and, in several cases, accounted for significant increases in yields of winter wheat. Although the lowest rate resulted in reduced plant height and lodging, a significant decrease in yields of spring oats was observed.	Brown and Earley (1973)

Yields were reduced in part by a reduction in ear leaf area (Konsler and Grabau, 1989) or ear elongation (Norberg et al., 1989). It appears that maize is more sensitive to damage by ethephon than small grains so the management of its use on maize is more critical. Ethephon is also registered for use on sugarcane in Hawaii to prevent photoperiod-sensitive flowering (Table 8.1). Moore and Osgood (1989) reported that application of ethephon at 0.56 kg ha^{-1} prevented flowering and increased cane yields by 7.5% and sugar yields by 10%. However, in other countries flowering was reduced without an increase in yields.

The action of ethephon on potato tuber production has also been evaluated. Although an initial trial was unsuccessful (Singh, 1970), later tests revealed that ethephon increased the number and decreased the size of tubers produced by potato plants (Garcia-Torres and Gomez-Campo, 1972; Langille, 1972; Schrader, 1987). These small tubers are desirable as seed potatoes. Koller and Hiller (1988) observed that ethephon also decreased two physiological disorders of potatoes, i.e., brown center and stem-end hollow heart. The causes of these disorders are not known; however, ethephon increased yields of marketable tubers by decreasing these disorders. Ethephon treatment also increased frost tolerance and the survival rate of tomato seedlings after transplanting (Liptay et al., 1982). Recently, Atta-Aly et al. (1999) reported that early application of ethrel increased the fruit yield of tomato by 15% over the control with a pronounced ripening delay. Transplanting of tobacco seedlings is occasionally delayed by unfavorable weather. Thus, ethephon could be used beneficially to delay growth and development of tobacco seedlings for 10 days to reduce the negative effect of weather (Kasperbauer and Hamilton, 1978).

Crop harvest can be enhanced by promoting abscission, dehiscence, and ripening. Effects on ripening include the development of color, flavor, and aroma. As shown in Table 8.1, ethephon is registered to promote ripening of cherries, table and raisin grapes, peppers, blackberries, tomatoes, apples, boysenberries, and pineapples. The effects of ethephon on the growth and yield of various plants, including cotton, sugar cane, bermudagrass, pea, soybean, tomato, and peach have been summarized comprehensively in Table 8.4.

8.2.4. Retprol and Its Applications

As discussed in Chapter 5 (Section 5.4), scientists of the All-Russian Scientific Research Institute of Agricultural Biotechnology and St. Petersburg Institute of the Basic Chemical Industry jointly developed a new C_2H_4-releasing preparation under the trade name of Retprol (Muromtsev et al., 1990, 1991, 1993, 1995; Bibik et al., 1995). This product is a preparative form of calcium carbide (CaC_2) that decomposes in soil. Upon addition to soil, Retprol decomposes into calcium hydroxide and acetylene (C_2H_2) under the influence of soil moisture. The released C_2H_2 is reduced to C_2H_4 by soil microorganisms. Retprol, the first commercial C_2H_4–promoting soil additive, is safer in storage than other commercially prepared C_2H_4-releasing products. Retprol is characterized by slow decomposition in soil and, correspondingly, by a prolonged period of moderate to high C_2H_4 levels in soil air. The efficiency of Retprol as an C_2H_4-releasing compound to improve the growth and yield of various crops was investigated in several greenhouse and field trials by using tomato, potato, cucumber, and diocious hemp as test crops.

Table 8.4. Effects of ethephon on various growth processes of different plants/crops.

Plant	Concentration applied	Physiological responses	Reference
Bermudagrass (Tifton 85)	35 mM	Treated plants showed 22% reduction in plant height, and a 118% and 101% increase in leaf/stem fresh and dry weight ratios, respectively. Data of glasshouse experiments revealed 112% more roots in treated plants grown under stressed conditions than untreated, 8 days after cutting removal and produced 10-fold higher number of tillers at 6 d after planting in soil than untreated cuttings. But differences in tillers disappeared after 21 days.	Shatters et al. (1998)
Bluegrass	1000-10,000 ppm	Applications resulted in marked inhibition of leaf elongation, but concomitantly stimulated stem growth. The highest concentration reduced leaf elongation by 75% and increased stem length by more than 20-fold in bluegrass.	Poovaiah and Leopold (1973)
Cotton	1.12 kg a.i. ha^{-1} (281 L ha^{-1})	Ethephon significantly reduced seed cotton yield when applied to cotton with 48-62% opened bolls. No detrimental effect on fiber quality was observed.	Smith et al. (1986)
Cotton	0.38-1.40 kg a.i. ha^{-1}	Application reduced the number of pink bollworm-infested bolls on early-season fruiting branches. High rates of application consistently reduced yields. In most cases, ethephon treatments delayed flowering.	Henneberry et al. (1992)
Fig	125-2000 ppm	High concentrations caused extensive tree defoliation and 500 ppm treatment resulted in early harvest of figs.	Ferguson et al. (1990)
Flax	0.90 kg a.i. ha^{-1}	Treatment significantly increased yield during the first year.	Lay and Dybing (1983)
Macadamia	100-1500 ppm	Treatment stimulated fruit abscission.	Nagao and Sakai (1987)

Table 8.4 (continued)

Crop	Rate	Description	Reference
Pea	2-250 ppm	Root treatment resulted in severe inhibition of nodulation; root growth was also slightly reduced.	Drennan and Norton (1972)
Peach	300 ppm	Treatment almost completely defruited Harbrite peach trees.	Byers and Lyons (1981)
Peach	480 ppm	Treatment caused thinning in Harken peach, and was equivalent in effectiveness to 150 ppm of GAF for fruit thinning.	Byers and Lyons (1981)
Peach	600 ppm	Treatment caused both fruit and leaf abscission.	Byers and Lyons (1981)
Peach	100 ppm	Treatment did not cause fruit thinning or leaf phytotoxicity to Cresthaven peach trees.	Byers and Lyons (1981)
Pepino	500 ppm	Ethephon has no effect on yield and quality character; however, in combination with salinity, it shortened the growth cycle by one month.	Prohens et al. (1999)
Soybean	0.28-84 kg ha^{-1}	Foliar application at low rates resulted in significant increases in the number of lateral branches and development of a shorter plant, whereas seed size was significantly reduced with the application of higher rates.	Sagaral and Foy (1989)
Sugarcane	1.12 kg ha^{-1}	Soil treatment significantly increased sucrose content of the immature internodal tissues, but reduction in glucose and fructose levels was observed.	Dill and Martin (1981)
Sunflower	400 ppm	Soil-applied ethephon caused reduction in stem height, chlorophyll breakdown, epinasty in leaves, and hypocotyl hypertrophy.	Kawase (1974)
Tomato	792 ppm (3 times)	Soil drenched with ethephon elicited a response similar to that of flooding (reduction in stem growth and chlorophyll content of the lower leaves and promotion in epinastic curvature of the leaf petiole and adventitious roots).	Kuo and Chen (1980)

Table 8.4 (continued)

Tomato	0-500 ppm	Soil application of ethephon markedly retarded the growth; however, branching and fruit yield were enhanced in some cases. Ethephon persisted for at least 6 weeks in soil.	Abdallah et al. (1986)
Tomato	100 ppm	Fruit receiving Ethrel treatment were higher in C_2H_4 and ACC levels than control. Early Ethrel-treated fruits became significantly larger in size and heavier in weight with a ripening delay of about 10 and 15 days, compared with those of control and aminoisobutyric acid-treated ones, respectively. The early application of ethrel increased the fruit yield by 15% over control with a pronounced ripening delay.	Atta-Aly et al. (1999)

Field trials during 1984-87 with a CaC$_2$ preparation revealed that its addition to soil at 120 kg ha^{-1} in a dry form during the growth stage of 3-4 true leaves of common tomatoes (Lebyazhinski variety) increased the yield by 35-50% (Table 8.5) along with accelerated ripening and improvement in quality (Muromtsev et al., 1990). Under greenhouse conditions, the CaC$_2$ preparation (60-80 kg ha^{-1}) in dry form or with irrigation water caused more flowering and accelerated tomato ripening. The harvest of tomatoes and cucumbers increased by 20-25% with gross yield increases of 10-15% (Muromtsev et al., 1990). Further field studies during 1990-1992 with tomato crop revealed that Retprol significantly improved the quality. Retprol application rates of 30 and 60 kg ha^{-1} increased the cluster number, the mean fruit weight and proportion of ripe fruit by 20-50% compared to the control. The ripening period was reduced by 6-9 days and its uniformity improved by 20%. Application of CaC$_2$ at 90 kg ha^{-1} was relatively less effective than Retprol in improving the growth parameters of tomato. The commercial tests indicated that Retprol applied at 30 kg ha^{-1} increased the tomato yield by 14 t ha^{-1} or by 33% compared to the control (Table 8.6) (Muromtsev et al., 1993). A similar kind of response was observed with mandarin fruits upon the application of CaC$_2$ to soil (Muromstev et al., 1991). Both fruit yield and quality of mandarin were increased substantially along with accelerated ripening.

The ability of C$_2$H$_4$ to change the sex ratio in plants in the direction of ferminization was observed in stands of dioceous hemp treated with Retprol. Experiments during 1988-92 with Retprol showed that the preparation (30, 60 and 90 kg ha^{-1}) significantly affected this physiological process as the ratio of male to female plants in the population changed in favor of female, while in controls this ratio was 1:1. The increased proportion of female plants resulted in a higher seed yield (by 0.2-0.3 t ha^{-1}) compared to the non-treated control. Similarly, studies with various cucumber cultivars revealed that Retprol promoted the shift of the plant sex toward the female side, resulting in increased number of female flowering ovaries and fruits. The cucumber yield was increased by 0.21 to 0.66 t ha^{-1} or by 16-38% compared to the controls (Table 8.7), depending upon the cultivars and the growth conditions (Muromtsev et al., 1993, 1995).

A two year trial with potato revealed that the application of Retprol and CaC$_2$ to soil resulted in increased number of tubers and tuber yield. The degree of response to both compounds depended on the varietal features of the potatoes. The application of various amounts of the C$_2$H$_4$ generating products (Retprol and CaC$_2$) increased the number of tubers by 25-80% and tuber yield by 29-121% for both varieties tested. Retprol was more effective than CaC$_2$ (Bibik et al., 1995).

8.2.5. Soil Ethylene and Its Ecological Significance

Ethylene production and catabolism in soil have been discussed in detail in Chapter 5. It is well documented that the production of C$_2$H$_4$ is widespread among the rhizosphere and soil microflora including both pathogenic and beneficial microorganisms. Moreover, organic amendments to soil enhance C$_2$H$_4$ accumulation in soil both biotically (major contribution) and abiotically (minor contribution). Plant roots can also be a potential source of soil C$_2$H$_4$ particularly, when exposed to any stress. All these natural sources contribute to the soil pool of C$_2$H$_4$. The addition of certain compounds can stimulate C$_2$H$_4$ synthesis in soil either by serving as substrates or as C and energy source for C$_2$H$_4$–producing microorganisms (see Chapters 4 and 5). Moreover, the combined application of an C$_2$H$_4$-producing microbial culture

Table 8.5. Effect of various doses of calcium carbide (dry form) on the productivity of common tomatoes of the Lebyazhinskii variety (Kherson Oblast, UkrNIIOZ, field experiments 1984-1986).*

Experiment	Commercial yield (t/ha) per year				Increase in yield	
	1984	1985	1986	Average	t/ha	% of control
Control	60.7	47.6	54.5	54.2	--	100
CaC$_2$						
30 kg/ha	62.3	56.6	59.2	59.4	5.2	110
90 kg/ha	76.4	60.1	70.0	68.8	14.6	127
120 kg/ha	--	71.4	76.4	73.9	19.7	136
150 kg/ha	--	66.2	66.7	66.5	12.3	123
Dextrel, 2 kg/ha**	63.7	50.0	56.1	56.6	24	104
HCP$_{05}$, t/ha	41.5	62.8	28.0			
Experimental precision, %	2.8	5.0	4.5			

*Soil: dark chestnut, clayey loamy, medium solonetzic; **, the synthetic C_2H_4 producer.
Source: Muromtseva et al. (1990).

Table 8.6. Tomato yields in response to applications of CaC$_2$ and Retprol.

Variant	Preparation rate (kg/ha)	Yield (t/ha)	Yield surplus (t/ha)	% of control
1990 Field Experiment				
Control	–	54	–	–
Control with optimal rate of calcium carbide	90	63	9	17
Retprol	15	60	6	11
	30	66	12	22
	60	66	12	22
	90	65	11	20
LSD$_{05}$		2.2		
1991 Field Experiment				
Control	–	70	–	–
Control with optimal rate of calcium carbide	90	73	3	4
Retprol	15	76	6	9
	30	86	16	23
	60	86	16	23
	90	75	5	7
LSD$_{05}$		1.1		
1992 Field Experiment				
Control	–	43	–	–
Control with optimal rate of calcium carbide	90	–	–	–
Retprol	15	57	14	33
	30			
	60			
	90			

Source: Muromtsev et al. (1993).

Table 8.7. Cucumber yields in response to applications of CaC$_2$ and Retprol.

		Field experiment						Commercial tests		
		Podarok (Odessa region)			Elektron (Moscow region)			Vodolei (Mordovia)		
			Yield surplus			Yield surplus			Yield surplus	
Variant	Preparation rate (kg/ha)	Yield (t/ha)	(t/ha)	% of control	Yield (t/ha)	(t/ha)	% of control	Yield (t/ha)	(t/ha)	% of control
Control	--	173	--	--	128	--	--	194	--	--
Calcium carbide	90	207	34	20	161	33	26	228	34	18
Retprol	30	238	65	38	161	33	26	--	--	--
	60	238	65	38	155	27	21	231	37	19
	90	239	66	38	149	21	16	--	--	--
LSD$_{05}$		14.0			13.9					

Source: Muromtsev et al. (1993)

and a suitable substrate can further increase C_2H_4 accumulation in soil. Since C_2H_4 is biologically active within an extremely low concentration range, it is most likely that C_2H_4 present in the soil atmosphere surrounding the roots can affect plant growth and crop productivity. This hypothesis is strongly supported by the plant response to an exogenous source of C_2H_4 applied to the roots (see Sections 8.2.1 and 8.2.4). Freytag et al. (1972) reported that soil-injected C_2H_4 improved the yields of cotton and sorghum. Similarly, Eplee (1975) showed that C_2H_4 injected into soil triggered suicidal germination of witchweed up to 90%, suggesting a role for C_2H_4 in weed eradication. The unequivocal proof that C_2H_4 uptake by plant roots can move to the shoots was demonstrated by rapid movement of labeled C_2H_4 in tomato and broad bean plants (Jackson and Campbell, 1975; El-Beltagy et al., 1986; 1990). High concentrations of C_2H_4 (1-10 ppm) in the rhizosphere of barley (Crossett and Campbell, 1975; Smith and Robertson, 1971), rice (Konings and Jackson, 1979; Smith and Robertson, 1971), tobacco (Smith and Robertson, 1971), tomato (Konings and Jackson, 1979; Nakayama et al., 1973; Smith and Robertson, 1971), soybean (Nakayama et al., 1973), pea (Goodlass and Smith, 1979; Konings and Jackson, 1979), white clover (Goodlass and Smith, 1979), citrus (Ishii and Kadoya, 1984b), and grapevine (Perret and Koblet, 1979) have been shown to markedly alter the growth of these plants. Thus, by altering the C_2H_4 concentrations in the soil atmosphere surrounding the roots, plant growth and development can be modified in a desired direction.

As indicated earlier, exogenous application of various compounds or organic materials can enhance C_2H_4 accumulation in soil and the rhizosphere biologically or nonenzymatically which can affect plant growth upon exposure of plant roots to the increased C_2H_4 levels. Exogenous additions of L-methionine (L-MET), its analogue, L-ethionine and ACC have been shown to increase C_2H_4 accumulation in the soil atmosphere (Arshad and Frankenberger, 1990a,b; Frankenberger and Phelan, 1985a,b; see Section 5.3.1). Arshad and co-workers investigated the effects of these amendments as a source of C_2H_4 in the rhizosphere on the growth and yield of various crops (Arshad and Frankenberger, 1990a; Arshad et al., 1993, 1994; Zahir and Arshad, 1998). Arshad and Frankenberger (1990a) found that soil application of L-MET affected vegetative growth and resistance to stem breaking (lodging) of two cultivars of corn (Table 8.8). Similarly, soil treatment with L-ethionine resulted in a significant epinastic response, enhanced fruit yield, and early fruit formation and ripening in tomato (Table 8.9). In another study, we found that *Albizia lebbeck* L. responded positively to low to moderate concentrations of L-MET (10^{-9} to 10^{-3} g kg^{-1} soil) applied to soil (Arshad et al., 1993). Growth parameters, including plant height, girth, dry weight of roots, total biomass, number of nodules, and dry weight of nodules of *A. lebbeck* were promoted significantly in response to various L-MET concentrations (Table 8.10). Plant N, P, and K concentrations and their uptake were also enhanced by treatment with L-MET. A significant quadratic dose-response relationship was found in all cases when each individual parameter was regressed against log [L-MET] excluding the control. Similarly, a significant growth and yield response in soybean was observed upon exposure to L-MET applied to soil (Arshad et al., 1994). We found that plant height, plant dry weight, root weight and total biomass, pods/plant and grain yield of soybean were significantly increased in response to various levels of L-MET application (Table 8.11). Recently, Zahir and Arshad (1998) investigated the response of sarsoon (*Brasica carinata*) and lentil (*Lens culinaris*) to C_2H_4 precursors added to the rhizosphere and

Table 8.8. Influence of L-methionine (MET) as an ethylene precursor on corn growth.

L-MET (mg kg⁻¹ soil)	Shoot height (cm)		Shoot fresh weight (g)		Shoot dry weight (g)		Stem diameter (mm)		Resistance to stem breaking (relative unit)	
	Kandy Korn	Miracle	Kandy Korn	Miracle	Kandy Korn	Miracle	Kandy Korn	Miracle	Kandy Korn	Miracle
Control	134a[a]	121a	159a	221a	26.1a	33.2a	15.4a	18.6a	3.41a	5.30b
0.0185	159b	134b	185ab	226a	28.7ab	30.4a	16.7ab	18.4a	3.67ab	3.41a
0.185	160b	137b	206bc	230ab	31.8ab	30.5a	17.0ab	19.1a	3.66ab	3.43a
1.85	173b	140b	231c	258b	34.5b	32.3a	17.4b	20.1b	4.35b	4.35ab

[a]Values sharing same letter(s) do not differ significantly at $p < 0.05$ level, according to Duncan's multiple range test.
Source: Arshad and Frankenberger (1990a).

Table 8.9. Influence of L-ethionine (L-ETH) as an ethylene precursor on tomato growth.

L-ETH (mg kg^{-1} soil)	Fresh fruit yield (g)	Avg. wt. of fresh fruit (g)	Total no. of ripe fruit	Epinastic movement (degrees) 72 h after treatment
Control	261a[a]	37.3a	11a	4.8a
0.2	477b	55.0ab	15ab	9.0b
2.0	445b	62.1b	16ab	9.8b
20	351ab	50.1ab	19b	12.3c

[a]Values sharing same letter(s) do not differ significantly at p < 0.05 level, according to Duncan's multiple range test. Source: Arshad and Frankenberger (1990a).

reported that the treatments had significant effects on growth and yield parameters of the test plants. The soil application of L-MET at 0.1 to 6 mg kg^{-1} caused a significant increase in plant height, pod length, number of seeds per pod, and straw and grain yield of sarsoon compared to untreated controls while, the 1000 grain weight was maximum with highest dose of L-MET applied. Similarly, the lowest level of ACC applied to soil increased grain yield and number of pods per plant of lentil significantly compared to the untreated controls, while other parameters (root and shoot weights, and 1000 grain weight) were not affected (Table 8.12). The highest dose of ACC resulted in significantly lower plant height. The proposed mechanism of action of these studies was substrate-dependent C_2H_4 production by the indigenous soil microflora. These studies enforce the possibility of using C_2H_4 precursors for biotransformations into C_2H_4 for the betterment of crop production. The combination of specific substrates and inoculum can release copious amounts of C_2H_4 in the rhizosphere and exposed roots which can lead to a physiological response. By using a C_2H_4 bioassay, the classical "triple" response in etiolated pea seedlings, Arshad and Frankenberger (1988) demonstrated that C_2H_4 of microbial origin can affect plant growth. We showed that MET-dependent C_2H_4 produced by an inoculum, *Acremonium falciforme* (Fig. 8.1) or by soil-indigenous microflora (Table 8.13) caused the classical "triple" response in etiolated pea seedlings, including reduction in elongation, swelling of the hypocotyl, and a change in direction of growth (horizontal). A similar response was observed by direct application of C_2H_4 gas.

Increased C_2H_4 concentrations in the soil air in response to other organic materials have been shown to evoke physiological responses upon exposed plants. In an attempt to investigate the reason(s) of sparse undergrowth in a heavy litter layer in pure stands of *Pinus radiata*, Lill and Waid (1975) found that under laboratory conditions, the litter produced a volatile factor which inhibited the shoot growth of white clover, ryegrass and *P. radiata* seedlings. They speculated that if this volatile factor accumulates in the field, it could be at least partially responsible for the sparse undergrowth in the *P. radiata* forests. Later, Lill and McWha (1976) identified this volatile factor as C_2H_4 by gas chromatography. They also observed that the inhibitory effects of the vapor from incubated *P. radiata* litter, on clover hypocotyl growth was completely removed by $KMnO_4$ and -195°C traps, and this effect was mimicked by direct application of C_2H_4. Similarly, plumule growth of rice seedlings was stimulated by litter vapor, and this

Table 8.10. Effect of L-methionine applied to soil on various growth parameters of *Albizia lebbeck* L. 32 weeks after treatments.

L-MET (g kg⁻¹ soil)	Plant height (cm)	Plant girth (cm)	Dry weight of shoot (g plant⁻¹)	Dry weight of root (g plant⁻¹)	No. of nodules (g plant⁻¹)	Dry weight of nodules (g plant⁻¹)	Total biomass
Control	309**	4.83	64.5	42.5	19.2	2.1	109
10^{-9}	445**	6.87**	68.0	44.8	21.8	3.2	117*
10^{-8}	422**	7.13**	74.5	45.0	29.0	5.3**	125**
10^{-7}	505**	7.30**	79.0	49.5**	35.0	8.1**	137**
10^{-6}	525**	7.83**	81.5	48.8*	36.0*	9.4**	140**
10^{-5}	423**	6.09**	75.8	46.0	32.0*	6.9**	129**
10^{-4}	350**	5.64**	71.0	40.8	32.5*	5.8**	118*
10^{-3}	380**	6.18**	68.0	38.2	29.2	5.8**	112
10^{-2}	341	5.39**	67.2	35.8	19.2	5.0**	108
10^{-1}	327	5.09*	64.0	37.5	12.8	2.2	102

*,** Means significantly different from control at $p \leq 0.05$(*) and $p \leq 0.01$(**), according ito Dunnett's Test.
Source: Arshad et al. (1993).

Table 8.11. Effect of soil-applied L-methionine (L-MET) on growth of soybean.

L-MET	Plant height	Straw weight	Root weight	Total biomass	Pods/ plant	Grain yield
(mg kg⁻¹ soil)	(cm)	----------------------(g/plant)----------------			(no.)	(g/plant)
Control	29.7	8.98	0.650	14.55	27.8	4.92
.0002	28.4	9.47	0.928*	13.85	22.5	3.45
.002	29.1	10.56*	0.890*	15.38	26.3	3.93
.02	30.8	10.34*	0.943*	17.38*	34.0*	6.10*
0.2	31.4	14.52*	0.950*	23.87*	42.3*	8.40*
2.0	34.3*	13.45*	1.035*	21.54*	42.5*	7.05*
20.0	31.9*	11.24*	1.003*	18.48*	35.8*	6.25*

*Means differed significantly with control at $p < 0.05$, according to Dunnett test.
Source: Arshad et al. (1994).

Table 8.12. Effect of ACC on growth and yield of lentil plants (average of 6 replicates).

Treatment (ACC) (mg kg⁻¹ soil)	Plant height (cm)	Dry root wt. (g)	Dry shoot wt. (g)	No. of pods plant⁻¹	1000 grain weight (g)	Grain yield (g plant⁻¹)
Control	27.5ab*	3.6a	8.1a	101.5b	30.3a	3.3b
0.5	28.7ab	4.2a	9.3a	135.8a	33.7a	4.9a
1.0	31.1a	3.9a	9.4a	115.7b	29.3a	3.9ab
2.0	24.7bc	3.8a	8.5a	104.7b	26.3a	3.1b
4.0	22.6c	3.4a	8.2a	103.3b	26.8a	3.1b

*Means sharing similar letter(s) do not differ significantly at $p < 0.05$.
Source: Zahir and Arshad (1998).

Table 8.13. Influence of L-methionine-derived C_2H_4 produced by indigenous soil microflora on etiolated pea seedlings.

Treatment[a]	Seedling length (cm)	Seedling diameter (mm)
Control	6.56 b [b]	1.87 a
AgNO₃ (240 mgL⁻¹)	13.50 d	1.93 ab
L-Methionine (5 mM)	5.14 ab	2.49 c
L -Methionine (5 mM) + AgNO₃ (240 mg L⁻¹)	11.10 c	2.06 ab
L -Methionine (10 mM)	3.90 a	2.75 d
L -Methionine (10 mM) + AgNO₃ (240 mg L⁻¹)	10.10 c	2.11 b

[a]Samples that did not receive AgNO₃ received NaNO₃ (240 mg L⁻¹).
[b]Values followed by the same letter were not significantly different at the 0.05 level according to Duncan's multiple-range test.   Source: Arshad and Frankenberger (1988).

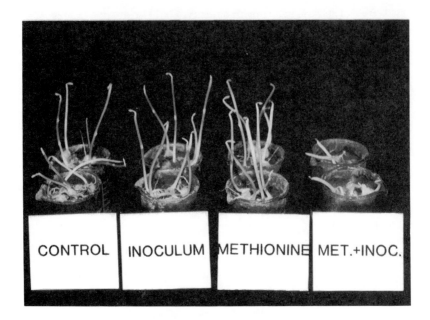

Fig. 8-1. Response of etiolated pea seedlings to the interaction between
L-methionine and inoculated Acremonium falciforme in sterile soil.
(From Arshad and Frankenberger, 1988.)

stimulation was also eliminated by $KMnO_4$ traps. Moreover, the addition of C_2H_4 to the vapor which had been passed through the $KMnO_4$ trap fully restored inhibition. They concluded that C_2H_4 released from the litter of *P. radiata* was responsible for the observed physiological and ecological effects on the growth of test plants and the same is applicable in the field where heavy litter layers of *P. radiata* had the potential of C_2H_4 accumulation in soil air at a level of physiologically active concentrations. They speculated that this high C_2H_4 level may be released as a result of microbial activity or senescence tissues (Lill and McWha, 1976). Likewise, the application of either composted or non-composted broiler litter to paddy rice has been found to significantly increase rice grain yield on some soils in eastern Arkansas (Miller and Wells, 1992). Soil and plant analysis of added nutrients revealed that the nutrients were not responsible for this yield response. However, considering the fact that rice is known to respond positively to C_2H_4 (Smith and Robertson, 1971; Metraux and Kende, 1983; Ku et al., 1970), led Miller and Wells (1992) to suggest that enhanced accumulation of C_2H_4 in the flooded, waste-amended soil may have been a factor in yield response. Later, Tang and Miller (1993) evaluated the effectiveness of poultry litter in stimulating C_2H_4 generation in soil. They confirmed that the application of poultry litter in either composted or non-composted form to soil enhanced C_2H_4 concentrations in soil air. They also concluded that enhanced C_2H_4 production in waste-amended soil may be an important factor in the positive response of lowland rice to waste application (Tang and Miller, 1993).

A number of studies were conducted by Ishii and co-workers to evaluate the significance of C_2H_4 released in soil from organic amendments on the growth of various plants (Ishii and Kadoya, 1984a,b; Iwasaki et al., 1981). Addition of dead or fresh plant materials obtained from grape, citrus, Japanese pear, peach, persimmon and rice greatly increased C_2H_4 evolution in soil (Ishii and Kadoya, 1984a). The application of dead grape leaves which caused the greatest C_2H_4 evolution markedly inhibited the growth of 'Muscat of Alexandria' grapevine cuttings and succinate dehydrogenase activity in the roots. In another study, they investigated the response of citrus plants exposed to various levels of C_2H_4 injected into the rhizosphere and compared the response evoked by C_2H_4 released from organic materials as applied by soil amendments (Ishii and Kadoya, 1984b). They found that the plant responses to organic materials were evoked by C_2H_4 released from the added material. Any material which did not stimulate C_2H_4 generation in soil did not have an affect on plant growth. They concluded that C_2H_4 evolution in waste-amended soils could be of great ecological significance. Iwasaki et al. (1981) reported that in citrus orchards, C_2H_4 concentration in soil of low productivity was higher than soil of high productivity. The physiological responses of plants evoked by higher C_2H_4 concentrations in waterlogged soils have already been discussed in Chapter 5 (Section 5.9).

8.3. ALTERATIONS IN ENDOGENOUS ETHYLENE BIOSYNTHESIS

Any factor/stimulus which causes a change in the endogenous levels of C_2H_4 in a plant could result in modified growth and development. With the advancement of knowledge on the physiology and biochemistry of C_2H_4, molecular approaches have led to the development of transgenic plants with low or high C_2H_4 production capacity. These approaches include inhibition of C_2H_4 biosynthesis by using antisense RNA of either ACC synthase (Oeller et al., 1991) or ACC oxidase (Hamilton et al., 1990) or by deamination of ACC (Klee et al., 1991) or by using bacterial *efe* genes encoding for the C_2H_4 –forming enzyme (Araki et al., 2000). Martin-Tanguy et al. (1993) reported that genetic transformation with a derivative of *rolC* from *Agrobacterium rhizogenes* and treatment with an inhibitor of C_2H_4 production, α-aminoisobutyric acid produced similar phenotypes and reduced C_2H_4 production in tobacco flowers. Likewise, potato plants expressing antisense and sense S-adenosylmethionine decarboxylase transgenes exhibited altered levels of C_2H_4 and polyamines while antisense plants also displayed abnormal phenotypes (Kumar et al., 1996). Inoculation of seeds or roots with specific inoculants could suppress the endogenous C_2H_4 synthesis, which subsequently creates a physiological response (Glick et al., 1998). These studies have opened up new opportunities for improving crop production, by altering the endogenous C_2H_4 levels in favor of a desired direction.

8.3.1. Transgenic Plants with ACC Oxidase/ACC Synthase Antisense

By employing molecular genetic approaches, transgenic plants with impaired biosynthesis of C_2H_4 have been developed (Hamilton et al., 1990; Klee et al., 1991; Oeller et al., 1991; Picton et al., 1993). In most of these studies, tomato has been used as a model system. Fruit from these transgenic plants were reported to have reduced C_2H_4 synthesis and to ripen abnormally. Hamilton et al. (1990) identified a cDNA clone for the tomato ACC oxidase gene (pTOM13) that inhibited C_2H_4 synthesis when expressed as an antisense gene in a transgenic plant. Cooper et al. (1998) reported that fruit of a transgenic tomato with ACC oxidase antisense transgenes exhibited reduced C_2H_4 biosynthesis (Fig. 8.2) and delayed ripening. Likewise, Picton et al. (1993) have demonstrated reduced ripening in ACC oxidase antisense tomato fruit. Reduction in C_2H_4 synthesis in transgenic plants using these approaches did not cause any apparent vegetative phenotypic abnormalities other than delayed leaf senescence (Picton et al., 1993). Fruit from these plants exhibited significant delays in ripening and the mature fruit typically remained firm longer than the non-transgenic control fruit.

By using an antisense ACC oxidase gene, Henzi et al. (1999a) studied C_2H_4 production in fully open flowers of transgenic lines of broccoli and demonstrated the feasibility of down-regulating C_2H_4 biosynthesis. Reduction up to 91% in C_2H_4 production after 96 h in comparison to the non-transgenic control were observed. In another study, Henzi et al. (1996b) demonstrated that broccoli flowers from all the transgenic plants exhibiting the tomato antisense ACC oxidase gene showed significantly reduced C_2H_4 production from 50 h and onward. Very recently, Henzi et al. (2000) investigated the morphological and agronomic characteristics of 12 transgenic broccoli lines containing a tomato antisense ACC oxidase gene. They observed that all lines showed reduced C_2H_4 production, particularly, significant reduction in C_2H_4 production from stalks of four field-grown transgenic lines of Green Beauty broccoli compared to the control, as observed 98 h after harvest. The transgenic plants also exhibited morphological changes to varying degrees. The results suggest that two enzyme systems may be involved in broccoli senescence, giving two bursts of C_2H_4 production, with only the second burst inhibited by the antisense ACC oxidase gene (Henzi et al., 2000).

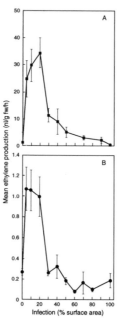

Fig. 8-2. Ethylene biosynthesis in fruit inoculated with *Colletotrichum gloeosporioides*. A) in ripe wild-type fruit (■) and B) ripe transgenic tomato with ACC oxidase antisense transgenes (●). Results are nL g fwt^{-1} h^{-1} in relation to symptom severity. Bars indicate the standard error of the mean. Source: Cooper et al. (1998).

Similarly, Ayub et al. (1996) generated transgenic cantaloupe melons in which C_2H_4 production was almost inhibited by the expression of an antisense ACC oxidase. Later they tested the chilling effect on fruits of these transgenic plants (Ben-Amor et al. , 1999). They observed that the sensitivity of cantaloupe melon to chilling injury can be considerably reduced by C_2H_4 suppression through expression of an antisense ACC oxidase gene in the transgenic plant.

Likewise, antisense RNA of ACC synthase has also been used to develop transgenic plants with reduced C_2H_4 synthesis. Oeller et al. (1991) observed that the expression of antisense ACC synthase RNA inhibited fruit ripening in transgenic tomato plants (Oeller et al., 1991). Similarly, a gene from bacteriophage T3 that encodes the enzyme, S-adeno-sylmethionine hydrolase (SAMase or AdoMetase, EC 3.3.1.2) was employed to inhibit C_2H_4 biosynthesis in transgenic plants (Hughes et al., 1987a,b; Studier and Movva, 1976). SAMase catalyzes the conversion of SAM to methylthioadenosine (MTA). Expression of SAMase in plants lowers the concentration of metabolic precursors (SAM and ACC) of C_2H_4 . Good et al. (1994) generated a transgenic tomato plant with the expression of T3 (encoding SAMase). The transgenic tomato plants produced fruits with a reduced capacity to synthesize C_2H_4 (Table 8.14). In another unpublished study, they observed delayed and modified ripening in tomatoes with expression of SAMase. They reported that the time required for fruit to develop to their final ripened state was approximately two-fold longer, the level of lycopene production was reduced, and the fruit demonstrated increased firmness and a delay in senescence for as long as three months after harvest (Good et al., unpublished data, cited by Good et al., 1994).

8.3.2. Transgenic Plants with Bacterial Ethylene-Forming Enzyme

By using the information upon sequencing of a bacterial *efe* gene encoding for the ethylene-forming enzyme (EFE) from *Ps. syringae* pv. *phaseolicola* PK2 (2-oxoglutarate-dependent dioxygenase), and its limited similarity with plant ACC oxidase (Fukuda et al., 1992, 1993), a transgenic tobacco plant was developed by Araki et al.

Table 8.14. Constitutive expression of SAMase reduces C_2H_4 production in transgenic tomato plants.

Sample	C_2H_4 (nl/g/h)	% reduction
Control – MS	2.9 + 0.15	
Control + NAA	11.4 + 4.33	
pAG-5120 – MS	1.22 + 0.05	58
pAG-5120 + NAA	3.51 + 1.66	69

Leaf disk assays were performed using untransformed controls and pAG-5120 transgenic tomato plants (cv. Large Red Cherry). MS samples are leaf disks placed on filter paper saturated with Murishige and Skoog basal medium. NAA samples had 10 μM naphthalene acetic acid added to the MS medium to stimulate C_2H_4 production. Samples for GC/FID analysis were taken after 20 h of ethylene accumulation ($n=3$).
Source: Good et al. (1994).

(2000) with the expression of bacterial EFE. They reported that two lines of transgenic plants produced C_2H_4 at consistently higher rates than the untransformed plants and their β-glucuronidase activities were expressed in different tissues. A significant dwarf morphology was observed in the transgenic tobacco displaying the highest C_2H_4 production which resembled the phenotype of a wild-type plant exposed to excess C_2H_4. This study for the first time unequivocally demonstrated the potential use of bacterial EFE to supply C_2H_4 as a hormonal signal via an alternate route (different from that of the MET-ACC route) using an ubiquitous C_2H_4 substrate, 2-oxoglutarate in plant tissues.

8.3.3. Transgenic Plants with Bacterial ACC-Deaminase

The plant growth promotion observed in response to inoculation with bacteria containing ACC-deaminase (see Section 8.3.4) provoked scientists to develop transgenic plants with the expression of ACC deaminase genes. These transgenic plants produce less C_2H_4 compared to wild-type parental plants, which subsequently affects the physiological processes and growth of plants, as summarized in Table 8.15. Klee and coworkers reported that transgenic tomato plants transformed with ACC-deaminase genes showed significant reduction in C_2H_4 synthesis (Fig. 8.3), delay in fruit ripening and improvement in fruit quality (Klee et al., 1991; Klee, 1993; Klee and Kishore, 1992). Similar kinds of observations in transgenic tomato plants expressing ACC deaminase (Figs. 8.4 and 8.5) were noted by Reed et al. (1995).

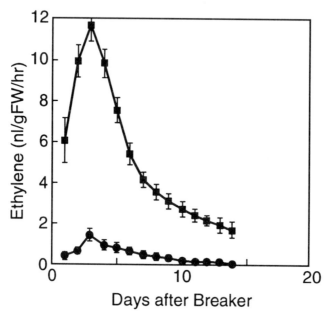

Fig. 8-3. Ethylene generation by 5673 transgenic with bacterial ACC-deaminase (●) and UC82B control (■) ripening fruit. Bars represent means ±SE at specific time points. Fruits were detached at breaker stage, and C_2H_4 generation was measured daily.
Source: Klee et al. (1991).

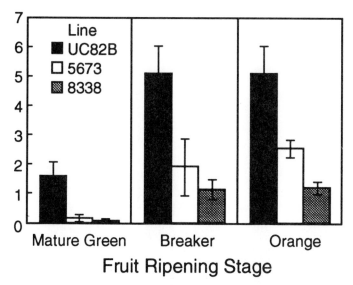

Fig. 8-4. Ethylene synthesis by tomato fruit at different
ripening stages for DR lines 8338 and 5673 (transgenic
tomato plants expressing ACC deaminase) and control line
UC82B. Vertical lines around the bars represent + SE.
Source: Reed et al. (1995).

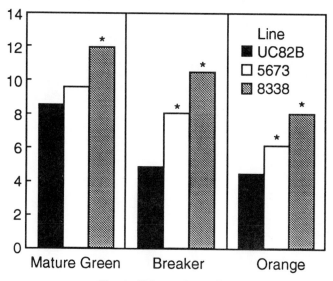

Fig. 8-5. Time-to-red for tomato fruit at different ripening
stages for DR lines 8338 and 5673 (transgenic tomato plants
expressing ACC-deaminase) and control line UC82B. Asterisk
shows statistically different from the UC82B control at the 5%
level. Source: Reed et al. (1995).

Table 8.15. Physiological responses evoked in plants transformed with ACC-deaminase genes.

Transgenic plants	ACC-deaminase genes source	Plant responses	Reference
Tomato	*Enterobacter cloacae* UW4	In general, transgenic tomato plants expressing ACC deaminase acquired greater amount of metals (Cd, Co, Cu, Ni, Pb and Zn) within the plant tissues and were less subject to the deleterious effects of metals on plant growth than were non-transgenic plants.	Grichko et al. (2000)
Tomato	*Pseudomonas* sp. 6G5	The transgenic plants showed a reduction in C_2H_4 synthesis up to 90%. Fruits from these plants exhibited a significant delay in ripening and mature fruits remained firm for at least 6 weeks longer than the nontransgenic control fruit.	Klee et al. (1991)
Tomato	*Ps. chloroaphis* strain 6G5	Two transformed tomato lines with the expression of ACC deaminase showed a significant reduction in C_2H_4 production compared to the parental control. The time for fruit to ripen was also extended for both the lines compared to the parental wild type.	Reed et al. (1995)
Tomato	*Pseudomonas* sp. strain 6G5	Fruits from the transgenic line 5673 ripened significantly slower than the control fruit when removed from the vine early in the ripening stage. In contrast, fruit that remained attached to the plants ripened much more rapidly, exhibiting little delay relative to the control. There was significantly more internal C_2H_4 in attached than detached fruit which was consistent with the ripening behavior. The transgenic fruit were significantly firmer than the control. The authors concluded that an enzyme other than ACC oxidase might also have a significant role in fruit softening.	Klee (1993)

Table 8.15 (continued)

Petunia, tobacco, tomato	*Ps. chloroaphis* strain 6G5	Senescence of flowers of tobacco plants expressing ACC deaminase was delayed. Likewise both initiation and duration of the ripening process were delayed in transformed tomato plants. C_2H_4 synthesis in transgenic plants was greatly reduced up to 77% in all tissue where ACC-deaminase was expressed.	Klee and Kishore (1992)
Petunia hybrida	*Pesudomonas*	Several transgenic plants expressing the ACC deaminase gene exhibited significant reduction in ACC content of pollen. Two independent transformants contained only trace amounts of ACC in pollen.	Lindstrom et al. (1996); Lei et al. (1996)

8.3.4. Inoculation with ACC-Deaminase Containing Plant Growth-Promoting Rhizobacteria

Microbial inoculation has been shown to alter the endogenous levels of C_2H_4 which subsequently leads to changes in plant growth (Table 8.16). Glick and his co-workers have conducted a series of studies to determine the mechanism responsible for the observed growth-promoting effects of a plant growth-promoting rhizobacterium (PGPR), *Pseudomonas putida* GR12-2 (Glick et al., 1994a,b; Hall et al., 1996; Xie et al., 1996; Mayak et al., 1999). They discovered that *Ps. putida* GR12-2 contained the enzyme ACC deaminase (Jacobson et al., 1994; Glick et al., 1994a,b) which hydrolyzes ACC into NH_3 and α-ketobutyrate (see Section 3.3) resulting in lowering of endogenous C_2H_4 production. Glick and his associates investigated the role of ACC deaminase in the growth-promoting activity of this bacterium by using its ACC deaminase minus and IAA overproducing mutants (Glick et al., 1994a,b; Hall et al., 1996; Mayak et al., 1999). It was reported that only the wild-type cells of *Ps. putida* GR12-2, but not any of the ACC deaminase minus or IAA overproducing mutants, promoted root growth of developing canola seedlings under gnotobiotic conditions. Similar observations were made in the case of lettuce, tomato, and wheat, in addition to canola (Hall et al., 1996). Later, other strains of bacteria (such as *Enterobacter cloacae*) which carry ACC deaminase activity were found to promote the growth of inoculated roots (Shah et al., 1997; Li et al., 2000). Glick et al. (1997) assessed the effects of seed inoculation with *Ps. putida* GR12-2 or its mutant *Ps. putida* GR12-2 *acd*68 (lacking ACC deaminase activity) on early growh of canola seedlings. Root and shoot lengths, fresh and dry weight, and the chlorophyll and protein contents of shoots were monitored in unstressed control seedlings and in seedlings growing in a saline soil or exposed to cold night temperatures. Under all conditions including salt or temperature stress, the wild-type bacterium, but not the mutant, promoted root growth. In soil inoculation experiments, the wild-type bacterium not only increased root growth, but was also effective in promoting shoot growth. Recently, Mayak et al. (1999) compared the effect of inoculation with *Ps. putida* GR12-2 (wild type), *Ps. putida* GR 12-2/acd36 (an ACC deaminase minus mutant) and *Ps. putida* GR12-2/aux1 (an IAA overproducing mutant) on the rooting of mung bean cuttings. Ethylene production in the inoculated cuttings was a result of the combined effects of ACC deaminase localized in the bacteria and bacterial produced IAA because IAA is known to stimulate endogenous C_2H_4 production via enhanced ACC synthase activity. Inoculation with the wild type and the ACC deaminase minus mutant increased the roots of cuttings significantly compared to the control, whereas the IAA overproducing mutant had no effect. The results imply that C_2H_4 is involved in the initiation and elongation of adventitious roots in mung bean cuttings and the PGPR (*Ps. putida* GR12-2) stimulates rooting of mung bean cuttings by lowering endogenous C_2H_4 production. All of these results have led Glick and co-workers to postulate a model (as illustrated in Fig. 8.6) to explain the mechanism of growth promotion by *Ps. putida* GR12-2. According to this model, the bacterial enzyme, ACC deaminase stimulates plant growth (in particular, root elongation) by sequestering and then hydrolyzing ACC released from inoculated germinating seeds. Hydrolysis of ACC results in lowering of C_2H_4 in seeds, which subsequently results in growth promotion. Glick and coworkers reported the following evidence in support of this model: (i) C_2H_4 has previously been found to be an inhibitor of plant root elongation in several different systems (Abeles et al., 1992). However, soil

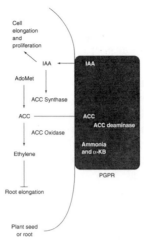

Fig. 8-6. Schematic representation of how a plant growth-promoting rhizobacteria (PGPR) bound to either a seed or plant root lowers the C_2H_4 concentration and thereby prevents C_2H_4 inhibition of root elongation. The arrows indicate a chemical or physical step in the mechanism and the symbol indicates inhibition (in this case the inhibition of root elongation by C_2H_4). Key: IAA, indoleacetic acid, ACC; 1-aminocyclopropane-1-carboxylic acid; AdoSAMMet, S-adenosyl-methionine; α-kB, α-ketobutyrate. Source: Glick et al. (1998).

bacteria that contain the enzyme, ACC deaminase can lower C_2H_4 levels by decreasing the plant ACC content (Glick et al., 1995; Penrose et al., unpublished results). Thus, the roots of C_2H_4 sensitive plants are invariably longer when ACC deaminase-containing bacteria are present (Hall et al., 1996); (ii) Three separate mutants of PGPR *Ps. putida* GR12-2 that were devoid of any ACC deaminase activity, unlike wild-type *Ps. putida* GR12-2, were unable to promote the elongation of canola roots in growth pouches under gnotobiotic conditions (Lifshitz et al., 1987; Glick et al., 1994a,b). Similarly, with tomato, lettuce and wheat, as well as canola, wild-type *Ps. putida* GR12-2, but not an ACC deaminase minus mutant of *Ps. putida* GR12-2 promoted root elongation under gnotobiotic conditions (Hall et al., 1996), a result which indicates that mutants devoid of ACC deaminase can no longer act as PGPR; (iii) The ACC level in extracts of roots from canola seedlings as measured by HPLC were lower when the seeds were treated with *Ps. putida* GR12-2 than in seedlings from untreated seeds. In addition, the ACC levels in root extracts from the seedlings in which the seeds were treated with an ACC deaminase minus mutant of *Ps. putida* GR12-2 were similar to those from the untreated seedlings (Table 8.16); (iv) The increase in the length of the roots of young (five- to seven-day-old) seedlings of canola, tomato, lettuce, wheat, oats and barley following treatments of the seeds with wild-type *Ps. putida* GR12-2 was similar to the response of these plants when their seeds were treated with the C_2H_4 inhibitor, AVG (Hall et al., 1996). Similarly, plants that are stimulated to the greatest extent by treatment with either *Ps. putida* GR12-2 or AVG are those that are the most sensitive to root length inhibition by ethephon (Hall et al., 1996). Thus, as far as root length is concerned, only C_2H_4 sensitive plants respond to the presence of PGPR that contain ACC deaminase; (v) Following transposon mutagenesis and selection for IAA overproducing mutants of *Ps.*

putida GR12-2, it was found that the mutant that overproduced IAA to the greatest extent was inhibitory to root elongation (Xie et al., 1996). The simplest explanation for this inhibitory effect on root growth of the IAA overproducing mutant of *Ps. putida* GR12-2 is that the increased level of IAA secreted by the mutant bacterium is taken up by the plant and therein interacts with the enzyme ACC synthase, stimulating the synthesis of excess ACC which is in turn converted to C_2H_4 (Yang and Hoffman, 1984; Kende, 1993); (vi) It was observed that with the transgenic tomato plants, the expression of the bacterial enzyme ACC deaminase dramatically lowered C_2H_4 levels in ripening tomato fruit (Klee et al., 1991; Sheehy et al., 1991). Although, strictly speaking, this situation is different from root-bacterial interaction, it confirms that this enzyme can in fact lower C_2H_4 levels in plants (see Section 8.5.4); and (vii) It was recently observed that when canola seeds were treated with *Escherichia coli* cells expressing a cloned *Enterobacter cloacae* ACC deaminase gene, these *E. coli* cells were able to promote root elongation (Shah et al., 1998). This demonstrates that enhanced root elongation was a direct consequence of the presence of the enzyme, ACC deaminase.

8.4. CONCLUDING REMARKS

As discussed earlier, the exogenous sources of C_2H_4 can be used for the benefit of agriculture production if employed carefully by keeping in view the specific hormone (C_2H_4)-plant response(s) and interactions. The availability of liquid C_2H_4 (e.g., ethephon) has provided a breakthrough in commercial use of C_2H_4 although its application is restricted mostly to the foliar spray. Preparations like Retprol have made it possible to use soil additives for increasing C_2H_4 concentrations in the root zone. Many other organic compounds/materials can be used as soil amendments to enhance C_2H_4 in the rhizosphere to a level of physiologically active concentrations. On the other hand, where the endogenous C_2H_4 could be inhibitory to certain plant growth processes, this inhibition can be overcome by using microbial isolates carrying ACC deaminase activity which converts ACC into NH_3 and α-ketobutyrate. Molecular approaches have successfully demonstrated the possibility of increasing or decreasing endogenous C_2H_4 synthesis in transgenic plants. There are many exciting opportunities of regulating the endogenous C_2H_4 synthesis in favor of plant growth and development. In the future, scientists will manipulate microbial sources of C_2H_4, and ACC deaminase via molecular approaches for the regulation of endogenous C_2H_4 in plants to benefit the agriculture industry.

Table 8-16. Physiological responses of plants to inoculation with bacteria containing ACC-deaminase activity.

Plant	Bacterial strains	Plant responses to inoculation	Reference
Canola	*Ps. putida* GR 12-2 and its ACC-deaminase minus mutants	Only wild type strain promoted root growth of developing canola seedlings under gnotobiotic conditions while its mutants lacking ACC-deaminase did not evoke any such response.	Glick et al. (1994b)
Canola	*Enterobacter cloacae* UW4 (wild type and an ACC-deaminase minus mutant)	Wild type parent strain possessing ACC deaminase activity stimulated the elongation of canola roots under gnotobiotic conditions while its ACC-deaminase minus mutant had no effect, confirming that mechanism action was lowering of C_2H_4 production.	Li et al. (2000)
Canola	*Ps. putida* GR 12-2 (wild type and over-producing IAA mutants)	Root and shoot growth were promoted in response to inoculation with wild type, but not with the mutant.	Glick et al. (1997)
Canola	*Enterobacter cloacae* *Escherichia coli* *Ps. putida* *Ps. fluorescens*	Only those strains exhibiting ACC-deaminase activity promoted the elongation of canola roots under gnotobiotic conditions.	Shah et al. (1998)
Canola	*Ps. fluorescens* strains CHA0 and CHO96 (ACC deaminase lacking strain)	Transformed strains with ACC deaminase activity increased root length of canola plants under gnotobiotic conditions, whereas, strains without this activity had no effect. The transformed strain also exhibited improved ability to protect cucumber against *Pythium* damping-off and potato tubers against *Erwinia* soft rot in a hermetically sealed container.	Wang et al. (2000)
Canola, lettuce, tomato, wheat	*Ps. putida* GR 12-2 and its ACC-deaminase mutants	Only wild type strain showed a positive effect on the root growth of inoculated plants. The results were comparable to those obtained with AVG treatment, an inhibitor of C_2H_4 biosynthesis in plants.	Hall et al. (1996)

Table 8.16 (continued)

Canola/tomato	*Kluyvera ascorbata* SUD165	The bacterium with ACC deaminase activity partially protected the plants against nickel toxicity under gnotobiotic conditions most likely via lowering the level of stress C_2H_4 induced by the nickel.	Burd et al. (1998)
Flower petals (zinnia, white sin carnation, maltese cross)	*Ps. putida* GR12-4 *Ps. putida* UW1	Inoculation with ACC-deaminase containing PGPR was effective in prolonging the petal life of C_2H_4 –sensitive flowers.	Nayani et al. (1998)
Mung bean	*Ps. putida* GR 12-2 (wild type) GR12-2/acd36 (ACC deamainase minus mutant) GR12-2/aux1 [indole acetic acid (IAA) overproducing mutant]	In all cases, inoculation caused changes in C_2H_4 production rates by the cuttings in parallel to the combined effect of ACC deaminase and bacterial produced IAA. The wild type and ACC-deaminase minus mutant treated cuttings had a significantly higher number of roots compared with cuttings rooted in water. Moreover, only the wild type influenced the development of longer roots. IAA overproducing mutants did not influence the number of roots. The results were inconsistent with the model that bacterially produced IAA stimulated the synthesis of ACC and hence C_2H_4, which affected the initiation and elongation of adventitious roots in mung bean cuttings.	Mayak et al. (1999)

8.5. REFERENCES

Abeles, F. B., Morgan, P. W., and Saltveit, M. E., Jr. , 1992, *Ethylene in Plant Biology*, 2nd Edition, Academic Press, San Diego.

Abdallah, M. M. F., El-Beltagy, A. S., Maksoud, M. A., Smith, A. R., and Hall, M. A., 1986, Effect of soil application of chloroethylphosphonic acid upon the growth and yield of tomatoes, *Acta Hortic.* **190**:383-387.

Agar, I. T., Massantini, R., Hess-Pierce, B., and Kader, A. A., 1999, Postharvest CO_2 and ethylene production and quality maintenance of fresh-cut kiwifruit slices, *J. Food Sci.* **64**:433-440.

Anonymous, 1967, Amchem-66-329, a new plant growth regulator, Amchem Products Inc., Information Sheet 37.

Araki, S., Matsuoka, M., Tanaka, M., and Ogawa, T., 2000, Ethylene formation and phenotypic analysis of transgenic tobacco plants expressing a bacterial ethylene-forming enzyme, *Plant Cell Physiol.* **41**:327-334.

Arshad, M., and Frankenberger, W. T., Jr., 1988, Influence of ethylene produced by soil microorganisms on etiolated pea seedlings, *Appl. Environ. Microbiol.* **54**:2728-2732.

Arshad, M., and Frankenberger, W. T., Jr., 1990a, Response of *Zea mays* and *Lycopersicon esculentum* to the ethylene precursors, L-methionine and L-ethionine, applied to soil, *Plant Soil* **122**:219-227.

Arshad, M., and Frankenberger, W. T., Jr., 1990b, Ethylene accumulation in soil in response to organic amendments, *Soil Sci. Soc. Am. J.* **54**:1026-1031.

Arshad, M., Javed, M., and Hussain, A., 1994, Response of soybean (*Glycine max*) to soil-applied precursors of phytohormones, *PGRSA Quarterly* **22**:109-115.

Arshad, M., Hussain, A., Javed, M., and Frankenberger, W. T., Jr., 1993, Effect of soil applied methionine on growth, nodulation and chemical composition of *Albizia lebbeck* L., *Plant Soil* **148**:129-135.

Atta-Aly, M. A., Riad, G. S., Lacheene, Z. El.-S., and El-Beltagy, A. S., 1999, Early application of ethrel extends tomato fruit cell division and increases fruit size and yield with ripening delay, *J. Plant Growth Regul.* **18**:15-24.

Ayub, R., Guis, M., Ben Amor, M., Gillot, L., Roustain, J.-P., Latche, A., Bouzayen, M., and Pech, J.-C., 1996, Expression of ACC oxidase antisense gene inhibits ripening of cantaloupe melon fruits, *Nature Biotech.* **14**:862-866.

Banno, K., Hayashi, S., and Tanabe, K., 1986, Promotion of flower bud formation and increase of pollen yield by application of ethephon and BA in 'Chojuro' pear (*Pyrus serotina*, Rehd.), *J. Japan Soc. Hort. Sci.* **55**:33-39.

Beaudry, R. M., and Kays, S. J., 1988, Flux of ethylene from leaves treated with a polar or non-polar ethylene-releasing compound, *J. Amer. Soc. Hort. Sci.* **113**:784-789.

Bebawi, F. F., and Eplee, R. E., 1986, Efficacy of ethylene as a germination stimulant *of Striga hermonthica* seed, Weed Sci. 34:694-698.

Ben-Amor, M., Flores, B., Latche, A., Bouzayen, M., Pech, J. C., and Romojaro, F., 1999, Inhibition of ethylene biosynthesis by antisense ACC oxidase RNA prevents chilling injury in *Charentais* cantaloupe melons, *Plant Cell Environ.* **22**:1579-1586.

Bergner, C., and Teichmann, C., 1993, A role for ethylene in barley plants responding to water shortage, *J. Plant Growth Regul.* **12**:67-72.

Bhatt, J. R., 1987, Gum-tapping in *Anogeissus latifolia* (Combretaceae) using ethephon, Curr. Sci. 56:936-930.

Bibik, N. D., Letunova, S. V., Druchek, E. V., and Muromtsev, G. S., 1995, Effectiveness of soil-acting ethylene producer in obtaining sanitized seed potatoes, *Russian Agri. Sci.* **9**:19-21.

Block, M. J., and Young, D. C., 1971, Conversion of β-hydroxethylhydrazine to ethylene, *Nature (London) New Biol.* **231**:288.

Bratsch, A. D., and Mack, H. J., 1990, Ethephon and mechanical topping influence growth, yield, and lodging of sweet corn, *HortSci.* **25**:291-293.

Bridger, G. M., Klinck, H. R., and Smith, D. L., 1995, Timing and rate of ethephon application to two-row and six-row spring barley, *Agron. J.* **87**:1198-1206.

Brown, C. M., and Earley, E. B., 1973, Response of one winter wheat and two spring oat varieties to foliar applications of 2-chloroethyl phosphonic acid (Ethrel), *Agron. J.* **65**:829-832.

Bukovac, M. J., Zucconi, F., Wittenbach, V. A., Flore, J. A., and Inoue, H., 1971, Effects of (2-chloroethyl)phosphonic acid on development and abscission of maturing sweet cherry *(Prunus avium* L.) fruit, *J. Amer. Soc. Hort. Sci.* **96**:777-781.

Bulman, P., and Smith, D. L., 1993, Yield and grain protein response of spring barley to ethephon and triadimefon, *Crop Sci.* **33**:798-803.

Burd, G. I., Dixon, D. G., and Glick, B. R., 1998, A plant growth-promoting bacterium that decreases nickel toxicity in seedlings, *Appl. Environ. Microbiol.* **64**:3663-3668.

Byers, R. E., and Lyons, C. G., 1981, Peach thinning results with ethylene release agents, in: *Proceedings Eighth Annual Meeting Plant Growth Regulator Society of America*, St. Petersburg, FL., Aug. 3-6, 1981, pp. 179-180.

Caldwell, C. D., Mellish, D. R., and Norris, J., 1988, A comparison of ethephon alone and in combination with CCC or DPC applied to spring barley, *Can. J. Plant Sci.* **68**:941-946.

Chace, E. M., and Church, C. G., 1927, Effect of ethylene on the composition of fruits, *Indust. Eng. Chem.* **19**:1135-1138.

Chace, E. M., and Sorber, D. G., 1936, Treating fruits and nuts in atmospheres containing ethylene, *Food Indust.* **8**:292-294.

Cooke, A. R., and Randall, D. L., 1968, 2-Haloethanephosphonic acids as ethylene releasing agents for the induction of flowering in pineapple, *Nature* **218**:974-975.

Cooper, W., Bouzayen, M., Hamilton, A., Barry, C., Rossall, S., and Grierson, D., 1998, Use of transgenic plants to study the role of ethylene and polygalacturonase during infection of tomato fruit by *Colletotrichum gloeosporioides*, *Plant Pathol.* **47**:308-316.

Cousins, H. H., 1910, Annual Report of the Jamaican Department of Agriculture 7:15.

Cox, W. J., and Andrade, H. F., 1988, Growth, yield, and yield components of maize as influenced by ethephon, *Crop. Sci.* **28**:536-542.

Crocker, W., 1948, *Growth of Plants*, Chapter 4, Reinhold Publ. Corp., New York, pp. 139-171.

Crossett, R. N., and Campbell, D. J., 1975, The effects of ethylene in the root environment upon the development of barley, *Plant Soil* **42**:453-464.

Dahnous, K., Vigue, G. T., Law, A. G., Konzak, C. F., and Miller, D. G., 1982, Height and yield response of selected wheat, barley and triticale cultivars to ethephon, *Agron. J.* **74**:580-582.

d'Andria, R., Chiaranda, F. Q., Lavini, A., and Mori, M., 1997, Grain yield and water consumption of ethephon-treated corn under different irrigation regimes, *Agron. J.* **89**:104-112.

De Munk, W. J., and Duineveld, T. L. J., 1986, The role of ethylene in the flowering response of bulbous plants, *Biol. Plant (Prague)* **28**:85-90.

Dennis, F. G., Jr., 1976, Trials of ethephon and other growth regulators for delaying bloom in tree fruits, *J. Amer. Soc. Hort. Sci.* **101**:241-245.

Denny, F. E., 1923, Method of coloring citrus fruits, U.S. Patent 1,475,938. December, 1923.

Denny, F. E., 1924, Hastening the coloration of lemons, *J. Agr. Res.* **27**:757-768.

Devlin, R. M., 1966, Plant Physiology, Reinhold Publ. Corp., New York.

DeWilde, R. C., 1971, Practical applications of (2-chloroethyl) phosphonic acid in agricultural production, *HortSci.* **6**:364-370.

Dill, G. M., and Martin, F. A., 1981, The effect of ethephon on endogenous sucrose and reducing sugar levels of immature internodal tissues of sugarcane, in: *Proc. Eighth Annual Meeting Plant Growth Regulator Society of America*, St. Petersburg, FL, Aug. 3-6, 1981, pp. 1.

Dilley, D. R., 1989, Air separator technology development for controlled atmosphere storage, in: *Proc. Fifth Intl. Controlled Atm. Res. Conf.*, Wenatchee, WA, pp. 409-418.

Dilley, C. L., and Dilley, D. R., 1985, New technology for analyzing ethylene and determining the onset of the ethylene climacteric of apples, in: *Controlled Atmospheres for Storage and Transport of Perishable Agricultural Commodities*, S. M. Blankenship, ed., Proc. 4th Natl. Cont. Atm. Res. Conf., pp. 353-362.

Dilley, D. R., Lange, E., and Tomala, K., 1989, Optimizing parameters for controlled atmosphere storage of apples, in: *Proc. Fifth Intl. Controlled Atm. Res. Conf.*, Wenatchee, WA, pp. 1-16.

Dollwet, H. H. A., and Kumamoto, J., 1972, The conversion of 2-hydroxethylhydrazine to ethylene, *Plant Physiol.* **49**:696-699.

Draber, W., 1977, Naturiche und synthetische Wachstumsregulatoren, in: *Chemie der Pflanzenschutz: und Schadlingsberkampfungsmittel, Band 4*, R. Wegler, ed., Springer-Verlag, Berlin, p. 1.

Drennan, D. S. H., and Norton, C., 1972, The effect of Ethrel on nodulation in *Pisum sativum* L., Plant Soil **36**:53-57.

Earley, E. B., and Slife, F. W., 1969, Effect of Ethrel on growth and yield of corn, *Agron. J.* **61**:821-823.

Edgerton, L. J., and Blanpied, G. D., 1968, Regulation of growth and fruit maturation with 2-chloroethanephosphonic acid, *Nature* **219**:1064-1065.

Egley, G. H., and Dale, J. E., 1970, Ethylene, 2-chloroethylphosphonic acid, and witchweed germination, *Weed Sci.* **18**:586-589.

Egley, G. H., 1982, Ethylene stimulation of weed seed germination, *Agric. Forest Bull.* **5**:13-18 (Univ. Alberta).

El-Beltagy, A. S., Madkour, M. A., and Hall, M. A., 1986, Uptake and movement of ethylene in tomatoes in relation to waterlogging, *Acta Horticul.* **190**:355-370.

El-Beltagy, A. S., Madkour, M. A., and Hall, M. A., 1990, ^{14}C-Ethylene movement and uptake in waterlogged broad bean (*Vicia faba* L.) plants, *Egypt. J. Hort.* **17**:171-180.

Eplee, R. E., 1975, Ethylene, a witchweed seed germination stimulant, *Weed Sci.* **23**:433-436.

Fairey, D. T., and Stoskopf, N. C., 1975, Effects of granular ethephon on male sterility in wheat, *Crop. Sci.* **15**:29-32.

Ferguson, L., Shorey, H., and Wood, D., 1990, Ethrel effects on fig harvest, in: *Proc. 17th Annual Meeting Plant Growth Regulator Society of America*, St. Paul, MN, Aug. 5-9, 1990, pp. 153-156.

Foster, K. R., and Taylor, J. S., 1993, Response of barley to ethephon: Effects of rate, nitrogen, and irrigation, *Crop. Sci.* **33**:123-131.

Foster, K. R., Reid, D. M., and Pharis, R. P., 1992, Ethylene biosynthesis and ethephon metabolism and transport in barley, *Crop Sci.* **32**:1345-1352.

Foy, C. L., 1983, Ethephon as an anti-lodging agent for small grains, in: *Proceedings Tenth Annual Meeting, Plant Growth Regulator Society of America*, East Lansing, MI, June 19-23, 1983, pp. 293-301.

Foy, C. L., and Witt, H. L., 1984, Ethephon for prevention of lodging of wheat and barley in Virginia: A three-year summary, in: *Proc. 11th Annual Meeting, Plant Growth Regulator Society of America*, Boston, MA, July 29-Aug. 1, 1984, pp. 68-69.

Frankenberger, W. T., Jr., and Phelan, P. J., 1985a, Ethylene biosynthesis in soil I. Method of assay in conversion of 1-aminocyclopropane-1-carboxylic acid to ethylene, *Soil Sci. Soc. Am. J.* **49**:1416-1422.

Frankenberger, W. T., Jr., and Phelan, P. J., 1985b, Ethylene biosynthesis in soil II. Kinetics and thermodynamics in the conversion of 1-aminocyclopropane-1-carboxylic acid to ethylene, *Soil Sci. Soc. Am. J.* **49**:1422-1426.

Freytag, A. H., Wendt, C. W., and Lira, E. P., 1972, Effects of soil-injected ethylene on yields of cotton and sorghum, *Agron. J.* **64**:524-526.

Fukuda, H., Ogawa, T., Ishihara, K., Fujii, T., Nagahama, K., Omata, T., Inoue, Y., Tanase, S., and Morino, Y., 1992, Molecular cloning in *Escherichia coli*, expression and nucleotide sequence of the gene for the ethylene-forming enzyme of *Pseudomonas syringae* pv. *phaseolicola* PK2, *Biochem. Biophys. Res. Commun.* **188**:826-832.

Fukuda, H., Ogawa, T., and Tanase, S., 1993, Ethylene production by microorganisms, *Adv. Microbiol. Physiol.* **35**:275-306.

Garcia-Torres, L., and Gomez-Campo, C., 1972, Increased tuberization in potatoes by ethrel (2-chloro-ethyl-phosphonic acid), *Potato Res.* **15**:76-80.

Gaska, J. M., and Oplinger, E. S., 1987, Influence of ethephon on sweet and field corn, *Bull. Plant Growth Regul. Soc. Am.* **16**:1988. (Abstract).

Gaska, J. M., and Oplinger, E. S., 1988a, Yield, lodging, and growth characteristics in sweet corn as influenced by ethephon timing and rate, *Agron. J.* **80**:722-726.

Gaska, J. M., and Oplinger, E. S., 1988b, Use of ethephon as a plant growth regulator in corn production, *Crop Sci.* **28**:981-986.

Gianfagna, T. J., Marini, R., and Rachmiel, S., 1986, Effect of ethephon and GA$_3$ on time of flowering in peach, *HortSci.* **21**:69-70.

Glick, B. R., Karaturovic, D. M., and Newell, P. C., 1995, A novel procedure for rapid isolation of plant growth-promoting pseudomonads, *Can. J. Microbiol.* **41**:533-536.

Glick, B. R., Penrose, D. M., and Li, J., 1998, A model for the lowering of plant ethylene concentrations by plant growth-promoting bacteria, *J. Theor. Biol.* **190**:63-68.

Glick, B. R., Jacobson, C. B., Schwarze, M. M. K., and Pasternak, J. J., 1994a, Does the enzyme 1-aminocyclopropane-1-carboxylate deaminase play a role in plant growth promotion by *Pseudomonas putida* GR12-2? in: *Improving Plant Productivity with Rhizosphere Bacteria*, M. H. Ryder, Stephens, P. M., and Bowen, G. D., eds., CSIRO, Adelaide, pp. 150-152.

Glick, B. R., Jacobson, C. B., Schwarze, M. M. K., and Pasternak, J. J., 1994b, 1-Aminocyclopropane-1-carboxylic acid deaminase mutants of the plant growth promoting rhizobacterium *Pseudomonas putida* GR12-2 do not stimulate canola root elongation, *Can. J. Microbiol.* **40**:911-915.

Glick, B. R., Liu, C., Ghosh, S., and Dumbroff, E. B., 1997, Early development of canola seedlings in the presence of the plant growth-promoting rhizobacterium *Pseudomonas putida* GR12-2, *Soil Biol. Biochem.* **29**:1233-1239.

Gonzalez, L. G., 1924, The smudging of mango trees and its effects, *Philipp Agric.* **12**:15-28.

Good, X., Kellogg, J. A., Wagoner, W., Langhoff, D., Matsumura, W., and Bestwick, R. K., 1994, Reduced ethylene synthesis by transgenic tomatoes expressing S-adenosylmethionine hydrolase, *Plant Molec. Biol.* **26**:781-790.

Goodlass, G., and Smith, K. A., 1979, Effects of ethylene on root extension and nodulation of pea (*Pisum sativum* L.) and white clover (*Trifolium repens* L.), *Plant Soil* **51**:387-395.

Green, C. F., Chalmers, I. F., and Packe-Drury-Lowe, S. J., 1988, Enhancing the performance of ethephon with mepiquat chloride on barley (*Hordeum distichon* cv. Panda) using an adjuvant comprising acidified sayal phospholipid, *Ann. Appl. Biol.* **113**:177-188.

Grichko, V. P., Filby, B.,, and Glick, B. R., 2000, Increased ability of transgenic plants expressing the bacterial enzyme ACC deaminase to accumulate Cd, Co, Cu, Ni, Pb and Zn, *J. Biotechnol.* **81**:45-53.

Gull, D. D., 1981, Ripening tomatoes with ethylene: Vegetable crops fact sheet, Florida Coop. Ext. Serv. VC-29.

Hall, J. A., Peirson, D., Ghosh, S., and Glick, B. R., 1996, Root elongation in various agronomic crops by the plant growth promoting rhizobacterium *Pseudomonas putida* GR12-2, *Isr. J. Plant Sci.* **44**:37-42.

Hamilton, A. J., Lycett, G. W., and Grierson, D., 1990, Antisense gene that inhibits synthesis of the hormone ethylene in transgenic plants *Nature* **346**:284-287.

Harvey, R. B., 1925, Blanching celery, Minn. Agr. Expt. Sta. Res. Bull. 222.

Harvey, R. B., 1928, Artificial ripening of fruits and vegetables, Minn. Agr. Sta. Res. Bull. 247.

Henneberry, T. J., Bariola, L. A., Chu, C. C., Meng, T., Jr., Deeter, B., and Jech, L. F., 1992, Early-season ethephon applications: Effect on cotton fruiting and initiation of pink bollworm infestations and cotton yields, *Southwestern Entomologist* **17**:135-147.Henzi, M. X., Christey, M. C., and McNeil, D. L., 2000, Morphological characterization and agronomic evaluation of transgenic broccoli (*Brassica oleracea* L. var. *italica*) containing an antisense ACC oxidase gene, *Euphytica* **113**:9-18.

Henzi, M. X., Christey, M. C., McNeil, D. L., and Davies, K. M., 1999a, *Agrobacterium rhizogenes*-mediated transformation of broccoli (*Brassica oleracea* L. var. *italica*) with an antisense 1-aminocyclopropane-1-carboxylic acid oxidase gene, *Plant Sci.* **143**:55-62.

Henzi, M. X., McNeil, D. L., Christey, M. C., and Lill, R. E., 1999b, A tomato antisense 1-aminocyclopropane-1-carboxylic acid oxidase gene causes reduced ethylene production in transgenic broccoli, *Aus. J. Plant Physiol.* **26**:179-183.

Hill, D. M., Joice, R., and Squires, N. R. W., 1982, Cerone: Its use and effect on the development of winter barley, in: *Chemical Manipulation of Crop Growth and Development*, J. S. McLaren (ed.). Butterworths, London, pp. 391-397.

Hills, L. D., and Haywood, E. H., 1946, *Rapid Tomato Ripening*, Faber & Faber, London. pp. 143.

Hughes, J. A., Brown, L. R., and Ferro, A. J., 1987a, Expression of cloned coliphage T3 S-adenosylmethionine hydrolase gene inhibits DNA methylation and polyamine biosynthesis in *E. coli*, *J. Bacteriol.* **169**:3625-3632.

Hughes, J. A., Brown, L. R., and Ferro, A. J., 1987b, Nucleotide sequence analysis of the coliphage T3 S-adenosylmethionine hydrolase gene and its surrounding ribonuclease III processing sites, *Nucl. Acid Res.* **15**:717-729.

Inaba, A., and Nakamura, R., 1988, Numerical expression for estimating the minimum ethylene exposure time necessary to induce ripening in banana fruit, *J. Amer. Soc. Hort. Sci.* **113**:561-564.

Irving, D. E., 1987, "Fantasia" nectarine: Effects of autumn-applied ethephon on blossoming and cropping, *New Zealand J. Expt. Agr.* **15**:67-72.

Ishii, T., and Kadoya, K., 1984a, Ethylene evolution from organic materials applied to soil and its relation to the growth of grapevines, *J. Japan. Soc. Hort. Sci.* **53**:157-167.

Ishii, T., and Kadoya, K., 1984b, Growth of citrus trees as affected by ethylene evolved from organic materials applied to soil, *J. Japan Soc. Hort. Sci.* **53**:320-330.

Iwahori, S., Lyons, J. M., and Sims, W. L., 1969, Induced femaleness in cucumber by 2-chloroethanephosphonic acid, *Nature* **222**:271-272.

Iwasaki, K., Mizutani, F., and Ishii, T., 1981, Soil aeration and the growth of citrus trees, in: *Systematic Studies for the Improvement of Citrus Growing on Steep Slope Orchards*, K. Kadoya, ed., Monbus-sho Sogo Kenkyo (A), Ehime Univ., Japan, pp. 42-46.

Jackson, M. B., and Campbell, D. J., 1975, Movement of ethylene from roots to shoots, a factor in the responses of tomato plants to waterlogged soil conditions, *New Phytol.* **74**:397-406.

Jacobson, C. B., Pasternak, J. J., and Glick, B. R., 1994, Partial purification and characterization of 1-aminocyclopropane-1-carboxylate deaminase from the plant growth promoting rhizobacterium *Pseudomonas putida* GR12-2, *Can. J. Microbiol.* **40**:1019-1025.

Kabachnik, M. I., and Rosiiskaya, P. A., 1946, Organophosphorus compounds. 1. Reaction of ethylene oxide with phosphorous trichloride, *Izv. Akad. Nauk. SSSR Khim. Nauk.* **406**:295.

Kasele, I. N., Nyirenda, F., Shanahan, J. F., Nielsen, D. C., and d'Andria, R., 1994, Ethephon alters corn growth, water use, and grain yield under drought stress, *Agron. J.* **86**:283-288.

Kasmire, R. F., Pratt, H. K., and Chacon, F., 1970, Honeydew melon maturity and ripening guide, Calif. Agr. Ext. Serv. MA-26.

Kasperbauer, M. J., and Hamilton, J. L., 1978, Ethylene regulation of tobacco seedling size, floral induction, and subsequent growth and development, *Agron. J.* **70**:363-366.

Kasperbauer, M. J., and Hamilton, J. L., 1978, Ethylene regulation of tobacco seedling size, floral induction, and subsequent growth and development, *Agron. J.* **70**:363-366.

Kawase, M., 1974, Role of ethylene in induction of flooding damage in sunflower, Physiol. Plant. **31**:29-38.

Kende, H., 1993, Ethylene biosynthesis, *Annu. Rev. Plant Physiol. Plant Mol. Biol.* **44**:283-307.

Ketring, D. L., 1977, Physiology of oil seeds, VI. A means to break dormancy of peanut *(Arachis bypogaea* L.) seeds in the field, *Peanut Sci.* **4**:42-45.

Kim, G.-H., and Wills, R. B. H., 1995, Effect of ethylene on storage life of lettuce, *J. Sci. Food Agric.* **69**:197-201.

Klee, H. J., 1993, Ripening physiology of fruit from transgenic tomato (*Lycopersicon esculentum*) plants with reduced ethylene synthesis, *Plant Physiol.* **102**:911-916.

Klee, H. J., and Kishore, G. M., 1992, Control of fruit ripening and senescence in plants, International Patent No. WO92/12249. European Patent Office, World Intellectual Property Organization.

Klee, H. J., Hayford, M. B., Kretzmer, K. A., Barry, G. F., and Kishore, G. M., 1991, Control of ethylene synthesis by expression of a bacterial enzyme in transgenic tomato plants, *Plant Cell* **3**:1187-1193.

Knee, M., Proctor, F. J., and Dover, C. J., 1985, The technology of ethylene control: Use and removal in post-harvest handling of horticultural commodities, *Ann. Appl. Biol.* **107**:581-595.

Knight, L. I., and Crocker, W., 1913, Toxicity of smoke, *Bot. Gaz.* **55**:337-371.

Koller, D. C., and Hiller, L. K., 1988, Response of russet burbank potatoes to ethephon sprays, *Am. Pot. J.* **65**:529-534.

Konings, H., and Jackson, M. B., 1979, A relationship between rates of ethylene production by roots, and the promoting or inhibiting effects of exogenous ethylene and water on root elongation, *Z. Pflanzenphysiol. Biodenkd.* **92**:385-397.

Konsler, J. V., and Grabau, L. J., 1989, Ethephon as a morphological regulator for corn, *Agron. J.* **81**:849-852.

Konze, J., and Lürssen, K., 1983, Unterschiedliche Wirkungsspektren von Ethylen-Wachstummsregulatoren: Welche Rolle spielt die freigesetzte Ethylenmenge? in: *Regulation des Phytohormon-gehaltes*, F. Bangerth, ed., Hohenheimer Arbeiten Band 129, Verlag Eugen Ulmer, Stuttgart, Germany, pp. 145.

Ku, V. V. V., Wills, R. B. H., and Ben-Yehoshua, S., 1999, 1-Methylcyclopropene can differentially affect the postharvest life of strawberries exposed to ethylene, *HortSci.* **34**:119-120.

Ku, H. W., Suge, H., Rappaport, L., and Pratt, H. K., 1970, Stimulation of rice coleoptile growth by ethylene, *Planta* **90**:333-339.

Kumar, A., Tylor, M. A., Arif, S. A. M., and Davies, H. V., 1996, Potato plants expressing antisense and sense S-adenosylmethionine decarboxylase (SAMDC) transgenes show altered levels of polyamines and ethylene: antisense plants display abnormal phenotypes, *The Plant J.* **9**:147-158.

Kuo, C. G., and Chen, B. W., 1980, Physiological responses of tomato cultivars to flooding, *J. Am. Soc. Hort. Sci.* **105**:751-755.

Langan, T. D., and Oplinger, E. S., 1987, Growth and yield of ethephon treated maize, *Agron. J.* **79**:130-134.

Langille, A. R., 1972, Effects of (2-chloroethyl) phosphonic acid on rhizome and tuber formation in the potato, *Solanum tuberosum* L. J. Amer. Soc. Hort. Sci. **97**:305-308.

Lavee, S., and Martin, G. C., 1974, Ethephon (1,2-^{14}C-(2-chloroethyl)phosphonic acid in peach fruits, I. Penetration and persistence, *J. Amer. Soc. Hort. Sci.* **99**:97-99.

Lay, C., and Dybing, C. D., 1983, Effects of plant growth regulators which delay senescence on seed and oil yields in oilseeds, in: *Proc. Tenth Annual Meeting Plant Growth Regulator Society of America*, East Lansing, MI, June 19-23, 1983, pp. 270-276.

Lei, C.-H., Lindstrom, J. T., and Woodson, W. R., 1996, Reduction of 1-aminocyclopropane-1-carboxylic acid (ACC) in pollen by expression of ACC deaminase in transgenic petunias, *Plant Physiol.* **111**:149.

Li, J., Ovakim, D. H., Charles, T. C., and Glick, B. R., 2000, An ACC deaminase minus mutant of *Enterobacter cloacae* UW4 no longer promotes root elongation, *Curr. Microbiol.* **41**:101-105.

Lifshitz, R., Kloepper, J. W., Kozlowski, M., Simonson, C., Carlson, J., Tipping, E. M., and Zaleska, L., 1987, Growth promotion of canola (rapeseed) seedlings by a strain of *Pseudomonas putida* under gnotobiotic conditions, *Can. J. Microbiol.* **33**:390-395.

Lill, R. E., and McWha, J. A., 1976, Production of ethylene by incubated litter of *Pinus radiata*, *Soil Biol. Biochem.* **8**:61-63.

Lill, R. E., and Waid, J. S., 1975, Volatile phytotoxic substances formed by litter of *Pinus radiata*, *N. Z. J. For. Sci.* **5**:65-70.

Lindstrom, J. T., Lei, C.-H., and Woodson, W. R., 1996, Reduction of 1-aminocyclopropane-1-carboxylic acid (ACC) in pollen by expression of ACC-deaminase in transgenic petunias, *HortSci.* **31**:571.

Liptay, A., Phatak, S. C., and Jaworski, C. A., 1982, Ethephon treatment of tomato transplants improves frost tolerance, *HortSci.* **17**:400-401.

Liu, F. W., 1978, Effects of harvest date and ethylene concentration in controlled atmosphere storage on the quality of 'McIntosh' apples, *J. Amer. Soc. Hort. Sci.* **103**:388-392.

Lower, R. L., and Miller,, C. H., 1969, Ethrel (2-chloroethanephosphonic acid) a tool for plant hybridizers, *Nature* 222:1072-1073.

Lürssen, K., 1982, Manipulation of crop growth by ethylene and some implications of the mode of generation, in: *Chemical Manipulation of Crop Growth and Development*, J. S. McLaren, ed., Butterworth Scientific, London, pp. 67-78.

Lürssen, K., 1991, Ethylene and agriculture, in: *The Plant Hormone Ethylene*, A. K. Mattoo and J. C. Suttle, eds., CRC Press, Boca Raton, FL, pp. 315-326.

Lürssen, K., and Konze, J., 1985, Relationship between ethylene production and plant growth after application of ethylene releasing plant growth regulators, in: *Ethylene and Plant Development*, J. A. Roberts and Tucker, G. A., eds., Butterworths, London, pp. 363-372.

Ma, B. L., and Smith, D. L., 1992, Chlormequat and ethephon timing and grain production of spring barley, *Agron. J.* 84:934-939.

Ma, B. L., Leibovich, G., Maloba, W. E., and Smith, D. L., 1992, Response of spring barley cultivars to nitrogen fertilizer and ethephon in regions with a short crop growing season, *J. Agron. Crop Sci.* 165:151-160.

Macnish, A. J., Joyce,, D. C., Hofman, P. J., and Simons, D. H., 1999, Involvement of ethylene in postharvest senescence of *Boronia heterophylla* flowers, *Aust. J. Exptl. Agric.* 39:911-913.

Martin-Tanguy, J., Corbineau, F., Burtin, D., Ben-Hayyim, G., and Tepfer, D., 1993, Genetic transformation with a derivative of *rolC* from *Agrobacterium rhizogenes* and treatment with α-aminoisobutyric acid produce similar phenotypes and reduce ethylene production and the accumulation of water-insoluble polyamine-hydroxycinnamic acid conjugates in tobacco flowers, *Plant Sci.* 93:63-76.

Mayak, S., Tivosh, T., and Blick, B. R., 1999, Effect of wild-type and mutant plant growth-promoting rhizobacteria on the rooting of mung bean cuttings, *J. Plant Growth Regul.* 18:49-53.

Maynard, J. A., and Swan, J. M., 1963, Organophosphorus compounds. I. 2-chloroalkylphosphonic acids as phosphorylating agents, *Aust. J. Chem.* 16:596-608.

McMurray, A. L., and Miller, C. H., 1968, Cucumber sex expression modified by 2-chloroethanephosphonic acid, Science 162:1397-1398.

Metraux, J. P., and Kende, H., 1983, The role of ethylene in the growth response of submerged deep water rice, *Plant Physiol.* 72:441-446.

Milbrath, J. A., Hansen, E., and Hartman, H., 1940, The removal of leaves from rose plants at the time of digging, Oregon Agric. Exp. Sta. Bull. 385, pp. 11.

Miller, E. V., 1947, The story of ethylene, Sci. Monthly 65:335-342.

Miller, D. M., and Wells, B. R., 1992, Indentifying the causes of poor rice growth on preciusion-graded soils, in: *Arkansas Rice Research Studies, 1991*, B. R. Wells, ed., Res. Series 422, Arkansas Agric. Exp. Sta., Fayetteville, AR, pp. 108-114.

Mohan, R. H. Y., and Jaiswal, V. S., 1970, Induction of female flowers on male plants of *Cannabis sativa* L. by 2-chloroethanephosphonic acid, *Experientia* 26:214-216.

Moore, P. H., and Osgood, R. V., 1989, Prevention of flowering and increasing sugar yield of sugarcane by application of ethephon (2-chloroethylphosphonic acid), *J. Plant Growth Regul.* 8:205-210.

Morgan, P. W., 1969, Stimulation of ethylene evolution and abscission in cotton by 2-chloroethanephosphonic acid, *Plant Physiol.* 44:337-341.

Morgan, P. W., 1982, Ethylene as an agricultural chemical, *Agric. For. Bull.* 5:29-37.

Morgan, P. W., 1986, Ethylene as an indicator and regulator in the development of field crops, in: *Plant Growth Substances, 1985*, M. Bopp, ed., Springer-Verlag, Berlin, pp. 375-379.

Muromtsev, G. S., Letunova, S. V., Beresh, I. G., and Alekseeva, S. A., 1990, Soil ethylene as a plant growth regulator and ways to intensify its formation in soils, *Biol. Bull. Acad. Sci. USSR* 16:455-461.

Muromstev, G. S., Krasinskaya, N. P., Letunova, S. V., and Beresh, I. G., 1991, Use of ethylene producing soil-acting preparation on citrus crops, *Soviet Agric. Sci.* 2:24-26.

Muromtsev, G. S., Letunova, S. V., Reutovich, L. N., Timpanova, Z. L., Gorbatenko, I. Y., Shapoval, O. A., Bibik, N. D., Stepanov, G. S., and Druchek, Y. V., 1993, Retprol--New ethylene-releasing preparation of soil activity, *Russian Agri. Sci.* 7:19-26.

Muromtsev, G. S., Shapoval, O. A., Letunova, S. V., and Druchek, Y. V., 1995, Efficiency of new ethylene producing soil preparation retprol on cucumber plants, *Selskokhozia. Biologiia* 5:64-68.

Murray, M., Beede, B., Weir, B., and Williams, J., 1995, Agricultural applications of ethephon, *HortSci.* 30:854.

Nafziger, E. D., Wax, L. M., and Brown, C. M., 1986, Response of five winter wheat cultivars to growth regulators and increased nitrogen, *Crop Sci.* 26:767-770.

Nagao, M. A., and Sakai, W. S., 1987, Effect of ethephoon on abscission of macadamia fruits, in: *Proceedings 14th Annual Meeting Plant Growth Regulator Society of America*, Honolulu, HI, Aug. 2-6, 1987, pp. 70.

Nakayama, M., Shimura, K., and Ota, Y., 1973, Physiological action of ethylene in crop plants. IV. Effects of ethylene application to roots on the growth of tomato and soybean plants, *Proc. Crop Sci. Soc. Japan* **42**:493-498.

Nayani, S., Mayak, S., and Glick, B. R., 1998, Effect of plant growth-promoting rhizobacteria on senescence of flower petals, *Indian J. Exptl. Biol.* **36**:836-839.

Norberg, O. S., Mason, S. C., and Lowry, S. R., 1989, Ethephon alteration of corn plant morphology, *Agron. J.* **81**:603-609.

Oeller, P. W., Min-Wong, L., Taylor, L. P., Pike, D. A., and Theologis, A., 1991, Reversible ihibition of tomato fruit senescence by antisense RNA, *Science* **254**:437-439.

Parrish, D. J., Williams, J. T., and Alley, M. M., 1985, Ethephon in winter wheat, in: *Proc. 12th Annual Meeting Plant Growth Regulator Society of America*, University of Colorado, Boulder, CO, July 28-Aug. 1, 1985, pp. 203.

Pearson, C. H., Golus, H. M., and Tindall, T. A., 1989, Ethephon application and nitrogen fertilization of irrigated winter barley in an arid environment, *Agron. J.* **81**:717-719.

Perret, P., and Koblet, W., 1979, Untersuchungen under den Zusammendang zwischen Sauerstaff-kohlendioxyd- und Aethylengehalt der Bodenluft und den Auftreten der Rebenctilorose, *Wein-Wiss.* **34**:151-170.

Picton, S., Barton, S. L., Bouzayen, M., Hamilton, A. J., and Grierson, D., 1993, Altered fruit ripening and leaf senescence in tomato expressing an antisense ethylene-forming enzyme transgene, *Plant J.* **3**:469-481.

Poovaiah, B. W., and Leopold, A. C., 1973, Effects of ethephon on growth of grasses, Crop Sci. **13**:755-758.

Prohens, J., Ruiz, J. J., and Nuez, F., 1999, Yield, earliness and fruit quality of pepino clones and their hybrids in the autumn-winter cycle, *J. Sci. Food Agric.* **79**:340-346.

Ramos, J. M., Garcia del Moral, L. F., Molina-Gano, J. L., Siamanca, P., Roca de Tagores, F., 1989, Effects of an early application of sulphur or ethephon as foliar spray on the growth and yield of spring barley in a Mediterranean environment, *J. Agric. Crop Sci.* **163**:129-137.

Rao, K. P., and Fritz, C. D., 1987, Effects of ethephon plant regulator on rice yield, in: *Proc. 14th Annual Meeting Plant Growth Regulator Society of America*, Honolulu, HI, Aug. 2-6, 1987, pp. 489.

Reed, A. J., Magin, K. M., Anderson, J. S. , Austin, G. D., Rangwala, T., Linde, D. C., Love, J. N., Rogers, S. G., and Fuchs, R. L., 1995, Delayed ripening tomato plants expressing the enzyme 1-aminocyclopropane-1-carboxylic acid deaminase, 1. Molecular characterization, enzyme expression, and fruit ripening traits, *J. Agric. Food Chem.* **43**:1954-1962.

Robinson, R. W., Shannon, S., and de la Gardia, M. D., 1968, Regulation of sex expression in cucumber, *Bio. Sci.* **19**:141-142.

Rodriguez, A. G., 1932, Influence of smoke and ethylene on the fruiting of the pineapple (*Ananas sativus* shult), *J. Dept. Agr. Puerto Rico* **26**:5-18.

Rogers, O. M., 1987, Ethephon overcomes seed scarification requirements of *Pelargonium*, in: Manipulation of Ethylene Responses in Horticulture, M. S. Reid, ed., *Acta Hort.* **201**:165-167.

Rosa, J. T., 1925, Ripening of tomatoes, *Proc. Åmer. Soc. Hort. Sci.* **22**:315-322.

Rowell, P. L., and Miller, D. G., 1971, Induction of male sterility in wheat with 2-chloroethylphosphonic acid (Ethrel), *Crop Sci.* **11**:629-631.

Rudich, J., Halevy, A. H., and Kedar,, N., 1969, Increase in femaleness of three cucurbits by treatment with ethrel, an ethylene-releasing compound, *Planta* **86**:69-76.

Sagaral, E. G., and Foy, C. L., 1989, Effects of selected growth regulators on growth, seed size, and yield of soybeans, in: *Proc. 16th Annual Meeting Plant Growth Regulator Society of America*, Arlington, VA, Aug. 6-10, 1989, p. 31.

Sagaral, E. G., and Parrish, D. J., 1989, Response of maize to two growth regulators, in: *Proc. 16th Annual Meeting Plant Growth Regulator Society of America*, Arlington, VA, Aug. 6-10, 1989, p. 32.

Sagaral, E. G., and Parrish, D. J., 1990, Effects of ethephon and Trigger on the growth and yield of corn, in: *Proc. 17th Annual Meeting Plant Growth Regulator Society of America*, St. Paul, MN, Aug. 5-9, 1990, p. 123.

Schipper, J. A., 1982, A smoke treatment influencing flower formation in iris bulbs, *Neth. J. Agr. Sci.* **30**:173-177.

Schrader, W. L., 1987, Use of ethephon for control of potato tuber size, in: Manipulation of Ethylene Responses in Horticulture, M. S. Reid, ed., *Acta Hortic.* **201**:171-174.

Schröder, R., and Lürssen, K., 1978, Mitted zur Regulierung des Pflanzenwachstums, Dtsch. Offeniegungsschrift Nr. 28(24):517.

Schwartz, T. K., Coffin, R. H., Evans, W. F., Nash, R. L., and Rao, K. P., 1982, The effect of ethephon (2-chloroethyl)phosphonic acid on lodging and yield of small grain cereal crops, in: *Proc. Ninth Annual Meeting Plant Growth Regulator Society of America*, Monterey, CA, July 5-9, 1982, pp. 89-95.

Serek, M., Sisler, E. C., and Reid, M. S., 1995, 1-Methylcyclopropene, a novel gaseous inhibitor of ethylene action, improves the life of fruits, cut flowers and potted plants, *Acta Hort.* **394**:337-345.

Shah, S., Li, J. Moffatt, B. A. , and Glick, B. R., 1998, Isolation and characterization of ACC deaminase genes from two different plant growth-promoting rhizobacteria, *Can. J. Microbiol.* **44**:833-843.

Shah, S., Li, J., Moffatt, B. A., and Glick, B. R., 1997, ACC deaminase genes from plant growth promoting rhizobacteria, in: *Plant Growth-Promoting Rhizobacteria. Present Status and Future Prospects*, A. Ogoshi, Kobayashi, K., Homma, Y., Kodama, F., Kondo, N., and Adkino, S., eds., OECD, Paris, France, pp. 320-324.

Shatters, R. G., Jr., Wheeler, R., and West, S. H., 1998, Ethephon induced changes in vegetative growth of "Tifton85" bermudagrass, *Crop. Sci.* **38**:97-103.

Sheehy, R. E., Honma, M., Yamada, M., Sasaki, T., Martineau, B., and Hiatt, W. R., 1991, Isolation, sequence, and expression in *Escherichia coli* of the *Pseudomonas* sp. strain ACP gene encoding 1-aminocyclopropane-1-carboxylate deaminase, *J. Bacteriol.* **173**:5260-5265.

Sherman, M., and Gull, D. D., 1981, A flow-through system for introducing ethylene into tomato ripening rooms, Vegetable Crops Fact Sheet, Fla. Coop. Ext. Serv. VC-30.

Sherman, M., 1985, Control of ethylene in the post harvest environment, *HortSci.* **20**:57-60.

Simmonds, N. W., 1959, Bananas, Longman, Green and Co., Ltd., London.

Simmons, S. R., Oelke, E. A., Wiersma, J. V., Lueschen, W. E., and Warnes, D. D., 1988, Spring wheat and barley responses to ethephon, *Agron. J.* **80**:829-834.

Singh, G., 1970, Influence of ethrel on growth and yield of potatoes, *Res. Life Sci.* **18**:38-43.

Sisler, E. C., Serek, M., and Dupille, E., 1996, Comparison of cyclopropene, 1-methylcyclopropene, and 3,3-dimethylcyclopropene as ethylene antagonists in plants, *Plant Growth Regul.* **18**:169-174.

Smith, C. W., Cothren, J. T., and Varvil, J. J., 1986, Yield and fiber quality of cotton following application of 2-chloroethyl phosphonic acid, *Agron. J.* **78**:814-818.

Smith, K. A., and Robertson, P. D., 1971, Effect of ethylene on root extension of cereals, *Nature* **234**:148-149.

Sterry, J. R., 1969, Ethrel as ethylene-evolving plant growth regulator, *Meded. Rijksfac. Landbouwwet. Gent.* **34**:462.

Studier, F. W., and Movva, N. R., 1976, SAMase gene of bacteriophage T3 is resopnsible for overcoming host restriction, *J. Virol.* **19**:136-145.

Stuart, N. W., Asen, S., and Gould, C. J., 1966, Accelerated flowering of bulbous iris after exposure to ethylene, *HortSci.* **1**:19-20.

Tang, T., and Miller, D. M., 1993, Ethylene production in anaerobically incubated soils amended with poultry litter, *Soil Sci.* **156**:186-192.

Thankur, R. P., and Rao, V. P., 1988, Effectiveness of ethrel as a male gametocide in pearl millet and its influence on ergot, *Plant Breeding* **101**:107-113.

Thomas, T. H., 1982, Plant Growth Regulator Potential and Practice, BCPC Publications, Croydon, U.K., pp. 271.

Tindall, T. A., Pearson, C. H., and Olsen, S., 1989, Ethephon application to spring barley under variable nitrogen and moisture regimes, *J. Fertilizer Issues* **6**:77-82.

Traub, H. P., Cooper, W. C., and Reece, P. C., 1940, Inducing flowering in the pineapple, *Ananas sativus, Proc. Amer. Soc. Hort. Sci.* **37**:521-525.

Uyemura, S., and Imanishi, H., 1983, Effects of gaseous compounds in smoke on dormancy release in *Freesia* corms, *Sci. Hortic.* **20**:91-99.

Van Andel, O. M., and Verkerke, D. R., 1978, Stimulation and inhibition by ethephon of stem and leaf growth of some gramineae at different stages of development, *J. Exp. Bot.* **29**:639-651.

Wang, C., Knill, E., Glick, B. R., and Defago, G., 2000, Effect of transferring 1-aminocyclopropane-1-carboxylic acid (ACC) deaminase genes into *Pseudomonas fluorescens* strain CHA0 and its *gac* A derivative CHA96 on their growth-promoting and disease-suppressive capacities, *Can. J. Microbiol.* **46**:898-907.

Warner, H. L., and Leopold, A. C., 1969, Ethylene evolution from 2-chloroethylphosphonic acid, *Plant Physiol.* **44**:156-158.

Webster, J. R., and Jackson, L. F., 1993, Management practices to reduce lodging and maximize grain yield and protein content of fall-sown irrigated hard red spring wheat, *Field Crops Res.* **33**:249-259.

Wiersma, D. W., Oplinger, E. S., and Guy, S. O., 1986, Environment and cultivar effects on winter wheat response to ethephon plant growth regulator, *Agron. J.* **78**:761-764.

Wills, R. B. H., and Kim, G. H., 1995, Effect of ethylene on postharvest life of strawberries, *Postharvest Biol. Technol.* **6**:249-255.

Wills, R. B. H., and Kim, G. H., 1996, Effect of ethylene on postharvest quality of green beans, *Aust. J. Exptl. Agric.* **36**:335-337.

Wills, R. B. H., and Wong, T., 1996, Effect of low ehylene levels on the storage life of the Asian leafy vegetables bak choi (*Brassica chinensis*), choi sum (*Brassica parachinensis*) and gai lan (*Brassica alboglabra*), *ASEAN Food J.* **11**:145-147.

Wills, R. B. H., Ku, V. V. V., Shohet, D., and Kim, G. H., 1999, Importance of low ethylene levels to delay senescence of non-climacteric fruit and vegetables, *Aust. J. Exptl. Agric.* **39**:221-224.

Wills, R. B. H., McGlasson, W. B., Graham, D., Lee, T. H., and Hall, E. G., 1989, Postharvest: An Introduction to the Physiology and Handling of Fruits and Vegetables, 3rd Ed., New South Wales Univ. Press, Sydney, Australia.

Xie, H., Pasternak, J. J., and Glick, B. R., 1996, Isolation and characterization of mutants of the plant growth-promoting rhizobacterium *Pseudomonas putida* GR12-2 that overproduce indoleacetic acid, *Curr. Microbiol.* **32**:67-71.

Yamaguchi, M., Chu, C. W., and Yang, S. F., 1971, The fate of ^{14}C(2-chloroethyl) phosphonic acid in summer squash, cucumber, and tomato, *J. Amer. Soc. Hort. Sci.* **96**:606-609.

Yang, S. F., 1969, Ethylene evolution from 2-chloroethylphosphonic acid, Plant Physiol. 44:1203-1204.

Yang, S. F., and Hoffman, N. E., 1984, Ethylene biosynthesis and its regulation in higher plants, *Annu. Rev. Plant Physiol.* **35**:155-189.

Zahir, Z. A., and Arshad,, M., 1998, Response of *Brassica carinata* and *Lens culinaris* to the ethylene precursors, L-methionine and 1-aminocyclopropane-1-carboxylic acid, *Soil Biol. Biochem.* **30**:2185-2188.

INDEX

Abiotic production, of ethylene in soil, 141, 170, 172
Abiotic stress, 195
Abscisic acid (ABA), 1, 23
Abscission, 242, 243, 244, 247, 258, 265, 271, 289, 301, 302
ACC deaminase, 56, 58, 59, 60–63, 90
  expression in microorganisms, 61, 324, 325
  expression in plants, 61, 318–321
  genes, 61, 318–321, 324
  induced by ACC, 60
  microorganisms containing, 62, 322–326
  properties, 63
  reaction catalyzed, 3, 60–61
ACC *N*-malonyltransferase, 23–25
  biochemical aspects, 24–25
  reaction catalyzed, 23
ACC oxidase, 15, 18–23, 202, 206, 207, 208, 218, 253, 259, 268, 317
  active site, 21
  biochemical and molecular aspects, 20–23
  genes, 26–27, 58, 263
  induction by, 19–20
  inhibitors, 19, 22
  location in plant tissues, 20
  mechanism of catalyzing reaction, 22–23
  reaction catalyzed, 18–20
ACC synthase, 14–18, 259, 268, 322
  biochemical and molecular aspects, 16–18
  genes, 26–27, 58, 263
  inducers, 17–18
  location in plant tissues, 16
  mechanism of reaction, 16
  reaction catalyzed, 14–15
Accelerated production, of ethylene by
  pathogen-infected tissues, 241–249
  Rhizobium-infected roots, 202–206
S-Adenosylmenthionine (SAM), precursor of ethylene
  in plants, 12, 13–15, 249, 253

Aeration, effect on ethylene production
  in soil, 147–153
  in vitro microbial production, 118, 119–122
Agricultural sources, of ethylene, 6, 290
Agriculture, role of ethylene, 289–326
  inoculation, ACC deaminase bacteria, 322–326
  sources and applications, 290–315
    ethylene gas, 290–291
    ethylene-releasing compounds, 291–305, 306–308
    soil ethylene, 305–315
    transgenic plants, altered in endogenous biosynthesis, 315–322
Air pollutant, ethylene, 6, 290
Algal-fungal symbiosis, 227; *see* Lichen
1-Aminocyclopropane-1-carboxylic acid (ACC)
  effect on nodulation, 196, 197, 198, 199, 200, 202, 203, 204, 208
  effect on plant growth, 311, 313
  precursor of ethylene in lichens, 229, 231, 232, 233, 234
  role in mycorrhizae, 217, 226
  role in pathogenesis, 242, 243, 244, 245, 247, 248, 250, 260, 266, 267
  precursor of ethylene in plants, 5, 12, 14–15, 18–20, 250, 251, 253
  precursor of ethylene in soil, 143, 144, 159, 162, 170, 171
Ammonium monooxygenase, 89, 182
Antifungal agents
  lubimin, 270
  phyrotuberin, 270
  phytuberol, 270
  pisatin, 270
  rishitin, 270
  scoporone, 256
  trifluralin, 254
  tulipalin, 256